ELECTRICITY AND ELECTRONICS

by

Howard H. Gerrish

William E. Dugger, Jr.
Director
Technology for all
Americans Project

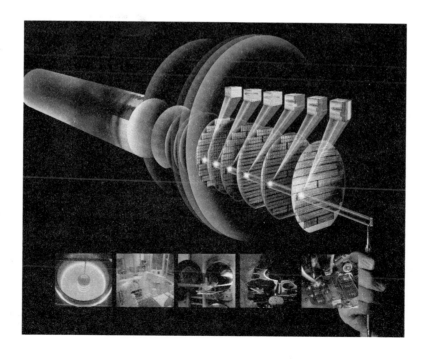

South Holland, Illinois
The Goodheart-Willcox Company, Inc.
Publishers

Copyright 1996

by

THE GOODHEART-WILLCOX COMPANY, INC.

Previous Editions Copyright 1989, 1980, 1977, 1975, 1968, 1964

All rights reserved. No part of this book may be reproduced, stored in a retrieval system, or transmitted in any form or by any means, electronic, mechanical, photocopying, recording, or otherwise, without the prior written permission of The Goodheart-Willcox Company, Inc. Manufactured in the United States of America.

Library of Congress Catalog Card Number 94-25667
International Standard Book Number 1-56637-078-7

1 2 3 4 5 6 7 8 9 10   96   99 98 97 96 95

Materials used for the cover art courtesy of Knight Electronics.

Library of Congress Cataloging in Publication Data

Gerrish, Howard H.
   Electricity and electronics / by Howard H. Gerrish, William E. Dugger, Jr.
     p. cm.
   Includes index.
   ISBN 1-56637-078-7
   1. Electric engineering. 2. Electronics.
I. Dugger, William. II. Title
TK146.G44 1996
621.3--dc19   94-25667
                                                CIP

INTRODUCTION

As a student interested in the rapidly changing and expanding field of electricity and electronics, you will find that *Electricity and Electronics* provides the fundamentals in easy-to-understand language.

Electricity and Electronics employs a dual approach to learning:
1. Experimentation and demonstration to assure thorough understanding of the principles. The instructional material is arranged to provide continuity in learning and challenging learning situations.
2. Practical application of developed principles and skills by constructing inexpensive, professional-appearing projects. The projects have proven appeal. In addition, each project has been thoroughly shop and laboratory tested.

Electricity and Electronics is divided into six parts. Each part begins with an overview of the chapters to be covered and ends with a summary and summary questions. In addition, each chapter is divided into sections. At the end of each section, review questions are provided for the material just covered.

When you have successfully completed this textbook and your course, you should have a thorough understanding of the concepts of electricity and electronics.

A bright outlook for the electronics industry in the future translates into opportunities for specialized education and a lifetime of challenging and rewarding employment.

Howard H. Gerrish

William E. Dugger, Jr.

CONTENTS

INTRODUCTION: The Technology of the Information Age 9

Part I Fundamentals of Electricity and Electronics

1 Science of Electricity and Electronics 19
Nature of Matter, Molecule and Atom, Electrons-Protons-Neutrons, Ionization, Static Electricity, Law of Charges, The Coulomb, Electrostatic Fields, Electric Voltage and Current.

2 Conductors, Insulators, and Semiconductors 27
Voltage, Sources, Volts, Electric Current, Direction of Current, Conductors, Conductance, and Insulators, Resistance, Ohm's Law, Factors Affecting Resistance, Resistors, Color Coding, Semiconductors.

3 Sources of Electricity 41
Chemical Action, Defects in Primary Cells, Zinc-Carbon Cell, Mercury Cell, Silver Oxide Cell, Nickel-Cadmium Cell, The Battery, Secondary Cells, Lead-Acid Cells, Battery Capacity, Other Electrical Energy Sources, Electrical Energy from Light, Photoelectrical Control, Electrical Energy from Heat, Electrical Energy from Mechanical Pressure, Electricity from Magnetism.

4 Basic Circuits 57
Circuit Fundamentals, Loads, Conductive Pathways, Controls, Basic Circuits, Polarity and Meters, Connecting Meters in a Circuit, Ohm's Law Equations, Power and Watt's Law, Combining Ohm's and Watt's Law.

5 Series, Parallel, and Combination Circuits 71
Series Circuits, Ohm's Law in Series Circuits, Kirchhoff's Voltage Law, Power in Series Circuit, Parallel Circuits, Kirchhoff's Current Law, Combination (Series-Parallel) Circuits.

Part II Applied Electricity

6 Magnetism 85
Basic Magnetic Principles, Laws of Magnetism, Magnetic Recording Tape, Magnetic Flux, Third Law of Magnetism, Electric Current and Magnetism, The Solenoid, Magnetic Circuits, Electromagnets, Solenoid Sucking Coil, The Relay, The Reed Relay, Circuit Breaker, Door Bell and Buzzer, Magnetic Shields.

7 Generators 97
Electrical Energy from Mechanical Energy, DC versus AC, Construction of a Generator, Generator Losses, Types of Generators, Independently Excited Field Generator, Shunt Generator, Series Generator, Compound Generator, Voltage and Current Regulation, Alternating Current, Phase Displacement, Alternating Current Generator, The Alternator, Practical Generator Applications.

8 **Instruments and Measurements** **111**
Basic Analog Meter Movement, Ammeter, Voltmeter, Voltmeter
Sensitivity, Ohmmeters, Multimeter, The VOM, Digital Meters,
Liquid Crystal Displays, Lighting Emitting Diodes, Other Meters,
Wheatstone Bridge, Iron Vane Meter Movement, The Wattmeter,
AC Meters, Loading a Circuit, How to Use a Meter, The Oscillo-
scope, Oscilloscope Familiarization.

Part III Alternating Current Circuits

9 **Inductance and RL Circuits** **131**
Inductance, Transient Responses, Mutual Inductance, The
Transformer, Turns Ratio and Voltage Ratio, Taps, Transformer
Power, Autotransformer, Transformer Losses, Reasons for Using
a Transformer, The Induction Coil, The Ignition System, Phase
Relationship in Transformers, Series and Parallel Inductance,
Inductance in AC Circuits, Induced Current and Voltage, Reactive
Power, Resistance and Inductance in an AC Circuit, Ohm's Law
for AC Circuits, Parallel RL Circuit.

10 **Capacitance and RC Circuits** **149**
Capacitance and the Capacitor, Types of Capacitors, Transient
Response of the Capacitor, RC Time Constant, Parallel and Series
Circuits and Computing Capacitance, Capacitance in AC Circuits,
Reactive Power, Resistance and Capacitance in AC Circuits,
Parallel RC Circuit.

11 **Tuned Circuits and RCL Networks** **165**
RCL Networks, Resonance, The Acceptor Circuit, The Tank
Circuit, The Reject Circuit, "Q" of Tuned Circuits, Loading
the Tank Circuit, Filtering Circuits, Filtering Action, Bypassing,
Low-Pass Filters, High-Pass Filter, Tuned Circuit Filters, The
Nomograph.

12 **Electric Motors** **179**
Motor Operation Principles, Counter EMF, Commutation and
Interpoles, Speed Regulation, The Shunt DC Motor, The Series
DC Motor, Compound DC Motors, Motor Starting Circuits,
Thyristor Motor Controls, The Universal Motor, Induction
Motors, Three-phase Induction Motor, Single-phase Induction
Motors, Repulsion Induction Motor, Shaded Pole Motor.

Part IV Electronic Devices and Applications

13 **Basic Electronic Devices** **199**
Semiconductors, Atomic Characteristics, Conduction of
Electricity, Doping, Semiconductor Diodes, Point Contact
Diodes, Silicon Rectifiers, Diode Characteristics and Ratings,
Series and Parallel Rectifier Arrangements, Testing Diodes,
Lighting Emitting Diodes, Transistors, FETs, Vacuum Tubes,
Thermionic Emitters, Cathodes, Diodes, Triodes, Tetrodes,
Pentodes, Cathode Ray Tubes.

14 Integrated Circuits **217**
History of the Integrated Circuit, IC Construction, Resistors,
Capacitors, Putting It Together, Common types of ICs.

Part V Basic Electronic Circuits

15 Power Supplies **231**
Power Supply Functions, The Power Transformer, Half-Wave
Rectification, Full-Wave Rectification, The Bridge Rectifier,
Filters, Voltage Regulation, Load Resistor, Zener Diode, Voltage
Regulator Circuit, Voltage Doublers, AC-DC Supply, Floating
Ground, Power Supply Construction.

16 Amplifiers and Linear Integrated Circuits **247**
Amplifier circuits, NPN and PNP Transistors, Biasing, Single
Battery Circuit, Methods of Bias, Amplifier Operation, Computing
Gain, Transistor Circuit Configurations, Classes of Amplifiers,
Thermal Considerations, Coupling Amplifiers, Transformer Coupling,
RC, Direct, and Push-Pull Coupling, Linear Integrated Circuits,
Voltage Regulators and References, Operational Amplifiers,
Linear IC Construction Projects.

17 Digital Circuits **273**
Binary Numbering System, Voltage Logic Levels, Bits, Nibbles,
and Bytes, Logic Gates, Logic Families, Digital IC Construction,
Projects.

18 Oscillators **293**
Oscillator Basics, Oscillator Types, Oscillator Projects.

Part VI Electronic Communication and Data Systems

19 Radio Wave Transmission **305**
Communication Model, Frequency Spectrum, Radio Transmitter,
Microphones, Modulation, Amplitude Modulation, Modulation
Patterns, Sideband Power, Transistorized Transmitters, Frequency
Modulation, Narrow Band FM, Modulation Index, The Radio Wave,
Radio Wave Travel, Ground Waves, Sky Waves, Project.

20 Radio Wave Receivers **323**
The AM Receiver, Tuning Circuit, RF Amplification, Diode
Detector, Superheterodyne Receiver, Mixer, IF Amplifiers,
Detector, Automatic Gain Control, Audio Preamplifier, Audio
Amplifier, Using ICs in Superheterodyne Receivers, Transducers,
Headphones, Tone Controls, Alignment, FM Receiver, FM
Detection, Noise Limiting, Projects.

21 Television **349**
Cameras, Scanning, Composite Video Signals, CRT Controls,
TV Receivers, The Television Channel, TV Innovations, Satellite
and Cable TV.

22 Computers 367
History, Microcomputers VS Minicomputers, Computer Operation, Input-Output Devices, Central Processing Unit, Storage Devices, Software, Computers in the Future.

23 Career Opportunities in Electronics 387
Careers and Types of Industries, Entrepreneurs, Career Information Sources, Education, Job Search Ideas.

Reference Section 395

Appendix 1, **395**, Appendix 2, **396**, Appendix 3, **397**, Appendix 4, **398**, Appendix 5, **400**, Appendix 6, **404**.

Dictionary of Terms 407

Acknowledgements 425

Index 426

SAFETY PRECAUTIONS FOR THE ELECTRICITY/ELECTRONICS SHOP

There is always an element of danger when working with electricity. Observe all safety rules that concern each project. Be particularly careful not to contact any live wire or terminal, regardless of whether it is connected to either a low voltage or a high voltage. Projects do not specify dangerous voltage levels. However, keep in mind at all times that it is possible to experience a surprising electric shock under certain circumstances. Even a healthy person can be injured or seriously hurt by the shock or what happens as a result of it. Do not fool around. Working with electricity can be fun, but it can also be dangerous!

PROJECTS AND ACTIVITIES

STATIC ELECTRICITY EXPERIMENTS 21, 22, 23	TRIAL MOTOR 181
	SERIES MOTOR 181
ELECTRICITY FROM A GRAPEFRUIT 41, 42	SHUNT MOTOR 181
	AC MOTOR 189
ELECTRICITY FROM NICKELS AND PENNIES 41, 42, 54	POWER SUPPLY 241, 242
VOLTAIC CELL 41, 42	TRICKLE CHARGER FOR AUTOMOBILE BATTERY 243
PHOTOELECTRIC CONTROL 52	LOW-COST MULTIRANGE VOLTMETER 262
THERMOCOUPLE 53, 54	1.5 VOLT FLASHER CIRCUIT 263
LAWS OF MAGNETISM 86	3 VOLT FLASHER CIRCUIT 263
MAGNETIC LINES OF FORCE 86	PARALLEL FLASHER CIRCUIT 263
ELECTROMAGNET 90	CONTINUITY CIRCUIT BUZZ BOX 263
SOLENOID SUCKING COIL 90, 91	INCANDESCENT BULB FLASHER 263
ELECTRODEMONSTRATOR 91, 132, 138, 404, 405, 406	VARIABLE FLASHER 263
CIRCUIT BREAKER 93	PORTABLE STEREO AMPLIFIER 264
DOOR BELL 93	TELEPHONE AMPLIFIER 268
BUZZER 93	PAGING AMPLIFIER 268
TACHOMETER 107	BINARY BINGO 284, 285, 286, 287
RPM METER 107	CODE PRACTICE OSCILLATOR 297
SELF-INDUCTION OF COIL 132	SQUAWKER HORN 298, 299
TRANSFORMER ACTION 139	AM TRANSMITTER 318, 319
INDUCTION COIL SHOCKER 139	CRYSTAL RADIO 343
RC BLINKER PROJECT 157, 158	FOUR TRANSISTOR RADIO 344
NOMOGRAPH 175, 176	SUPERHETERODYNE RADIO 344, 345, 346

THE TECHNOLOGY OF THE INFORMATION AGE

We live in the age of a third revolution. It is known as the Information Revolution. This is a time when information and computers greatly influence our lives.

The first revolution was one of agricultural advances. This revolution began hundreds of years ago. Society, from its very beginning, was agricultural. Through the Agricultural Revolution, more food and better farming and production methods were developed.

The second revolution began in the mid 19th Century. Industry caused the economic base to shift from farm to mill, from country to city. This was called the Industrial Revolution. The key result of the Industrial Revolution, however, was not its products. It was the increased use of science and technology.

The study of science meant knowledge gained by study and practice. This scientific knowledge was then used to study ways of changing or working with the environment to make life better. This is known as technology.

Science and technology used during the Industrial Revolution created a strong base. This base is used in the current Information Revolution. The items produced now are moving toward supplying services, such as communications systems and transportation. The Information Revolution has given us a computer-based economy. The technology of electricity and electronics made computers possible.

In the past 25 years, we have witnessed some outstanding progress in electronics. Fig. 1 shows the past, present, and future of invention and innovation in electronics from 1920 up to the year 2000. Great change has taken place and, most likely, will continue to do so.

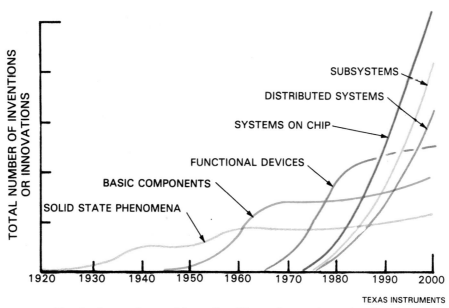

Fig. 1. Innovation and invention life cycle are shown by decades.

It is now possible to produce integrated circuits with the space to store more than a million bits of data on a single microchip. Fifth generation computers run at amazing speeds and may one day emulate human reasoning. Semiconductors now switch electrical signals on and off in 5.8 trillionths of a second. Lasers can be used to transmit 20 billion bits of information per second over a 42 mile length of fiber optic cable. That is roughly equal to sending the entire text of ten 30-volume encyclopedias in one second. See Fig. 2.

Personal microcomputers now have more computing power than the first electronic computer, ENIAC. In addition, they are 20 times faster than ENIAC. They have a much larger memory. They are thousands of times more reliable. They consume the power of a small light bulb, rather than that of a locomotive. They occupy 1/30,000 of the space. And they cost 1/10,000 the price of ENIAC, Fig. 3.

ELECTRICITY AND ELECTRONICS IN TECHNOLOGY

There are many ways electricity and electronics are used in our world. Three areas in which both are used to further technology are information systems, communication systems, and transportation systems.

INFORMATION SYSTEMS

The computer has become the current counterpart of the steam engine. Like the steam engine that brought on the Industrial Revolution, the computer has been a key force in the Information Revolution. This is because the computer is an information machine.

Businesses use computers to obtain and send out a great deal of information. Computers can help businesses provide more complete and responsive services, Fig. 4. Computers can also assist hospitals in providing better health care. Manufacturers seeking lower costs, while improving the quality of their products, use computers, Fig. 5. City, county, state,

Fig. 3. Desktop computers are smaller, quicker, and cheaper than the first electronic computer, ENIAC.

Fig. 2. The information this woman is researching could be sent to her computer in less than one second. This makes future research less time consuming.

Fig. 4. Touch screen monitors increase worker productivity. This allows product orders to be filled and shipped in less time.

The Technology of the Information Age

and federal governments keep costs down by using information generated by computers. Schools use computers in teaching. Farmers use computers to raise more healthy crops and livestock, while making the best use of feeds and equipment. Individuals use computers to manage personal finances, improve education, and play games, Fig. 6. For these and many other needs, computers can be a key part of the answer.

COMMUNICATION SYSTEMS

In its most basic sense, communication is the process of transmitting and receiving messages so that the receiver (destination) understands the message the sender (source) sent. Electronics provides a medium for communication. The telegraph, developed in the 1830s, started communication by electrical signals rather than by the five senses used until then.

In 1948, Claude E. Shannon proposed that communication systems contain a few basic elements. Warren Weaver expanded on Shannon's theory in 1949. A diagram of the Shannon-Weaver communication model is shown in Fig. 7. It is the basis for modern electronic communication.

We communicate today using many electronic devices, such as telephones, televisions, radios, phonograph recordings, computers, and recording tapes, Fig. 8. Electronic communication will be explored in more detail later in this text.

Communication brings us closer. Without it, we become isolated. Electronics makes modern communication possible.

EMERSON ELECTRIC

Fig. 5. These computers are produced using the just-in-time manufacturing process. With this process, only the parts needed for one day of work are kept on hand. Information systems help keep a record of what is needed and when.

MAXIS

Fig. 6. Computers can teach us and entertain us at the same time. This program shows the complexities of running a city.

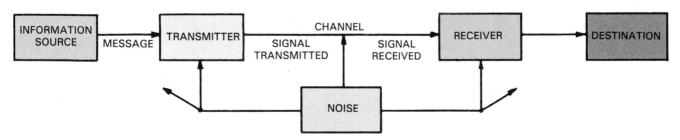

Fig. 7. Shannon-Weaver communication model.

Electricity and Electronics

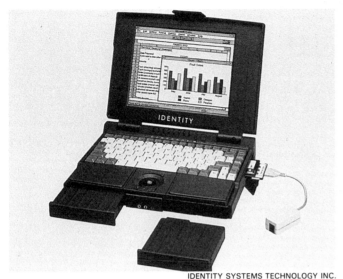

Fig. 8. Communication through computers is the fastest growing sector in the communication field.

IDENTITY SYSTEMS TECHNOLOGY INC.

TRANSPORTATION SYSTEMS

Transportation involves moving goods or people from one place to another. Electricity and electronics help make a great deal of transportation possible. Electronic communication systems aid pilots of aircraft when communicating with those on the ground, Fig. 9. Pilots also use electronic navigational systems found on modern aircraft to plan the safest flight routes. Computers are used to provide automatic pilot options when the jet is in the air.

Automobiles have detailed electronic devices that work everything from air-conditioners to fuel systems. See Fig. 10. Elevators use electronics to operate safely. Other ground transportation systems such as trains and subways could not operate without using electricity and electronics.

These three systems are only a few of the many uses for electricity and electronics in the Information Age. As technology expands, the uses for electricity and electronics will also expand.

MILESTONES AND EVENTS IN ELECTRICITY AND ELECTRONICS

There have been many inventions and developments in electricity and electronics over the years. Scientists, engineers, technicians, and teachers have contributed to the constant progress.

A chart showing the major discoveries, inventions, and developments in electricity and electronics over the past two centuries is shown in Fig. 11. This chart does not list everything, but it does attempt to highlight the major events.

Fig. 9. In efforts as delicate as space travel, superior communication systems are a must.

NASA

The Technology of the Information Age

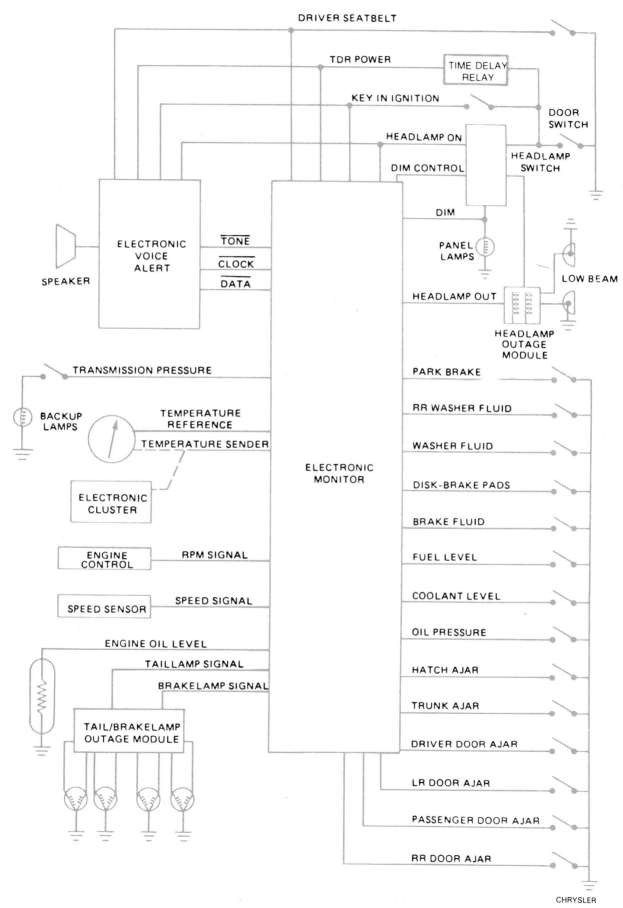

Fig. 10. This is a diagram of a voice alert system in a modern car. If a condition is not right, a signal goes to the voice alert. The voice alert circuit can then "talk" to the driver of the car.

Electricity and Electronics

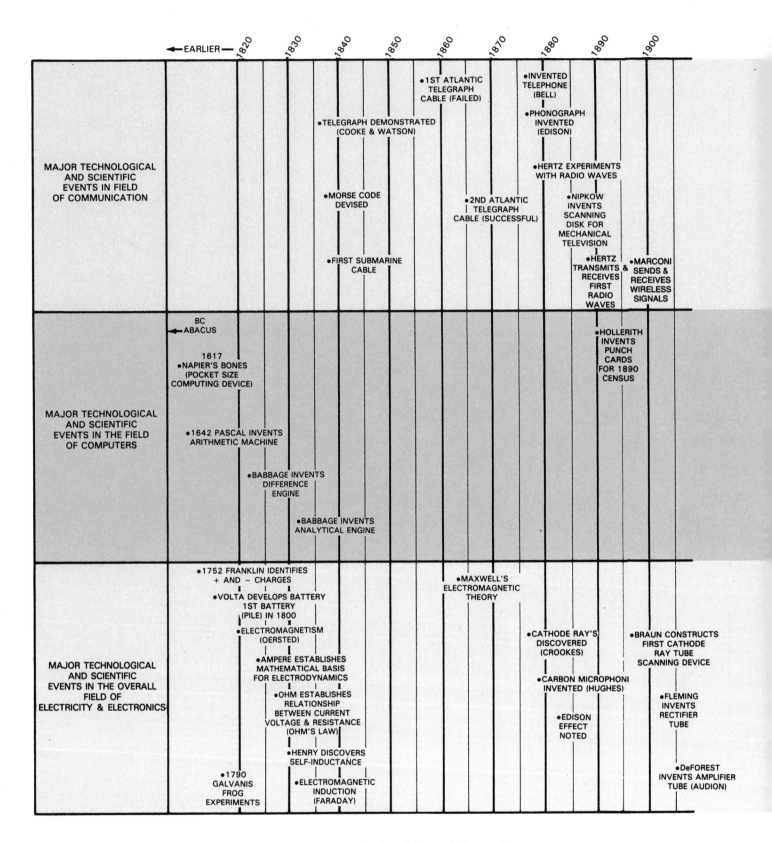

Fig. 11. Milestones in electricity and electronics.

The Technology of the Information Age

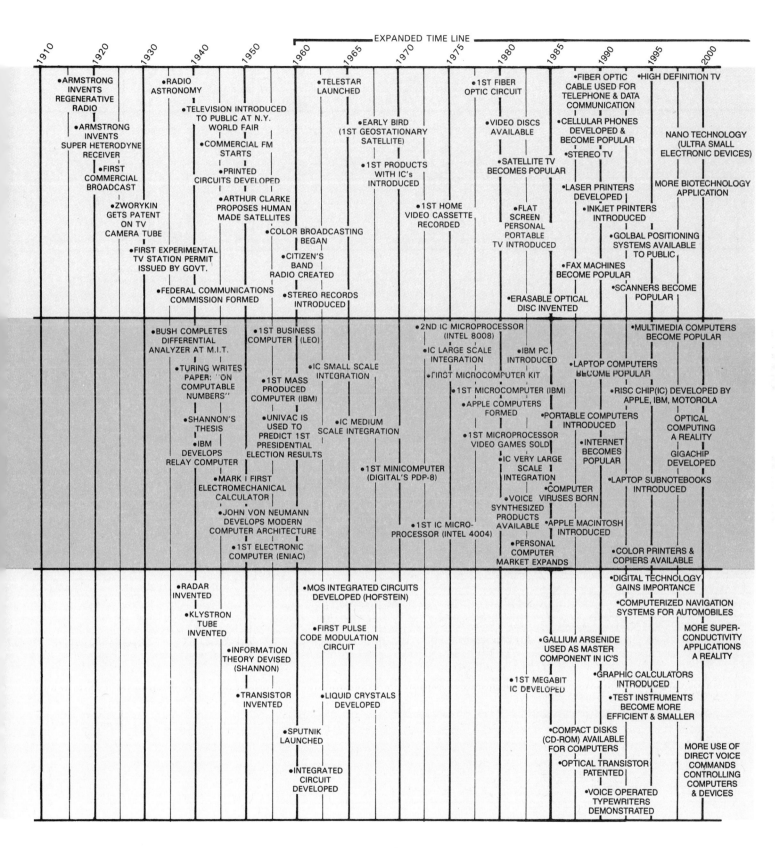

Fig. 11. Continued.

Part I

FUNDAMENTALS OF ELECTRICITY AND ELECTRONICS TECHNOLOGY

1 Science of Electricity and Electronics
2 Conductors, Insulators, and Semiconductors
3 Sources of Electricity
4 Basic Circuits
5 Series, Parallel, and Combination Circuits

The science and technology of electricity is not quite two centuries old. Yet in this short time period, much has been developed and accomplished. The founders of the field of electricity, Oersted, Faraday, Maxwell, Ampere, and Volta, would be astonished if they could observe the current marvels of their creative inquiry and curiosity.

The first five chapters of ELECTRICITY AND ELECTRONICS were developed to give the reader a broad scientific and technological knowledge and historical understanding of the fundamentals of direct current electricity. (Concepts, circuits and theory relating to alternating current electricity are presented in Chapters 6 through 12. These, then, are followed by a study of solid-state electronics.)

Chapter 1 is the *Science of Electricity and Electronics*. It explores the origin of the electron, how it can be moved along in a circuit or remain at rest, as in static electricity. In addition, voltage and electric current will also be explained.

Chapter 2 presents the basic building blocks of electricity. It is titled *Conductors, Insulators, and Semiconductors*. The theory of current flow is presented, along with an in-depth discussion of resistance. Insulators and semiconductors are also introduced.

There are six basic sources where electricity is generated. Most of these are explained in Chapter 3, *Sources of Electricity*. The process of generating electricity from one of these sources produces the important movement of electrons in a circuit.

Basic circuits are covered in Chapter 4, along with Ohm's Law and Watt's Law. You will also learn about electrical circuit controls, loads, and conductive pathways in this chapter.

Chapter 5 explores the three fundamental circuits: series, parallel, and combination.

Electricity and Electronics

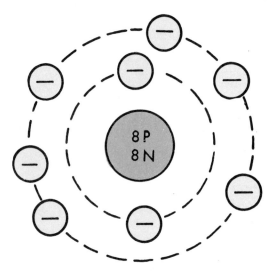

Oxygen atom with equal numbers of protons, neutrons, and electrons.

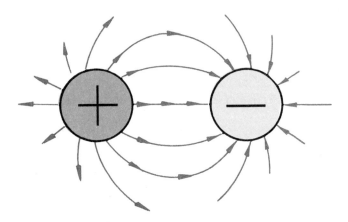

Unlike charges attract each other.

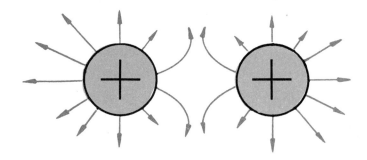

Like charges repel each other.

The whole science and technology of electricity revolves around the nature of matter and a few principles. All matter is made up of electrical particles. Some are positively charged, some negatively charged and some are neutral. Like particles repel each other and unlike charges attract.

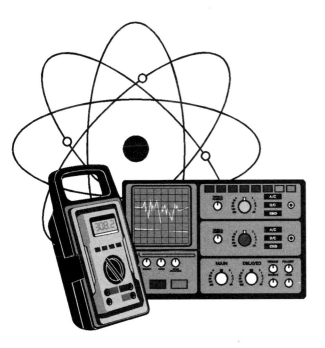

Chapter 1

SCIENCE OF ELECTRICITY AND ELECTRONICS

After studying this chapter, you will be able to:
- *Identify the relationship between elements and compounds.*
- *Construct a model of an atom.*
- *Discuss the concepts of atomic weight and atomic number.*
- *State the Law of Charges and explain it using examples.*
- *Explain what is meant by electric current.*

You are fortunate to live in an age in which the opportunity exists to study the electron. New discoveries, developments, and applications in electronics occur almost daily. These open a promising vista of unlimited opportunities for the creative scientist, as well as for the skilled technician. We are living in a truly electronic age.

1.1 THE NATURE OF MATTER

The scientist tells us that everything is made up of matter. MATTER may be defined as anything which occupies space or has mass.

Some matter differs from other matter by frequent observable characteristics or properties. It may differ in color, taste, or hardness. These characteristics permit us to identify the various forms of matter. Sometimes observable characteristics will not permit identification of a substance. It is then necessary to make a chemical analysis to determine its nature and behavior. Most materials we use are made of a combination of various kinds or mixtures of matter.

LESSON IN SAFETY: The experimenter, the technician, the scientist, and the engineer, must respect the power of electricity. Although voltages used in projects and experiments described in this book are not dangerous to a normal, healthy individual, there is never an excuse for carelessness.

In the basic study of the electron theory, it must be understood that some types of elementary matter exist as mixtures. Further effort to break this matter into parts by chemical decomposition will produce no further change in the characteristics of the matter. These simple forms of material have been called ELEMENTS. Scientists have made further subdivisions of the elements by physical forces, such as those generated in the "atom smasher."

Our concern, however, is the nature of an element. There are many of them. Some familiar elements are iron, copper, gold, aluminum, carbon, and oxygen. Chemists have isolated over ninety different kinds of elements existing in nature and several more found in their laboratories. A COMPOUND is a mixture of two or more elements.

MOLECULE AND THE ATOM

If you were to take a crystal of table salt and cut it in half, you would have two parts, but both would be common salt. Salt is a chemical compound of sodium (Na) and chlorine (Cl) or NaCl. In your imagination you might conceive the further division of the crystal of salt until it could not be divided again. This question has disturbed scientists and philosophers for hundreds of years. Can a particle of matter be divided to the point that if it is divided again it will disappear?

In science, the smallest particle that can exist and still retain the properties of the original compound or element is called a MOLECULE. Referring again to the crystal of salt, even though the salt has been divided to the smallest particle that can exist by itself, it is still a compound of sodium and chlorine. So this tiny particle must consist of two parts; one of sodium and one of chlorine. The molecule of salt would no longer be salt if it were divided. However, if a molecule of an element was subdivided, the parts would be all alike. These tiny particles of an element are named ATOMS.

The smallness of just one molecule will stagger your imagination. If you began filling a small match box with molecules at the rate of ten million per second, it would take you over a billion years to fill the box!

The word *atom* is derived from the Greek word meaning indivisible. It was not until recent years that the atom was divided into parts. It is not within the scope of this text to describe the chemical and physical reactions that take place when an atom is "split." However, it is important to learn some things about the structure of an atom. See Fig. 1-1 for a chart showing the relationship of matter, elements, compounds, atoms, and molecules.

ELECTRONS—PROTONS—NEUTRONS

Physicists have discovered that atoms are composed of minute particles of electricity. The center of each atom is the NUCLEUS. It contains positively charged particles of electricity called PROTONS, and also neutral (neither positive or negative) particles. Most of the mass or weight of the atom is in its nucleus. Surrounding the nucleus, in rapidly revolving orbits, are negatively charged particles of electricity called ELECTRONS.

The structure of the atom may be compared to the solar system, where the sun is at the center, like the nucleus. The planets (Earth, Venus, Mars, etc.) revolving around the sun may be compared to the whirling electrons in orbit. The whirling electrons or planetary electrons are kept from falling into the center by centrifugal force. Normally for each proton in the nucleus there is one electron in orbit. The positive and negative charges cancel, leaving the atom electrically neutral. The atomic structure of each element may be described as having a fixed number of electrons in orbit. Examples of the atomic structure of two common elements are displayed in Fig. 1-2.

All elements are arranged in order in the Periodic Table of Elements according to their basic ATOMIC NUMBER. That is the number of electrons in the orbits. Or they may be arranged by their ATOMIC WEIGHT, which is the approximate number of protons and neutrons in the nucleus. Referring again to Fig. 1-2, the atomic weight of hydrogen is one (scientifically 1.008), and its atomic number is one. The atomic weight of oxygen is sixteen; its atomic number is eight.

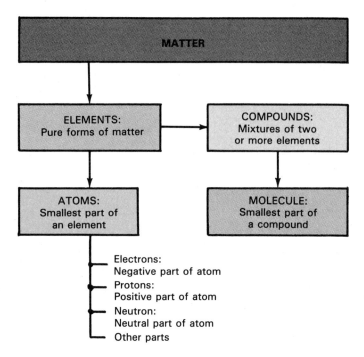

Fig. 1-1. Relationship between matter, elements, compounds, atoms, and molecules.

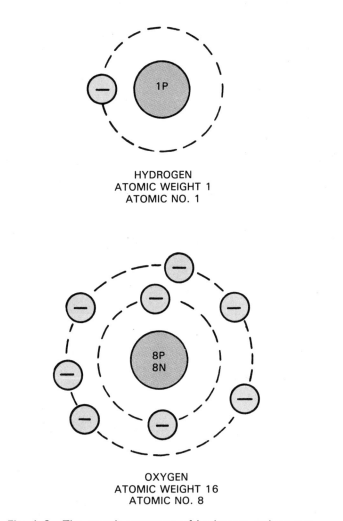

Fig. 1-2. The atomic structure of hydrogen and oxygen.

IONIZATION

Usually an atom remains in its normal state unless energy is added to it by some exterior force such as heat, friction, or bombardment by other electrons. When energy is added to the atom, it will become excited. Electrons which are loosely bound to the atom in the outer rings or orbits may leave the atom. If electrons leave the balanced or neutral atom, then there is a deficiency of electrons and the atom is no longer in electrical balance. The atom is then said to be IONIZED. If the atom has lost electrons, which are negative particles of electricity, the atom then would be a positively charged or POSITIVE ION. If the atom gained some electrons, then the atom would be negatively charged or a NEGATIVE ION. This is an important lesson in electronic theory.

REVIEW QUESTIONS FOR SECTION 1.1

1. Define the following terms:
   a. Matter.
   b. Elements.
   c. Compounds.
   d. Atom.
   e. Molecule.
2. Name some common elements.
3. What is atomic weight?
4. Define atomic number.
5. What is the meaning of ionization?

1.2 STATIC ELECTRICITY

The word *static* means at rest. Electricity can be at rest. The generation of static electricity can be demonstrated in many ways. By stroking the fur of a cat, you will notice that its fur tends to be attracted to your hands as you bring your hand back over the cat. If this is done at night, you may see tiny sparks and hear a crackling sound.

This is caused by the discharge of static electricity. What is happening? It is simply that the friction between your hand and the cat's fur is exciting the atoms. The atoms are becoming unbalanced through the movement of electrons from one atom to another. The sparks are created as the atoms attempt to neutralize themselves. You may have experienced the building up of a static charge in your body while walking across a carpet or sliding across a car seat. Sometimes a spark discharge may result when you touch another person or the door handle of the car.

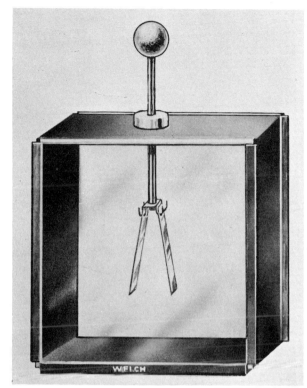

WELCH SCIENTIFIC CO.

Fig. 1-3. An electroscope. The gold leaves in the glass enclosure will indicate an electric charge.

LAW OF CHARGES

One of the fundamental laws in the study of electricity is the Law of Charges. That is: LIKE CHARGES REPEL EACH OTHER. UNLIKE CHARGES ATTRACT EACH OTHER.

This law will be used many times in experiments described in this text. It is important to remember that ONE ELECTRON WILL REPEL ANOTHER ELECTRON. A negatively charged mass will be attracted by a positively charged mass. The Law of Charges may be demonstrated in the electrical laboratory by means of an electroscope or pith balls, Fig. 1-3.

Materials needed:

1. Electroscope.
2. Vulcanite Rod and Fur.
3. Glass Rod and Silk.
4. Single and Double Pith Balls on Stands.

EXPERIMENT I. Quickly rub the vulcanite rod with the fur. There is now a negative charge on the rod as electrons were transferred to the rod from the fur by friction. Bring the charged rod close to the ball on top of the electroscope. You will notice that the gold leaves in the electroscope will expand.

Remove the rod and the leaves will close. The electroscope is used in this case to show the presence of an electrical charge. The negatively charged rod repels the electrons on the ball of the electroscope and forces them down to the leaves. The leaves, both being negatively charged, will repel each other and expand, Fig. 1-4.

EXPERIMENT II. Once again, rub the vulcanite rod in order to charge it. Now touch the ball on the electroscope with the rod. You will notice that the leaves expand and remain that way after the rod is removed. The rod has shared its charge with the electroscope and the electroscope remains charged, Fig. 1-5. Now touch the ball with your finger. The leaves will close, showing that the electroscope has been discharged through your body.

EXPERIMENT III. Repeat Experiments I and II, using the glass rod and the silk. The glass rod will charge positively. However, the results will be the same, with opposite polarity.

EXPERIMENT IV. Repeat Experiment I by bringing the charged rod close to, but not touching, the electroscope. In this position, touch the electroscope with your other hand. Remove your hand and then the rod. You will observe that the electroscope will remain charged. As no actual contact was made between the rod and the electroscope, the electroscope was charged by INDUCTION. The nearness of the negative rod forced the electrons on the electroscope to travel away, through your hand. Of course, when your hand was removed, there was no way for the electroscope to regain its lost electrons so it must remain positively charged as indicated by the expanded leaves.

EXPERIMENT V. Once again, rub the vulcanite rod and charge it negatively. Bring the rod close to a hanging pith ball, Fig. 1-6. The ball will at first be attracted to the rod due to unlike charges. When the pith ball touches the rod, it assumes the charge of the rod and is immediately repelled. Now both the ball and rod are negative. Try once again to touch the ball with the rod. Notice that the ball is always repelled.

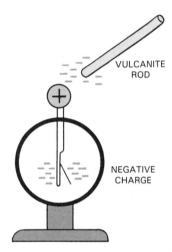

Fig. 1-4. The negatively charged rod forces electrons down to the leaves of the electroscope. The leaves expand because they are both negative and, therefore, repel each other.

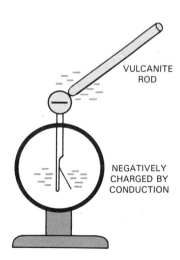

Fig. 1-5. The electroscope becomes charged when the rod touches the ball and will remain charged.

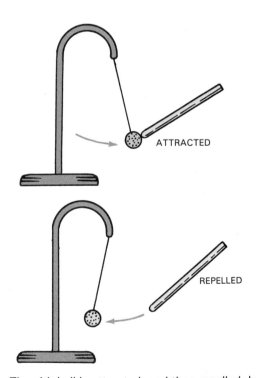

Fig. 1-6. The pith ball is attracted, and then repelled, by the charged rod.

EXPERIMENT VI. Using a second pith ball, charge it in the same manner as Experiment V. You will now have two pith balls charged negatively. Try to bring the balls close together and notice that they will not touch. There is a repulsive force existing between them, Fig. 1-7.

EXPERIMENT VII. Leave one pith ball negative. Charge the second pith ball positively by using the glass rod and the silk. As you bring the two balls close to each other, you will notice that they are attracted to each other, Fig. 1-8.

These seven experiments prove the Law of Charges. You must know this law, otherwise lessons in electricity will be hard to understand.

THE COULOMB

It would be impractical to describe the difference in charge between two bodies as one body having one, or even a dozen, more electrons than the other body. This is because an atom and parts of an atom (electrons) are very small particles.

The force of attraction and repulsion of charged particles was studied by the French scientist, Charles A. Coulomb. He developed a practical unit for measurement of an amount of electricity. It is known as the COULOMB. It represents approximately 6.24×10^{18} electrons (6,240,000,000,000,000,000).

ELECTROSTATIC FIELDS

Once again, perform Experiment IV. When the vulcanite rod was brought close to, but not touching, the electroscope, why did it affect the electroscope? This is due to the invisible lines of force that exist around a charged body. These lines of force are called the ELECTROSTATIC FIELD or DIELECTRIC FIELD. The field is strongest very close to the charged body. It diminishes at a distance inversely proportional to the square of the distance.

Refer to Fig. 1-9. Two charged balls are shown, with lines representing the electrostatic fields of opposite polarity and the attractive force existing between them. In Fig. 1-10, two charged balls are shown with like polarities and the repulsive force exisitng between the balls due to the electrostatic fields. Once again, the electrostatic fields add further proof to the Law of Charges. Like charges repel. Unlike charges attract.

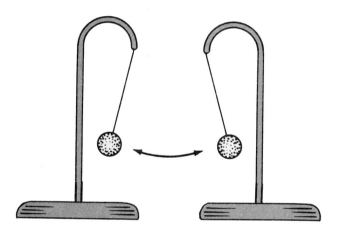

Fig. 1-7. The like charged pith balls repel each other.

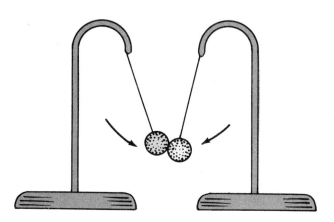

Fig. 1-8. Unlike charged pith balls attract each other.

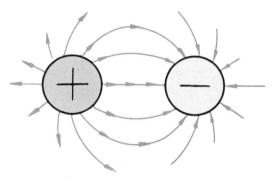

Fig. 1-9. The electrostatic fields of unlike charged bodies, showing attractive force.

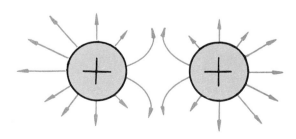

Fig. 1-10. The electrostatic fields of like charged bodies, showing repulsive force.

LESSON IN SAFETY: Horseplay has no place where machinery is operating and tools are being used. Do not pay the penalty for carelessness or poor working habits. It may be painful and very expensive!

REVIEW QUESTIONS FOR SECTION 1.2

1. Define static electricity.
2. What is a compound?
3. Explain what is meant by electrostatic field.
4. Like charges _____ each other.
5. What is meant by a charged body?

1.3 ELECTRIC VOLTAGE AND CURRENT

At this point, you should have a thorough understanding of static electricity and the electrostatic fields existing around charged bodies. Consider the illustration in Fig. 1-11. It is very similar to Fig. 1-9, except that a short piece of copper wire has been connected between the balls, A and B. Again, you should realize what is meant by a negatively charged body, such as Ball A. It actually has an excess of negative charges of electrons. In the case of Ball B, it will have a lack of electons because of its positive charge. When the wire, or CONDUCTOR, is connected between the two balls, then the excess electrons on Ball A will follow along the conductor to Ball B. Electrons will continue to flow until the number of electrons on both balls are equal. At this point, the potential difference between them is zero. This movement of electrons is called ELECTRIC CURRENT. The force which caused these electrons to move is called electromotive force (emf) or difference in potential. It is known as VOLTAGE.

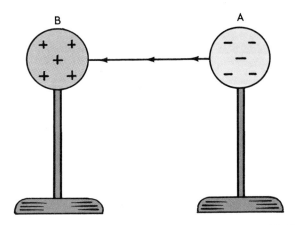

Fig. 1-11. Electrons flow from the negatively charged ball to the positive ball.

REVIEW QUESTIONS FOR SECTION 1.3

1. What is meant by a negatively charged body?
2. The movement of electrons is known as _____ _____.
3. What is electromotive force?

SUMMARY

1. Matter is anything that occupies space or has mass.
2. Elements are basic or pure forms of matter.
3. Compounds are mixtures or combinations of one or more elements.
4. Atoms are the simplest form of an element still having the unique characteristics of that element.
5. Molecules are the simplest form of a compound, still having the unique characteristics of that compound.
6. The negatively charged particle of an atom is the electron and the positively charged particle is the proton.
7. Like charges repel each other while unlike charges attract each other.
8. The coulomb is a quantity of electrons (6,240,000,000,000,000,000 or 6.24×10^{18} electrons).
9. Current is the movement of electrons in a conductor.
10. Voltage is the force behind the electron moving them along to produce current.

TEST YOUR KNOWLEDGE, Chapter 1

Please do not write in the text. Place your answers on a separate sheet of paper.

1. A negative charge of electricity is called an _____; a positive charge, a _____; a neutral particle, a _____.
2. What is the name for a unit of measurement for a quantity of electricity? How many electrons does it represent?
3. What are the two laws of electrostatic charges?
4. A negative rod will charge an electroscope by contact. The leaves will both be:
   a. Positive.
   b. Negative.
   c. Neutral.
5. A positive rod will charge an electroscope by induction. The leaves will both be:
   a. Positive.
   b. Negative.
   c. Neutral.

6. Electron movement is called _____ _____.
7. When a charged rod is brought close to a pith ball, explain the action. Why does this action occur?
8. Use a reference book in the library to discover the atomic weight and atomic number of the following:
   a. Cu (Copper).
   b. Al (Aluminum).
   c. Ge (Germanium).
   d. As (Arsenic).
9. The invisible lines of force that surround a charged body are called the:
   a. Electron force.
   b. Electrostatic field.
   c. Electric field.
   d. None of the above.

FOR DISCUSSION

1. How do lightning rods protect a building during a storm?
2. What great scientist discovered the relationship between lightning and electricity? Describe his experiment.
3. A police car seems to have some unusual radio interference while driving along the highway. This interference is not present when the car is not moving. Suggest a possible cause and explain.
4. Electrostatic fields are used in industry to reclaim by-products and to prevent air pollution by smoke and particles. Explain how this procedure will scrub the exhaust and emissions. Also, how will this cleansing impact the "greenhouse" effect on the earth?

Electricity and electronics provide us with many wonderful tools.

DEVRY INSTITUTES

Electricity and Electronics

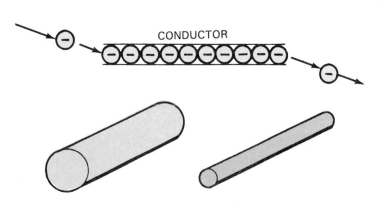

1. Conductor of electricity.

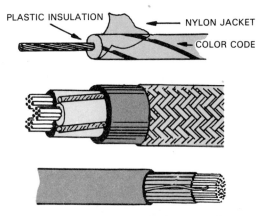

2. Insulators stop electricity from traveling where it can cause harm or damage.

3. Individual resistors line the center of this circuit board. Many new electronic components have resistors built into them.

Chapter 2

CONDUCTORS, INSULATORS, AND SEMICONDUCTORS

After studying this chapter, you will be able to:
- *Explain the meaning of voltage and how it relates to volts.*
- *Discuss the theory of electric current.*
- *Describe the two theories of current flow direction.*
- *Distinguish between conductors, insulators, and semiconductors.*
- *State and explain Ohm's Law.*
- *List the factors affecting resistance.*
- *Determine the value of various color coded resistors.*

All electricity and electronics is based on the fundamental concept of moving electrons in some type of material. In Chapter 1, we discussed the concept of electron action in certain materials that causes electrical current flow. In this chapter, we will discuss three fundamental terms of electricity and electronics: voltage, current, and resistance. Conductors, insulators and semiconductors will also be explained.

2.1 VOLTAGE

The force or difference in potential which causes electrons to flow is called VOLTAGE. Whenever there is a negative charge and a positive charge there is a difference in potential. If a material, such as a conductor, is connected between the two different charges, the electrons will flow from the negative charge to the positive charge. See Fig. 2-1. The electrons will continue to flow until there is no difference in potential or voltage.

Voltage is also called ELECTROMOTIVE FORCE or EMF. The term ELECTRICAL PRESSURE is also used to refer to voltage. In any case, all of the terms mean the same thing, which is the push or pull, or force, that moves electrons along in a material. The abbreviation for voltage in electrical formulas is E. This represents electromotive force.

VOLTAGE SOURCES

There are six basic sources of voltage. One of these sources was presented in Chapter 1 in the discussion of static electricity. To review, when certain materials are rubbed together, or when friction is created, static electricity is created. This type of electricity is often referred to as "at rest," since these charges are stationary, or static. These two different charges have a difference in potential or a voltage. If a piece of wire or conductor is placed between them, an electron flow will occur.

Electric voltage can also be caused by chemical energy created by the chemical action of certain materials in cells and batteries. Energy from light can be converted to electrical energy, or voltage, by a solar cell. Mechanical pressure can be converted to electrical voltage by the use of certain kinds of pressure sensitive materials. Voltage can be created by thermal energy or heat.

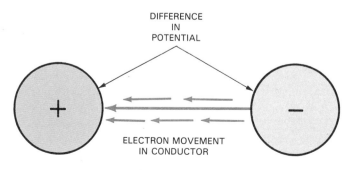

Fig. 2-1. Electrons flow due to a difference in potential.

27

These four sources of voltage will be discussed in Chapter 3. A sixth source of voltage, and the one which is most used today, is the conversion of magnetic energy to electrical energy. This source will be discussed in detail in Chapter 6.

VOLTS

The unit for measuring electrical pressure or voltage is the VOLT. This is a very common unit. Many items we come in contact with every day are rated in volts, such as 117 volts in the home, 12 volts in the car battery, or 1.5 volts in a "D" cell used in a flashlight.

There is sometimes a need to change volt references to smaller or larger units. To do this, PREFIXES are added to the base word, "volt." These prefixes denote an increase or decrease in value.

Units which are larger than the basic volt unit are the KILOVOLT (kV) and the MEGAVOLT (MV). The kilovolt is one thousand times larger than the volt. The megavolt is one million times larger than the volt. Units which are smaller than the volt include the MILLIVOLT (mV), and the MICROVOLT (μV). The millivolt is one thousandth the size of the volt. The microvolt is one millionth the size of the volt. Fig. 2-2 shows the relationship of all these prefixes.

A sample conversion from one volt group to another is shown in Fig. 2-3. In this example, 32 millivolts is being converted to volts. The first step is to locate the decimal. Since it is not shown in the example, it is assumed to be after the last digit (2). The next step is to move the decimal to the decimal location for the value to which you are converting. Generally, if the number to which you are converting is larger, the decimal will be moved left. If it is smaller, the decimal will be moved right. In this example, the decimal point will be moved three spaces to the right. This is the decimal location for the microvolt. Finally, add zeros in the spaces left between the decimal point and the number. Refer again to Fig. 2-3.

REVIEW QUESTIONS FOR SECTION 2.1

1. Define voltage.
2. Give three other terms for voltage.
3. The unit of measurement for voltage is the _____.
4. What is a prefix?
5. Convert the following voltage values:
   a. 5.6 V = _____ mV.
   b. 23 μV = _____ mV.
   c. 1200 kV = _____ V.
   d. .5 MV = _____ kV.
   e. 12.6 mV = _____ μV.
   f. 16,000 V = _____ kV.

2.2 ELECTRIC CURRENT

You will recall that the flow of electrons is called ELECTRIC CURRENT. It is necessary to measure the flow of electrons or current in a circuit, so units of measurements have been standardized. The abbreviation for electric current in formulas is I. This represents current intensity or amperage. Comparing the electricial circuit to a water pipe, one would measure the flow of water as so many gallons per minute.

Electric current is measured by the number of electrons flowing each second past a given point in the conductor. This is where the term coulomb is useful to represent a quantity of electrons. The standard unit for the measurement of electric current is the AMPERE, or AMP. This represents the flow of one coulomb per second or 6.24×10^{18} electrons moving past a fixed point in the circuit. See Fig. 2-4. Mathematically,

$$I = \frac{Q}{t} \text{ or } Q = It$$

where Q = coulombs
I = current in amperes
t = time in seconds

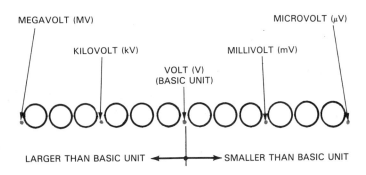

Fig. 2-2. Prefixes used with the volt unit.

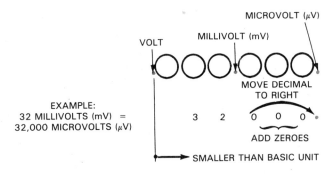

Fig. 2-3. Conversion from millivolts to microvolts.

Conductors, Insulators, and Semiconductors

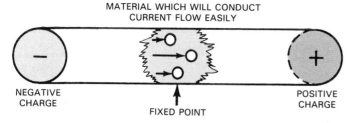

Fig. 2-4. One ampere is one coulomb (6.24 x 10¹⁸) moving past a fixed point in one second.

Example: If 40 coulombs of electrons passed a given point in 10 seconds, then a current of 4 amperes would flow.

$$I = \frac{40 \text{ coulombs}}{10 \text{ seconds}} = 4 \text{ amperes}$$

As with voltage, prefixes may be added to the basic unit of ampere to change the basic value. Since the ampere is a fairly large unit of measure, only the smaller MILLIAMP (mA) and MICROAMP (μA) are used. The milliamp is one thousand times smaller than the amp. The microamp is one million times smaller than the amp. Fig. 2-5 shows the relationship of the prefixes to the basic unit.

DIRECTION OF CURRENT

Electrons have a negative charge and protons have a positive charge. When there is a large enough concentration of either charge, a difference in a potential, or a voltage, is created. This difference in potential is what forces the electrons along in a material.

There are two theories about the direction of current flow. The ELECTRON THEORY OF CURRENT FLOW defines the direction of electron flow as moving from NEGATIVE (−) to POSITIVE (+) in the external circuit. Another theory, the CONVENTIONAL THEORY OF CURRENT FLOW, states that current flows from POSITIVE (+) to NEGATIVE (−) in the external circuit. In both cases, the current always flows in the opposite direction of the current in the voltage source. Throughout this textbook we will be using the Electron Theory of Current Flow (− to +).

REVIEW QUESTIONS FOR SECTION 2.2

1. Define current.
2. I is the abbreviation for what?
3. What is an ampere?
4. The electron theory for current flow states that the current in the external circuit flows from _____ to _____.
5. Convert the following current values:
   a. 12 A = _____ mA.
   b. 4295 mA = _____ A.
   c. 16,210 μA = _____ mA.
   d. .9 mA = _____ μA.
   e. 1.5 A = _____ μA.
   f. .005 A = _____ mA.

2.3 CONDUCTORS, CONDUCTANCE, AND INSULATORS

When a source of electrical voltage is connected to a device, such as a lamp, with copper wire, the electrons will flow, or be conducted, from the negative terminal of the source, through the lamp, and back to the positive terminal of the source. This copper wire is the path along which electric current will flow. The wire is called a CONDUCTOR.

Some materials are good conductors of electricity. These materials have a large number of free electrons. Copper conductors are used in most electrical circuits. Copper is an excellent conductor. Silver is even better than copper. However, it is too expensive for general use. Gold is also a good conductor. Many high-power lines are aluminum for added strength and light weight. Brass, zinc, and iron are also fair conductors of electricity.

The actual conduction of electricity is done by transferring electrons from one atom to the next in the conductor. Assume that a piece of copper wire is neutral. If an electron is forced into one end of the wire, an electron is forced out from the other end, Fig. 2-6. The original electron did not flow through the conductor. Yet the energy was transferred by the interaction between the electrons in the conductor. The actual transfer of electrical energy occurs at an amazing speed.

Fig. 2-5. Prefixes used with the ampere unit. Kiloamp is seldom used.

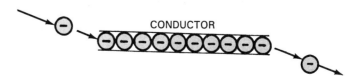

Fig. 2-6. The transfer of energy between atoms causes current to flow through a conductor.

29

It has been accurately measured and approaches the speed of light: 186,000 miles per second! In terms of the metric system, this would be 300,000 kilometers or 300,000,000 meters per second.

The unit for measuring electrical conductance is the SIEMENS. This unit was named after the German inventor Ernst von Siemens who did a great deal of work in the development of telegraphy use. The abbreviation for the siemens is S.

The unit formula for computing electrical conductance is as follows:

$$G = \frac{1}{R}$$

where G is conductance (in siemens) and R is resistance (in ohms).

LESSON IN SAFETY: Your body is a good conductor of electricity. Never touch a circuit wire or component unless you are sure that it is not energized. An electric current flowing from one hand across your chest and heart and out of your other hand, for example, can be dangerous, even fatal. The wise technician uses only one hand when working on high voltage circuits. The other hand is kept in a pocket! The operators of switchboards and power stations stand on insulated platforms. Then there is no direct path to ground in case of accidental contact with a high voltage wire.

Conductors are arranged according to size by the American Wire Gauge System, Fig. 2-7. The larger gauge numbers have smaller diameters and cross-sectional areas. For example, a No. 14 wire is larger than a No. 20 wire.

A common way of expressing cross-sectional area of a round wire is in circular mils (cmil). One MIL is 0.001 inches. The number of CIRCULAR MILS in an area of wire (conductor) is equal to the square of the diameter (D^2), where D is in mils. Therefore, a wire with a diameter of 1 mil has an area of 1 cmil ($1 \times 1 = 1$). A wire with a diameter of 2 mils has an area of 4 cmils ($2 \times 2 = 4$). A diameter of 15 mils gives an area of 225 cmils ($15 \times 15 = 225$). See Fig. 2-8.

The circular mil is a more convenient method of expressing the diameter of a conductor than the fractional inch. An example will help illustrate this point. According to our equation, a wire with a diameter of 50 mils (0.05 inches [50×0.001]) has an area of 2500 cmil. You will recall from geometry that the area of a circle is πr^2. Computing the area of this wire in square inches:

$$A = 3.14 \times (0.025 \text{ in.})^2$$
$$= 3.14 \times 0.000625 \text{ in.}^2$$
$$= 0.0019625 \text{ in.}^2$$

No doubt, you will find it easier to work with 2500 cmils than 0.0019625 in.^{2}.

There are other ways to express the formula for area of a circle (πr^2). Note the following variations:

$$A = \pi \left(\frac{D}{2}\right)^2 = \pi \frac{D^2}{4} = \frac{3.14}{4} D^2 = 0.7854 D^2$$

Area of a wire, given in SQUARE MILS, is equal to $0.7854 D^2$, where D is in mils. To determine the number of square mils in one circular mil, recall that 1 circular mil is the area of a wire 1 mil in diameter. Using the previous equation, $A = 0.7854 D^2$, gives:

$$A = 0.7854 \times (1 \text{ mil})^2$$
$$= 0.7854 \text{ mil}^2$$

This is the square mil area in one circular mil. In order, then, to find the circular mil area of a conductor whose area is given in square mils, it is necessary to divide by 0.7854.

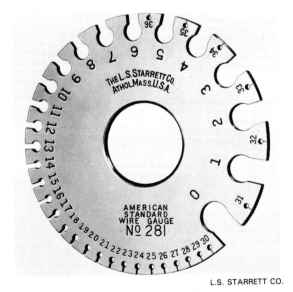

L.S. STARRETT CO.

Fig. 2-7. A gauge is used to find wire size.

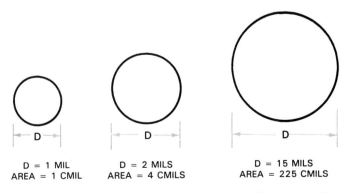

| D = 1 MIL | D = 2 MILS | D = 15 MILS |
| AREA = 1 CMIL | AREA = 4 CMILS | AREA = 225 CMILS |

Fig. 2-8. Circular mil is found by squaring the diameter in mils.

Conductors, Insulators, and Semiconductors

GAUGE NO.	DIAM. MILS	CIRCULAR MIL AREA	RESISTANCE OHMS PER 1,000 FT. OF COPPER WIRE AT 25°C	GAUGE NO.	DIAM. MILS	CIRCULAR MIL AREA	RESISTANCE OHMS PER 1,000 FT. OF COPPER WIRE AT 25°C
1	289.3	83,690	0.1264	21	28.46	810.1	13.05
2	257.6	66,370	0.1593	22	25.35	642.4	16.46
3	229.4	52,640	0.2009	23	25.57	509.5	20.76
4	204.3	41,740	0.2533	24	20.10	404.0	26.17
5	181.9	33,100	0.3195	25	17.90	320.4	33.00
6	162.0	26,250	0.4028	26	15.94	254.1	41.62
7	144.3	20,820	0.5080	27	14.20	201.5	52.48
8	128.5	16,510	0.6405	28	12.64	159.8	66.17
9	114.4	13,090	0.8077	29	11.26	126.7	83.44
10	101.9	10,380	1.018	30	10.03	100.5	105.2
11	90.74	8,234	1.284	31	8.928	79.70	132.7
12	80.81	6,530	1.619	32	7.950	63.21	167.3
13	71.96	5,178	2.042	33	7.080	50.13	211.0
14	64.08	4,107	2.575	34	6.305	39.75	266.0
15	57.07	3,257	3.247	35	5.615	31.52	335.0
16	50.82	2,583	4.094	36	5.000	25.00	423.0
17	45.26	2,048	5.163	37	4.453	19.83	533.4
18	40.30	1,624	6.510	38	3.965	15.72	672.6
19	35.89	1,288	8.210	39	3.531	12.47	848.1
20	31.96	1,022	10.35	40	3.145	9.88	1,069.

Fig. 2-9. Copper wire table.

The resulting equation is:

$$A = \frac{.7854D^2 \text{ (square mil area)}}{.7854 \text{ (square mil area in one cmil)}}$$

or $\frac{.7854D^2}{.7854}$

then $A = D^2$

Therefore, the circular mil cross-sectional area of a conductor equals the square of its diameter in mils.

Finally, in order to change a circular mil wire size to square mils, multiply by .7854. This information is important when comparing current carrying capacity of round and square wires, Fig. 2-9.

INSULATORS

Materials with only a few free electrons do not conduct electrons well. They are called INSULATORS. Insulation is used on wires to protect people from coming in accidental contact with circuits. It is also used to avoid electrical contact between wires and supporting devices. Common materials used for insulation are rubber, glass, air, bakelite, mica, and asbestos.

It is apparent then, that some materials resist the flow of electrons, while others have less resistance to the flow of electrons. This resistance must always be considered an important characteristic of electrical circuits.

REVIEW QUESTIONS FOR SECTION 2.3

1. What is a conductor?
2. Name three good conductors.
3. Define an insulator.
4. Name four good insulators.
5. At what speed does current flow?
6. What is the unit for conductance?

2.4 RESISTANCE

The opposition to the flow of electrical current is called RESISTANCE. The abbreviation for resistance in electrical formulas is R. The unit for measuring resistance is the OHM and its symbol is Ω. As with voltage and current, prefixes can be used to increase or decrease the basic value of the ohm. Fig. 2-10 shows a chart of common prefixes used with the ohm unit.

OHM'S LAW

There is an important relationship between voltage, current, and resistance. First, one volt is needed to force one ampere through one ohm of resistance. Next,

Electricity and Electronics

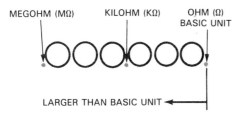

Fig. 2-10. Prefixes used with the ohm unit. Milli and micro are seldom used.

voltage (E) and current (I) in a circuit are DIRECTLY PROPORTIONAL. This means that the more voltage there is, the more current there will be (when resistance remains the same). Finally, current (I) and the resistance (R) are INVERSELY PROPORTIONAL. This means that as resistance increases, current will decrease (when voltage remains constant).

The mathematical relationship between voltage, current, and resistance is called OHM'S LAW. It will be covered in more detail in Chapter 5.

FACTORS AFFECTING RESISTANCE

There are four factors which cause one material or conductor to have more resistance than another conductor. These factors are:
1. SURFACE AREA OF CONDUCTOR. The larger the surface area or diameter, the less the resistance.
2. KIND OF MATERIAL. Certain materials have more resistance than others.
3. LENGTH OF CONDUCTOR. The longer the conductor, the greater the resistance.
4. TEMPERATURE OF MATERIAL. The hotter the material, the greater the resistance.

Surface area of conductor

The flow of electron traffic can be easily understood by comparing it to automobile traffic. Consider the size of the road. A large, smooth, four lane highway will carry more traffic than a narrow, rough country road, Fig. 2-11. This is also true of the electrical conductor.

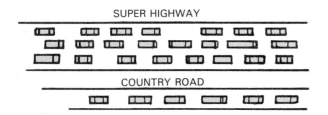

Fig. 2-11. A superhighway can carry more traffic than a country road. Because it is smoother, it also has less resistance to the flow of automobiles.

RESISTANCE is the opposition to electric current in a conducting wire or material. A large wire can carry more electrons than a small wire. It has less resistance to the flow, because of its larger size, Fig. 2-12. Resistance to the flow of electricity by a wire or a conductor will vary inversely with the size or cross-sectional area of the conductor.

Not every conductor is a solid wire. Frequently, it is necessary to use several smaller wires twisted together to form a cable. This method allows greater flexibility and ease of installation. If a wire is subject to movement or vibration, such as occurs in a lamp cord, a stranded wire produces greater strength and less danger of breakage.

Stranded wires are usually made up of 7, 19, or 37 separate wires twisted together. A further improvement is made by twisting several stranded wires into one cable, see Fig. 2-13. Do not confuse this stranded cable with the multi-wire conductor used in many electronic applications. The multi-conductor has several wires, each insulated from the other. These wires are color-coded so that correct connections can be made.

Fig. 2-12. A large conductor has less resistance than a small conductor. A large conductor also can carry a larger current.

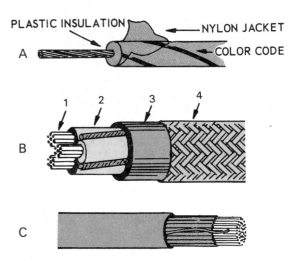

Fig. 2-13. A—Stranded hookup wire used in electronic circuits. B—Power cable. (1) Three stranded conductors. (2) Insulation and jute fillers. (3) Tough, flexible jacket. (4) Bronze protective armor. C—Wire for a telephone switchboard. Numerous conductors are insulated from each other and are color coded.

LESSON IN SAFETY: When making connections with stranded wires, the stranded ends should be tightly twisted together and coated with solder. This is called TINNING. Fine wires that are separated from the connection are a mark of poor quality and may cause dangerous short circuits.

Kind of material

It was mentioned in Chapter 1 that some materials are good conductors of electricity and others are poor conductors. The kind of material used in the manufacturing of wire also affects conductance or resistance.

LESSON IN SAFETY: Wires used to conduct electricity throughout living and working areas are selected according to size. They must be capable of carrying the correct amount of electricity needed for a particular use. Most circuits for lighting use either No. 12 or No. 14 wire, rated at 15 amperes. The wires for a range or water heater may be No. 6 to No. 10. Always use wires of the correct size to avoid trouble.

Length of conductor

A third consideration in the selection of a conductor is its length. If one foot of wire has a certain resistance, then ten feet of the same wire will have ten times more resistance; fifty feet will have fifty times more. Remember that this resistance uses up power, creating losses in the line. Devices connected to long conductors may not have the required voltage for proper operation, due to line losses.

LESSON IN SAFETY: When using long extension cords to operate lights and tools, be sure that they have sufficient size. In this way, there will be little line loss or voltage drop. Small wire extension cords may heat up and burn. Motor driven tools at the end of a small, long extension cord will heat up and operate inefficiently. Also, it is wise to equip all tools with a three wire cord. The third wire is a ground and is connected to the case and handle of the tool. In case of a short circuit, the third wire will conduct the electricity harmlessly into the ground, instead of into the body of the user. This precaution has saved many lives.

Temperature of material

A fourth consideration is temperature. Most metals used in conductors, such as copper and aluminum, increase in resistance as the temperature increases. In many electronic circuits, careful design is necessary to insure proper ventilation and radiation of heat from current carrying devices. When wires are enclosed in metallic coverings or conduit, the National Electric Code specifies that larger wires be used. In this case, heat due to losses in the line will not change the current carrying ability of the conductors.

REVIEW QUESTIONS FOR SECTION 2.4

1. Define resistance.
2. Explain the relationship known as Ohm's Law.
3. Name four factors that influence resistance.
4. Convert the following resistance units:
   a. .5 MΩ = ________ Ω.
   b. 2.7 kΩ = ________ Ω.
   c. 3300 Ω = ________ kΩ.
   d. 8.2 MΩ = ________ kΩ.

2.5 RESISTORS

Resistance units permit the design engineer to use electricity in many interesting and profitable ways. Because resistance is part of a circuit that uses power, most useful and labor saving electrical devices use some form of resistance.

Resistance may purposely be put into a circuit to reduce voltages and to limit current flow in electronic devices. Fig. 2-14 shows several forms of molded composition fixed resistors. They are manufactured in a number of shapes and sizes. The symbol for a fixed value resistor is shown in Fig. 2-15.

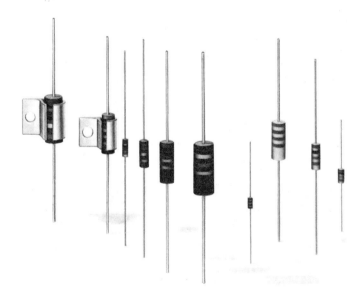

ALLEN-BRADLEY

Fig. 2-14. Group of carbon composition resistors.

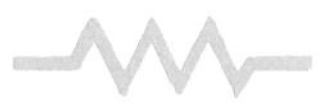

Fig. 2-15. Fixed resistor symbol.

Electricity and Electronics

The chemical make-up that causes resistance is accurately controlled in the lab. Resistors can be purchased in a range of sizes, from one ohm to several million ohms. The physical size of the resistor is rated in WATTS. This refers to the ability of the resistor to dissolve heat caused by resistance. Common sizes used in electronic work are the 1/2 watt and 1 watt sizes.

Resistors, then, are grouped by ohms and watts. For example, a 1000 ohm resistor can be purchased in a 1/2 watt, 1 watt, or 2 watt size. In each size the resistance would be the same. Electrical power, or wattage, will be explained in more detail in Chapter 5. See Fig. 2-16 for the construction of a composition resistor.

Another type of small wattage fixed value resistor is the thin film resistor. It is similar to the molded composition resistor in appearance and function. However, the thin film resistor is made from depositing a resistance material on a glass or ceramic tube. Leads with caps are fitted over each end of the tube to make the body of the resistor. Thin film resistors are usually color coded. See Fig. 2-17.

For higher current uses, resistance units are wirewound. A wire is wound on a ceramic core. The wire has a specific fixed-value resistance. The entire resistance component is insulated by a coat of vitreous (opaque) enamel. Several of these resistors are shown in Fig. 2-18. These are manufactured in sizes from 5 watts to 200 watts. The wattage used depends on the heat dissipation required during operation. Metal oxide resistors are also used for high voltage and wattage applications. See Fig. 2-19.

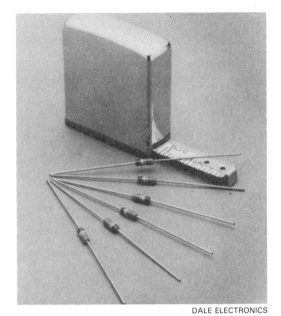

DALE ELECTRONICS

Fig. 2-17. Thin film resistors.

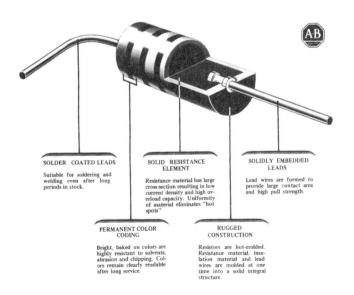

ALLEN-BRADLEY

Fig. 2-16. Cutaway of a carbon composition resistor.

Fig. 2-18. Resistors are made in many types, shapes, and sizes to meet specific circuit requirements.

Conductors, Insulators, and Semiconductors

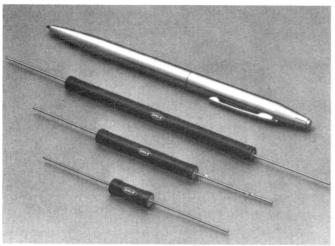

Fig. 2-19. Metal oxide resistors.

Fig. 2-21. Adjustable resistor (potentiometer) symbol.

Fig. 2-22. Precision thick film chip resistors.

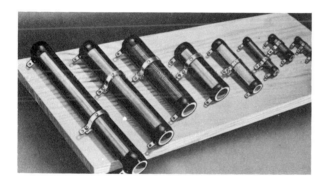

Fig. 2-20. These adjustable resistors provide a sliding tap for voltage divider uses.

Another type of wirewound resistor is the adjustable resistor. Unlike the wirewound resistor, however, the adjustable resistor is not entirely covered by enamel material. Instead, a portion of one side of the wire is exposed. An adjustable sliding tap is attached across the exposed surface. The tap is moved to adjust the resistance. Adjustable resistors, sometimes with two or more adjustable taps, are used a great deal as voltage dividers in power supplies and other devices. Adjustable resistors are shown in Fig. 2-20. The symbol for an adjustable resistor is shown in Fig. 2-21.

Precision thick film chip resistors are new in the electronics field. Fig. 2-22 shows these small resistors.

Most electronic equipment requires the use of variable resistance parts. TVs and radios, for example, have volume controls. These controls are often knobs, located on the front panel of the device. They are generally rotary type knobs; that is, they revolve around an axis. The resistive element is a molded composition ring located inside the knob. The variation in resistance is provided by a sliding contact arm attached to the ring. Connections may be made through terminals to both ends of the resistance, as well as to the sliding contact arm. This part is called a POTENTIOMETER, Fig. 2-23. Its symbol is shown in Fig. 2-24.

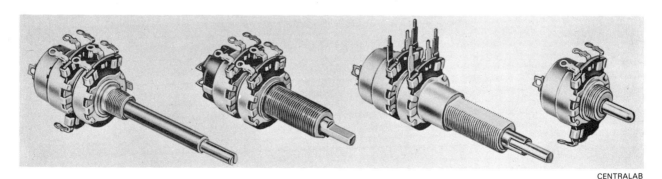

Fig. 2-23. Potentiometers used in electronic circuitry for fixed and variable resistance.

35

Electricity and Electronics

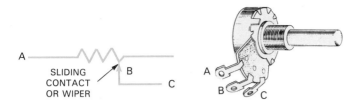

Fig. 2-24. Potentiometer symbol.

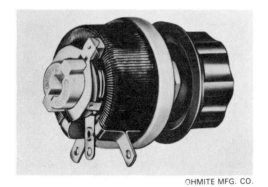

Fig. 2-25. A wirewound potentiometer for higher current and power uses.

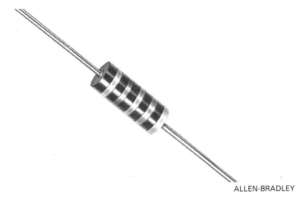

Fig. 2-26. Color code bands encircle the resistor.

For higher current and power uses, potentiometers may be wirewound resistance wire instead of molded composition, Fig. 2-25.

There are also many types and sizes of resistors designed for specific uses. These include very accurate precision resistors and heavy, rugged resistors for high power uses.

One new form of resistor that is rapidly becoming popular in transistor circuitry is the THERMISTOR. It is used to offset the effect of temperature change in a circuit. It is constructed so that as temperature increases, resistance will decrease.

OHMMETERS

Ohmmeters are used to measure the value of resistance in a component or a circuit. The power for ohmmeter operation is provided in its case. In using an ohmmeter, isolate the resistance to be tested. There should be no power in the circuit when using the ohmmeter. Some ohmmeters require zeroing on each scale before using.

More will be presented on the design and use of ohmmeters in Chapter 8.

COLOR CODING RESISTORS

Most low wattage resistors, up to 2 watts, are color coded. The colors are used worldwide. The universal color code has been adopted by the Electronics Industries Association (EIA) and the United States Armed Forces (MIL).

Refer to Fig. 2-26. Note how the color codes are printed all around the body of the resistor. In this way, value can be easily determined in any position. To see how to read color coded resistors, refer to the standard color code chart shown in Fig. 2-27.

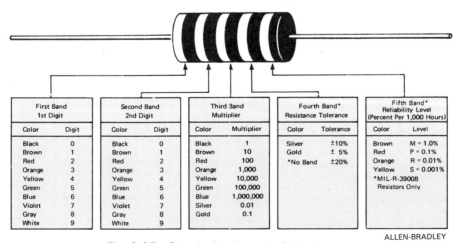

Fig. 2-27. Standard color code for resistors.

Conductors, Insulators, and Semiconductors

Resistors often have four color bands without a fifth reliability band. If this is the case, read the resistor with the first four colors. A few examples of color coded resistor values are shown in Fig. 2-28.

REVIEW QUESTIONS FOR SECTION 2.5

1. What is electrical resistance?
2. Give symbols for the following:
   a. Fixed resistor.
   b. Tapped resistor.
   c. Potentiometer.
   d. Ammeter.
3. Name four types of resistors.
4. Give the color codes (first color, second color, multiplier, tolerance) for the following resistors:
   a. 2 400 Ω, 10%.
   b. 680 Ω, 5%.
   c. 91,000 Ω, 20%.
   d. 27 Ω, 10%.
   e. 3 600 Ω, 5%.
   f. 100 Ω, 10%.
   g. 5.1 Ω, 5%.
   h. 9.1 meg Ω, 10%.

2.6 SEMICONDUCTORS

Conductors have very low resistance to electron or current flow, while insulators have very high resistance to current flow. SEMICONDUCTORS, lie in between conductors and insulators in their ability to conduct or resist current flow. The most commonly used semiconductor materials are silicon (Si), germanium (Ge), and selenium (Se). A new material, gallium arsenide (GaAs), is now being used more as a basic semiconductor material. A chart with the resistance comparison of conductors, insulators, and semiconductors is shown in Fig. 2-29.

Semiconductors are the basic building materials for solid-state electronic devices, such as transistors and integrated circuits. They are used in nearly every electronic circuit, from stereo radios to computers. Semiconductors will be explained more in Chapter 13.

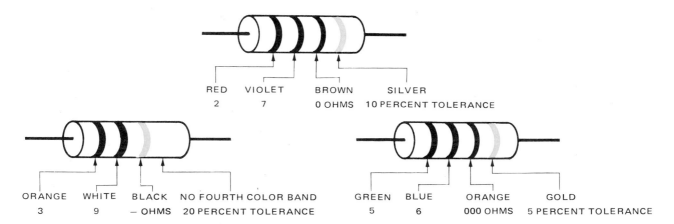

Fig. 2-28. Examples of standard color coded resistors.

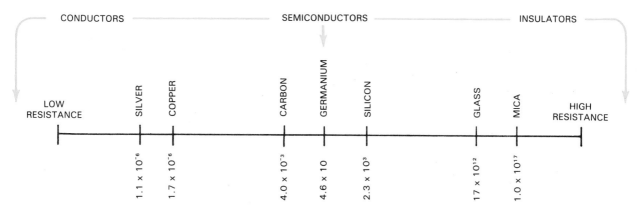

Fig. 2-29. Resistance of some conductors, semiconductors, and insulators (in ohms/cm).

Electricity and Electronics

REVIEW QUESTIONS FOR SECTION 2.6

1. What is a semiconductor?
2. Name three elements that work well as semiconductor materials.
3. Name a new semiconductor material.
4. Conductors have _____ resistance while semiconductors have a relatively _____ resistance.

SUMMARY

1. Voltage is the force or pressure which moves electrons along in a conductor. This movement of electrons is called current. The opposition to current flow is resistance. When resistance is constant, the more voltage there is, the larger the current. An increase in resistance at that voltage level will cause a decrease in current flow.
2. Prefixes are used to change the value of a basic unit, such as the volt, the amp, and the ohm.
3. Using the Electron Theory of Current Flow, electrons flow from negative (−) to positive (+) in an external circuit (not in the voltage source).
4. Four factors affecting resistance are the length, surface, and temperature of the conductor, and the kind of material from which the conductor is made.
5. Color codes are used to identify values of small wattage (2 watts and under) resistors.
6. Conductors are materials which have low resistance to current flow while insulators have high resistance to current flow. Semiconductors lie in between conductors and insulators in their resistance to current.

TEST YOUR KNOWLEDGE, Chapter 2

Please do not write in the text. Place your answers on a separate sheet of paper.

MATCHING QUESTIONS: Match the following definitions with the correct terms.

    a. Volt.   c. Circular mil.
    b. Amp.   d. Siemens.

1. Standard unit of measurement for electric current.
2. Standard unit of measurement for electrical conductance.
3. Standard unit of measurement for electrical pressure of force.
4. Standard unit of measurement for cross-sectional area of round wire.
5. The prefixes KILO and MEGA denote units that are (larger, smaller) than the basic unit.
6. Give the abbreviations for the following when used in formulas:
   a. Electric current.
   b. Resistance.
   c. Voltage.
7. Briefly explain the two theories on the direction of current flow.
8. Why are insulators used?
9. _____ is the opposition to the flow of electrical current.
10. Physical size of a resistor is rated in:
    a. Amperes.
    b. Volts.
    c. Watts.
    d. None of the above.
11. What is an ohmmeter used to measure?
12. Make the following conversion:
    a. 4 300 Ω = _____ kΩ.
    b. 16,000 V = _____ mV.
    c. 9.3 mA = _____ μA.
    d. 1,270,000 V = _____ kV.
    e. .05 kΩ _____ Ω.
    f. 28 μA = _____ mA.
    g. 22.5 V = _____ kV.
13. Give the values for the following color coded resistors:

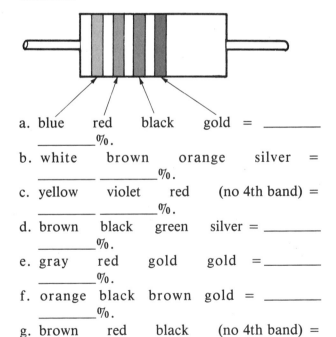

  a. blue   red   black   gold = _____ _____ %.
  b. white   brown   orange   silver = _____ _____ %.
  c. yellow   violet   red   (no 4th band) = _____ _____ %.
  d. brown   black   green   silver = _____ _____ %.
  e. gray   red   gold   gold = _____ _____ %.
  f. orange   black   brown   gold = _____ _____ %.
  g. brown   red   black   (no 4th band) = _____ _____ %.
  h. blue   gray   gold   silver = _____ _____ %.

FOR DISCUSSION

1. Discuss the idea of electron current flowing one way in an external circuit (− to +) and the other way

(+ to −) in the power source or battery.
2. A good understanding of the relationship between conductors, insulators, and semiconductors is important in the study of electricity and electronics.

Discuss the reasons why this is so.
3. Relate the concept of resistance to other areas of your daily life. In what other ways do you see its effect?

Resistors are scattered among other electrical components in this stereo receiver.

Electricity and Electronics

Some dry batteries, like these, are rechargeable. To allow this, the chemical reaction causing the electrical current must be reversed.

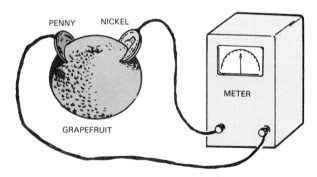

Basic principle of a battery: when two unlike elements are placed in a chemical, the reaction of the elements to the chemical creates an electrical potential (charge) between the elements.

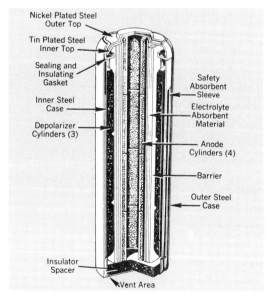

Modern dry battery in cutaway. Center rod is the positive electrode (terminal). The case is the negative electrode. A chemical "paste" surrounds the center rod and is in electrical contact with the battery case too.

Batteries are an important source of electrical power. We use them widely in portable electrical/electronic units. Among these are flashlights, watches, radios, cameras, automobiles, toys, and calculators.

Chapter 3

SOURCES OF ELECTRICITY

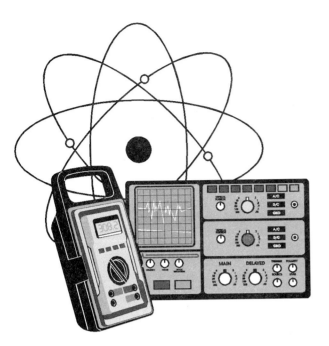

After studying this chapter, you will be able to:
  List the six basic sources of electricity.
  Explain the chemical action that creates electricity in various types of cells.
  Define polarization.
  Distinguish between series and parallel connections in batteries.
  Calculate the output voltage of batteries connected in series.
  Demonstrate proper use of a hydrometer and explain its use.
  Calculate the theoretical capacity of a battery.

Every student knows the story of Benjamin Franklin and his kite. For centuries before Franklin, scientists and philosophers had observed lightning. It was through the experimentation and research of Dr. Franklin that the relationship between lightning and static electricity was confirmed. What is electricity and where does it come from? Years before the discovery of the electron theory by J. J. Thomson, it was suggested by Dr. Franklin that electricity consisted of many tiny particles or electric charges. He further theorized that electrical charges were created by the distribution of electrical particles in nature.

We have learned that a potential difference or electromotive force (emf) is created when electrons are redistributed. A body might assume a charge; its polarity is determined by the deficiency or excess of electrons. People since have turned their scientific interests and research to the development of machines and processes which will cause an electrical imbalance and an electrical pressure.

There are six basic sources of electricity or electromotive force. They are friction, chemical action, light, heat, pressure, and magnetism.

In this chapter, we will discuss producing electricity from chemical action or batteries. You will also learn how electricity is produced by light, by solar batteries, by pressure, and from heat.

3.1 CHEMICAL ACTION

One of the more familiar sources of an electrical potential or voltage is the battery. In 1790, the Italian scientist, Galvani, observed a strange phenomena during the dissection of a frog supported on copper wires. Each time he touched the frog with his steel scalpel, its leg would twitch. Galvani reasoned that the frog's leg contained electricity.

As a result of these experiments, Alessandro Volta, another Italian scientist, invented the electric cell, named in his honor, called the Voltaic Cell. The unit of electrical pressure, the VOLT, is also named in his honor. Volta discovered that when two dissimilar elements were placed in a chemical which acted upon them, an electrical potential was built up between them. Thus, electricity can be produced by chemical action.

The student may construct several voltaic cells to demonstrate this action. Cut a one inch square of blotting paper and soak it in a strong salt solution. Place the wet paper between a penny and a nickel as shown in Fig. 3-1. If a sensitive meter is connected to the coins, it will indicate that a small voltage is present.

In Fig. 3-2 electricity is created by a grapefruit. Make small cuts in the skin of a grapefruit. In one cut, place the penny; in the other, the nickel. Once again, a meter will indicate a voltage. A better cell may be made by placing a carbon rod (these may be removed from an old dry cell) and a strip of zinc in a glass jar containing an acid and water solution, Fig. 3-3. Follow all safety precautions when performing experiments.

Electricity and Electronics

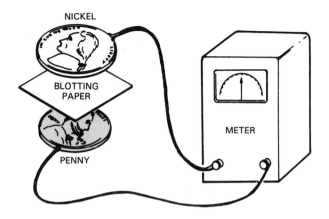

Fig. 3-1. A simple cell is produced by a nickel, a penny, and a salt solution.

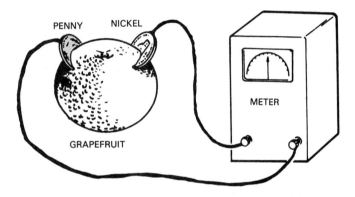

Fig. 3-2. A grapefruit will produce enough electricity to operate a small transistor radio.

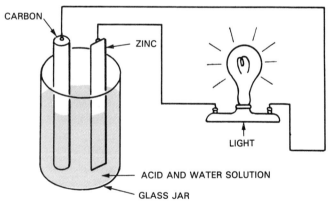

Fig. 3-3. An experimental cell, made with zinc, carbon, and acid.

LESSON IN SAFETY: When mixing acid and water, always pour acid into water. NEVER pour water into acid. Acid will burn your hands and your clothing. Wash your hands at once with clear water if you spill acid on them. Acid may be neutralized with baking soda. See your instructor for first aid treatment!

When the polarity of the carbon rod is tested, it will be positive. The zinc will be negative. If a wire is connected between these elements or electrodes, a current will flow. A VOLTAIC CELL can be described as a way of converting chemical energy into electrical energy.

In the above mentioned example of a voltaic cell, the sulfuric acid (H_2SO_4) and water (H_2O) solution is also known as ELECTROLYTE. When the electrodes are placed in this acid electrolyte, a chemical action takes place. The sulfuric acid breaks down into positive ions ($2H^+$) and negative ions (SO_4^{2-}). The negative ions move toward the zinc electrode, and combine with it by making zinc sulfate ($ZnSO_4$). The positive ions move toward the carbon electrode. This action creates a potential difference between the electrodes. The zinc will be negative. The carbon will be positive. This cell will develop about 1.5 volts.

If a load, such as a light, is connected to the cell, a current will flow and the light will glow, Fig. 3-3. As the cell is used, the chemical action continues until the zinc electrode is consumed. The chemical equation for this action would be:

$$Zn + H_2SO_4 + H_2O \rightarrow ZnSO_4 + H_2O + H_2\uparrow$$

Zinc plus sulfuric acid plus water chemically reacts to form zinc sulfate and water and free hydrogen gas. This cell cannot be recharged because the zinc has been consumed. It is called a PRIMARY CELL. The chemical action cannot be reversed.

DEFECTS IN PRIMARY CELLS

One might think that the chemical action of the voltaic cell would continue to produce a voltage as long as the active ingredients of the cell were present. In studying the equation for the discharge of the cell, you will observe the formation of free hydrogen gas. Since the carbon electrode does not enter into chemical action, the hydrogen forms gas bubbles. These collect around the carbon electrode. As the cell continues to discharge, an insulating blanket of bubbles will form around the carbon. This reduces the output and terminal voltage of the cell. The cell is said to be POLARIZED. The action is called POLARIZATION.

To overcome this defect in the simple voltaic cell, a DEPOLARIZING AGENT may be added. Compounds which are rich in oxygen, such as manganese dioxide (MnO_2), are used for this purpose. The oxygen in the depolarizer combines with the hydrogen bubbles and forms water. This chemical action appears as:

$$2MnO_2 + H_2 \rightarrow Mn_2O_3 + H_2O$$

The free hydrogen has been removed, so the cell will continue to produce a voltage.

Sources of Electricity

One might assume that when current is not being used from the cell, the chemical action would also stop. However, during the smelting of zinc ore, not all impurities are removed. Small particles of carbon, iron, and other elements remain. These impurities act as the positive electrode for many small cells within the large cell. But this chemical action adds nothing to the electrical energy produced at the cell terminals. This action is called LOCAL ACTION. It may be reduced by using pure zinc for the negative electrode, or by a process called AMALGAMATION. A small quantity of mercury is added to the zinc during manufacturing. As mercury is a heavy liquid, any impurities in the zinc will float on the surface of the mercury, causing them to leave the zinc surface. This process increases the life of a primary cell.

THE ZINC-CARBON CELL

Although the primary cell has been described as a liquid cell, the liquid type is not in common use. Rather, the primary cell is often a DRY CELL. A dry cell averts the danger of spilling liquid acids.

Flashlight batteries (cells) are examples of a dry cell. The dry cell consists of a zinc container which acts as the negative electrode. A carbon rod in the center is the positive electrode. Surrounding the rod is a paste made of ground carbon, manganese dioxide, and sal ammoniac (ammonium chloride), mixed with water. The depolarizer is the MnO_2. The ground carbon increases the effectiveness of the cell by reducing its internal resistance. During discharge of the cell, water is formed.

You may recall having difficulty removing dead cells from a flashlight. The water caused them to expand. Although this problem has been solved by improved manufacturing techniques, it is still not advisable to leave cells in your flashlight for long periods of time. You should keep fresh cells in your flashlight, so it will be ready for emergency use.

LESSON IN SAFETY: Improper battery use may cause leakage and explosion. Therefore, obey the following precautions.
1. Install the batteries with the positive (+) and negative (−) polarities in the proper direction.
2. Do not use new and old batteries together.
3. Do not use carbon zinc batteries with other types of batteries.
4. Never attempt to short circuit, disassemble, or heat batteries. Do not throw batteries in fire.

Carbon zinc batteries are not rechargeable. If recharged, they may leak and explode.

THE ALKALINE CELL

The alkaline battery is composed of manganese dioxide for the positive activating substance, zinc powder for the negative activating substance, and caustic alkali for the electrolyte. Recent progress in electronic product design has demanded more compact supply sources. The number of products needing a large current and a large capacity have increased. This requires the development of more advanced batteries. Cylindrical alkaline batteries are now widely used to supply power for electronic products. They can be used with common manganese dioxide batteries, Figs. 3-4 and 3-5.

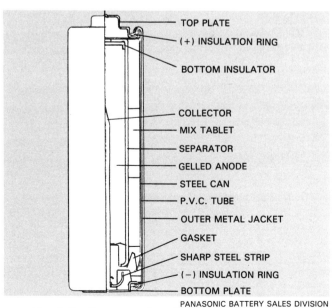

MALLORY BATTERY CO.

Fig. 3-4. Popular sizes of alkaline cells.

PANASONIC BATTERY SALES DIVISION

Fig. 3-5. Cutaway of an AA size alkaline cell.

THE MERCURY CELL

A relatively new type of dry cell is shown in Fig. 3-6. It is called a mercury cell. It creates a voltage of 1.34 volts by the chemical action between zinc (−) and mercuric oxide (+). It is costly to make. However, the mercury cell is better in that it creates about five times more current than the conventional dry cell. It also maintains its terminal voltage under load for longer periods of operation. The mercury cell has found wide use in powering field instruments and portable communications systems.

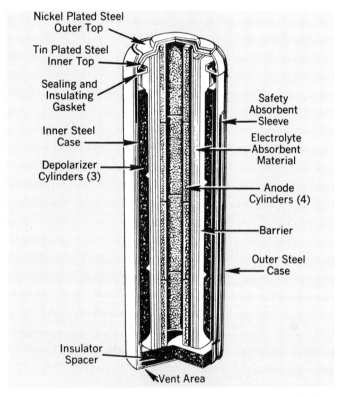

Fig. 3-6. Mercury cell. It creates voltage by chemical action between zinc and mercuric oxide.

LITHIUM CELL

Lithium has the highest negative potential of all metals. It is, therefore, the best substance for an anode.

Many battery make-ups are possible by mixing lithium with various cathode substances.

Energy densities of these batteries can be computed by RESPECTIVE REACTION EQUATIONS. Fig. 3-7 shows energy densities of lithium batteries compared with those of conventional batteries. Even if high voltage batteries are desired, batteries using water-based (aqueous) electrolyte cannot be used in lithium batteries. Lithium batteries have high energy density due to a non-aqueous organic solvent for electrolyte.

Features of lithium batteries, such as voltage and discharge capacity, are determined by the type of cathode substance used. Fluorocarbon is an INTERCALATION (inserted between or among existing elements) compound. It is produced through reaction of carbon powder and fluorine gas. It is expressed in $(CF)_n$.

Lithium is the lightest metal, with an electric capacity of 3.86 Ah/g. The electrode reaction is expressed by $Li \rightarrow Li^+ + e$ where the lithium is dissolved during discharge. Lithium is also the most negative metal with a standard electromotive force of $E_o = -3.045$ V. Therefore, lithium is the most suitable anode for production of high voltage and light weight batteries.

If an alkalimetal salt is dissolved in a non-aqueous organic solvent, an electrolyte with ionic conduction will be produced. An aprotic organic solvent with a high dielectric constant, low viscosity, and a high boiling point is used for the electrolyte. Refer to Fig. 3-8.

SILVER OXIDE CELLS

Silver oxide cells have several advantages over other types of cells. These advantages include:
- Very stable discharge voltage.
- Excellent high discharge characteristics.

Reaction		E*	(Wh/kg)	
$nLi + (CF)_n$	$\rightarrow nLiF + nC$	3.2*	2,260	Poly-carbonmonofluoride Lithium Battery
$8Li + 3SOCl_2$	$\rightarrow 6LiCl + Li_2SO_3 + 2S$	3.61*	1,877	Thionyl Chloride Lithium Battery
$2Li + CuF_2$	$\rightarrow 2LiF + Cu$	3.54	1,646	—
$2Li + NiF_2$	$\rightarrow 2LiF + Ni$	2.83	1,370	—
$2Li + 2SO_2$	$\rightarrow Li_2S_2O_4$	2.95*	1,114	Sulfur Dioxide Lithium Battery
$2Li + 2MnO_2$	$\rightarrow Li_2O + Mn_2O_3$	2.69	768	Manganese Dioxide Lithium Battery
$2Li + Ag_2CrO_4$	$\rightarrow Li_2CrO_4 + 2Ag$	3.35*	520	Silver Chromate Lithium Battery
Conventional batteries				
$Zn + 2MnO_2$	$\rightarrow Zn^{2-} + 2MnOOh$	1.7	234	Carbon Zinc Battery
$Zn + HgO$	$\rightarrow Zn(OH)_4^{2-} + Hg$	1.4	266	Mercury Battery
$Zn + Ag_2O$	$\rightarrow ZnO + 2Ag$	1.6	287	Silver Oxide Battery

PANASONIC BATTERY SALES DIVISION

Fig. 3-7. Theoretical energy densities of lithium batteries compared with conventional batteries.

Sources of Electricity

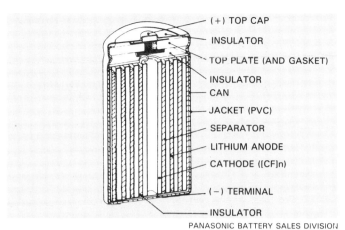

Fig. 3-8. Cross-sectional view of a cylindrical shaped lithium battery.

- High energy density per unit volume.
- Wide range of operating temperatures.
- Compact, thin size.

Compact silver oxide batteries have the highest electrical volume and leakage resistance of any battery of that size. This is particularly true when used in watches. Two types of silver oxide batteries are made for use in watches. One type uses caustic potash for electrolyte. The other uses caustic soda. The caustic potash battery has the symbol W on the bottom of the battery. It is for high drain use. It is used in wristwatches with LCD lamps and multi-function analog watches. The caustic soda battery has the symbol SW on the bottom of the battery. It is for low drain use, and is used mostly in single function analog watches. Fig. 3-9 shows a cutaway of a silver oxide cell.

NICKEL-CADMIUM RECHARGEABLE CELL

Another development in the field of dry cells is the rechargeable cell. Basically, these are nickel-cadmium alkaline batteries with paste rather than liquid for the electrolyte. Among the many advantages of these cells is their ability to be recharged. Other advantages include long life, high efficiency, compactness, and light weight. The nickel-cadmium cell will produce a high discharge current due to its low internal resistance. A flashlight using a rechargeable cell is shown in Fig. 3-10. Other uses include the powering of transistor radios. burglar alarm systems, photo-flashes, and aircraft instruments.

A nickel-cadmium "D" cell with a built-in solar cell charger is shown in Fig. 3-11. This cell can be installed in flashlights, radios, communications equipment, etc. It takes up the same amount of space that a standard battery requires. Yet, it provides solar battery regeneration for years, without the use of any outside charging device.

Fig. 3-10. This battery powered flashlight can be recharged by plugging it into any 115 volt outlet.

Fig. 3-9. Cutaway view of a silver oxide cell.

Fig. 3-11. Sunlight can be used to recharge this solar nickel-cadmium cell.

One type of nickel-cadmium cell uses positive and negative plates, a separator, alkaline electrolyte, a metal case, and a sealing plate with self-resealing safety vent. It is shown in Fig. 3-12.

The positive plate is a porous, sintered nickel base plate. It is filled with nickel hydroxide. The negative plate is a punched plate of thin steel, coated with cadmium active material. The separator is made of a polyamide fiber. For high temperature uses, it is made of a nonwoven polypropylene fiber. The positve plate, separator, and negative plate are pressed together, wound into a coil, and inserted in the metal case.

The electrolyte is an alkaline aqueous solution. It is totally absorbed into the plate and separator. The metal case is constructed of nickel-plated steel. It is welded on the inside to the negative plate. It becomes the negative pole. The sealing plate uses a special liquid sealing agent to form a perfect seal. The positive plate is welded on the inside to the sealing plate. It becomes the positive pole. The self-resealing safety vent permits the discharge of gas in the event of an abnormal increase of internal pressure. This prevents against rupture or other damage. The vent is made of a special alkaline and oxidation resistant rubber. This insures that operating pressure and safety features will be retained over a long period of time.

The electromechanical processes of a nickel-cadmium alkaline cell are outlined below:

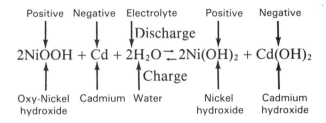

$$\underset{\text{Oxy-Nickel hydroxide}}{2\text{NiOOH}} + \underset{\text{Cadmium}}{\text{Cd}} + \underset{\text{Water}}{2\text{H}_2\text{O}} \underset{\text{Charge}}{\overset{\text{Discharge}}{\rightleftarrows}} \underset{\text{Nickel hydroxide}}{2\text{Ni(OH)}_2} + \underset{\text{Cadmium hydroxide}}{\text{Cd(OH)}_2}$$

In this process, charging and discharging are reversed in a very efficient manner. The electrical energy used during discharge is regained during recharge.

During the final charging stage, an oxygen gas is created with the reaction occurring at the positive:

Positive $\quad \underset{\text{Hydroxide ions}}{4\text{OH}^-} \rightarrow \underset{\text{Oxygen}}{\text{O}_2\uparrow} + \underset{\text{Water}}{2\text{H}_2\text{O}} + \underset{\text{Electrons}}{4\text{e}^-} \ldots (1)$

This oxygen passes through the separator to the negative. After this, an absorption reaction takes place at the negative and absorption occurs.

Negative $\quad \underset{\text{Oxygen}}{\text{O}_2} + \underset{\text{Water}}{2\text{H}_2\text{O}} + \underset{\text{Electrons}}{4\text{e}^-} \rightarrow \underset{\text{Hydroxide ions}}{4\text{OH}^-} \ldots (2)$

THE BATTERY

Often, a single cell is called a battery. By strict definition, however, a battery consists of two or more cells connected together. These cells are enclosed in one case.

In the study of electricity, it is important to understand the purpose and results of connecting cells in groups. First, consider the series connection. In this method the positive terminal of one cell is connected to the negative terminal of the second cell. In Fig. 3-13,

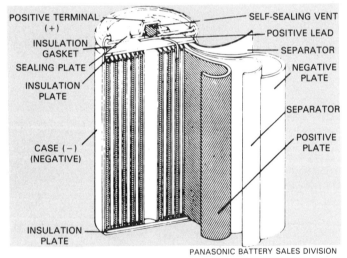

Fig. 3-12. Construction of a nickel-cadmium cell.

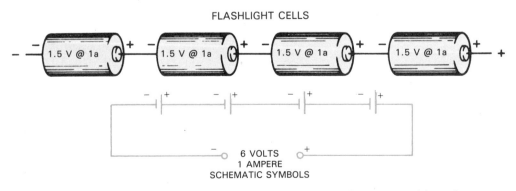

Fig. 3-13. Pictorial and schematic diagram of four cells connected in series.

Sources of Electricity

four cells are connected in SERIES. The output voltage will equal

$$E_{out} = E_{one\ cell} \times n,$$

where n equals the number of cells. So,

$$E_{out} = 1.5\ volts \times 4 = 6\ volts.$$

Notice that the voltage has increased four times. However, the capacity of the battery to supply a current is the same as one cell. Cells are connected in this manner to supply higher voltages for many uses. A flashlight may use two or more cells in series. B batteries for portable radios have many cells in series to produce 22 1/2, 45, or 90 volts. The amperage does not increase by connecting cells in series.

In Fig. 3-14, the positive terminals have been connected together, and the negative terminals have been connected together. These cells are connected in PARALLEL. The total voltage across the terminals of the battery is the same as one cell only. Although the voltage has not increased, the life of the battery has been increased because the current is drawn from all cells instead of one. The amperage is added by connecting cells in parallel. Generally speaking, if the load applied to the battery is a low resistance load, it is better to use the parallel connection. For a high resistance load, use the series connection.

Cells can also be connected in a mixed grouping. In Fig. 3-15, two groups of batteries with 6 volts terminal voltage are connected in parallel. Total voltage is still 6 volts. But the capacity has been increased by this SERIES-PARALLEL method of connecting cells. The total current is added for 2 amperes.

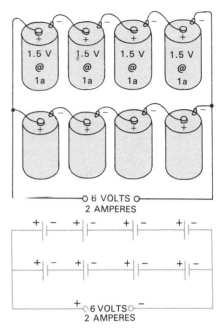

Fig. 3-15. Pictorial and schematic diagram of two groups of cells connected in series and those two groups connected in parallel.

SECONDARY CELLS

A secondary cell can be recharged or restored. The chemical reaction which occurs on discharge may be reversed by forcing a current through the battery in the opposite direction. This charging current must be supplied from another source, which may be a generator or a power company. Fig. 3-16 shows one type of battery charger used by garages and gas stations for recharging automobile batteries. An alternating current, which will be studied in a later chapter,

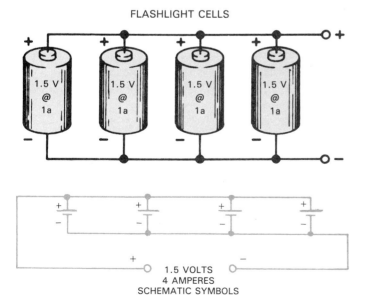

Fig. 3-14. Pictorial and schematic diagram of four cells connected in parallel.

CENTURY MFG. CO.

Fig. 3-16. This type of quick charger is used on car batteries. It is found in many service stations.

must be RECTIFIED to a direct current for charging the battery.

LEAD ACID CELLS

A common type of lead acid cell is the car storage battery. A storage battery does not store electricity. Rather, it stores chemical energy which produces electrical energy. The active ingredients in a fully charged battery are lead peroxide (PbO_2) which acts as the positive plate, and pure spongy lead (Pb) for the negative plate. The liquid electrolyte is sulfuric acid (H_2SO_4) and water (H_2O). The positive plates are a reddish-brown color. Negative plates are gray.

The chemical reaction is rather involved. However, study the information given in Fig. 3-17. Notice that during discharge, both the spongy lead and the lead peroxide (also called lead dioxide) plates are being changed to lead sulfate and the electrolyte is being changed to water. When the cell is recharged, the reverse action occurs. The lead sulfate changes back to spongy lead and lead peroxide; the electrolyte to sulfuric acid.

LESSON IN SAFETY: During the charging process of a storage battery, highly explosive hydrogen gas may be present. Do not smoke or light matches near charging batteries. Charge only in a well ventilated room. Batteries should be first connected to the charger, before the power is applied. Otherwise, the sparks made during connection might ignite the hydrogen gas and cause an explosion.

The electrolyte of a fully charged battery is a solution of sulfuric acid and water. The weight of pure sulfuric acid is 1.835 times heavier than water. This is called its specific gravity.

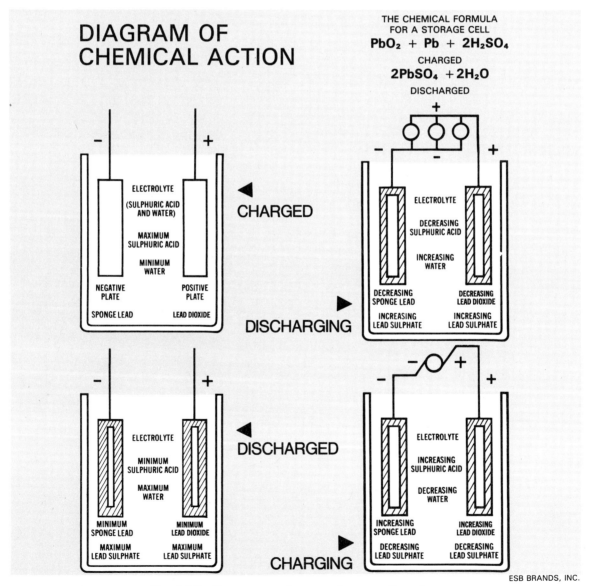

Fig. 3-17. How a secondary, or storage cell works.

Sources of Electricity

SPECIFIC GRAVITY is the weight of a liquid as it compares to water. The specific gravity of water is 1.000. The acid and water mixture in a fully charged battery has a specific gravity of approximately 1.300 or less. As the electrolyte changes to water when the cell discharges, the specific gravity becomes approximately 1.100 to 1.150. Therefore, the specific gravity may be used to determine the state of charge of a cell. The instrument used to measure the specific gravity is a HYDROMETER, Fig. 3-18.

Readings may be taken from the floating bulb and scale. The principle of the hydrometer is based on Archimedes principle in physics. This states that a floating body will displace an amount of liquid equal to its own weight. If the cell is in a fully charged state, the electrolyte liquid is heavier, so the float in the hydrometer will not sink as far. The distance that it does sink, is calibrated in specific gravity on the scale. This can be read as the state of charge of the cell.

The plates of a modern lead acid cell are made of a grid of special lead-antimony alloy. A paste of the chemically active material is pressed on the grid, Fig. 3-19. The grid holds the material in place and adds mechanical strength. When the paste on the plate grids is hard and dry, each plate is formed to its correct chemical material by placing it in an electrolyte, and passing a current through it in the proper direction. One direction makes a positive plate; the opposite direction, a negative plate. Refer again to Fig. 3-19. It shows the construction of a cell, which consists of alternate negative and positive plates grouped together. All negative plates and all positive plates are connected together by a lead plate strap at the top. The plates are insulated from each other by separators. These separators are thin, porous layers of insulation. They are usually made of wood, rubber, or spun glass.

LESSON IN SAFETY: Sulfuric acid used in automotive storage batteries will burn your hands, eat holes in your clothes and generally destroy other materials which it contacts. When testing a cell with a hydrometer, be certain that small drops of acid at the end of the rubber tube do not drip on you or on nearby objects of value. Hold the hydrometer over the battery filler hole while taking readings. Store hydrometer in hanging position in glass or rubber jar.

LESSON IN SAFETY: Expensive storage batteries may be destroyed by excessive vibration and rough handling. Chemicals may break off from the plates, and cause internal short circuits and dead cells. Handle a battery gently and be sure it is securely clamped and bolted in your car.

In the 12 volt automotive battery, six of these cells are placed in a molded hard rubber case. Each cell has its own individual compartment. At the bottom of each compartment a space sediment chamber is provided. This is where particles of chemicals broken from the plates due to chemical action or vibration, may collect. Otherwise, these particles might short out the plates and make a dead cell.

The individual cells are connected in series by lead alloy connectors. The entire battery is then sealed with a battery-sealing compound. Each cell is provided with a filler cap, so that the distilled water may be added to the electrolyte as it evaporates. Periodically check the liquid level in each cell. It should be about 1/4 in. above the top edges of the plates. If the electrolyte does not completely cover the plates, the ends of the plates exposed to air will become seriously sulfated or coated with a hard inactive chemical compound. Badly sulfated batteries have a short life. Maintenance free batteries are also available.

LESSON IN SAFETY: Storage batteries are heavy. Use the proper carrying strap when moving a battery. Get help from a fellow worker when lifting a battery into a car. Do not strain yourself by improper lifting.

DELCO-REMY DIV., GENERAL MOTORS CORP.

Fig. 3-19. Cutaway view of a typical lead acid battery. This battery is designed for maintenance free use in automobiles. Note the arrangement of plates and intercell connectors.

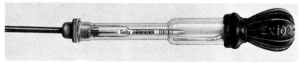

EXIDE SALES, ELECTRIC STORAGE BATTERY CO.

Fig. 3-18. A hydrometer is used to check the state of charge of a battery. This tool is used in many service stations to check the charge of car batteries.

Electricity and Electronics

BATTERY CAPACITY

It is important that you understand the term CAPACITY as it relates to batteries. The capacity of a battery is its ability to produce a current over a certain period of time. It is equal to the product of amperes supplied by the battery and the time. Capacity is measured in AMPERE-HOURS (AH). The description of an automotive battery might indicate a capacity of 100 ampere-hours. This would mean that the battery could supply:

100 amps for 1 hr. $100 \times 1 = 100$ amps hrs.
50 amps for 2 hrs. $50 \times 2 = 100$ amps hrs.
10 amps for 10 hrs. $10 \times 10 = 100$ amps hrs.
1 amp for 100 hrs. $1 \times 100 = $ amp hrs.

A battery will not perform exactly by this schedule, as the RATE OF DISCHARGE must always be considered. A rapidly discharged battery will not give its maximum ampere-hour rating. A slowly discharged battery may exceed its rated capacity. The Society of Automotive Engineers (SAE) has set standards for the rating of automotive batteries. A manufacturer must meet these standards in order to advertise a battery as a specific ampere-hour capacity.

Several factors determine the capacity of a storage battery:

1. The number of plates in each cell. An increased number of plates provides more square inches of surface area for chemical action. Automotive batteries are commonly made with 13, 15, and 17 plates per cell. The number of plates is a determination of the life and quality of a battery.
2. The kind of separators used will have some effect on the capacity and life of a battery.
3. The general condition of the battery with respect to its state of charge, age, and care will influence the capacity rating of any given battery.

To compare primary and secondary cells and batteries, refer to Fig. 3-20. Note that their most common uses are given.

Category	Type		Configuration			(V)					Storage life	Number of times it can be recharged	Typical applications:
			Positive activating substance	Electrolyte	Negative-activating substance	Nominal voltage	Strong current	Voltage stability	Low temperature	High temperature			
Primary batteries	Carbon Zinc batteries		MnO_2	$ZnCl_2$ NH_4Cl	Zn	1.5	C B during intermittence	C	B	B	A	Non-rechargeable	Lighting equipment, radios, tape recorders, electronic calculators, hearing aids, toys, TVs
	Alkaline-manganese dioxide batteries	Cylinder	MnO_2	KOH(ZnO)	Zn	1.5	B	B	A	A	A	Non-rechargeable	Electric shavers, camera strobes, tape recorders, electronic calculators
		Button									AA	Non-rechargeable	Clocks, watches, toys, electronic calculators, cameras, lighters
	Mercury batteries		HgO	KOH(ZnO)	Zn	1.35	B	AA	B	A	A	Non-rechargeable	Hearing aids, electronic calculators, exposure meters, EE cameras, wireless microphones, measuring instruments
			HgO	NaOH(ZnO)	Zn	1.35	C	AA	C	A	AA	Non-rechargeable	Watches
			HgO+MnO_2	KOH(ZnO)	Zn	1.4	B	A	A	A	A	Non-rechargeable	Medical equipment, radios, cameras, electronic shutters, hearing aids, pagers
	Silver oxide batteries		Ag_2O	KOH(ZnO)	Zn	1.55	B	AA	B	A	A	Non-rechargeable	Cameras, watches, electronics calculators, lighters, hearing aids, radios, thermometers, remote control units
			Ag_2O	NaOH(ZnO)	Zn	1.55	C	AA	C	A	AA	Non-rechargeable	Watches
	Lithium batteries		(CF)n	Organic tetrachloride solvent	Li	3.0	C~B	A	AAA	AA	AAA	Non-rechargeable	Memory backup power for cameras, illuminated buoys, watches, communication equipment, water and gas meters, computers
			MnO_2	Organic tetrachloride solvent	Li	3.0	C~B	B	A	A	AA	Non-rechargeable	Electronic calculators, memory backup power sources, cameras, watches, toys
Secondary batteries	Sealed nickel-cadmium (Ni-Cd) batteries		NiOOH	KOH	Cd	1.2	AA	A	AA	A	B	300~1000	Lighting equipment, electric shavers, electric tools, electronic calculators, ECRs, emergency lights, guide lights
	Sealed lead-acid batteries		PbO_2	H_2SO_4	Pb	2.0	A	A	A	A	A	100~300	VTRs, TVs, measuring instruments, ECRs, emergency equipment, ups, Engin starting, Tools, HHC

PANASONIC BATTERY SALES DIVISION

Fig. 3-20. Comparison of primary and secondary batteries and their uses.

Sources of Electricity

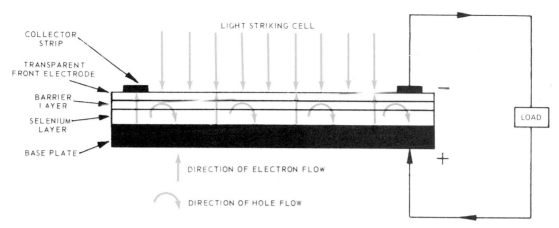

Fig. 3-21. Typical structure of a photovoltaic cell.

REVIEW QUESTIONS FOR SECTION 3.1

1. Explain how a primary cell operates.
2. Name four primary cells.
3. What are the advantages of a silver oxide cell?
4. Name two types of rechargeable secondary cells.
5. Connecting cells in series increases their _____ rating.
6. Give some typical uses for zinc carbon and alkaline batteries.

3.2 OTHER SOURCES OF ELECTRICAL ENERGY

ELECTRICAL ENERGY FROM LIGHT

For many years scientists have attempted to transform light energy from the sun into useful amounts of electrical power. Although certain experimental equipment has been successful, widespread use of solar power is still in its infancy.

Just as light may be produced by an electric current running through a resistance or filament in a light bulb, light may also produce an electric current. The PHOTOVOLTAIC CELL is one device which makes this conversion. This device will develop an electrical potential between its terminals, when the surface of the cell is exposed to light.

A selenium photovoltaic cell is constructed of a layer of selenium compound, a barrier layer, and transparent front electrode, Fig. 3-21. When light strikes the surface, electrons from the selenium compound cross the barrier. This creates a potential difference between the base plate and the front electrode. The electrons cannot return to the selenium compound because the barrier permits conduction in one direction only. It is a UNIDIRECTIONAL CONDUCTOR. Fig. 3-22 shows the construction of a solar photovoltaic cell.

Fig. 3-23 shows sun batteries used by experimenters. The batteries may be connected in series and parallel for higher power requirements.

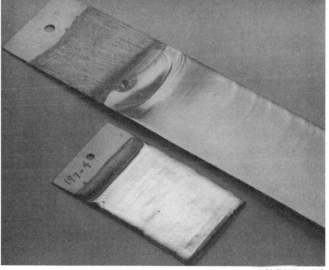

Fig. 3-22. Solar photovoltaic cell construction.

Fig. 3-23. Selenium photovoltaic cells, mounted and unmounted, with and without leads.

Other uses for photocells include light meters used in photography. In this application, a sensitive meter responds to the current generated by the cell. The brighter the scene, the greater deflection of the meter.

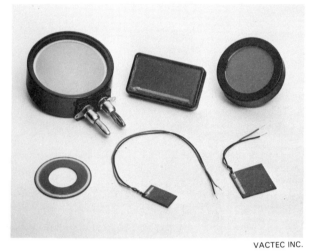

Fig. 3-24. Photocells and their symbol. Various sizes are shown; both hermetic packages and plastic coated.

Fig. 3-24 shows some photosensitive cells used to control the amount of light reaching the film. Photocells are used in industry for counting, sorting, and automatic inspection.

PHOTOELECTRIC CONTROL

A photoelectric control device using a photocell is shown in Fig. 3-25. When light shines on the cell, the resistance of the circuit changes. A cadmium-sulphide photocell is a light variable resistor. It is most sensitive in the green to yellow portion of the light spectrum. With it you can use light to control many electronic devices. Photocells are used in counting operations, burglar alarms, door opening mechanisms, and in many other devices.

ELECTRICAL ENERGY FROM HEAT

A device used to indicate and control the heat of electric ovens and furnaces is shown in Fig. 3-26. It is called a THERMOCOUPLE. When two dissimilar metals in contact with each other are heated, a potential difference will develop between the metals, Fig. 3-27.

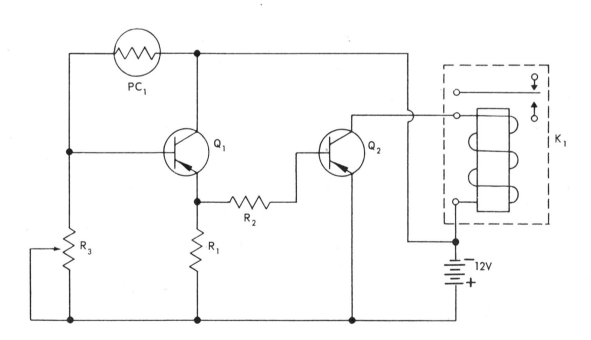

PHOTOELECTRIC CONTROL
PARTS LIST

R_1, R_2 — 100 Ohms, 1/2 W
R_3 — 1000 Ohms, Potentiometer
Q_1, Q_2 — 2N408, Transistors
K_1 — 4.6 mA, Relay
PC_1 — Photocell, (Sylvania 8143)

Fig. 3-25. Schematic and parts list for photoelectric control.

Sources of Electricity

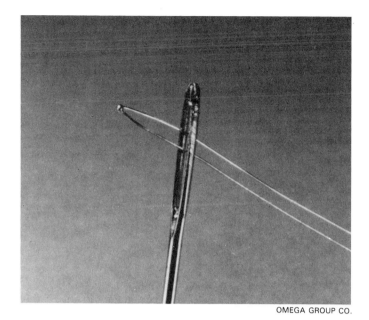

Fig. 3-26. Small thermocouple.

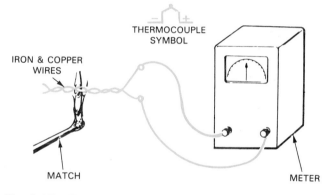

Fig. 3-27. The basic principle of the thermocouple can be demonstrated by heating two dissimilar wires twisted together.

Fig. 3-28. Temperature is measured with this instrument by pointing it at the heat.

Fig. 3-29. Attaching a thermocouple module to a digital multimeter allows the thermocouple to be used to measure temperature.

In the demonstration of Fig. 3-27, an iron stovepipe wire and a copper wire are twisted tightly together. Their ends are connected to a sensitive meter such as a GALVANOMETER. A galvanometer is a device that is capable of measuring very small currents. When the flame of a lit match heats the twisted joint, a reading can be observed on the meter indicating that an electromotive force is present. Output voltage of a thermocouple may be strengthened and used to work large motors, valves, controls, and recording devices.

Commercial types of thermocouples employ various kinds of dissimilar metals and alloys such as nickel-platinum, chromel-alumel, and iron-constantan. These unfamiliar names apply to special alloys developed for thermocouples.

The combination indicating device including meter and thermocouple is a PYROMETER. See Figs. 3-28 and 3-29. For an instrument which must be sensitive to temperature change, a large number of thermocouples may be joined in series. Such a group is known as a THERMOPILE.

Electricity and Electronics

ELECTRICAL ENERGY FROM MECHANICAL PRESSURE

Many crystalline substances such as quartz, tourmaline, and Rochelle salts have a peculiar characteristic. When a voltage is applied to the surfaces of the crystal, the crystal will become distorted. The opposite is also true. If a mechanical pressure or force is applied to the crystal surface, a voltage will be developed. The crystal microphone, Fig. 3-30, is a familiar example of this process. Sound waves striking a diaphragm mechanically linked to the crystal surfaces, cause distortion in the crystal. This develops a voltage across its surfaces. Thus, sound waves are converted to the electrical energy.

Another familiar application is the phono cartridge installed in the pick-up arm of a record player. The phonograph needle is moved by the sound grooves in the record. The moving needle applies mechanical pressure to the crystal in the cartridge and an electrical potential is developed. The potential will vary in frequency and magnitude according to the movement of the needle. The varying potential is amplified and connected to a speaker. You hear the voice or music.

Creating electricity by the mechanical distortion of a crystal is know as PIEZOELECTRIC EFFECT. Crystals may be cut for particular operating characteristics. In a later chapter, the use of crystals as frequency controls for radio transmitters will be discussed.

ELECTRICITY FROM MAGNETISM

A current primary source of electrical energy is the dynamo or generator. Generators prove that magnetism can produce electricity.

A GENERATOR is a rotating machine that converts mechanical energy into electrical energy. This source of electricity requires detailed study. Chapter 7 is devoted to generators, generator types, and controls.

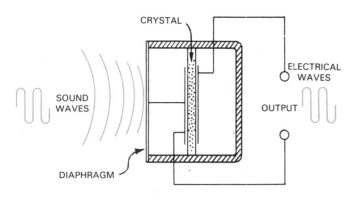

Fig. 3-30. Crystal microphone converts sound waves to electrical energy.

REVIEW QUESTIONS FOR SECTION 3.2

1. The _____ produces electricity from light.
2. A light variable resistor is called a _____.
3. What device is used to produce electricity from heat?
4. The effect of producing electricity from pressure is _____.
5. Sketch and explain how a photovoltaic cell operates.

EXPERIMENT I. Construct a cell using a penny, a nickel, and blotting paper soaked in salt solution. Measure the voltage with a multimeter.

EXPERIMENT II. Construct a thermocouple by twisting the ends of a copper and an iron wire together. Connect the wires to a voltmeter. Then heat the junction with a match. Observe the voltage developed.

EXPERIMENT III. Obtain these materials: 4 flashlight cells, 4 cell holders, and 1 voltmeter. Connect the cells in series, as in Diagram 1.

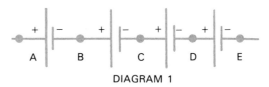

DIAGRAM 1

What is the voltage between A and B? A and C? A and D? A and E? What conclusion do you draw from this experiment?

EXPERIMENT IV. Using the same materials as used in Experiment III, connect the cells in parallel, as in Diagram 2.

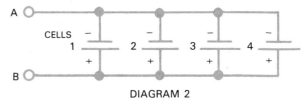

DIAGRAM 2

What is the voltage between A and B? Remove cell No. 4. Measure voltage from A to B. What is the voltage now? Remove cells No. 3 and 4. Measure the voltage from A to B. Remove cells No. 2, 3, and 4. Measure the voltage from A to B. What conclusion do you draw from this experiment?

EXPERIMENT V. Secure a sun battery. Connect voltmeter leads to battery terminals. Observe meter reading. Next, shine a flashlight on the cell and note the higher meter reading. Finally, place the cell in sunshine and note the still higher meter reading.

SUMMARY

1. Electricity can be produced by chemical action.
2. Cells are the basic unit for producing electricity by chemical action. Batteries are two or more cells connected together.
3. Primary cells cannot be recharged while secondary cells can be recharged.
4. Some popular primary cells include the zinc-carbon cell, the alkaline cell, the mercury cell, the lithium cell, and the silver oxide cell.
5. Nickel-cadmium cells and lead-acid cells are secondary cells.
6. Connecting cells in series (− to +) increases their voltage rating while connecting them in parallel (− to − and + to +) increases their current rating.
7. Battery capacity is a current producing rating measured in ampere-hours (AH).
8. A photovoltaic cell produces electricity from light.
9. Photoresistive cells are light sensitive resistors.
10. The device used to produce electricity by heat is the thermocouple.
11. Electricity can also be produced by applying pressure to certain objects such as quartz, tourmaline, and Rochelle salts.

TEST YOUR KNOWLEDGE, Chapter 3

Please do not write in the text. Place your answers on a separate sheet of paper.

1. Name six basic sources of electricity or electromotive force.
2. What safety precautions should be taken when mixing acid and water?
3. A _____ _____ is a way of converting chemical energy into electrical energy.
4. A cell that cannot be recharged is a:
   a. One-way cell.
   b. Secondary cell.
   c. Primary cell.
   d. None of the above.
5. What is polarization?
6. List five advantages silver oxide batteries have over other types of batteries.
7. What is the major advantage of the nickel-cadmium cell?
8. In a _____ connection, the positive terminal of one cell is connected to the negative terminal of the second cell.
9. In a _____ connection, all positive terminals are connected and all negative terminals are connected.
10. What is the composition of the electrolyte in a lead acid storage battery?
11. Specific gravity:
    a. Can be measured with a hydrometer.
    b. Is used to determine the state of charge of a cell.
    c. Is the weight of a liquid as it compares to water.
    d. All of the above.
12. What is the measurement of battery capacity?
13. A _____ is a device used to indicate and control the heat of electric ovens and furnaces.
14. What is piezoelectric effect?

FOR DISCUSSION

1. How does chemical action operate our nervous system? Research this question. Write a brief report on your findings.
2. Do research on one of the following topics. Give a report in class.
   a. The electrical system of a lightning bug.
   b. Electric eel.
3. How can electricity from light be used to solve some of our energy problems? What are the restrictions?
4. Why do materials such as quartz have a piezoelectric effect?

Electricity and Electronics

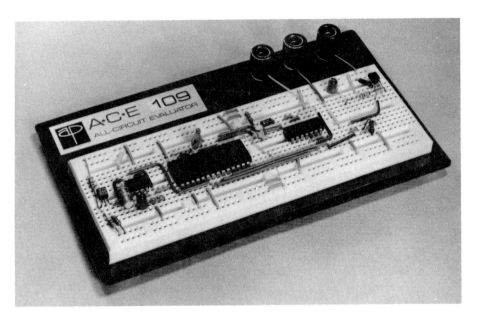

A "breadboard" such as this is a good way to check whether a circuit design will work.

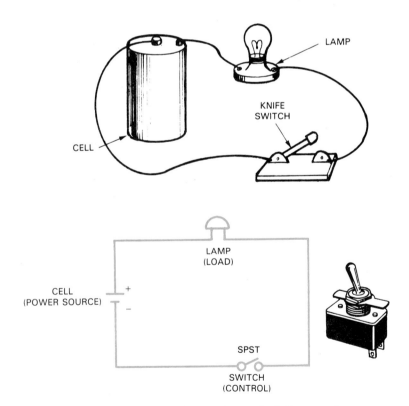

A simple circuit must have all of these parts.

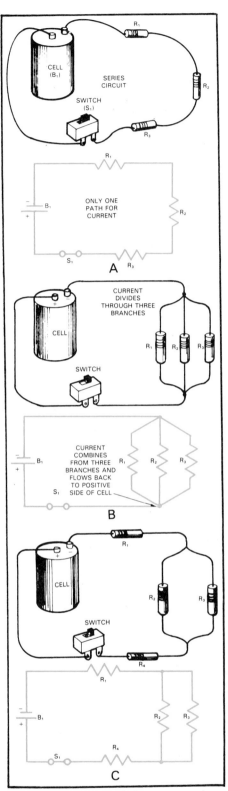

All circuits are one of three types: A—Series (only one path for current). B—Parallel (has more than one path for current). C—Combination (has both series and parallel paths for current).

Electrical power can only be used when electrical current can travel along a pathway and eventually end up where it began. This pathway is called a circuit because the current must complete the entire "circuit." In addition to a path, every circuit has a power source, a load, and some type of control.

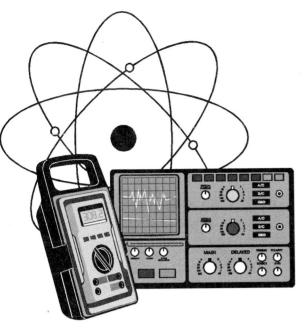

Chapter 4
BASIC CIRCUITS

After studying this chapter, you will be able to:
  Identify the fundamentals of a circuit.
  Explain the manufacture of printed circuits.
  Describe the three types of basic circuits.
  Define polarity.
  Explain how to connect meters in a circuit.
  Demonstrate understanding of both Ohm's Law and Watt's Law by solving a variety of problems.

4.1 FUNDAMENTALS OF A CIRCUIT

ELECTRICAL CIRCUITS are complete pathways on which electric current flows. Three factors are needed for a circuit to operate. Many circuits have a fourth factor that provides some type of control of current flow. These factors are voltage source, load, conductive pathway, and control. See Fig. 4-1.

VOLTAGE SOURCE

As was mentioned in Chapter 3, there are six basic voltage sources. These are friction, chemical action, heat, light, pressure, and magnetism. Friction is rarely used to provide a direct source of voltage in a circuit. The other five sources are used in various ways.

Symbols are used in electrical and electronic circuits to present various devices or components. Fig. 4-2 shows the symbols for voltage sources used to power electrical and electronic circuits.

LOADS

Every electrical circuit must have some type of load. A LOAD is the electrical device which is connected to the output of the voltage source. Current flows through the load and is consumed by the load. Refer back to Fig. 4-1. Some types of loads also provide resistance to current flow. These are called RESISTIVE LOADS. Without this resistance, an uncontrolled current would progress to a point where it would overload the circuit.

Fig. 4-1. Basic factors needed for operation of an electrical circuit.

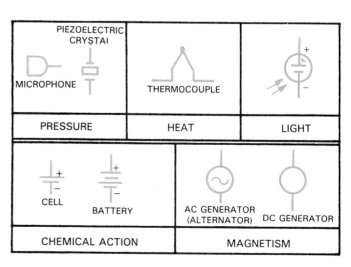

Fig. 4-2. Symbols for basic voltage sources.

Electricity and Electronics

If a circuit has no resistance, then the circuit is said to have a SHORT. Resistance is zero, in this case. If there is more than one resistor in a circuit, and some are not working while others are, that is a PARTIAL SHORT. There is reduced resistance in the circuit. See Fig. 4-3 for an illustration of these two conditions.

When the pathway for current flow is broken, an OPEN CIRCUIT occurs. The circuit is incomplete and no current flows.

As has been stated, loads are electrical and electronic devices or parts that consume current flow. Common resistive loads include lamps, resistors, computers, and heating elements. Other loads, such as inductors, capacitors, solenoids, transformers, and relays will be explained in Chapters 11 and 12. They are known as inductive and capacitive loads. Symbols for common loads are given in Fig. 4-4.

CONDUCTIVE PATHWAY

The actual pathway used in circuits to provide for electron flow is a conductor of some type. The conductor may be wire or it may be a thin piece of copper adhered to thin plastic. This type of conductor is called a PRINTED CIRCUIT. A large current carrying pathway is called an ELECTRICAL BUS.

SCHEMATICS are electrical drawings which show the correct wiring of component parts in a circuit. In a schematic, the conductors are usually shown as lines. Fig. 4-5 shows symbols for connected and disconnected conductors.

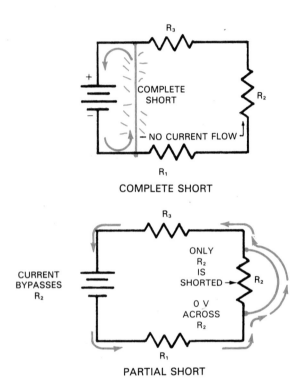

Fig. 4-3. Shorts and partial shorts in a circuit.

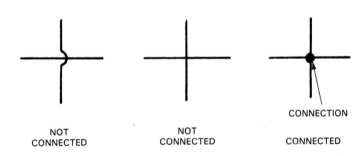

Fig. 4-5. Schematic of wires.

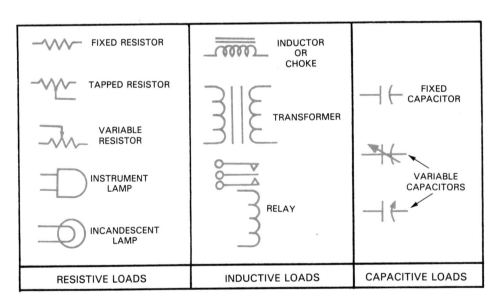

Fig. 4-4. Symbols for common electrical loads.

Basic Circuits

To show how a schematic relates to an actual wiring job, refer to Figs. 4-6 and 4-7. Study these carefully.

A BREADBOARD is a useful device for learning about circuits, Fig. 4-8. A circuit can be laid out on the breadboard to learn if it will work. Parts are mounted on the board. Temporary connections are made. Then the circuit is tested. By using a breadboard, you can be sure that the wiring of the circuit is correct. You can also be sure that parts with the correct values have been used. When you know the circuit will work, it can be soldered.

Printed circuits

Printed circuits are made from a thin layer of conducting material, such as copper foil. The conducting material is bonded to an insulated board. The copper can be bonded on one side or on both sides.

The actual circuit is made by dipping the board in an etchant. Circuit element areas are protected by covering with a resist material. Exposed areas are dissolved by the etchant. After the board is removed from the etchant, the resist material is removed and the board is cleaned. It is now ready to be drilled.

Printed circuit boards can also be made using a photographic method. Light-sensitive material covers the surface of the board. A negative of the circuit board layout is placed over the board. The board is then exposed to light and developed in a manner similar to that method used for film. During developing, the unexposed areas are washed away. Then the board is etched, as was just described.

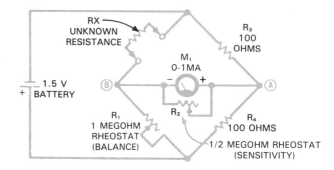

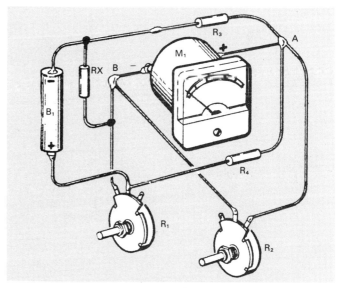

Fig. 4-7. Relationship of actual wiring job to its schematic.

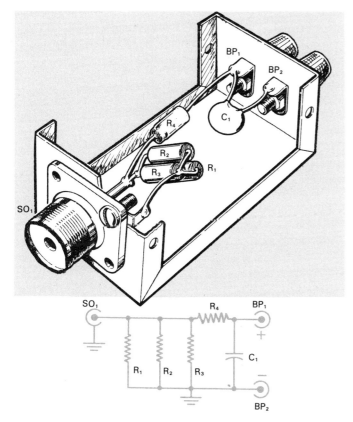

Fig. 4-6. Relationship of actual wiring job to its schematic.

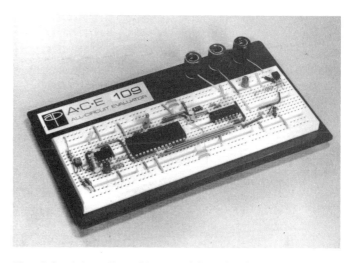

Fig. 4-8. A breadboard is a useful device for learning about circuits.

Fig. 4-9 shows a simple printed circuit and its schematic.

CONTROLS

SWITCHES are electrical devices or components that turn circuits on and off. The simplest switch is the single-pole, single-throw (SPST) switch. Fig. 4-10, of a knife switch, shows how a SPST switch works.

There are many more types of switches. For example, when both sides of a line are to be switched, then a double-pole, single-throw (DPST) switch is used. Fig. 4-11 shows the circuit diagrams for various switches.

It is sometimes desirable to switch a circuit from two different locations. In this case, a three-way switch is used. This type of switch permits a device to be turned on and off from two different places. For example, some lights can be turned on from different switches in the same room.

The schematic of this hookup is shown in Fig. 4-12. The light is on. But is can be turned off by moving either switch A or B. In Fig. 4-13, the light is off. But it can be turned on by either switch A or B. Follow the circuit through the switches in each position.

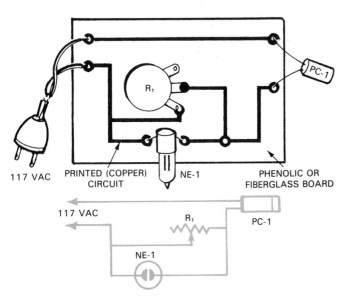

Fig. 4-9. Relationship of a printed circuit and its schematic. Study this carefully.

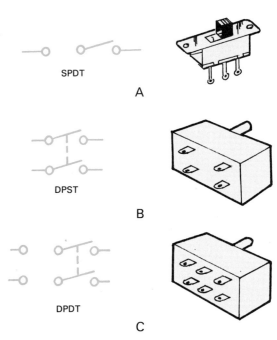

Fig. 4-11. Types of switches. A—Single-pole, double-throw (SPDT) switches from one point to another. B—Double-pole, single-throw switch. C—Double-pole, double-throw (DPDT) switches a double line to two other points.

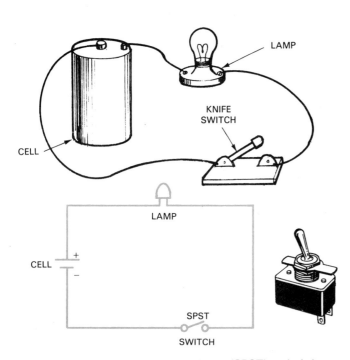

Fig. 4-10. A single-pole, single-throw (SPST) switch in a simple circuit.

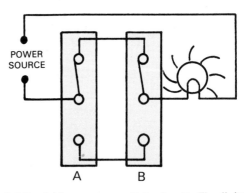

Fig. 4-12. A three-way switch circuit. The light is on.

Basic Circuits

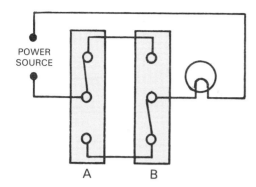

Fig. 4-13. A three-way switch circuit. The light is off.

Push button switches, Figs. 4-14 and 4-15, also come in assemblies. Fig. 4-16 shows one of these.

Many electronic devices now use rotary switches. The channel selector switch on a television set is a rotary switch, Fig. 4-17. There are many different types of rotary switch circuits. The single-pole rotary switch symbol and a standard section circuit diagram are shown in Fig. 4-18.

Fig. 4-14. Push button, NO (normally open). It is used to switch bells or alarms or to momentarily close a circuit.

Fig. 4-15. Push button, NC (normally closed). It is used to momentarily open a circuit.

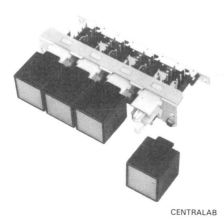

Fig. 4-16. Push button switch assemblies.

Each kind and size of wire has the ability to conduct a specified amount of electric current. All conductors have some resistance. When a current overcomes this resistance, heat is produced. If a wire is operated within its bounds, this heat is sent out into the surrounding air. The temperature does not rise too high. At times, however, too much current is forced through a conductor. Then, the temperature rises to a point where the wire becomes hot and destroys itself. This is known as OVERLOADING. If the wire is near combustible material, such as that found in the wall of a structure, a fire might result.

Overloading a circuit can be caused by:
1. Excessive load which draws more than a safe amount of current.
2. A direct short circuit.

Circuits and appliances are usually protected from overload by a fuse or circuit breaker. A FUSE is a thin strip of metal which melts at a low temperature. Those used in the home are usually designated 15 and 20 amperes. If a current exceeds the rating of the fuse, the fuse will melt and open the circuits. This prevents damage to equipment and the danger of fire. The symbol for a fuse, as it appears in electrical circuit diagrams, appears in Fig. 4-19.

Fig. 4-17. Front view of a typical rotary switch.

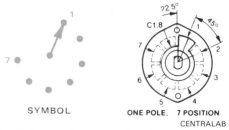

Fig. 4-18. Rotary switch symbol and standard section circuit diagram.

Fig. 4-19. Fuses and symbol for a fuse.

Electricity and Electronics

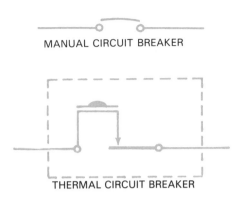

Fig. 4-20. Symbols for circuit breakers.

A CIRCUIT BREAKER is a magnetic or thermal device that automatically opens the circuit when there is too much current flow, Fig. 4-20. It must be manually reset before the circuit can be used again. Needless to say, the cause of the overload should be found and removed before the current is turned on again.

REVIEW QUESTIONS FOR SECTION 4.1

1. _____ _____ are complete pathways on which electric current flows.
2. What is a load?
3. Explain the difference between a short circuit and an open circuit.
4. Electrical drawings which show the correct wiring of component parts in a circuit are:
   a. Printed circuits.
   b. Electrical buses.
   c. Schematics.
   d. None of the above.
5. _____ are electrical devices or components that turn circuits on and off.

4.2 BASIC CIRCUITS

Although it may seem that there are many types of circuits, there are only three basic types. They are the series circuit, the parallel circuit, and the combination (series-parallel) circuit.

A SERIES circuit has only one pathway on which current can flow. Since there is only one pathway, if a break or open occurs, all current will stop. Fig. 4-21 shows a series circuit.

A PARALLEL circuit has more than one pathway on which current can flow. There are BRANCHES where electrons can divide to flow through the load. Remember, the maximum current ALWAYS flows through the path of least resistance. Therefore, the

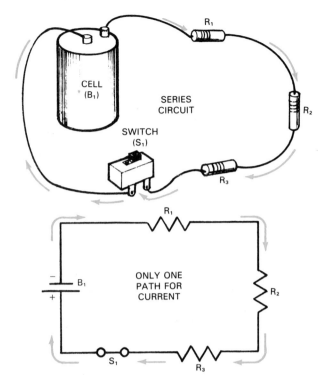

Fig. 4-21. A simple series circuit.

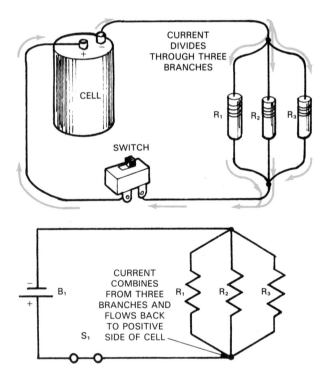

Fig. 4-22. A simple parallel circuit.

lower resistance path in a parallel circuit has the most current. In all cases, the amount of current that leaves the negative lead of the voltage source will be the same amount as that which returns to the positive lead. Fig. 4-22 shows a simple parallel circuit.

Basic Circuits

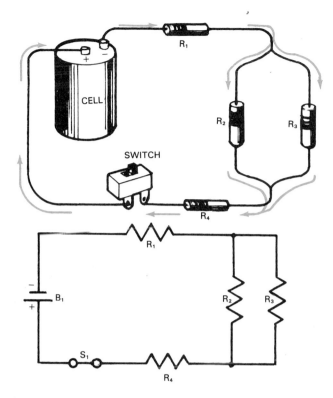

Fig. 4-23. A combination circuit. Note that R_1 and R_4 are in series with the cell and R_2 and R_3 are in parallel.

A COMBINATION circuit has components connected in both series and parallel. The series section has all of the current flowing through it. The parallel section has branch circuits through which the current divides and flows. A combination circuit is often called a SERIES-PARALLEL circuit, Fig. 4-23.

Series, parallel, and combination circuits will be discussed in greater detail in Chapter 5.

REVIEW QUESTIONS FOR SECTION 4.2

1. What is a series circuit?
2. A parallel circuit has (more than, only) one path for current.
3. Current will always take the path of least _____.
4. Describe what happens to the current in a combination circuit.

4.3 POLARITY AND METERS

The property of a device or components in a circuit to have a negative or positive value is called POLARITY. The correct value is important when connecting cells or batteries in a circuit or when measuring voltage or current. Also, some components, such as electrolytic capacitors, have a positive and negative polarity. They must be correctly placed in the current.

Fig. 4-24 shows the polarity of certain components as well as the polarity created by current flow. Note that resistors (R_1 & R_2) do not normally have a negative or positive polarity. But current flowing through them causes a voltage drop. This causes one end to have a negative charge and the other end a positive charge. Polarity is very important when connecting voltmeters or ammeters in a dc circuit.

CONNECTING METERS IN A CIRCUIT

VOLTMETERS are used to measure voltage in a circuit. They are always connected in parallel, across what is to be measured. Correct polarity must be observed in order to obtain the correct reading and to prevent damage to the meter. If the voltage is not known, it is a good idea to start on the highest range of the meter and move down. Some of the newer meters are autoranging making this unnecessary. Fig. 4-25 shows the symbol for a voltmeter and the correct polarity connection in a series circuit. Note that the black lead of a meter is negative polarity and the red lead is positive.

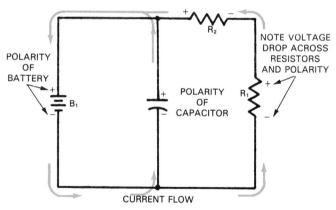

Fig. 4-24. Polarity in a circuit.

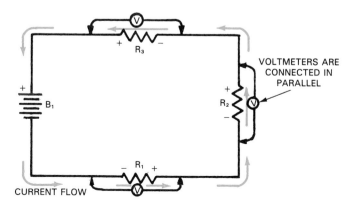

Fig. 4-25. Voltmeters are connected in parallel in a circuit.

AMMETERS are used to measure current. They are always connected in series in the circuit. The circuit must be broken open to measure the current flowing in that part of the circuit. Again, observing correct polarity is vital in the proper use of an ammeter. Fig. 4-26 shows how to measure current in a parallel circuit.

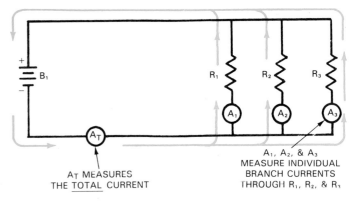

Fig. 4-26. Ammeters are connected in series in a circuit.

REVIEW QUESTIONS FOR SECTION 4.3

1. What is polarity?
2. Ammeters are connected in _____ in a circuit; voltmeters are connected in _____ in a circuit.
3. In Fig. 4-26, how would a voltmeter be connected to measure the voltage across R_2? How would an ammeter be connected to measure the current flowing through R_3? Show polarities.
4. How would meters be connected to measure total current and total voltage in the circuit shown in Question 3?

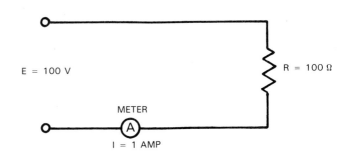

Fig. 4-27. E = 100 V, R = 100 Ω. What is I?

4.4 OHM'S LAW

One of the basic laws of electrical circuits was derived from experimentation done by George Simon Ohm, the German scientist and philosopher, during the 19th century. Ohm's Law states that the current in amperes in a circuit is equal to the applied voltage divided by the resistance, or:

$$I = \frac{E \text{ (in volts)}}{R \text{ (in ohms)}}$$

Notice the letters used in this equation:
- I = intensity of current in amperes.
- E = electromotive force in volts.
- R = resistance in ohms.

In non-mathematical language, this formula means:
- As voltage is increased, current increases.
- As voltage is decreased, current decreases.
- As resistance is increased, current decreases.
- As resistance is decreased, current increases.

A certain device which has a resistance of 100 ohms is connected to a 100 V source of electricity. How much current will flow in the circuit? See Fig. 4-27.

Solution:
$$I = \frac{E}{R} \text{ or } I = \frac{100 \text{ V}}{100 \text{ Ω}} = 1 \text{ ampere}$$

If voltage is increased to 200 volts, how much current will flow? See Fig. 4-28.

Solution:
$$I = \frac{200 \text{ V}}{100 \text{ Ω}} = 2 \text{ amperes}$$

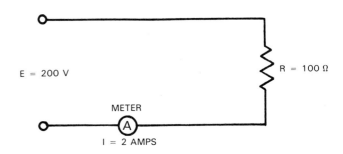

Fig. 4-28. E = 200 V, R = 100 Ω. What is I?

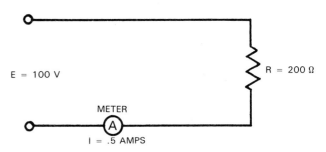

Fig. 4-29. E = 100 V, R = 200 Ω. What is I?

Return the voltage to 100 volts, but increase the resistance to 200 ohms. See Fig. 4-29.

Solution:
$$I = \frac{100 \text{ V}}{200 \text{ Ω}} = .5 \text{ amperes}$$

Finally, decrease the resistance to 50 ohms. See Fig. 4-30.

Solution:
$$I = \frac{100\text{ V}}{50\text{ }\Omega} = 2 \text{ amperes}$$

Compare these computations to the previous discussion, and notice that these statements have been proved true. By transposition of symbols in Ohm's Law, it is possible to solve for any one unknown quantity, if the other two quantities are known.

Ohm's Law may be written three ways:
$$I = \frac{E}{R}, \text{ or } E = I \times R, \text{ or } R = \frac{E}{I}$$

Problem. A current of .5 amperes is flowing in a circuit which has 100 ohms resistance. What is the applied voltage? See Fig. 4-31.
$$E = IR = .5 \times 100 = 50 \text{ volts}$$

Problem. A circuit has an applied voltage of 50 volts and the current is measured at .5 amperes. What is the resistance of the circuit shown in Fig. 4-32?
$$R = \frac{E}{I} \text{ or } R = \frac{50}{.5} = 100 \text{ }\Omega$$

Practicing problems is the best way to learn Ohm's Law. A device to help you solve Ohm's Law problems is shown in Fig. 4-33.

When working problems involving Ohm's Law, be certain that all quantities are in the simplest form: volts in volts, current in amperes, resistance in ohms. If one or more of these quantities appears with a prefix (kilo, mega, milli, etc.) they must be converted before solving the equation.

Problem. A circuit has a resistance of 10 kilohms and the current is measured as 100 milliamperes. What is the applied voltage?
$$E = I \times R \text{ or } E = .1 \text{ amperes} \times 10{,}000 = 1000 \text{ V}$$

For methods of making conversions to basic electrical units, see the Appendix.

REVIEW QUESTIONS FOR SECTION 4.4

1. State Ohm's Law in your own words.
2. In the formula, $I = \frac{E}{R}$, which symbol represents current?
3. In a circuit, as voltage increases, _____ increases.
4. What is the resistance of a circuit that has 20 volts and 2 amperes?
5. When resistance is 100 ohms and the current is .5 amperes, what is the voltage?

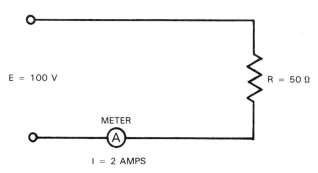

Fig. 4-30. E = 100 V, R = 50 Ω. What is I?

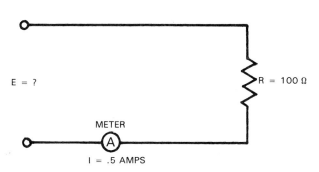

Fig. 4-31. R = 100 Ω, I = .5 amperes. What is the voltage?

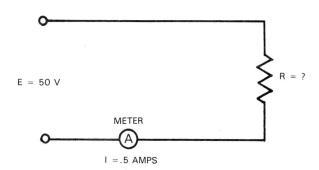

Fig. 4-32. E = 50 V, I = .5 amperes. What is the resistance?

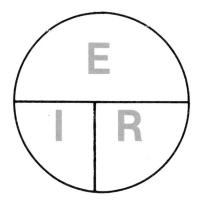

Fig. 4-33. A memory device to use when solving Ohm's Law problems. Cover the unknown quantity with a finger. The remaining letters give the correct equation. For example, E is unknown. Cover E and I R remains. Thus, E = I × R.

4.5 POWER AND WATT'S LAW

The only component in any circuit that uses electrical power is resistance. POWER is the time rate of doing work. In the physics of machines, it was discovered that when a force moved through a distance, work was done. For example, if a 10 pound weight is lifted one foot, the work done equals,

F (force) × D (distance) or
10 × 1 = 10 foot-pounds (ft.-lb.)
of work.

See Fig. 4-34.

No reference is made to time in this equation. One might take five seconds or 10 minutes to lift the 10 pound weight. However, if one lifted the 10 pound weight once each second, then the power expended would be 10 ft.-lb. per second. To carry the example one step further, if one lifted the 10 pound weight in one-half, or .5 second, then the power expended would equal:

$\frac{10}{.5}$ or 20 ft.-lb. per second.

A machine that works at a rate of 550 ft.-lb./second is equivalent to ONE HORSEPOWER. Also, 33,000 ft.-lb./min. equals one horsepower.

In electricity, the unit of power is the WATT. It was named in honor of James Watt, who is credited with the invention of the steam engine. The symbol for a watt is P.

When one volt of electrical pressure moves one coulomb of electricity in one second, the work accomplished is equal to one watt of power. Recall the definition of one ampere: the movement of one coulomb of electrons past a given point in a circuit is one second. So, power in an electrical circuit is equal to:

P (watt) = E (volts) × I (amperes)

Use these figures for comparison of electrical power to mechanical power:

746 watts = 1 horsepower

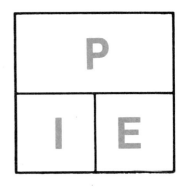

Fig. 4-35. A memory device to help solve power problems. Cover the unknown quantity and the remaining letters give the correct equation. For example, E is unknown. Cover E and P/I remains. Thus, E = P ÷ I.

The power formula, sometimes called Watt's Law, can be arranged algebraically. If two quantities are known, the third unknown may be found. Use the device in Fig. 4-35.

$P = I \times E$ or $I = \frac{P}{E}$, or $E = \frac{P}{I}$

Example. A circuit with an unknown load has an applied voltage of 100 volts. The measured current is 2 amperes. How much power is consumed?

P = I × E or 2 amps × 100 volts
= 200 watts

Example. An electric toaster rated at 550 watts is connected to a 117 volt source. How much current will this appliance use?

$I = \frac{P}{E}$ or $\frac{550 \text{ W}}{117 \text{ V}}$ = 4.7 amps

LESSON IN SAFETY: An electric current produces heat when it passes through a resistance. Heaters and resistors will remain hot for some time after the power is removed. Handle them carefully. Burns should be treated immediately. See your instructor.

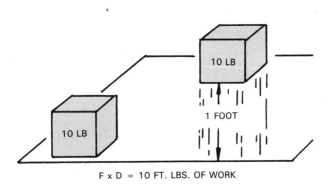

F x D = 10 FT. LBS. OF WORK

Fig. 4-34. When a 10 pound weight is lifted one foot, 10 foot-pounds of work is done.

REVIEW QUESTIONS FOR SECTION 4.5

1. Define power.
2. State Watt's Law. Use it in a formula with a given voltage and current.
3. The unit for power is the _____.
4. Compute the voltage and the power for the following figure.

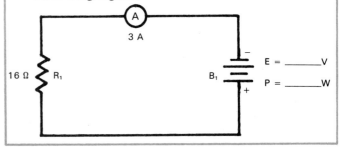

Basic Circuits

4.6 OHM'S LAW AND WATT'S LAW

It is possible to combine Ohm's Law and Watt's Law to produce simple formulas which will permit you to solve any unknown, if two quantities are known. In Fig. 4-36, these equations are given, with an explanation of each.

These formulas are arranged in a wheel-shaped memory device for ready reference. Refer to Fig. 4-37. To use the device, find the equation that uses the two known quantities in your circuit. For example, if current and resistance are known and power is unknown, use Equation 11. If power and voltage are known and current is unknown, use Equation 5.

1.	$E = I \times R$	Ohm's Law
2.	$E = \frac{P}{I}$	Watt's Law
3.	$E = \sqrt{PR}$	By transposing equation 12 and taking the square root.
4.	$I = \frac{E}{R}$	Ohm's Law
5.	$I = \frac{P}{E}$	Watt's Law
6.	$I = \sqrt{\frac{P}{R}}$	By transposing equation 9 and taking the square root.
7.	$R = \frac{E}{I}$	Ohm's Law
8.	$R = \frac{E^2}{P}$	By transposing equation 12.
9.	$R = \frac{P}{I^2}$	By transposing equation 11.
10.	$P = I \times E$	Watt's Law
11.	$P = I^2 \times R$	By substituting $I \times R$ from equation 1, for E.
12.	$P = \frac{E^2}{R}$	By substituting $\frac{E}{R}$ from equation 4, for I.

Fig. 4-36. This table states a variety of basic formulas that are needed to solve problems.

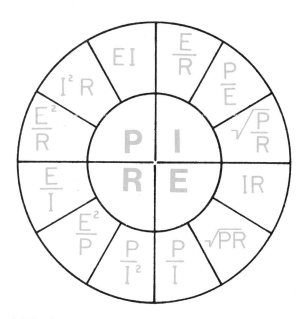

Fig. 4-37. A memory device combining Ohm's Law and Power Laws. It can be a great help when solving problems.

REVIEW QUESTIONS FOR SECTION 4.6

1. Give the correct missing symbol in the formulas below:

$$E = \frac{P}{?} \quad I = \sqrt{\frac{P}{?}} \quad P = \frac{?}{R} \quad R = \frac{P}{?}$$

2. Solve for the unknown values below.

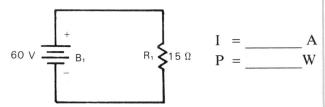

I = _____ A
P = _____ W

3. Solve for the unknown values below.

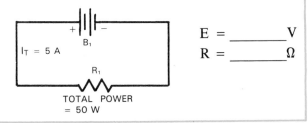

E = _____ V
R = _____ Ω

SUMMARY

1. Circuits are complete pathways for current. All circuits have a voltage source, some type of a load, a conductive pathway, and some type of control.
2. There are three basic types of circuits. They are the series, the parallel, and the combination circuits.
3. The property of a circuit or component to have a negative and positive value is called polarity.
4. Ohm's Law is a mathematical formula stating the relationship between voltage, current, and resistance in a circuit. It is $E = I \times R$.
5. Electrical power is the rate of doing electrical work.
6. Watt's Law is a mathematical formula stating the relationship of the power, voltage, and current in a circuit. It is $P = I \times E$.
7. There are definite relationships between Ohm's Law and Watt's Law, as shown in Figs. 4-36 and 4-37.

TEST YOUR KNOWLEDGE, Chapter 4

Please do not write in the text. Place your answers on a separate sheet of paper.

1. Four factors are necessary in order for a circuit to operate. Name them.
2. Name two devices that protect against circuit overload.
3. A _____ circuit has only one pathway on which current can flow. A _____ circuit has more than one pathway on which current can flow.
4. Which device is used to measure voltage in a circuit?
   a. Ammeter.
   b. Voltmeter.
   c. Circuit breaker.
   d. None of the above.
5. State Ohm's Law using three different equations.
6. A machine that works at 550 ft.-lb./second is equivalent to one _____.
7. Which of the following is Watt's Law?
   a. $P = I \times E$.
   b. $I = \frac{P}{E}$.
   c. $E = \frac{P}{I}$.
   d. All of the above.
8. Use Figs. 4-36 and 4-37 to solve the following problems.
   a. E = 100 V, I = 2 amps, R = _____.
   b. E = 50 V, R = 1000 ohms, I = _____.
   c. I = .5 amps, R = 50 ohms, E = _____.
   d. E = 10 V, I = .001 amps, R = _____.
   e. I = .05 amps, R = 1000 ohms, E = _____.
   f. P = 10 W, I = 2 amps, E = _____.
   g. E = 100 V, I = .5 amps, P = _____.
   h. P = 500 W, E = 250 V, I = _____.
   i. I = .01 amps, R = 100 ohms, E = _____.
   j. P = 100 W, I = 2 amps, R = _____.
   k. E = 10 V, P = 10 W, R = _____.
   l. E = 500 V, I = 2 amps, R = _____.
   m. E = 100 V, R = 1000 ohms, P = _____.
   n. I = .5 amps, R = 50 ohms, P = _____.
   o. I = 4 amps, R = 10 ohms, P = _____.
   p. I = 10 mA, E = 50 V, P = _____.
   q. I = 20 mA, E = 100 V, R = _____.
   r. P = 10 W, I = 1 amp, R = _____.

s. E = 1000 V, R = 1000 ohms,
    I = _____.
t. I = 100 mA, R = 100 ohms,
    E = _____.
u. I = 100 mA, R = 100 ohms,
    P = _____.
v. P = 500 W, E = 100 V,
    I = _____.
w. E = 100 V, R = 100 ohms,
    P = _____.
x. E = 50 V, R = 10 kilohms,
    I = _____.
y. P = 10 W, R = 10 ohms,
    E = _____.
z. P = 50 W, R = 2 ohms,
    I = _____.

FOR DISCUSSION

1. Discuss why different circuits are important in electrical devices. For example, why not have only a simple series circuit in an electrical device?
2. Explain the difference between a partial short and a complete short.
3. What is meant by polarity? Why is it so important in electricity and electronics?
4. State Ohm's Law and Watt's Law in your own words. What is the relationship that exists between these two laws?
5. Based on what you have learned in this chapter about series and parallel circuits, how would you design a security system for your room which would let you open a window for fresh air?

TEKTRONIX

Basic laws of electricity come into play when performing complex tasks.

Electricity and Electronics

PANASONIC

Even this small cellular phone will include all of the three basic types of circuits described in Chapter 5.

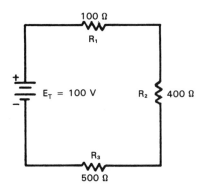

Only one path is provided for electric current.

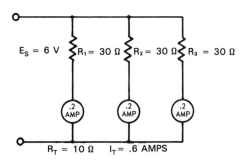

More than one path has been provided for current.

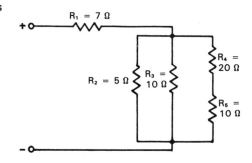

This is a more complicated type circuit.

All circuits can be included in the three basic types: series, parallel and combination (includes both series and parallel). Can you identify them?

Chapter 5

SERIES, PARALLEL, AND COMBINATION CIRCUITS

After studying this chapter, you will be able to:
- *Use Ohm's Law to compute the total resistance, total voltage, and total current of series, parallel, and combination circuits.*
- *Determine voltage drops for series and parallel circuits.*
- *Explain Kirchhoff's Voltage Law as it applies to a series circuit.*
- *Use Ohm's Law to compute the power in series, parallel, and combination circuits.*
- *Explain Kirchhoff's Current Law as it applies to a parallel circuit.*

As was mentioned in Chapter 4, there are three basic types of electrical circuits. They are the series circuit, the parallel circuit, and the combination circuit (series-parallel). In this chapter, these circuits will be explored in more detail. Also, Kirchhoff's Laws will be presented.

5.1 SERIES CIRCUITS

When components in a circuit are arranged so the current will pass through all the components, this is a SERIES CIRCUIT. The components are connected in-line or end-to-end in a series string, Fig. 5-1.

The total amount of current which leaves the negative pole of the battery will flow through R_1, R_2, and R_3 equally. Refer to the formula shown in Fig. 5-2. The total current leaving the battery is I_T. Current flow through individual resistor is expressed by I_{R_1}, I_{R_2}, and I_{R_3}.

When resistors are connected in series, the total resistance of the circuit is equal to the sum of individual values of all resistors, Fig. 5-3.

FORMULA FOR CURRENT IN SERIES CIRCUIT:
$I_T = I_{R_1} = I_{R_2} = I_{R_3} \ldots$

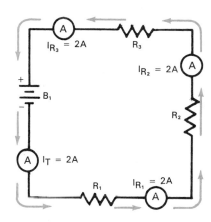

Fig. 5-2. Current flows equally through all resistors in a series circuit.

FORMULA FOR RESISTANCE IN SERIES:
$R_T = R_1 + R_2 + R_3 \ldots$

$R_T = 60 \Omega + 40 \Omega + 20 \Omega$
$R_T = 120 \Omega$

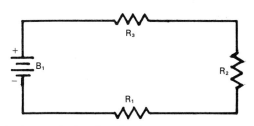

Fig. 5-1. In a typical series circuit, components are connected in one pathway, or end-to-end.

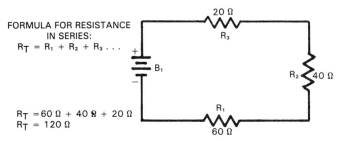

Fig. 5-3. Total resistance in a series circuit is the sum of the individual values of each resistor.

71

The total voltage in a series circuit is equal to the individual voltage drops across the separate loads. This is shown mathematically in Fig. 5-4. The total voltage is expressed as E_T. Individual voltage drops across the resistors are represented by E_{R_1}, E_{R_2}, and E_{R_3}.

Refer back to Fig. 5-4. E_{R_1} = 6 volts, E_{R_2} = 12 volts, E_{R_3} = 18 volts. R_1 has the least amount of voltage drop. One can assume it is the smallest resistor. The voltage drop in R_2 is twice as large as in R_1. The resistance value of R_2, then, must be twice as large as R_1. R_3 is three times the resistance of R_1, since the voltage drop in R_3 is three times as large as R_1. R_3 is one and one-half times as large as R_2, since the voltage drop in R_3 is 18 volts and in R_2 is 12 volts.

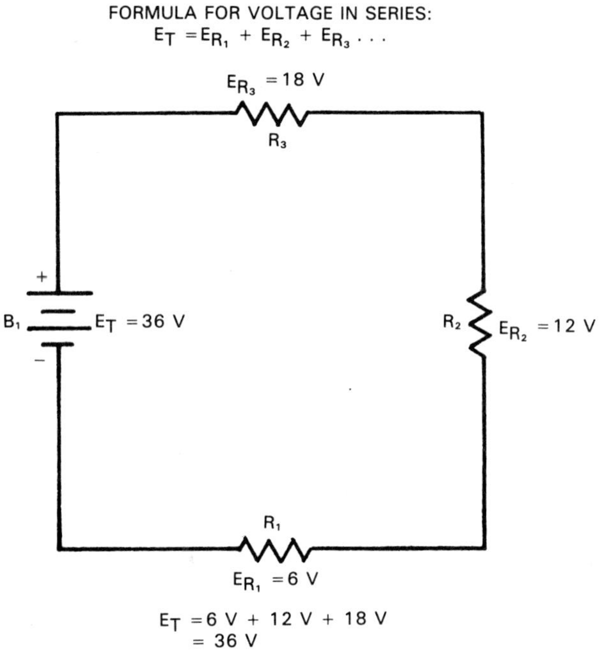

OHM'S LAW IN SERIES CIRCUITS

Ohm's Law mathematically describes the relationship between current, voltage, and resistance. Series circuits have all of these factors. Therefore, Ohm's Law applies to series circuits. Refer to Fig. 5-5.

If resistors are connected in series, the total resistance of the circuit is the sum of the resistors. In Fig. 5-5, R_1 = 100 Ω, R_2 = 400 Ω, and R_3 = 500 Ω. Then $R_T = R_1 + R_2 + R_3$ or 100 + 400 + 500 = 1000 Ω.

If 100 volts were applied to the terminals of this circuit, the current flowing in the circuit would equal,

$$I = \frac{E}{R} \text{ or } I = \frac{100 \text{ V}}{1000 \text{ Ω}} = .1 \text{ amp}$$

Recall that these resistors are connected in series. All the current flowing in the circuit must pass through each resistor. Therefore, current is the same in all parts of the circuit, Fig. 5-6.

As current passes through a resistor, a certain amount of energy is used and a certain amount of pressure, or voltage, is lost. The voltage loss across each resistor may be calculated by Ohm's Law, in the form, $E = I \times R$.

For R_1, E_{R_1} = .1 × 100 = 10 V
For R_2, E_{R_2} = .1 × 400 = 40 V
For R_3, E_{R_3} = .1 × 500 = 50 V

Again, the symbol E_{R_1} means the voltage across resistor R_1. Note that the sum of the voltage losses or VOLTAGE DROPS is equal to the source voltage. In this example,

$E_{R_1} + E_{R_2} + E_{R_3} = E_{SOURCE}$
or 10 V + 40 V + 50 V = 100 V

Because the voltage drop or loss across a resistor is computed by using the $E = I \times R$ form of Ohm's Law, this loss is often called the IR DROP. This means the same as VOLTAGE DROP.

Fig. 5-4. Total voltage in a series circuit is the sum of individual voltage drops across separate loads.

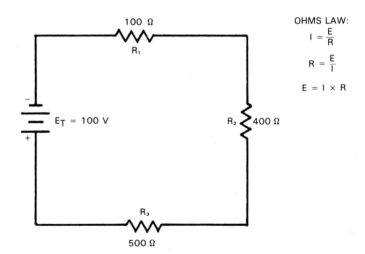

Fig. 5-5. Ohm's Law in a series circuit.

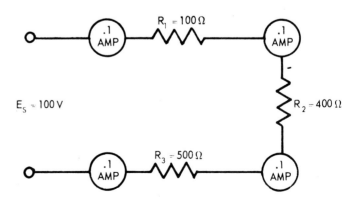

Fig. 5-6. The current measured at any point in the series circuit is the same.

KIRCHHOFF'S VOLTAGE LAW

In the mid-nineteenth century, German scientist Gustav R. Kirchhoff devised two basic laws of electricity relating to voltage and current. Kirchhoff's Voltage Law applies to series circuits. It states that the sum of voltage drops and voltage rises around any closed series loop (or circuit) is zero. In simpler terms, all of the voltage drops (E_{R_1}, E_{R_2}, and E_{R_3}, etc.) in a series circuit must add up to the total applied voltage (E_T). See Fig. 5-7.

POWER IN A SERIES CIRCUIT

As the voltage overcomes the resistance, work is done. Power is the time rate of doing work. You will recall that in electricity, $P = I \times E$. The energy lost in overcoming resistance takes the form of heat. This must be dissipated by the resistor into the surrounding air. This explains why some resistors are larger than others. They must be large enough to provide radiation surface for heat dissipation. The total power used in this circuit, as shown in Fig. 5-8, would equal:

$$P = I \times E \text{ or } .1 \times 100 \text{ V} = 10 \text{ W}$$

To calculate the power consumed by individual resistors, the formula $P = I^2R$ may be used.

For R_1, $P = (.1)^2 \times 100 = 1$ W
For R_2, $P = (.1)^2 \times 400 = 4$ W
For R_3, $P = (.1)^2 \times 500 = 5$ W
For R_T, $P = (.1)^2 \times 1000 = 10$ W

Notice that the sum of the individual powers used by each resistor is equal to the total power used by the circuit.

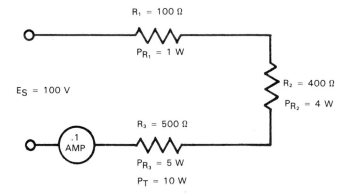

Fig. 5-8. The power used by each resistor may be computed by the formula, $P = I \times E$.

REVIEW QUESTIONS FOR SECTION 5.1

1. How does Kirchhoff's Voltage Law relate to series circuits?
2. In the circuit shown below, find the following:

   R_T = _____ Ω.
   I_T = _____ A.
   E_{R_1} = _____ V.
   E_{R_2} = _____ V.
   E_{R_3} = _____ V.
   P_T = _____ W.

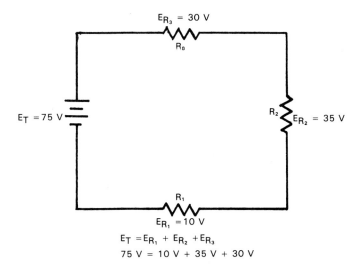

3. In the circuit shown below, find the following:

   R_T = _____ Ω.
   E_T = _____ V.
   E_{R_1} = _____ V.
   E_{R_2} = _____ V.
   E_{R_3} = _____ V.
   P_{R_1} = _____ W.
   P_T = _____ W.

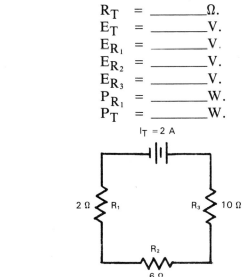

Fig. 5-7. An example of Kirchhoff's Voltage Law.

5.2 PARALLEL CIRCUITS

When several components are connected to the same voltage source, the components are connected in parallel or side-by-side. Multiple paths for current flow are provided by a parallel circuit. Each resistor is a path of its own. In Fig. 5-9, three separate paths are provided by R_1, R_2, and R_3. When components are connected in this manner, the total resistance of the circuit is DECREASED every time another component is added.

Compare a parallel circuit to a system of pipes carrying water. Two pipes will carry more water than one pipe. If a third pipe is added, the three pipes will carry more water than two pipes. As more pipes are added, the total resistance to the flow of water is DECREASED. Refer to Fig. 5-10.

In Fig. 5-11, resistor $R_1 = R_2 = R_3$. Each resistor will carry the same amount of current. The three resistors together will carry three times more current than only one. Therefore, the total resistance of the circuit must be one-third of the resistance of a single resistance. With values of 30 Ω at R_1, R_2, and R_3, then:

$$R_T = \frac{30 \, \Omega}{3} = 10 \, \Omega.$$

When EQUAL resistors are connected in parallel, the total resistance of the parallel group or network is equal to any one resistor divided by the number of resistors in the network.

Total resistance = R_T =

$$\frac{R}{N} = \frac{\text{value of one resistor}}{\text{number of resistors in network}}$$

In a parallel circuit, the applied voltage is the same for each resistor. This is because each is connected across the same voltage source. Therefore, the currents are, using Ohm's Law:

For R_1, $I = \dfrac{6 \, V}{30 \, \Omega} = .2$ amps

For R_2, $I = \dfrac{6 \, V}{30 \, \Omega} = .2$ amps

For R_3, $I = \dfrac{6 \, V}{30 \, \Omega} = .2$ amps

The total current flowing through the network would be the sum of the individual branch currents:

$$I_T = I_{R_1} + I_{R_2} + I_{R_3}$$
$$= .2 + .2 + .2 = .6 \text{ amps}$$

This may be proved by applying Ohm's Law, using the total resistance,

$$I_T = \frac{E}{R_T} = \frac{6 \, V}{10 \, \Omega} = .6 \text{ amps}$$

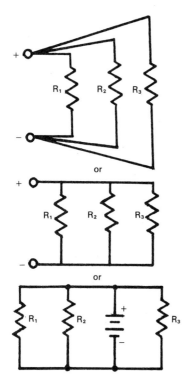

Fig. 5-9. Schematic diagrams of three resistors in parallel.

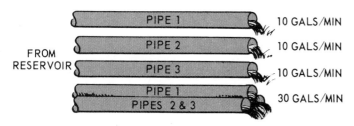

Fig. 5-10. Electricity may be compared to the flow of water. Increasing the number of pipes will decrease the resistance to water flow.

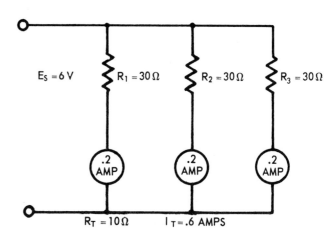

Fig. 5-11. Total current equals the sum of branch currents.

Series, Parallel, and Combination Circuits

Again, there are these laws concerning parallel circuits.
1. The voltage across all branches of a parallel network is the same.
2. The total current is equal to the sum of the individual branch currents.

The differences between these laws and those which apply to series circuits should be compared.

It is not unusual to find two unequal resistors connected in parallel. In this case, the total resistance may be found by using the formula,

$$R_T = \frac{R_1 R_2}{R_1 + R_2}$$

In Fig. 5-12, resistance $R_1 = 20\ \Omega$ and $R_2 = 30\ \Omega$. Apply the formula.

$$R_T = \frac{20 \times 30}{20 + 30} \text{ or } \frac{600}{50} = 12\ \Omega$$

With an applied voltage of $E_S = 6$ V, the total current in the circuit will be,

$$I_T = \frac{E}{R_T} = \frac{6}{12} = .5 \text{ amps}$$

This equation may be proved by finding the sum of individual branch current,

For R_1, $I = \frac{6\ V}{20\ \Omega} = .3$ amps

R_2, $I = \frac{6\ V}{30\ \Omega} = .2$ amps

$I_1 + I_2 = .3 + .2 = .5$ amps or I_T

Two observations should be carefully studied.
1. The TOTAL resistance of any parallel circuit must always be LESS than the value of any branch resistance. Note in Fig. 5-12, $R_T = 13\Omega$ which is less than R_1 which is 20Ω and R_2 which is 30Ω.
2. The branch of the circuit containing the greatest resistance conducts the least current.

Referring to Fig. 5-13, three unequal resistors are connected in parallel. $R_1 = 5\ \Omega$, $R_2 = 10\ \Omega$, and $R_3 = 30\ \Omega$. To find the total resistance, find the sum of the individual branch currents. Assuming a voltage of 30 volts then:

For Branch R_1, $I_1 = \frac{E}{R_1} = \frac{30}{5} = 6$ amps

For Branch R_2, $I_2 = \frac{E}{R_2} = \frac{30}{10} = 3$ amps

For Branch R_3, $I_3 = \frac{E}{R_3} = \frac{30}{30} = 1$ amp

$I_T = I_1 + I_2 + I_3 = 6 + 3 + 1 = 10$ amps

By Ohm's Law:

$$R_T = \frac{E}{I} \text{ so } R_T = \frac{30}{10} = 3\ \Omega$$

As I_T is the sum of all the currents, and is equal to $\frac{E}{R_T}$ (Ohm's Law) then,

$$\frac{E}{R_T} = \frac{E}{R_1} + \frac{E}{R_2} + \frac{E}{R_3}$$

Divide both sides of the equation by E, then

$$\frac{1}{R_T} = \frac{1}{R_1} + \frac{1}{R_2} + \frac{1}{R_3}$$

Therefore, to solve for the total resistance of a circuit in Fig. 5-13 use this formula:

$$\frac{1}{R_T} = \frac{1}{5} + \frac{1}{10} + \frac{1}{30} = \frac{10}{30}$$

Inverting both sides of the equation,

$$R_T = \frac{30}{10} = 3\ \Omega$$

This formula is used in the computation of the total resistance of parallel circuits. You should practice using it by completing the problems at the end of this section and chapter.

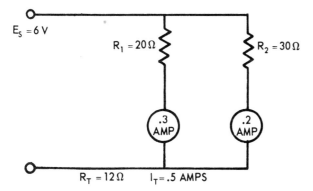

Fig. 5-12. Schematic of two unequal resistors in parallel.

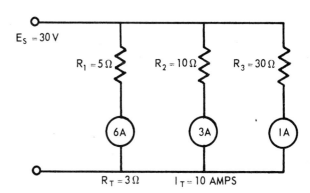

Fig. 5-13. Three unequal resistors connected in parallel.

KIRCHHOFF'S CURRENT LAW

Kirchhoff's Current Law states, the algebraic sum of the currents entering and leaving any point in a (parallel) circuit must equal zero. In a simpler form, the amount of current entering a parallel circuit must be the same as the current leaving that circuit. To see how this works, refer to Fig. 5-14.

Kirchhoff's Current Law provides the basis for the operation of a parallel circuit. It states that the total line current equals the sum of the individual branch currents.

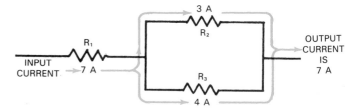

Fig. 5-14. Kirchhoff's Current Law.

REVIEW QUESTIONS FOR SECTION 5.2

1. Explain Kirchhoff's Current Law.
2. Solve for the unknown values in the parallel circuit shown.

   R_T = _____ Ω.
   I_T = _____ A.
   E_{R_1} = _____ V.
   I_{R_1} = _____ A.
   I_{R_2} = _____ A.
   I_{R_3} = _____ A.
   P_T = _____ W.

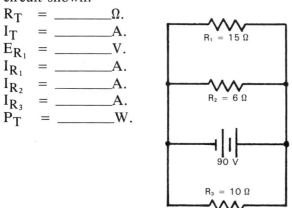

3. Solve for the unknown quantities in the circuit shown.

   R_T = _____ Ω.
   E_T = _____ V.
   I_{R_1} = _____ A.
   I_{R_2} = _____ A.
   I_{R_3} = _____ A.
   P_T = _____ W.

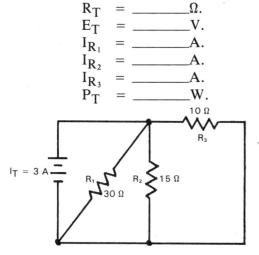

4. In a parallel circuit containing a 4 ohm, a 5 ohm, and a 6 ohm resistor, the current flow is:
   a. Highest through the 4 ohm resistor.
   b. Lowest through the 4 ohm resistor.
   c. Highest through the 6 ohm resistor.
   d. Equal through all three resistors.

5.3 COMBINATION CIRCUITS (SERIES-PARALLEL)

The technician and engineer are often required to compute the total resistance of combination (series-parallel) circuits. This total resistance is referred to as the EQUIVALENT RESISTANCE of the circuit. It can be solved in logical order by using the series and parallel resistance formulas of the previous discussion. Referring to Fig. 5-15, $R_1 = 7\ \Omega$, $R_2 = 5\ \Omega$, $R_3 = 10\ \Omega$, $R_4 = 20\ \Omega$, $R_5 = 10\ \Omega$. The applied voltage, E_S, equals 30 volts.

Step 1. Combine R_4 and R_5, Fig. 5-16. They are in series, so $R_4 + R_5 = 20 + 10 = 30\ \Omega$.

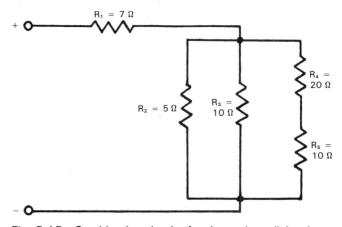

Fig. 5-15. Combination circuit of series and parallel resistors.

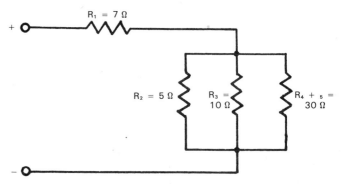

Fig. 5-16. Two series resistors in one branch of the parallel network have been combined.

Step 2. Find the total resistance of the parallel circuit, Fig. 5-17.

$$\frac{1}{R_T} = \frac{1}{R_2} + \frac{1}{R_3} + \frac{1}{R_{4\&5}} \text{ or}$$

$$\frac{1}{5} + \frac{1}{10} + \frac{1}{30} = \frac{10}{30},$$

$$R_T = \frac{30}{10} = 3 \, \Omega$$

Step 3. This equivalent resistance of 3 ohms is in series with R_1 of 7 ohms, Fig. 5-18. So the R_T of the circuit equals 3 ohms + 7 ohms or 10 ohms.

The current flowing in the total circuit is,

$$I_T = \frac{E}{R_T} = \frac{30}{10} = 3 \text{ amps}$$

The voltage drop across R_1 is,

$$E_{R_1} = IR \text{ or } 3 \times 7 = 21 \text{ V}$$

If 21 volts is lost by R_1, then the remaining part of the applied voltage appears across the network of parallel resistors. The remaining voltage equals 9 V.

The current through the three branches must equal the total current of 3 amps in the circuit, Fig. 5-19.

$$I_{R_2} = \frac{E}{R_2} = \frac{9 \text{ V}}{5 \, \Omega} = 1.8 \text{ amps}$$

$$I_{R_3} = \frac{E}{R_3} = \frac{9 \text{ V}}{10 \, \Omega} = .9 \text{ amps}$$

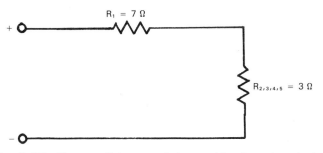

Fig. 5-17. The parallel network is combined to the single equivalent resistance.

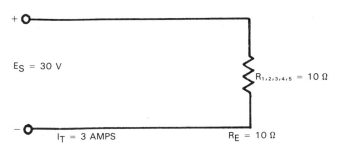

Fig. 5-18. Total resistance of the combination circuit is equivalent to a single resistance.

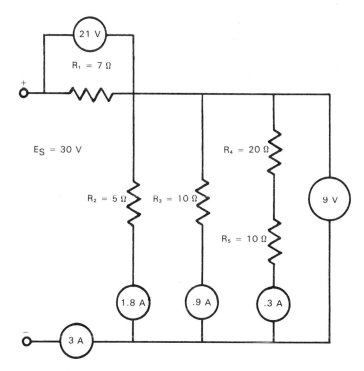

Fig. 5-19. The complete analysis of the combination circuit, showing voltages and currents.

$$I_{R_{4\&5}} = \frac{E}{R_{4\&5}} = \frac{9 \text{ V}}{30 \, \Omega} = .3 \text{ amps}$$

$$I_T = I_{R_2} + I_{R_3} + I_{R_{4\&5}}$$

$$= 1.8 + .9 + .3 = 3 \text{ amps}$$

Knowing the current and voltage of this circuit, the power consumed would be:

$$P = I \times E \text{ or } 3 \text{ amps} \times 30 \text{ V} = 90 \text{ W}$$

The sum of the powers dissipated by the individual resistors should equal 90 watts. Using the formula $P = I^2R$ then:

$$P_{R_1} = (3)^2 \times 7 = 63 \text{ W}$$
$$P_{R_2} = (1.8)^2 \times 5 = 16.2 \text{ W}$$
$$P_{R_3} = (.9)^2 \times 10 = 8.1 \text{ W}$$
$$P_{R_4} = (.3)^2 \times 20 = 1.8 \text{ W}$$
$$P_{R_5} = (.3)^2 \times 10 = .9 \text{ W}$$
$$P_T = 90 \text{ W}$$

The solution of series and parallel circuits involving the use of Ohm's and Watt's Laws requires practice. Practice is the only way to assure understanding. Quick recognition of known and unknown quantities and a logical approach to the solution will assure proper results. Always prove your answer by using other forms of Ohm's Law.

REVIEW QUESTIONS FOR SECTION 5.3

1. What is another name for a combination circuit?
2. Find the unknown values for the circuit shown.
   R_T = _____ Ω.
   E_T = _____ V.
   E_{R_1} = _____ V.
   E_{R_2} = _____ V.
   E_{R_3} = _____ V.
   E_{R_4} = _____ V.
   I_{R_1} = _____ A.
   I_{R_2} = _____ A.
   I_{R_3} = _____ A.
   I_{R_4} = _____ A.
   P_T = _____ W.

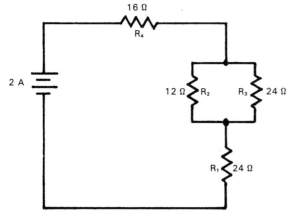

3. What is the voltage drop across the R_X (3 ohm) resistor?
   a. 2.4 volts.
   b. 24 volts.
   c. 9.6 volts.
   d. 0.96 volts.

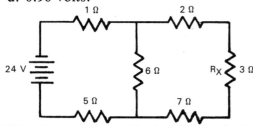

4. What is the value of I_T in the circuit shown?
   a. I_T = 1.14 amp.
   b. I_T = 0.4 amp.
   c. I_T = 0.667 amp.
   d. I_T = 1 amp.

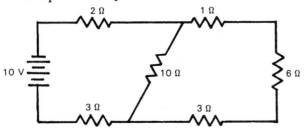

SUMMARY

1. A series circuit is one where there is only path for current flow.
2. In a series circuit, the following facts are known:
   a. Current is equal throughout
      ($I_T = I_{R_1} = I_{R_2} = I_{R_3}$).
   b. Resistances add up to the total resistance
      ($R_T = R_1 + R_2 + R_3$).
   c. The total voltage is equal to the individual voltage drops across the resistors ($E_T = E_{R_1} + E_{R_2} + E_{R_3}$).
3. Ohm's Law is:
   $I = \dfrac{E}{R}$ or $E = I \times R$ or $R = \dfrac{E}{I}$
4. Kirchhoff's Voltage Law states that all of the voltage drops (E_{R_1}, E_{R_2}, E_{R_3}, etc.) in a series circuit must add up to the total or applied voltage.
5. Power is the time rate for doing work.
6. The power formulas are:
   $P = I \times E$ or $P = I^2 R$ or $P = \dfrac{E^2}{R}$
7. Parallel circuits have more than one pathway for current to flow.
8. The formulas for voltage, current, and resistance in a parallel circuit are:
   a. $E_T = E_{R_1} = E_{R_2} = E_{R_3}$, etc.
   b. $I_T = I_{R_1} + I_{R_2} + I_{R_3}$, etc.
   c. $\dfrac{1}{R_T} = \dfrac{1}{R_1} + \dfrac{1}{R_2} + \dfrac{1}{R_3}$, etc.
   or $R_T = \dfrac{R_1 \times R_2}{R_1 + R_2}$
9. Kirchhoff's Current Law states that the amount of current entering a junction or parallel circuit must be the same as the current leaving that circuit.
10. Combination circuits are made of certain components connected in series and others connected in parallel.

TEST YOUR KNOWLEDGE, Chapter 5

Please do not write in the text. Place your answers on a separate sheet of paper.
1. When resistors are connected in _____, the total resistance of the circuit is equal to the sum of the individual values of all the resistors.
2. Voltage loss across a resistor is also known as voltage _____.
3. In a parallel circuit, the total resistance of the circuit is _____ with each component that is added.

Series, Parallel, and Combination Circuits

4. When equal resistors are connected in parallel, $R_T = $ _____.
5. When unequal resistors are connected in parallel, $R_T = $ _____.
6. In combination circuits, total resistance is referred to as the _____ resistance.
7. In a series circuit, $P_T = $ _____.
8. In a parallel circuit, $P_T = $ _____.
9. Using the circuit shown, solve for the following values.
   $R_T = $ _____.
   $E_T = $ _____.
   $E_{R_1} = $ _____.
   $E_{R_2} = $ _____.
   $E_{R_3} = $ _____.
   $P_T = $ _____.

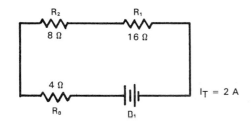

10. Using the circuit shown, solve for the following values.
    $R_T = $ _____.
    $E_T = $ _____.
    $E_{R_1} = $ _____.
    $I_{R_1} = $ _____.
    $I_{R_2} = $ _____.
    $I_{R_3} = $ _____.
    $P_T = $ _____.

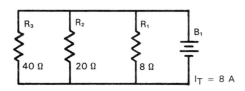

11. Using the circuit shown, solve for the following values.
    $R_T = $ _____.
    $I_T = $ _____.
    $E_{R_1} = $ _____.
    $E_{R_2} = $ _____.
    $E_{R_3} = $ _____.
    $I_{R_1} = $ _____.
    $I_{R_2} = $ _____.
    $I_{R_3} = $ _____.
    $P_T = $ _____.

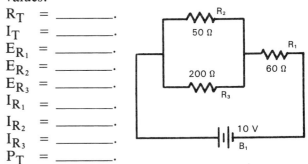

12. Using the circuit shown, solve for the following values.
    $R_T = $ _____.
    $E_T = $ _____.
    $E_{R_1} = $ _____.
    $E_{R_2} = $ _____.
    $E_{R_3} = $ _____.
    $E_{R_4} = $ _____.
    $I_{R_1} = $ _____.
    $I_{R_2} = $ _____.
    $I_{R_3} = $ _____.
    $I_{R_4} = $ _____.
    $P_T = $ _____.

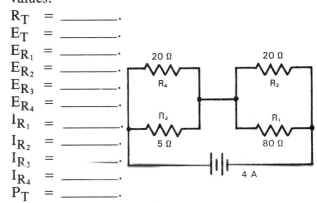

FOR DISCUSSION

1. What are the implications of Kirchhoff's two laws for electrical circuits?
2. Why does the resistance of a conductor decrease as its surface area increases?
3. You have a 6 volt car radio that uses 1.5 amps. You want to install it in a 12 volt car. Draw the circuit for and give the values of the components to do this.
4. Draw the circuit for working one light from two locations.

Part I Summary

FUNDAMENTALS OF ELECTRICITY AND ELECTRONICS TECHNOLOGY

IMPORTANT POINTS

1. Matter is anything that has size, weight, and mass. Elements are basic or pure forms of matter. Compounds are mixtures of elements.
2. Atoms are the smallest form of an element still having the characteristics of that element. Molecules are the smallest form of a compound still having the characteristics of that compound.
3. Electrons have a negative ($-$) charge and protons have a positive ($+$) charge.
4. Static electricity is electricity that is at rest or not moving.
5. Like charges repel each other while unlike charges attract each other.
6. A coulomb is 6.24×10^{18} electrons.
7. Voltage is the force or push that moves electrons along in a conductor. It is measured in volts.
8. Current is the movement of electrons in a conductor. It is measured in amperes.
9. One ampere is the rate measure of one coulomb of electrons (6.24×10^{18}) moving past a given point in a conductor in one second.
10. Prefixes are used to change the value of a basic unit. The following prefixes are used for current, voltage, and resistance:

    Mega (M) = 1,000,000 times larger
    Kilo (K) = 1,000 times larger
    Milli (m) = $\frac{1}{1,000}$ times smaller
    Micro (μ) = $\frac{1}{1,000,000}$ times as small

11. The Electron Theory of Current Flow states that current in the external circuit always flows from negative to positive.
12. A conductor has free electrons and low resistance. It permits current to flow easily. The unit for conductance is the siemen.
13. Insulators do not have any free electrons and have a high resistance to current.
14. Resistance is the opposition to current. The unit for resistance is the ohm (Ω).
15. Ohm's Law states the relationship between current, voltage, and resistance in a formula.
16. Semiconductors lie between conductors and insulators in their resistance to current.
17. Primary cells cannot be recharged. Secondary cells can be recharged.
18. Connecting cells in parallel increases their current rating. Connecting them in series increases their voltage rating.
19. Electrical energy can be created from light, by a solar cell or photovoltaic cell.
20. A thermocouple converts heat to electrical energy.
21. The effect of converting pressure to electrical energy is called piezoelectricity.
22. A circuit is a complete pathway on which current flows. It usually consists of a voltage source, a pathway, a load, and a control.
23. A series circuit has only one pathway on which current can flow.
24. A parallel circuit has more than one pathway on which current can flow.
25. A combination circuit has certain components connected in series and others connected in parallel.
26. Power is the time rate of doing electrical work. It is measured in watts.

SUMMARY QUESTIONS

1. Solve for the unknown quantities in the circuit shown.
   E_T = _____.
   P_T = _____.

2. Convert the following units.
   a. 610 μA = _____ mA.
   b. 42,000 μV = _____ V.
   c. .07 MΩ = _____ Ω.
   d. 29 kV = _____ V.
   e. 1.75 A = _____ mA.

3. The value for a resistor which has yellow, violet, brown, and silver color bands is _____ Ω, _____ %.

4. Solve for the unknown quantities in the circuit shown.
   R_T = _____.
   I_T = _____.
   E_{R_1} = _____.
   F_{R_2} = _____.
   E_{R_3} = _____.
   P_T = _____.

5. According to Kirchhoff's Law of Voltages, the algebraic sum of all the voltages in a series circuit is equal to:
   a. Zero.
   b. The source voltage.
   c. The total voltage drop.
   d. The sum of the IR drop of the circuit.

6. What is the value of R_3 in the circuit below?
   a. 8 ohms.
   b. 10 ohms.
   c. 20 ohms.
   d. 100 ohms.

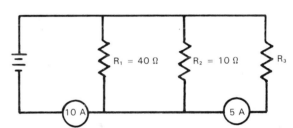

7. Solve for the unknown quantities in the circuit shown.
   R_T = _____.
   I_T = _____.
   I_{R_1} = _____.
   I_{R_2} = _____.
   I_{R_3} = _____.
   I_{R_4} = _____.
   E_{R_1} = _____.
   E_{R_2} = _____.
   E_{R_3} = _____.
   E_{R_4} = _____.
   P_T = _____.

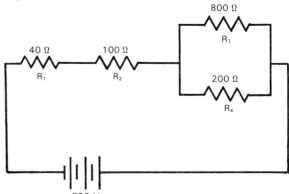

Part II

APPLIED ELECTRICITY

6 Magnetism
7 Generators
8 Instruments and Measurements

The greatest source of electricity is the conversion of mechanical energy to electrical energy by magnetic generators and alternators. Over 98 percent of all electricity generated comes from this source. The two basic types of electricity, according to the type of electron flow, are direct current (dc) and alternating current (ac).

Chapter 6, *Magnetism* discusses both permanent magnetic and electromagnetic principles. Applications of magnetic devices such as relays, solenoids, compasses, and recording tapes are explained.

The theory and operation of generators and alternators are presented in Chapter 7, *Generators*. Direct and alternating current are discussed.

Each student of electricity and electronics must be familiar with test instruments. Also the proper use of instruments in the measurement of voltage, current, resistance, wattage, or any other electrical unit is important to all studying this field. Chapter 8 presents an orientation to instruments and measurement, from the simple voltmeter to the impressive oscilloscope.

Electricity and Electronics

Floppy disks for computers have magnetic properties.

A demonstration of electricity being used to create a magnet.

A demonstration designed to show how an electromagnet can open and close an electrical switch.

Magnetism is closely linked to electricity. It is important to generation of electricity, communication, and to certain types of controls. Electricity can be used to create magnetism. Magnetism can be used to store information and to change electrical voltage, among other things.

84

Chapter 6

MAGNETISM

After studying this chapter, you will be able to:
- Discuss basic magnetic principles.
- State the three Laws of Magnetism mentioned in this chapter.
- Describe the link between electric current and magnetism.
- Explain Roland's Law.
- Discuss various types of relays and the manner in which they work.
- Describe the use of magnetic shields.

6.1 BASIC MAGNETIC PRINCIPLES

For centuries, magnetism has been of interest to scientists. Shepherds, tending their flocks in ancient days, were mystified by small pieces of stone, which attracted the iron tip on the shepherd's staff. The ancient Chinese navigators discovered that a small piece of this odd stone attached to a string would always turn in a northerly direction. These small stones were iron ore. They were called magnetite by the Greeks, because they were found near Magnesia in Asia Minor. Since mariners used these stones in the navigation of their ships, the stones became known as "leading stones" or LODESTONES. These were the first forms of natural magnets. Today, a MAGNET may be defined as a material or substance which has the power to attract iron, steel, and other magnetic materials.

Through laboratory experiments, it was discovered that the greatest attractive force appeared at the ends of a magnet. These concentrations of magnetic force are called MAGNETIC POLES. Each magnet has a NORTH POLE and a SOUTH POLE. It was also discovered that many invisible lines of magnetic force existed between poles. Each line was an independent line. It did not cross or touch its bordering line.

Fig. 6-1 shows the magnetic lines of force. This field picture was created by placing a bar magnet beneath a sheet of paper, and sprinkling it with fine iron filings. Note the pattern of lines existing between the poles. Note the concentration of lines at each end of the magnet or its poles. Each line of force from the north pole goes to the south pole through space and returns to the north pole through the magnet. These closed loops of the magnetic field may be described as MAGNETIC CIRCUITS. You may compare the magnetic circuit to the electrical circuit, where the magnetizing force may be compared to voltage, and the magnetic lines may be compared to current.

Further scientific investigation has proved that the earth acts as one enormous magnet, with its poles close to the north and south geographical poles. Referring

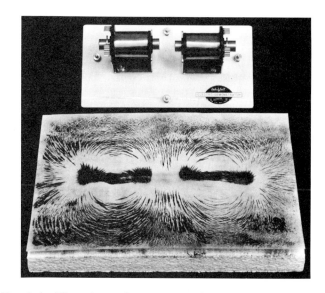

Fig. 6-1. The electrodemonstrator is used to show the magnetic fields of two permanent magnets.

to Fig. 6-2, you will observe that magnetic north and the north geographic pole do not coincide. A compass would not necessarily point toward true north. This angle between true and magnetic north is called the ANGLE OF DECLINATION OR VARIATION.

There is, however, an imaginary line around the earth where the angle of declination is zero. When standing on this line your compass would point to true north as well as magnetic north. At all other locations on the surface of the earth, the compass reading must be corrected to find true north.

LAWS OF MAGNETISM

The power of a magnet to attract iron has already been discussed. In Fig. 6-3, two bar magnets have been hung in wire saddles and are free to turn. Notice that when the north (N) pole of one magnet is close to the south (S) pole of the other, an attractive force brings the two magnets together. If the magnets are turned so that two N poles or two S poles are close to each other, there will be a repulsive force between the two magnets. This proves two of the Laws of Magnetism:

UNLIKE POLES ATTRACT EACH OTHER.
LIKE POLES REPEL EACH OTHER.

Place the magnets under a sheet of paper. Using the iron filings to detect the invisible fields, observe that in the attractive position (N and S together) the invisible field is very strong between the two poles, Fig. 6-4. However, when the magnets are placed with like poles together, the repulsive force is shown by an absence of lines between the poles.

CAUTION: Certain precautions should be observed during handling and storage of permanent magnets. Since magnetism is a result of molecular alignment, any rough handling, such as dropping or pounding, may upset the molecular alignment and weaken the magnet.

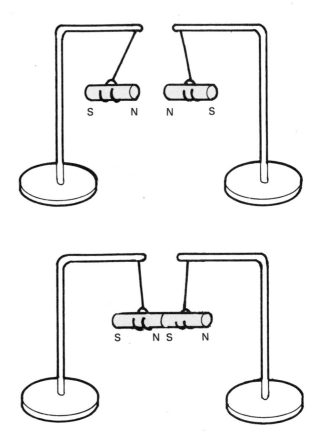

Fig. 6-3. These permanent magnets demonstrate the Laws of Magnetism.

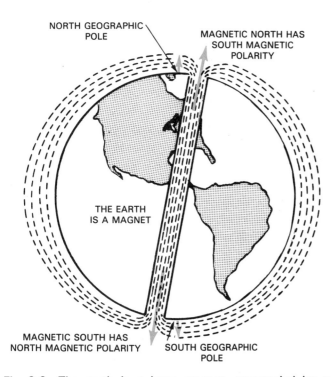

Fig. 6-2. The earth is a large magnet, surrounded by a magnetic field.

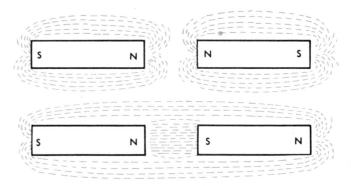

Fig. 6-4. These sketches show the magnetic fields of attracting and repelling magnets.

Magnetism

What causes a substance to become magnetized? One theory states that molecules in an iron bar act as tiny magnets. They are arranged in-line, so their North and South poles are joined. When the iron is demagnetized, the molecules are in random positions, Fig. 6-5. This fact can be proved by breaking a piece of magnetized iron into several pieces. Each piece is a separate magnet. Fig. 6-6 shows a broken magnet.

The conclusion is further supported by the way a magnet is made. For example, take an unmagnetized bar. Rub it a few times in the same direction with a permanent magnet. A test will show that the bar is now magnetized. Rubbing with the magnet lined up the molecules and caused the iron to become magnetized.

Magnets are made by placing the material to be magnetized in a very strong magnetic field.

CAUTION: Magnets should be stored in pairs with north and south poles together to insure long life. A single horseshoe magnet may be preserved by placing a small piece of soft iron, called "keeper," across its poles.

Permanent magnets are made in a number of shapes and sizes, Fig. 6-7. Flexible magnets may be used as a visual teaching aid, Fig. 6-8. The resistance symbols in this drawing are made of a flexible magnetic material. They are held in place on a sheet iron bulletin board by magnetism. Industry uses these "boards" for traffic control, production flow, temporary notices, and other announcements.

CAUTION: Heat will destroy a magnet. Heat energy causes an increase in molecular activity and expansion. This permits the molecules to return to their random positions on the unmagnetized piece of steel.

Magnetic recording tape

Recording tape used in cassettes, videotapes, and computers uses magnetic principles to store information. Magnetized iron oxide particles are adhered to a plastic tape or floppy disk using a binder. By magnetizing these with a recording head, information, music, or photographs can be stored. Likewise, the information can be "read" from a tape or floppy disk by a magnetic head. Fig. 6-9 shows a typical floppy disk used in a computer disk drive.

Fig. 6-5. In a permanent magnet, the molecules are in line.

Fig. 6-6. A long magnet may be broken into several smaller magnets.

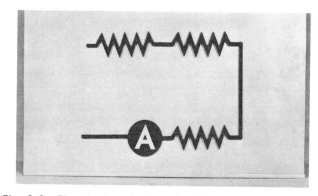

Fig. 6-8. Electrical symbols made of flexible magnets are attached to a metal board. This idea has hundreds of uses.

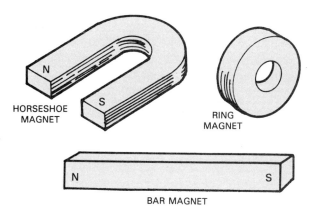

Fig. 6-7. Magnets are made in a number of styles, shapes, and sizes. Note that N and S poles of ring magnet cannot be identified.

MAXELL

Fig. 6-9. Magnetic floppy disks used in computers.

MAGNETIC FLUX

The many invisible lines of magnetic force surrounding a magnet are called the MAGNETIC FLUX. If a magnet is strong, the lines will be more dense. So, the strength of the magnetic field may be determined by its flux density, or the number of lines per square inch or per square centimeter. Flux density is expressed by the equation:

$$B = \frac{\Phi}{A}$$

where B equals flux density, Φ (phi) equals the number of lines, and A equals the cross-sectional area, in either square inches or square centimeters. If A is measured in square centimeters, then B is the number of lines per square centimeter or GAUSS. B can be given in webers per square meter.

THIRD LAW OF MAGNETISM

A simple experiment will help to show a third Law of Magnetism. Place one bar magnet on a table, then slowly approach it with a second bar magnet. If the polarities are opposite, they will attract each other when close enough together. This experiment helps to indicate that the ATTRACTIVE FORCE INCREASES AS THE DISTANCE BETWEEN THE MAGNETS DECREASES.

The magnetic force, either attractive or repelling, varies inversely as the square of the distance between the poles. For example, if the distance between two magnets with like poles is increased to twice the distance, the repulsive force reduces to one-quarter of its former value. This is valuable information. It explains the reason for accurately setting the gap in magnetic relays and switches. The term GAP is the distance between the core of the electromagnet and its armature.

REVIEW QUESTIONS FOR SECTION 6.1

1. The two poles of a permanent magnet are the _____ pole and the _____ pole.
2. Unlike poles _____ each other.
3. Name three common shapes for magnets.
4. Name six uses of magnets that you come in contact with each day.
5. The invisible lines of magnetic force are called _____ _____.
6. Flux density is measured in _____.

6.2 ELECTRIC CURRENT AND MAGNETISM

During the eighteenth and nineteenth centuries, a great deal of research was directed to the link between electricity and magnetism. The Danish physicist, Hans Christian Oersted, discovered that a magnetic field existed around a conductor carrying an electric current, Fig. 6-10, where a current carrying conductor is passed through a sheet of cardboard. Small compasses placed close to the conductor, point in the direction of the magnetic lines of force. Reversing the current will also reverse the direction of the compasses by 180 deg. This shows that the direction of the magnetic field depends upon the direction of current flow.

The LEFT HAND RULE FOR A CONDUCTOR may be used to reveal the direction of the magnetic field. Grasp conductor with your left hand, extending your thumb in direction of current flow. Your fingers around conductor will indicate circular direction of field.

In Fig. 6-11, the dot in the center of the conductor at left is the point of an arrow. It shows that current is flowing toward you. Circular arrows give the direction of the magnetic field. This principle is very important when electrical wires carry alternating currents. Then, placement of wires, or "lead dress," has a certain influence on the working of the circuit.

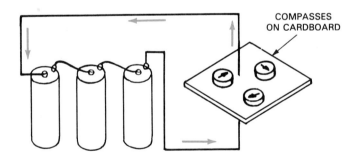

Fig. 6-10. Compasses line up to show circular pattern of magnetic field around current carrying conductor.

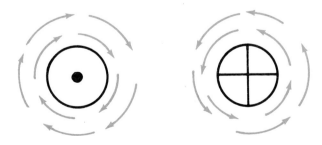

Fig. 6-11. These conventions are used to show the link between current flow and the magnetic field.

Magnetism

THE SOLENOID

When a current carrying conductor is wound into a coil, or SOLENOID, each of the magnetic fields circling the conductors tend to merge or join together. A solenoid will appear as a magnetic field with a N pole at one end, and a S pole at the opposite end. This solenoid is shown in Fig. 6-12.

The LEFT HAND RULE FOR A COIL may be used to learn the polarity of the coil. Grasp the coil with your left hand in such a manner that your fingers circle the coil in the direction of current flow. Your extended thumb will point to the N pole of the coil.

The strength of the magnetic field of a solenoid depends upon the number of turns of wire in the coil, and the current in amperes flowing through the coil. The product of the amperes and turns is called the AMPERE-TURNS of a coil. It is the unit of measurement of field strength. If, for example, a coil of 500 ampere-turns will produce a required field strength, any combination of turns and amperes totaling 500 will work, such as,

50 turns × 10 amps = 500 amp-turns
100 turns × 5 amps = 500 amp-turns
500 turns × 1 amp = 500 amp-turns

MAGNETIC CIRCUITS

A detailed study of magnetic circuits is beyond the scope of this text. However, you should know the terms used to describe quantities and characteristics of magnetic circuits.

It has already been mentioned that a likeness exists between magnetic circuits and electrical circuits. Comparing circuits, the number of lines of magnetic flux, Φ (in electricity this is current, I), is in direct proportion to the force producing them. This force (F) is in units of magnetomotive force called GILBERTS. (In electricity this is voltage, E.) Also, there is resistance in a magnetic circuit, which is called RELUCTANCE. Its unit of measure is the REL (R). The equation then becomes:

$$\Phi = \frac{F}{R} \text{ (Rowland's Law)}$$

where Φ = Total number of lines of magnetic force

F = Force which produces the field in gilberts

R = Resistance to the passage of lines of force (reluctance)

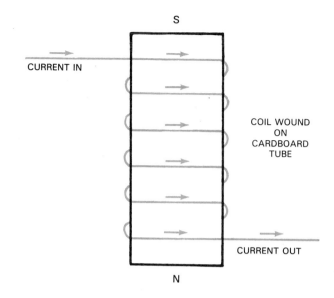

Fig. 6-12. A wire wound into a coil is a solenoid and has a polarity set by the direction of current flow.

The relationship between F and ampere-turns is F = 1.257 IN, where F is gilberts, I is the current, and N is the number of turns.

The term reluctance has been defined as the unit of measurement of the resistance to the passage of magnetic lines of force. This is also the resistance offered by one cubic centimeter of air. Experimentation proves that, as the flux path increases in length, the reluctance increases. Likewise, as the cross-sectional area of the flux path decreases, the reluctance increases.

Reluctance also depends upon the material used as the path. The ability of a material or substance to conduct magnetic lines of force is called PERMEABILITY. The permeability of air is considered as 1. The term is expressed by the Greek letter μ, (pronounced mew.) The factors determining the resistance in a magnetic circuit may be compared to the factors affecting the resistance of electron flow in a conductor. Refer to Chapter 4. For the magnetic circuit:

$$R = \frac{\ell}{\mu A}$$

where R equals rels, ℓ equals length of path, μ equals the permeability of the material in the path, and A equals the cross-sectional area of the path.

There is a distinct link between the density of a magnetic field, the force producing the field, and the permeability of the substance in which the field is produced. Mathematically,

$$\mu = \frac{B}{H} = \frac{\text{Flux lines per cm}^2 \text{ or gauss}}{\text{Gilberts per cm}}$$

The symbol H is the INTENSITY of a magnetizing force through a unit of length. It is commonly expressed in gilberts per centimeter. mathematically,

$$H = \frac{F}{\ell} = \frac{1.257 \text{ IN}}{\ell}$$

The term ℓ is, again, length of path expressed in centimeters. The B, H, and μ of common magnetic materials may be found in an electricity handbook.

ELECTROMAGNETS

The construction of a solenoid coil has already been discussed. In the case of the solenoid, air only is the conductor in the magnetic circuit. Other substances will conduct magnetic lines of force better than air. These materials would be described as having greater PERMEABILITY.

To prove this, a soft iron core may be inserted in the solenoid coil, Fig. 6-13. The strength of the magnetic field will be greatly increased. Two reasons may be given. First, the magnetic lines have been confined (concentrated) into the smaller cross-sectional area of the core. Secondly, the iron provides a far better path (greater permeability) for the magnetic lines. Such a device is known as an ELECTROMAGNET. The rules used to learn the polarity of an electromagnet are the same as those for the solenoid. Use the Left Hand Rule. You may wish to construct a simple electromagnet as shown in Fig. 6-14.

The electromagnet is made by selecting a prewound solenoid coil and putting the soft iron core in the center. Wire according to the diagram. Experiment with this coil by using a compass. Connect the ends of the coil to a 6-volt battery. Notice the polarity of the coil. Reverse the connections to the battery and observe the compass action. Remember the direction of electron flow in a circuit. Does this experiment prove the Left Hand Rule?

Carry on this experiment using the electromagnet to pick up several small nails. As the nails are held by magnetic force, open the switch. The nails will fall. Leaving the switch open, try to pick up some fine iron filings. Notice that some filings will be drawn to the core. In other words, the core has retained a small amount of its magnetism. This is called RESIDUAL MAGNETISM. As very little magnetism remained, the core would be considered as having LOW RETENTIVITY. Now, replace the core of the magnet with a steel core. Repeat the experiment. Once the steel is magnetized by the coil, it retains its magnetism. A permanent magnet has been made, because steel has a HIGH RETENTIVITY.

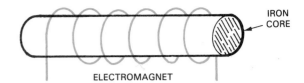

Fig. 6-13. The coil with an iron core is described as an electromagnet.

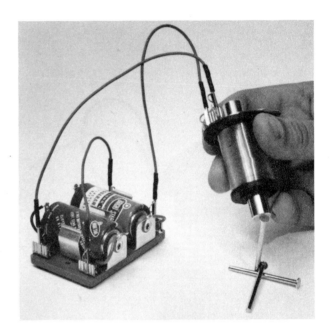

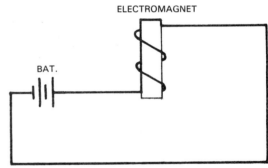

Fig. 6-14. An electrodemonstrator (see Appendix 6 for more information) is used to show principles of an electromagnet.

SOLENOID SUCKING COIL

Refer to Fig. 6-15. The experiment is still being conducted. This is to show the sucking action of the solenoid. First, energize the solenoid coil by closing the switch. Then bring the iron core close to one end of the coil. Notice the pulling magnetic force. Release the core from your fingers. It will be sucked into the center of the coil and come to rest at the center of the coil. Is this not converting magnetism into mechanical motion? The solenoid sucking coil has many uses in industry, for the electrical control of mechanical action.

Magnetism

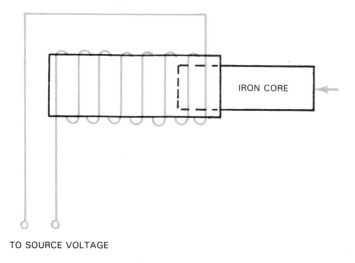

Fig. 6-15. When the solenoid is energized, the core is sucked into the center.

REVIEW QUESTIONS FOR SECTION 6.2

1. What is an electromagnet?
2. Explain the Left Hand Rule for a coil.
3. A coiled current carrying conductor is known as a _____.
4. What is permeability?
5. The amount of magnetism left in a magnetic core is called ___ ___ _____.
6. The unit of measure for magnetomotive force is the:
   a. Flux.
   b. Gilbert.
   c. Gauss.
   d. None of the above.

Refer to Fig. 6-16. It shows a simple, useful door chime. It is built by arranging the coil and parts as shown. The door chime uses the sucking coil principle. When the coil is energized, the core is sucked into the coil and strikes the chime.

6.3 THE RELAY

The relay is a device used to control a large flow of current by means of a low voltage, low current circuit. It is a magnetic switch.

Fig. 6-16. In this application, the electrodemonstrator is used as a door chime. The same principles and construction may be used for other worthwhile and useful projects.

Assemble the relay, as shown by the diagram, Fig. 6-17. A flashlight cell is used to energize the electromagnet. This circuit is controlled by the switch. When the coil is magnetized, its attractive force pulls the lever arm, called an ARMATURE, toward the coil. The contact points on the armature will open or close depending upon the arrangement, and control the larger high voltage circuit. In this circuit a light bulb is used which is connected to a 115 volt power source. The advantages of this device are clear.

1. From the safety point of view, the operator touches only a harmless, low-voltage circuit, yet controls perhaps several hundred volts by means of the relay.
2. Heavy current machines may be controlled from a remote location without any need to run heavy wires to the controlling switch.
3. Switching action by means of relays may be very rapid and positive.

There are hundreds of uses for relays at home and in industry to control motors and machines. In the automobile, the relay is used in the voltage regulator and in controlling the starting motor, the horn, and the headlights. In electronic equipment and radio transmitters, a relay is used in many high voltage circuits. It provides a sensitive, positive means of control. Relays of this type are shown in Fig. 6-18.

By looking through a radio parts catalog, you will notice a large number of relays designed for certain uses, Fig. 6-19. When choosing a relay for a special purpose, think about the number of switching contacts required and the current carrying ability of these contacts. Well-designed relays have points made of silver, silver alloys, tungsten, and other alloys.

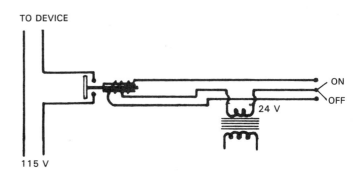

Fig. 6-18. Basic circuit using low voltage relay, as used to control lighting circuit in the home.

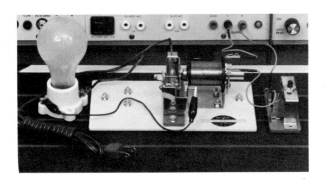

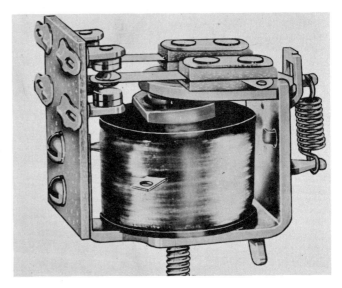

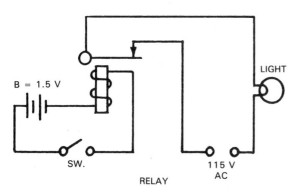

Fig. 6-17. This circuit uses the electrodemonstrator as a relay.

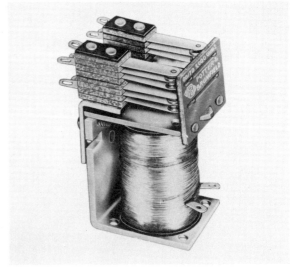

Fig. 6-19. Type of relay used in electronic control circuits.

Magnetism

A relay may be chosen for either opening or closing a circuit. Its "at rest" position would be "normally closed" or "normally open." The coil is the most vital specification. Relays are designed with coils which work on a direct current or an alternating current. Some are very sensitive. They require a milliampere or less to energize. The coil chosen should produce enough magnetic field at its rated voltage to insure positive contact of switch points at all times.

THE REED RELAY

The reed relay is another similar use of magnetism. Refer to Fig. 6-20. Two magnetically sensitive switch contacts are enclosed in a glass tube. If a permanent magnet is brought close to the glass tube, it will cause the switch contacts to close. The reed relay can also be worked by an electromagnet. The operating coil is placed around the reed relay.

CIRCUIT BREAKER

Referring to electromagnets, the strength of a magnetic field depends upon the AMPERE-TURNS. If a coil is made with a fixed number of turns, then its field strength may be varied by controlling the current in the coil. Upon this principle is devised the CIRCUIT BREAKER.

Assemble the coil and armature parts as in Fig. 6-21. Compare this device to the relay. Notice that, in the circuit breaker, the coil and armature points are in series and the whole device is connected in series in the line carrying the load. If the line current exceeds a preset value, the coil will build up enough magnetic force to overcome the spring tension of the armature. The armature releases the points and will open and "break" the circuit. The contact points must be reset or closed by hand.

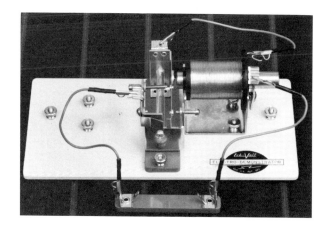
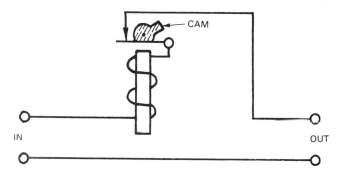

Fig. 6-21. Top. Electrodemonstrator assembled as a circuit breaker. Bottom. Circuit breaker circuit.

This device prevents a circuit from being overloaded beyond the safe carrying volume of the wires. Many homes and businesses use the circuit breaker in place of the fuse.

LESSON IN SAFETY: The circuit breaker is installed for safety. It prevents overload circuits and the danger of fire. The breaker should never be disabled. If the breaker opens the circuit over and over, learn the reason at once. Then take instant steps to solve the problem.

DOOR BELL AND BUZZER

A common device that uses electromagnets is the buzzer. Arrange the coils and armature as in Fig. 6-22. Notice that the coil and the contact points are again connected in series. When the coil is energized, the armature is attracted toward the coil. The contact points open and "break" the circuit. The attraction of the coil then falls to zero. The armature spring pulls the contact points closed. This once again energizes the coil and opens the contacts. This action continues as long as a voltage is applied. The device "buzzes" due to vibration of the armature. An extension may be placed on the vibrating armature on which a striker is attached. The striker hits a bell. This is the principle of the doorbell.

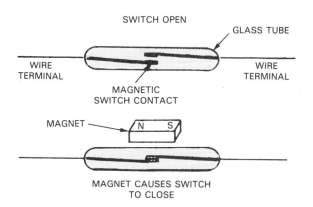

Fig. 6-20. A sketch of a reed relay. The magnet causes the reed contacts to close.

Electricity and Electronics

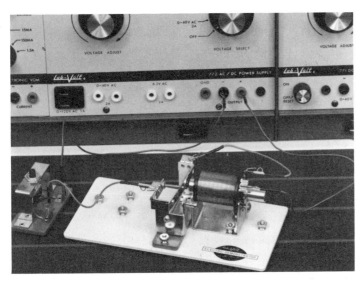
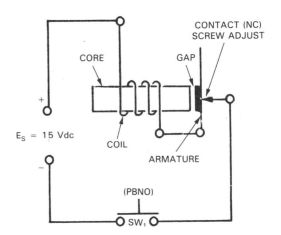

Fig. 6-22. Left. Electrodemonstrator assembled as a buzzer. Right. Buzzer circuit.

Clear-cut examples of electromagnetic uses will be discussed later in this text. A thorough understanding of the action of magnetic fields is vital. The principles of magnetism are the basis for many electrical and electronic events.

MAGNETIC SHIELDS

A magnetic field has no bounds. To the first-time student, it is hard to understand that the force of the magnet will pass through any kind of substance. It does not matter whether it be a concrete wall, glass, wood or any other substance. However, an instrument or a circuit may be shielded from magnetic lines of force. This is done by using the PERMEABILITY of some other substance.

If a piece of iron is placed in a magnetic field, the lines of force will follow through the iron rather than through the air. This is because iron has a much greater permeability. The iron acts as a low resistance path for the magnetic lines. In Fig. 6-23 this action may be observed. By this method, magnetic lines may be conducted around an object. This is called SHIELDING.

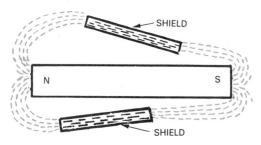

Fig. 6-23. Magnetic lines of force may be conducted around a device by using a high permeability material as a shield.

Many shields are used in electronic equipment to prevent magnetic fields from affecting the working of circuits.

REVIEW QUESTIONS FOR SECTION 6.3

1. What is a relay?
2. The strength of a magnetic field depends on the _____-_____.
3. A circuit breaker:
   a. Is installed for safety.
   b. Prevents overloaded circuits.
   c. Is often used in place of a fuse.
   d. All of the above.
4. Explain how a door buzzer operates.
5. What is shielding?

EXPERIMENT I. Set up the electrodemonstrator as shown in Fig. 6-1. The coils are used in this experiment only as a HOLDING DEVICE for the two permanent magnets. Suspend the cardboard over the magnets on long bolts fastened to the base. Fix the magnets in the coils by wedging them with a pointed match stick or toothpick.
   A. Place magnets so that poles attract. Sprinkle iron filings on cardboard and observe pattern.
   B. Place magnets so that poles repel. Observe pattern.
   C. Place magnets so that ends touch and observe pattern. What conclusions can you draw from these observations?

EXPERIMENT II. Make two wire saddles as shown in Fig. 6-3, and suspend permanent magnets in free space.

A. Place so that north poles are next to each other. Observe.
B. Place so that south poles are next to each other. Observe.
C. Place so that unlike poles are next to each other. Observe.

Make your conclusions and state the Laws of Magnetism in your own words.

EXPERIMENT III. Construct and use the electromagnet, the buzzer, the door chime, and the circuit breaker. Use your electrodemonstrator. Ask your instructor to check each setup BEFORE power is applied.

SUMMARY

1. Magnets are materials or substances which have the power to attract iron, steel, or other magnetic material.
2. Magnets have poles. Every magnet has a north pole and a south pole.
3. The earth is a large magnet.
4. Like poles repel each other; unlike poles attract each other.
5. Temporary magnets lose their properties quickly. Permanent magnets keep their magnetic properties for a long period of time.
6. The invisible lines around a magnet are called magnetic flux lines.
7. Electromagnets are created when current flows through a coil of wire.
8. Magnetic resistance is reluctance.
9. Permeability is the ability of a material to conduct magnetic lines of force.
10. Residual magnetism is what is left in a core after the electromagnetic field is turned off.
11. The relay is an electromagnetic switch.
12. An electromagnetic circuit breaker can be used to prevent an overload of current.

TEST YOUR KNOWLEDGE, Chapter 6

Please do not write in the text. Place your answers on a separate sheet of paper.

1. Concentrations of magnetic force are called _____ _____.
2. What is the name of the angle between true north and magnetic north?
3. State the first two Laws of Magnetism.
4. What is the equation for flux density? What do each of the letter symbols in the equation mean?
5. State the third Law of Magnetism.
6. Compare a magnetic circuit to an electrical circuit. Use Rowland's Law.
7. List three advantages of using relays.
8. In your own words, state the meaning of the following letter symbols.
   a. H.
   b. μ.
   c. IN.

FOR DISCUSSION

1. What are some uses of magnets in our everyday living at home?
2. Explain the differences and similarities between magnetism and gravitation.
3. How does a magnetic compass work in the northern hemisphere? How does it work in the southern hemisphere?
4. How can relays be used in elementary computers?

Electricity and Electronics

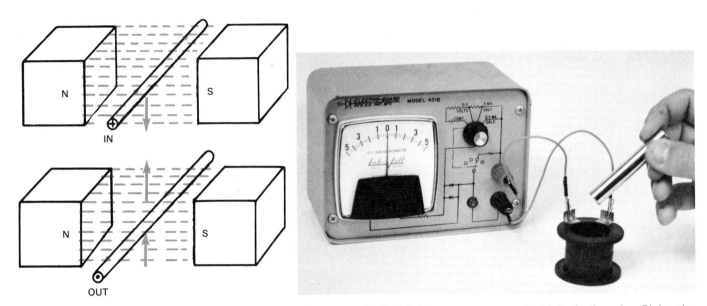

Experiments will prove that if you pass a wire through a magnetic field (left) you can cause electricity in the wire. Right, the galvanometer, a coil, and a magnet will let you observe this principle.

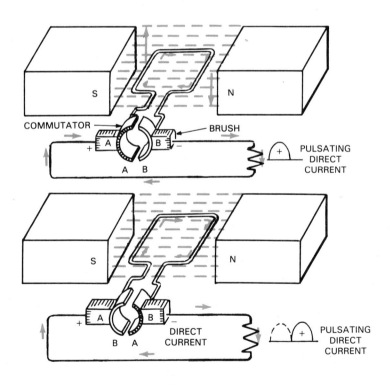

This shows the basic principle of the generator. As the wire coil turns, electricity flows through the wire coil, into the rings of the commutator and out through the brushes into the electrical circuit.

As you learned in Chapter 6, magnetism is closely linked to electricity. As you continue to study, you will see how magnetism is involved in generating electric power through generators.

Chapter 7
GENERATORS

After studying this chapter, you will be able to:
- *Explain how mechanical energy can produce electrical energy.*
- *Discuss the difference between direct current and alternating current.*
- *Explain the construction and function of a generator.*
- *Identify the types of generators.*
- *Discuss voltage and current regulation.*
- *Use your knowledge of alternating current to solve practical problems.*

7-1 ELECTRICAL ENERGY FROM MECHANICAL ENERGY

In Chapter 6, the discoveries of Hans Christian Oerstead were cited as proof that electric current produces a magnetic field. In 1831, scientist Michael Faraday wondered: If electricity produces magnetism, can magnetism produce electricity?

Based on his research, Mr. Faraday learned that magnetism could in fact produce electricity. If a conductor was moved through a magnetic field, a voltage was induced (brought about) in the conductor. This is known as MAGNETIC INDUCTION. In order to induce voltage, there must be a magnetic field, a conductor, and relative motion between the field and conductor, Fig. 7-1.

In the following experiment, you can see the principles of magnetic induction at work. Refer to Fig. 7-2. A coil is attached to a galvanometer. This is a meter that shows the flow and direction of an electric current. A permanent bar magnet is placed in the hollow coil. As the magnet moves into the coil, the needle on the meter shows current flowing in one direction. As the magnet is taken out of the coil, the needle shows current flowing in the other direction. When the magnet is still, no current is produced. These same movements occur when the magnet is held still, while the coil is moved over the magnet.

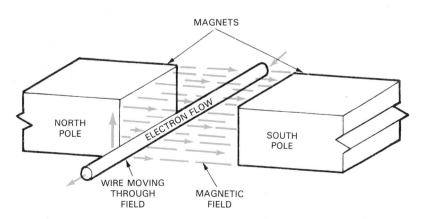

Fig. 7-1. Magnetic induction. A conductor passed through a magnetic field displaces electrons in conductor. The electrons then move through the conductor.

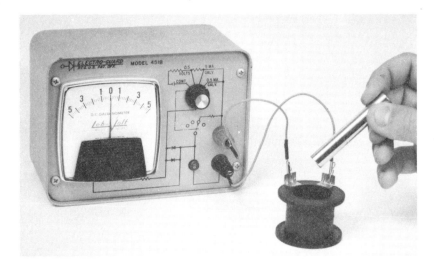

Fig. 7-2. Use a galvanometer, a coil, and a magnet to observe the principles of magnetic induction.

Fig. 7-3 shows the action of coil turning in a magnetic field. In position A, coil top moves parallel to field of magnetism. No voltage is produced. In position B, both sides of the coil are cutting the field at right angles. The highest voltage is produced at this angle. Position C is like position A. The voltage drops to zero. In position D, the coil is again cutting the field at right angles. Highest voltage is induced, but in the opposite direction of position B. The curve in Fig. 7-3 shows the voltage induced in the coil in one turn.

How is this induced voltage created? Fig. 7-4 shows single conductors passing through a magnetic field, but in opposite directions. Current flow is shown by the arrows. In each case, the induced current in the conductor forms a magnetic field around the conductor. This field opposes, and is repelled by, the fixed field. This is stated in LENZ'S LAW: "The polarity of an induced electromagnetic force is such that it sets up a current, the magnetic field of which always opposes the change in the existing magnetic field." More simply, Lenz's Law says the field induced around the conductor is opposed by the existing field. Therefore, in order to get electricity, some form of mechanical force must be applied to overcome this opposition. For example, water and steam supply the mechanical force to turn turbines in large power plants.

The strength of the induced voltage in a rotating coil depends on:

1. The number of magnetic lines of force cut by the coil.

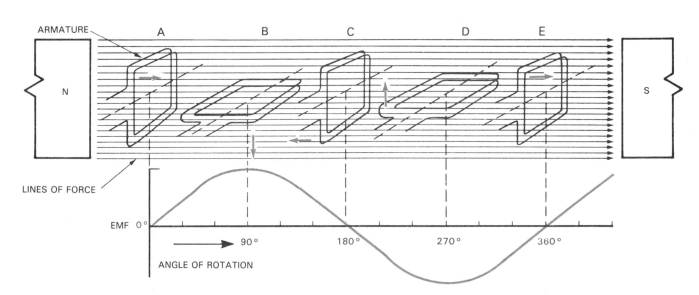

Fig. 7-3. Step-by-step development of induced voltage during one revolution of a coil.

Generators

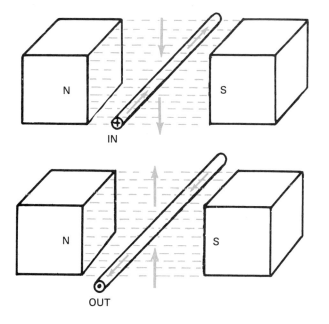

Fig. 7-4. The direction of current flow as a conductor cuts across a magnetic field.

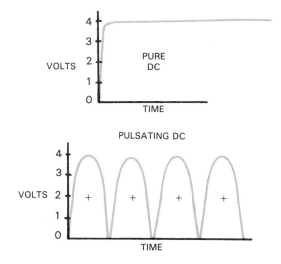

Fig. 7-5. Graphs showing pure dc and pulsating dc.

2. The speed at which the conductor moves through the field. Therefore, when a single conductor cuts across 100,000,000 (10^8) magnetic lines in one second, one volt of electrical pressure will be produced. The voltage can be increased by winding the armature with many turns of wire, and also by increasing its speed of rotation. Reviewing what was learned in Chapter 6, this link can be expressed by the mathematical equation:

$$E = \frac{\Phi N}{10^8},$$

where E equals the induced voltage, Φ equals the lines of magnetic flux, and N equals revolutions per second.

For example, if the fixed magnetic field consisted of 10^6 lines of magnetic flux and a single conductor cut across the field 50 times per second, the induced voltage would equal:

$$E = \frac{10^6 \times 50}{10^8} = 50 \times 10^{-2}$$
$$= .5 \text{ volts}$$

DC VERSUS AC

Direct current (dc) flows in one direction in a conductor. The flow of electrons may be constant. This is called PURE DIRECT CURRENT. Or the flow may be intermittent. This is called PULSATING DIRECT CURRENT. See Fig. 7-5.

Alternating (ac) current flows in alternating directions (back and forth) in a conductor within a given time period. Alternating current is shown in the graph, Fig. 7-3. This curve is called a SINE WAVE.

CONSTRUCTION OF A GENERATOR

A GENERATOR is a device which changes mechanical energy into electrical energy. You saw this change in Fig. 7-3. The revolution of the coil (mechanical energy) was changed to induced current (electrical energy). This action is an example of a very simple generator. It is not very useful in this form. It is not powerful enough to do work. It requires some improvements.

A stronger magnetic field can be made by replacing the permanent magnets with electromagnets. Field coils can be placed over pole pieces or shoes fastened to the steel frame or generator case. The revolving coil, or ARMATURE, is hung in the case with the proper bearings. The single coil is replaced by wire coils of many turns on the armature. Outside connections to the rotating armature are made through commutators or slip rings, and brushes. A COMMUTATOR is a device that reverses electrical connections. It is used on direct current generators, Fig. 7-6. Slip rings are used on alternating current generators, Fig. 7-7.

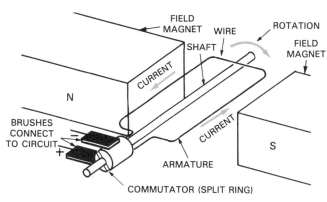

Fig. 7-6. Simple dc generator.

99

Electricity and Electronics

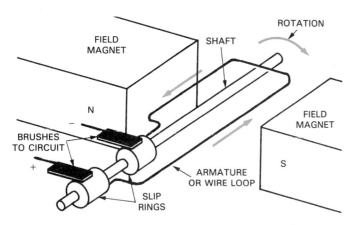

Fig. 7-7. Simple ac generator. Wire loop carries the induced current. Electrons flow out one brush, through circuit, and back in through other brush.

With an ac generator, the induced current in the armature flows first to highest voltage in one direction. The current then flows to highest voltage in the other direction. AC generators will be covered later in this chapter.

The dc generator is like the ac generator in construction. Refer back to Fig. 7-6. However, a direct current is needed at the output terminals of the dc generator. The alternating current in the armature is changed to a pulsating direct current in the outside circuit by a commutator. Study Fig. 7-8. The drawings show the action of the commutator and brushes.

Note that the current in the outside circuit is always flowing in one direction. The output of this generator is shown in Fig. 7-9. It rises and falls from zero to maximum to zero, but always in the same direction. Following the action in Fig. 7-8, brush A is in contact with commutator section A, and brush B is in contact with section B. The first induced wave of current flows through the armature out of brush B, around the external circuit and into brush A, completing the circuit. When the armature revolves one-half turn, the induced current will reverse its direction. But the commutator sections have also turned with the armature. The induced current flowing out of section A is now in contact with brush B and flows through the external circuit to brush A into section B, completing the circuit. The commutator has acted as a switch. It reversed the connections to the rotating coil when the direction of the induced current was reversed.

The current in the outside circuit is pulsating direct current. The output of this generator is not a smooth direct current. The weakness of pulsating dc can be improved two ways. The number of rotating coils on the armature can be increased and commutator sections can be supplied for each set of coils.

To help you understand how the coils are added to the armature, see Fig. 7-10. Each coil has its own induced current. As the current starts to fall off in one, it is replaced by the induced current in the next coil.

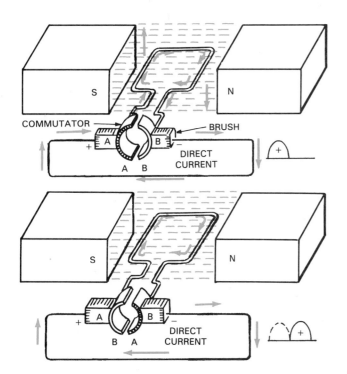

Fig. 7-8. The commutator changes the alternating current in the armature to a pulsating direct current in the outside circuit.

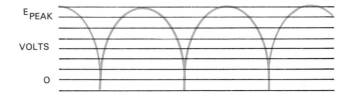

Fig. 7-9. Generator output in volts.

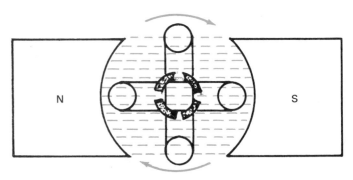

Fig. 7-10. A simple generator with two coils at right angles to each other. They are rotated in magnetic field.

Generators

The current is created as the coils cut across the magnetic field. A graph of the output is shown in Fig. 7-11. It is still a pulsating current. However, the pulses come twice as often and are not as large. The output of the two coil generator is much smoother. By increasing the number of coils, the output will closely duplicate a pure direct current with only a ripple variation.

GENERATOR LOSSES

All of the mechanical power used to turn the generator is not converted into useful electrical power. There are some losses. You will recall that all wires have some resistance, depending on their size, material, and length. Resistance uses power.

In the generator coils there are many feet of copper wire. The resistance of this wire must be overcome by the induced voltage. Voltage used in this manner gives nothing to the external circuit. It only creates heat in the generator windings. Power loss in any resistance is equal to:

$$P = I^2R$$

Of special concern is the fact that power loss increases as the square of the current. To be specific, if the current in the generator doubles, the power loss is four times, or 2^2, more. The limiting factor in generator output is usually the wire size, and current carrying capacity of the armature windings. Losses resulting from resistance in the windings are classified as COPPER LOSSES or the I^2R LOSS.

The armature windings in the generator are wound on an iron core, which is slotted to hold the coils. Like a conductor, as the solid iron core rotates, it induces voltages, causing an alternating current to flow in the core. This alternating current is known as an EDDY CURRENT. It produces heat, which is a loss of energy. To reduce this loss, the core is made of built up thin sections or laminations, Fig. 7-12. Each section is insulated from the next section by lacquer or, at times, only oxide and rust. Lamination reduces voltage in the core and increases resistance to eddy currents.

Eddy currents will increase in the core as the speed of rotation increases or the field density increases. When iron cores are used in rotating machines and transformers, they are laminated to reduce eddy current losses.

A third loss occurring in a generator is called HYSTERESIS LOSS. It is also called molecular friction. As the armature rotates in the fixed magnetic field, many of the magnetic particles in the armature core remain lined up with the fixed field. These particles then rotate against those not lined up with the field. This

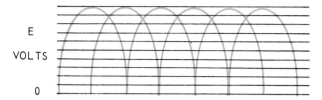

Fig. 7-11. Output of generator in Fig. 7-10.

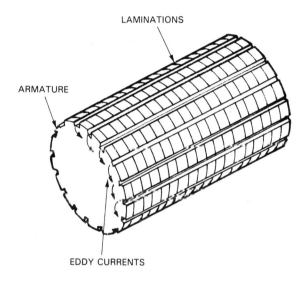

Fig. 7-12. Lamination of the metal core of an armature reduces the flow of eddy currents.

causes internal friction between the magnetic particles and creates a heat loss. Generator manufacturers now use a silicon steel which has a low hysteresis loss. Annealing the core further reduces this loss.

REVIEW QUESTIONS FOR SECTION 7.1

1. Voltage is highest when the coil moves (parallel, at a right angle) to the magnetic field.
2. In your own words, explain Lenz's Law.
3. On what does the strength of induced voltage depend?
4. Define direct current and alternating current.
5. A _____ is a device that reverses electrical connections.
6. Sketch a dc generator. Use arrows to show current flow.
7. Name three types of generator losses.

7.2 TYPES OF GENERATORS

Basic generator types include independently excited field, shunt, series, and compound.

INDEPENDENTLY EXCITED FIELD GENERATOR

Generator output is determined by the strength of the magnetic field and the speed of rotation. Field strength is measured in ampere-turns. So, an increase in current in the field windings will increase the field strength.

Output voltage is in direct proportion to field strength, times the speed of rotation. Therefore, most output regulating devices depend on varying the current in the field.

The field windings may be connected to an independent source of dc voltage, Fig. 7-13. With the speed constant, the output may be varied by controlling the exciting voltage of the dc source. This is done by inserting RESISTANCE IN SERIES with the source and field windings. This type of generator is called an independently excited field generator.

SHUNT GENERATOR

The inconvenience of a separate dc source for field excitation led to the development of the shunt generator. In this generator a part of the generated current is used to excite the fields. The field windings consist of many turns of small wire. They use only a small part of the generated current. The total current generated must, of course, be the sum of the field excitation current and the current delivered to the load. Thus, the output current can be thought of as varying according to the applied load. The field flux does not vary to a great extent. Therefore, terminal voltage remains constant under varying load conditions. This type of generator is considered a constant voltage machine.

If the generator excites its own field, how does the generator start? Beginning voltage is produced when the armature windings cut across a small magnetic field. This field is caused by magnetism left over in the pole shoes or field coil cores. A diagram of a shunt generator is shown in Fig. 7-14.

All machines are designed to do a certain amount of work. If overloaded, their lives are shortened. A generator is no exception. When overloaded, the shunt generator terminal voltage drops rapidly. Current flood causes the armature windings to heat and break down.

SERIES GENERATOR

The field windings of a generator may be placed in series with the armature and the load. Such a generator is sketched in Fig. 7-15. It is seldom used.

COMPOUND GENERATOR

The compound generator uses both series and shunt windings in the field. The series windings are often a few turns of large wire. They are mounted on the same poles with the shunt windings. Both windings add to the field strength of the generator. If both act in the

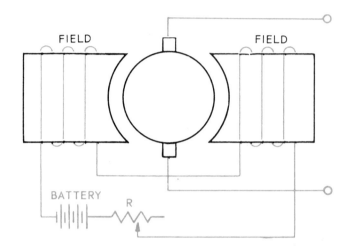

Fig. 7-13. An independently excited generator.

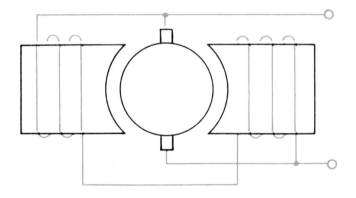

Fig. 7-14. A shunt generator.

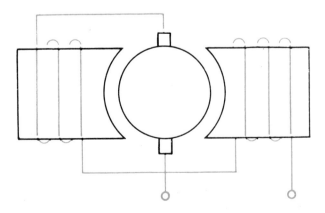

Fig. 7-15. A series wound generator.

same direction or polarity, an increase in load causes an increase of current in the series coils. This would increase the magnetic field and the terminal voltage of the output. The fields would be ADDITIVE; the resulting field would be the sum of both coils. However, the current through the series winding produces magnetic saturation of the core. This results in a decrease of voltage as the load increases.

The way terminal voltage behaves depends on the degree of COMPOUNDING. A compound generator which maintains the same voltage either at no-load or full-load is said to be FLAT-COMPOUNDED. An OVER-COMPOUNDED generator, then, will increase the output voltage at full-load, and an UNDER-COMPOUNDED generator will have a decreased voltage at full-load current.

A variable load may be placed in parallel with the series winding to adjust the degree of compounding. Fig. 7-16 shows schematic diagrams of the compound, the series, and the shunt generator.

VOLTAGE AND CURRENT REGULATION

The regulation of a power source, whether a generator or a power supply, may be defined as the percentage of voltage drop between no-load and full-load. Mathematically, it may be expressed as:

$$\frac{E_{\text{no-load}} - E_{\text{full-load}}}{E_{\text{full-load}}} \times 100 = \% \text{ Regulation}$$

To explain this formula, assume that the voltage of a generator with no-load applied is 100 volts. Under full-load the voltage drops to 97 volts, then,

$$\frac{100 \text{ V} - 97 \text{ V}}{97 \text{ V}} = \frac{3}{97} \times 100 = 3.1\% \text{ approx.}$$

In most uses, the output generator voltage should be maintained at a fixed value under varying load conditions. The output voltage of the generator depends upon the field strength. The field strength depends upon the field current.

Current, according to Ohm's Law, varies inversely with resistance. Therefore, a device which would vary the resistance in the field circuit would also vary the voltage output of the generator. This regulator is shown in Fig. 7-17. It is often used in automobiles.

The generator output terminal G is joined to the battery and the winding of a magnetic relay. The voltage produced by the generator causes a current to flow in the relay coil. If the voltage exceeds a preset value, the increased current will provide enough magnetism to open the relay contacts.

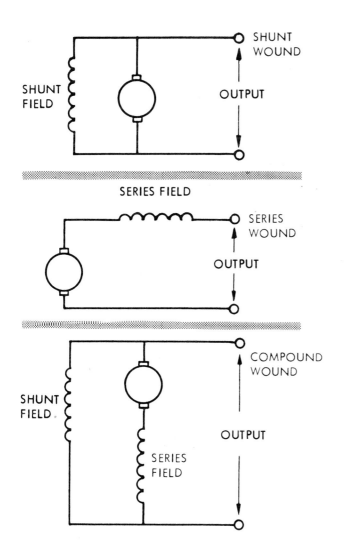

Fig. 7-16. Compare these wiring diagrams of the shunt, the series, and the compound generator.

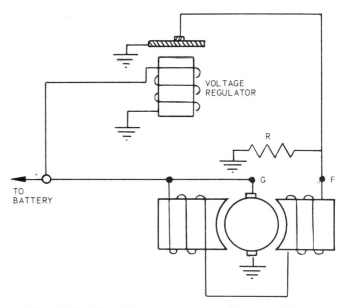

Fig. 7-17. Circuit for a generator voltage regulator.

Notice that the generator field is grounded through these contacts. When they open, the field current must pass through resistance R to ground. This reduces the current which reduces the field strength and reduces the terminal voltage. When the voltage is reduced, the relay contact closes, permitting maximum field current. Terminal voltage rises.

In operation, these contact points vibrate. They alternately cut resistance in and out of the field circuit and maintain a constant voltage output of the generator.

Mechanical-magnetic relays have served this purpose for many years. Now, however, electronic devices are being used on cars. An electronic regulator using an IC for the switching functions is shown in Fig. 7-18. A study of integrated circuits will be done in a later chapter of this text.

REVIEW QUESTIONS FOR SECTION 7.2

1. Generator output is determined by two items. Name these items.
2. This generator requires an external power source to provide voltage to the field windings:
   a. Series generator.
   b. Compound generator.
   c. Shunt generator.
   d. Independently excited field generator.
3. This generator is considered a constant voltage device:
   a. Series generator.
   b. Compound generator.
   c. Shunt generator.
   d. Independently excited field generator.
4. This generator has both series and shunt windings in the field:
   a. Series generator.
   b. Compound generator.
   c. Shunt generator.
   d. Independently excited field generator.
5. State the mathematical equation for regulation of a power source.

7.3 ALTERNATING CURRENT

Direct current flows in only one direction. Alternating current changes its direction of flow at times in the circuit. In dc, the source voltage does not change its polarity. In ac, the source voltage changes its polarity between positive and negative.

Fig. 7-19 shows the magnitude and polarity of an ac voltage. Starting at zero, the voltage rises to maximum in the positive direction. It then falls back to zero. Then it rises to maximum with the opposite polarity and returns to zero.

The current wave is also plotted on the graph. It shows the flow of current and the direction of flow. Above the zero line, current is flowing in one direction. Below the zero line, the current is flowing in the opposite direction.

The graph represents instantaneous current and voltage at any point in the cycle. But what is a cycle? A cycle is a sequence or chain of events occurring in a period of time.

An ac cycle may be described as a complete set of positive and negative values for ac. The alternating

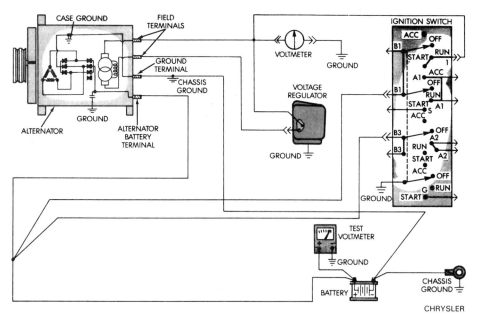

Fig. 7-18. Electrical voltage regulators are used on all new cars.

Generators

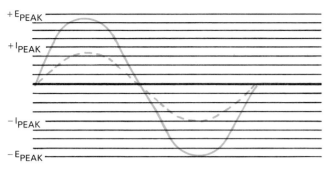

Fig. 7-19. Current and voltage of alternating current.

current in your home changes direction 120 times per second. It has a frequency of 60 cycles per second (60 cps). FREQUENCY measured in cycles per second, or hertz (Hz), is the number of complete cycles occurring per second. If 60 cycles occur in one second, then the time period for 1 cycle is 1/60 of a second, or .0166 seconds. This is the PERIOD of the cycle. Refer again to Fig. 7-19. The maximum rise of the waveform shows the AMPLITUDE of the wave, including the PEAK voltage and current.

We learned that induced current in a rotating wire in a magnetic field flowed first in one direction and then in the other direction. This was defined as an alternating current. Two points to remember are:
1. The frequency of this cycle of events increases as rotation speed increases.
2. The amplitude of the induced voltage depends on the strength of the magnetic field.

When solving problems involving alternating currents, vectors are used to depict the magnitude and direction of a force. A VECTOR is a straight line drawn to a scale that represents units of force. An arrowhead on the line shows the direction.

The development of an ac wave is shown in Fig. 7-20. This wave is from a single coil armature, represented by the rotating vector, making one revolution through a magnetic field. Assume that the peak induced voltage is 10 volts. Using a scale in which 1 inch equals 5 volts, the vector is two inches or 10 volts long. Vectors of this nature are assumed to rotate in a counterclockwise direction.

The time base in Fig. 7-20 is a line using any convenient scale. It shows the period of one cycle or revolution of the vector. The time base is grouped into segments that represent the time for certain degrees of rotation during the cycle. For example, at 90 degrees rotation, one quarter of the time period is used. At 270 degrees rotation, three-quarters of the time period is used.

The wave is developed by plotting voltage amplitude at any instant of revolution against the time segment. The developed wave is called a sine wave. The instantaneous induced voltages are proportioned to the sine of the angle θ (theta) that the vector makes with the horizontal. (Refer to Appendix for explanation of trigonometric functions and tables.) The instantaneous voltage may then be found at any point of the cycle by making use of the following equation:

$$e = E_{max} \text{SIN } \theta$$

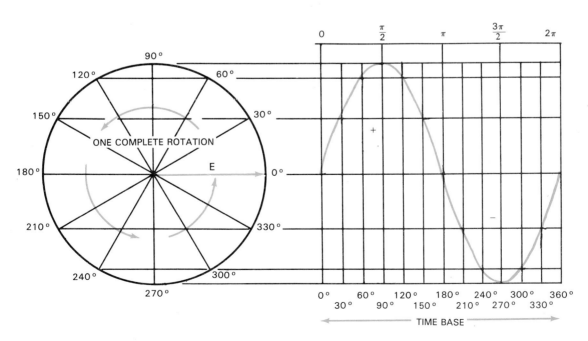

Fig. 7-20. The development of a sine wave.

To apply this equation, assume that an ac generator is producing a peak voltage of 100 volts. What is the instantaneous voltage at 45 degrees of rotation?

$$e = 100 \times \sin 45°$$

$$e = 100 \times .707 = 70.7 \text{ V}$$

A study of the differences between an ac wave and a direct current raises a key question. What is the actual value of the ac wave? The voltage and current vary constantly and reach peak value only twice during a cycle.

Often, the AVERAGE VALUE of the wave is needed. This is the mathematical average of all the instantaneous values during one half-cycle of the alternating current. The formula for computing the average value from the peak value (max) of the ac wave is:

$$E_{avg} = .637 \, E_{max}$$

If E_{avg} is known, the conversion to find E_{max} can be made using this equation:

$$E_{max} = 1.57 \, E_{avg}$$

A more useful alternating current value is found by comparing it to the heating effect equivalent to a direct current. This is the EFFECTIVE VALUE. Use this formula to find it:

$$E_{eff} = .707 \, E_{max}$$

If E_{eff} is known, the conversion to find E_{max} can be made by using this equation:

$$E_{max} = 1.414 \, E_{eff}$$

The effective value is also called the rms (root mean square) value. It represents the square root of the average of all currents squared, between zero and maximum of the wave. The currents are squared so the power produced may be compared to direct current. Watt's Law states: $P = I^2R$. Using the .707 factor, the value of a direct current may be found which will equal the alternating current. For example, a peak ac current of 5 amperes will produce the same heating effect in a resistance as a dc current of 3.53 amperes:

$$I_{eff} = .707 \times 5 = 3.53 \text{ amps dc}$$

Note that average and effective values may be applied to either voltage or current waves.

PHASE DISPLACEMENT

A few wave forms may be drawn on the same time base to show the phase relationship between them. In Fig. 7-21, E and I show the voltage and current in a given circuit. The current and voltage rise and fall at the same time. They cross the zero line at the same

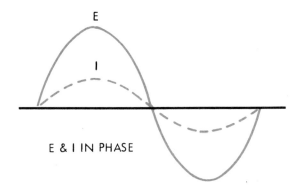

Fig. 7-21. Current and voltage are in phase.

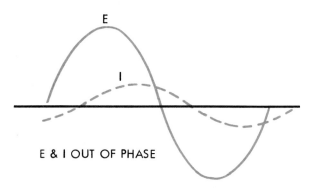

Fig. 7-22. Current and voltage are out of phase.

point. The current and voltage are IN PHASE. The "in phase" condition only exists in the purely resistive circuit. This will be discussed later.

Many times the current will lead or lag the voltage, Fig. 7-22. This creates a PHASE DISPLACEMENT between the two waves. Displacement is measured in degrees. It is equal to the angle θ between the two polar vectors. The waves are OUT OF PHASE. Several out of phase currents and voltages may be shown in this way.

ALTERNATING CURRENT GENERATOR

The ac generator is like the dc generator in many respects with one key exception. The commutator is omitted. The ends of the armature coils are brought out to SLIP RINGS. Brushes sliding on the slip rings provide connection to the coils at all times. The current in the externally connected circuit is an alternating current.

In large commercial generators, the magnetic field is rotated and the armature windings are placed in slots in the stationary frame or stator of the generator. This method allows for the generation of large currents in the armature while avoiding sending these currents through moving or siding rings and brushes.

Generators

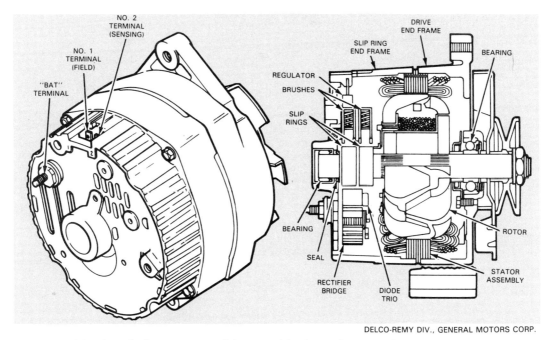

Fig. 7-23. A typical ac generator (alternator) is shown in external and cutaway views.

The rotating field is excited through slip rings and brushes by an externally connected dc generator. It is called the EXCITER. The dc voltage is needed for the magnetic field. Commercial power generators are turned by water power and steam.

THE ALTERNATOR

The ac generator (also called an alternator) is used in the charging system of all U.S. cars. Fig. 7-23 shows the inside of the unit, including a built-in voltage regulator to control output. The output is rectified from alternating current to direct current for charging the battery and other electrical devices in the car. Manufacturers say the alternator has some advantages over the dc generator. These include higher output at lower speeds and trouble-free service.

PRACTICAL GENERATOR APPLICATIONS

The principle of the generator may be shown in the useful and practical rpm meter shown in Fig. 7-24. It is simple to construct and provides the chance to design a case. A small dc motor, such as those found in toy automobiles, is used as the generator. The shaft of the motor is wrapped with friction tape or rubber. This friction wheel is held against a rotating shaft or wheel. The generated emf is measured on the 0-10 mA meter. This meter may be calibrated to read directly in rpms. The switch permits measurements in either clockwise or counterclockwise rotation. The "adjust screw" is used as a calibrator or range setting control.

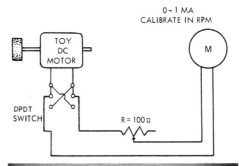

Fig. 7-24. A simple rpm meter (tachometer).

REVIEW QUESTIONS FOR SECTION 7.3

1. Define alternating current cycle.
2. _____ is the number of cycles per second.
3. What is a vector?
4. 1.75 volts eff = _____ volts max.
5. Another name for effective voltage is _____ _____.
6. Explain the terms, in phase and out of phase.

SUMMARY

1. To produce an induced current flow by magnetism, three factors must exist:
   a. There must be a magnetic field.
   b. There must be a conductor (or coil) in a closed circuit.
   c. There must be a relative movement between the field and the conductor.
2. A generator is a device which converts mechanical energy into electrical energy.
3. Lenz's Law states that the polarity of an induced emf is such that it sets up a current, the magnetic field of which always opposes the change in the existing field.
4. Direct current (dc) is current flowing in a single direction in a conductor. Alternating current (ac) is current flowing in more than one direction in a conductor.
5. Generators have copper losses, eddy currents, and hysteresis losses that reduce their efficiency.
6. Generator types include the independently excited field, shunt, series, and compound.
7. Regulation of a power source, whether it is a generator or a power supply, may be defined as the percentage of voltage drop between no-load and full-load.
8. A cycle is a complete set of positive and negative values for alternating current.
9. Frequency is the number of cycles that occur each second. It is measured in hertz.
10. The usable value for ac for doing actual work is referred to as the effective or root-mean-square value.
11. Formulas for converting the effective value to the peak value (and vice versa) are:

    $E_{eff} = .707\ E_{max}$
    $E_{max} = 1.414\ E_{eff}$

12. Two sine waves are in phase when they are of the same frequency and they go through the zero points at the same time.
13. Direct current generators have commutators while ac alternators have slip rings.

TEST YOUR KNOWLEDGE, Chapter 7

Please do not write in the text. Place your answers on a separate sheet of paper.

1. Highest induced voltages are produced at _____ degree angles.
2. The strength of induced voltage depends on: (choose all that apply)
   a. The number of magnetic lines of force cut by the coil.
   b. The size of the armature.
   c. The speed at which the conductor moves through the field.
   d. None of the above.
3. State the mathematical equation used to find the strength of induced voltage in a rotating coil.
4. Commutators are used on _____ current generators. Slip rings are used on _____ current generators.
5. In what two ways can a pulsating current be improved?
6. Losses resulting from resistance in the windings are classified as _____ losses.
7. An eddy current:
   a. Produces heat.
   b. Flows in the core of an armature winding.
   c. Causes a loss of energy.
   d. All of the above.
8. Explain the cause of hysteresis loss.
9. Identify each of the following generator diagrams.

a.

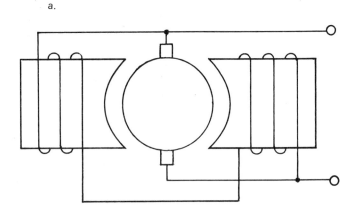

Generators

b.

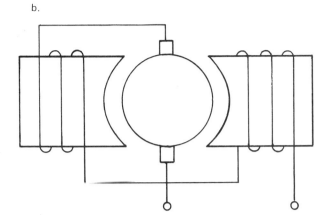

c.

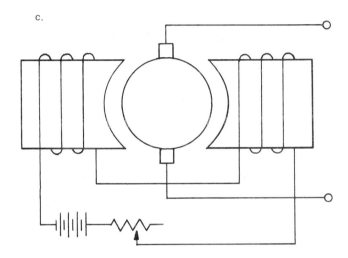

10. A generator has a no-load voltage of 25 volts. When load is applied, terminal voltage drops to 24 volts. What is the percent of regulation?
11. A generated voltage has a peak value of 240 volts. What is the instantaneous voltage at 60 degrees?
12. What is effective value of the generated voltage in Problem 11?
13. What is the peak value of the alternating current used in your home (117 volts)?

FOR DISCUSSION

1. Explain how sources of energy can be converted into electricity.
2. Why do we need both ac and dc electricity in electrical and electronic systems?
3. How does a generator differ from an alternator?
4. Why is ac not as efficient as dc?

GENERAC CORP.

Portable generators bring power to remote worksites.

Electricity and Electronics

TEKTRONIX

Technicians must know not only how to use instruments, but also how to use the instruments with other electronic systems. This chapter explains the construction of the instruments and how to connect them properly to test circuits.

Chapter 8

INSTRUMENTS AND MEASUREMENTS

After studying this chapter, you will be able to:
- Explain the operation of basic analog meter movements.
- Compute shunt resistor values.
- Compute multiplier resistor values.
- Discuss the concept of meter sensitivity.
- Identify various types of multimeters.
- Explain the operation of an oscilloscope.

Electricity and electronics students rely on instruments to precisely judge the action and traits of a circuit. Skillful use of instruments is the mark of a good technician. The technician must know what he or she is trying to measure and how to measure it. Experienced technicians are able to make correct measurements. These are valuable in the design of electronic equipment, as well as in the detection and correction of circuit failures.

8.1 BASIC ANALOG METER MOVEMENT

A common type of meter movement measures current and voltage. It is the D'Arsonval or stationary-magnet, moving-coil galvanometer, Fig. 8-1. The movement consists of a permanent-type magnet and a rotating coil in the magnetic field. An indicating needle is attached to the rotating coil, Fig. 8-2.

When a current passes through the moving coil, a magnetic field is produced. This field reacts with the stationary field and causes rotation (deflection) of the needle. This deflection force is proportional to the strength of the current flowing in the moving coil. The moving coil is returned to its "at rest" position by hair springs. These springs also connect to the coil. The deflecting force rotates the coil against the restraining force of these springs.

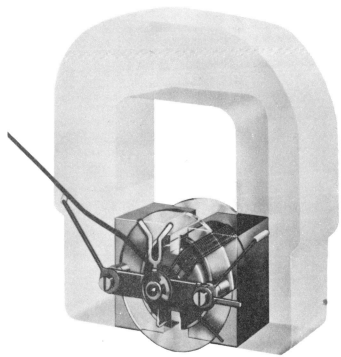

DAYSTROM, WESTON INSTRUMENTS DIV.

Fig. 8-1. A phantom view of the D'Arsonval meter movement.

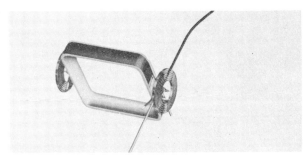

Fig. 8-2. On the D'Arsonval, the indicating needle is attached to a rotating coil of the meter.

IMPORTANT: Changing direction of current in the moving coil causes deflection of the needle in the opposite direction. Be sure that correct polarity of voltages and current are observed when connecting a meter to a dc circuit.

This simple meter can be improved using a dampening device. A dampener prevents the indicating needle from overshooting and oscillating when making a measurement. To do this, the moving coil is wound on an aluminum bobbin. Since the bobbin tends to oscillate in the magnetic field, an emf is induced in it (Lenz's Law). As a result, any impulse for the coil to oscillate is quickly dampened out. Analog meters have a scale marked off in specific increments, or values.

Meters are costly instruments. They should be used with skill and judgment. Precision types have jewel bearings like fine watches. The quality of the meter is judged by its sensitivity and accuracy.

AMMETER

An AMMETER measures current flow in a circuit in amperes, milliamperes, or microamperes, depending on the scale chosen.

The moving coil of the meter is wound with many turns of fine wire. If a large current is allowed to flow in the coil, it will quickly burn out. In order to measure larger currents with the basic movement, a SHUNT, or alternate path, is supplied. The major amount of current will flow through this, leaving only enough current to safely work the moving coil. This shunt is in the form of a precision resistor connected in parallel with the meter coil.

The use of shunts is illustrated in Fig. 8-3. Follow this problem. The specification of a certain meter movement requires .001 amps, or one milliampere, for full scale deflection of the needle. The ohmic resistance of the moving coil is 100 ohms. Compute the shunt resistor values so that the meter will measure currents from 0-1 mA, 0-10 mA, 0-50 mA, and 0-100 mA. Each range involves using a different shunt.

Step 1. Calculate the voltage which, if applied to the coil, will cause one milliampere flow of current and full scale deflection of needle.

$$E = I \times R = .001 \times 100 = .1 \text{ volt}$$

where I = amps for fullscale deflection, and R = resistance of meter coil.

The meter will read from 0-1 mA without a shunt.

Step 2. To convert this same meter to read from 0-10 mA, a shunt must be connected which will carry 9/10 of the current, or 9 mA, leaving the one millampere to operate the meter. Since .1 volt must be applied across the meter, and also across the shunt, the shunt resistance must be:

$$R_S = \frac{.1 \text{ V}}{.009 \text{ amps}} = 11.1 \text{ ohms}$$

The meter, with 11.1 ohms shunt, will measure from 0-10 mA.

Step 3. To convert this meter to read from 0-50 mA, a shunt must be used which will carry 49/50 of the current, or 49 mA. The computation is the same as in Step 2:

$$R_S = \frac{.1 \text{ V}}{.049 \text{ amps}} = 2.04 \text{ ohms}$$

Step 4. To convert the meter for the 0-100 mA scale, a shunt must be used which will carry 99/100 of the current, or 99 mA.

$$R_S = \frac{.1 \text{ V}}{.099 \text{ amps}} = 1.01 \text{ ohms}$$

The meter with the 1.01 ohms shunt will measure currents from 0-100 mA. Look again at the circuit in Fig. 8-3. Notice the switching device used to change the ranges of the meter. The correct scale on the dial must be used to correspond to the selected range.

IMPORTANT: An ammeter must always be connected in series with a circuit, never across or parallel with it. To make a series connection requires breaking the circuit to connect the meter. A parallel connection may instantly destroy the meter, Fig. 8-4.

REVIEW QUESTIONS FOR SECTION 8.1

1. Another name for a stationary-magnet, moving-coil meter is the _____ movement.
2. Explain how a moving-coil meter operates.
3. Used to measure current flow in amperes, milliamperes, or microamperes:
   a. Ammeter.
   b. Dampener.
   c. Shunt.
   d. None of the above.
4. Compute the value of the shunt in the circuit below.

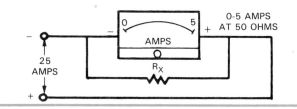

Instruments and Measurements

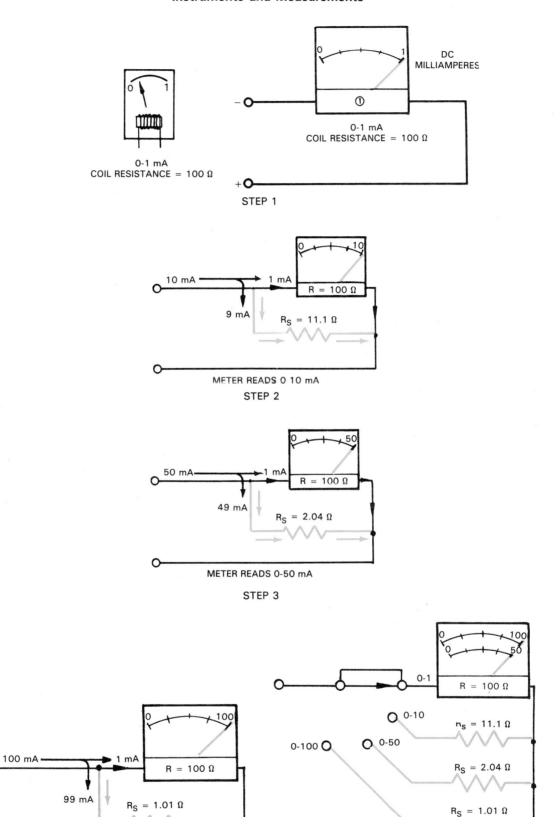

Fig. 8-3. Step 1. Left. Sketch of the basic ammeter. Right. The voltage which causes the full scale deflection current is computed. Step 2. The shunt carries 9/10 of the current. Step 3. The shunt carries 49/50 of the current. Step 4. The shunt carries 99/100 of the current.

Electricity and Electronics

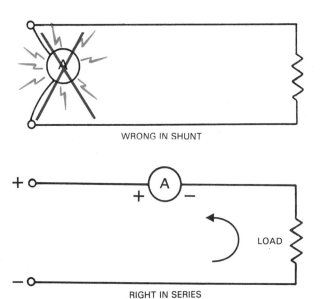

Fig. 8-4. Connecting an ammeter. Top. Incorrect. Bottom. Correct.

8.2 VOLTMETER

The same basic movement used for the ammeter may also be used to measure voltage, providing the impressed voltage across the coil never exceeds .1 volt as computed for full scale deflection. To arrange the meter to measure higher voltages, MULTIPLIER RESISTORS are switched in series with the meter movement.

Using the same meter as used for the ammeter, refer to Fig. 8-5. Follow the steps as the multipliers are computed so that the meter will measure voltages from 0-1 V, 0-10 V, 0-100 V, and 0-500 V.

Step 1. Remember that no more than .1 volt is allowed across the meter coil at any time. Therefore, a resistor must be placed in series with the meter that will cause a voltage drop of .9 V, if the meter is used to measure one volt. Also, the meter only allows .001 amperes for full scale deflection. This is the highest current allowed in the circuit. The multiplier resistor must produce a .9 volts drop when .001 amperes flow through it.

$$R_M = \frac{.9 \text{ V}}{.001 \text{ amps}} = 900 \text{ ohms}$$

Step 2. To convert to the 0-10 volt range, a resistor must be selected to produce a 9.9 volt drop.

$$R_M = \frac{9.9 \text{ V}}{.001 \text{ amps}} = 9900 \text{ ohms}$$

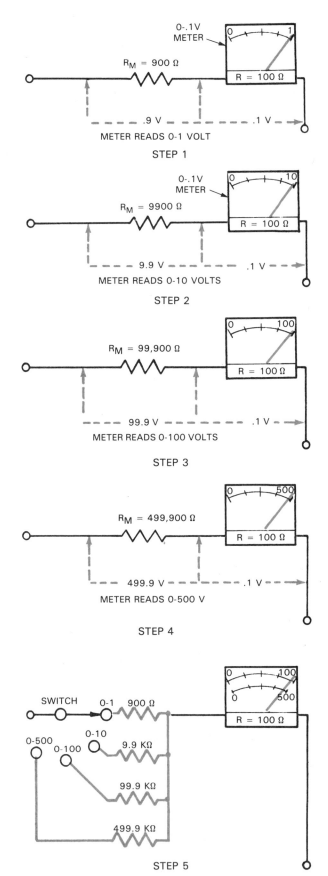

Fig. 8-5. Step 1. The multiplier causes an IR drop of .9 V. Step 2. The multiplier causes an IR drop of 9.9 V. Step 3. The multiplier causes an IR drop of 99.9 V. Step 4. The multiplier causes an IR drop of 499.9 V. Step 5. A switch selects the range.

Step 3. To convert to the 0-100 volt range, a resistor must be selected to produce a 99.9 volt drop.

$$R_M = \frac{99.9 \text{ V}}{.001 \text{ amps}} = 99,900 \text{ ohms}$$

Step 4. Finally, to use the 0-500 volt range, the resistor must cause a 499.9 volt drop.

$$R_M = \frac{499.9 \text{ V}}{.001 \text{ amps}} = 499,900 \text{ ohms}$$

Again, a switching device is used to select the correct multiplier resistor for the range in use. Read the scale on the dial which corresponds to the range selected.

IMPORTANT: A voltmeter is always connected in parallel or across the circuit, never in series. To measure voltage, the circuit does not have to be broken. See Fig. 8-6. Added precautions: when measuring an unknown voltage, always start with the meter on its highest range and adjust downward to the proper range to avoid damaging the meter. Be sure that the leads are connected with the correct polarity when measuring dc. BLACK is negative, RED is positive.

VOLTMETER SENSITIVITY

The sensitivity of a meter is a sign of quality. OHMS-PER-VOLT is the unit for measuring sensitivity. In Step 4 of the previous example, the total resistance of the meter and its multiplier resistance is:

499,900 ohms in R_M
+ 100 ohms meter resistance
500,000 ohms total resistance

This amount of resistance is used in the 500 volt range which is equivalent to:

$\frac{500,000 \text{ ohms}}{500 \text{ V}}$ or 1000 ohms per volt.

By Ohm's Law, I = E/R, and the reciprocal of I is R/E, which is the same as sensitivity.

Therefore, sensitivity is equal to the reciprocal of the current required for full scale deflection. For the meter used in the above example:

Sensitivity = $\frac{1}{I}$ or $\frac{1}{.001}$ or 1000 ohms/volt

This is an easy method of gauging sensitivity.

A quality meter will have a sensitivity of 20,000 ohms/volt. Precision laboratory meters go as high as 200,000 ohms/volt. Accuracy of the meter is commonly expressed as a percentage, such as 1 percent. The true value will be within one percent of the scale reading.

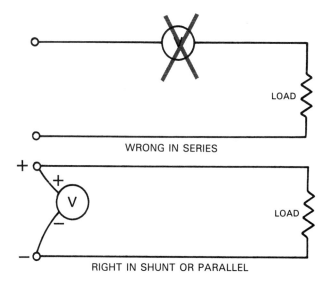

Fig. 8-6. Connecting a voltmeter. Top. Incorrect. Bottom. Correct.

OHMMETERS

A meter used to measure the value of an unknown resistance is called an OHMMETER. The same meter movement as used for the voltmeter and ammeter may be used for this purpose. A voltage source and a variable resistor are added to the circuit. The series type ohmmeter is shown in Fig. 8-7.

A three-volt battery is used as source, and is built into the meter case. The meter movement permits only .1 volt for a current of .001 amps for full scale deflection. Therefore, a multiplier resistor is placed in series with the meter to drop the voltage:

$$R_M = \frac{2.9 \text{ V}}{.001 \text{ amps}} = 2900 \text{ ohms}$$

This resistance plus the meter resistance is equal to 3000 ohms. Part of this resistance is made variable to make up for changes in battery voltage due to aging.

To use the meter, short the test terminals. This is equal to zero resistance and the meter deflects from left to right. The needle is adjusted to zero by the variable resistor marked, "OHMS ADJUST." An

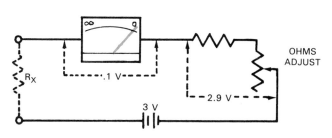

Fig. 8-7. Schematic diagram of a series ohmmeter.

Electricity and Electronics

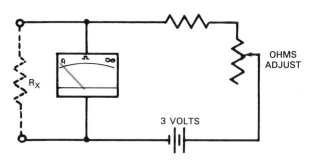

Fig. 8-8. Shunt ohmmeter.

unknown resistor, R_X, is now placed between the test terminals. The needle will deflect less than full scale depending upon the resistance value. The scale reads directly in ohms. Other ranges may note that the scale value should be multiplied by 10, by 1000, or by 100,000.

It is interesting to note that with this circuit, the meter scale reads right to left, which is opposite to the voltmeter or ammeter scale.

A shunt ohmmeter may be connected as in Fig. 8-8. In this circuit, the unknown resistance R_X is shunted across the meter. Low values of R_X cause lower currents through the meter. High values of R_X cause higher meter currents. Arranged in this way, the indicating needle deflects from right to left in the normal manner. Zero resistance is on the left. The scale increases from left to right.

REVIEW QUESTIONS FOR SECTION 8.2

1. Voltmeters are connected in _____ in a circuit.
2. The black lead of a meter has _____ polarity; the red lead has _____ polarity.
3. What is the unit for measuring sensitivity of a meter?
4. A _____ ohmmeter has zero resistance on the right side of the meter face.
5. Find the value of the multiplier resistor (R_X) in the following circuit. It should read 50 volts across points A to B.

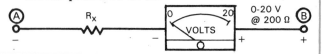

8.3 MULTIMETER

Multimeters, used by technicians, commonly include voltmeters, ammeters, and ohmmeters in one case, Fig.

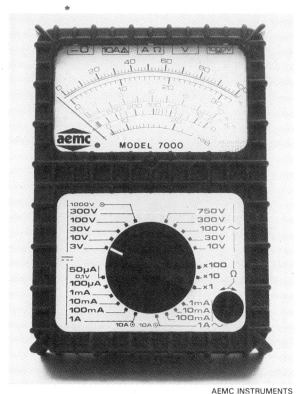

AEMC INSTRUMENTS
Fig. 8-9. Analog volt-ohm-milliammeter.

8-9. Ample switching is supplied to change function and range of the meter.

IMPORTANT: Before using the ohmmeter to test a circuit, be sure that the power is turned off on the equipment under test. Pull the plug, to be sure. Also discharge all electrolytic capacitors by shorting terminals to ground with an insulated screwdriver. Multimeters are expensive. Carelessness will ruin them.

THE VOLT-OHM MILLIAMMETER (VOM)

A simple multimeter to use in electronic circuits is the volt-ohm milliammeter (VOM). It consists of a circuit that has a meter, a power source, a variable resistor (for zeroing the meter), and a load. There are two basic types of VOMs: the series and the shunt. Fig. 8-10 shows a series VOM, Fig. 8-11 shows a shunt type VOM. A commercial VOM is shown in Fig. 8-12.

Volt-ohm-milliammeters have the advantage of being portable and low in cost. They do, however, have a low input resistance (in ohms per volt) on the lowest voltage range. This can cause accuracy problems.

When the field-effect transistor was developed, the VOM was redesigned to overcome the low input impedance problem. The FIELD EFFECT TRANSISTOR-VOM (FET-VOM) measures ac and dc voltage, ac and dc current, resistance, and decibel ratings. Fig. 8-13 shows a commercial FET-VOM.

Instruments and Measurements

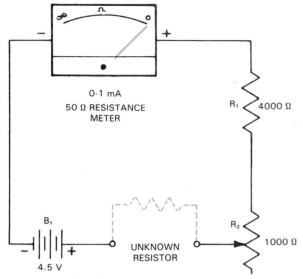

Fig. 8-10. A series VOM.

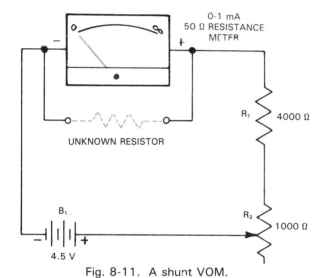

Fig. 8-11. A shunt VOM.

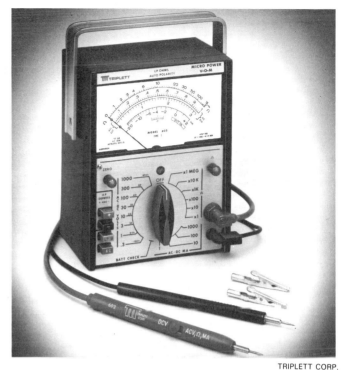

TRIPLETT CORP.

Fig. 8-13. Field effect transistor, volt-ohm-meter.

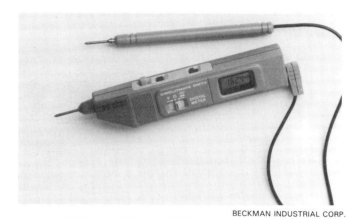

BECKMAN INDUSTRIAL CORP.

Fig. 8-14. A miniature DVM used for portable work.

DIGITAL METERS

Digital multimeters (DVM) are very popular in the electronics field. These meters precisely measure ac and dc current, ac and dc voltage, and resistance. Many are portable. Some, however, operate from ac line voltage.

The digital multimeter can provide instant visual display of unknown circuit values. The cost of these multimeters has decreased since their appearance in the test instrument market. Figs. 8-14 and 8-15 show typical digital multimeters.

Fig. 8-12. Volt-ohm-meter.

Electricity and Electronics

Fig. 8-15. Certain digital meters use accessories, such as the current clamp. Dangerous currents can be measured accurately and safely without breaking the circuit under test.

Fig. 8-16. Hand-held digital meter.

Fig. 8-17. A liquid crystal display chip.

Digital multimeter used for hand-held testing is shown in Fig. 8-16. They are self-contained and portable.

Liquid crystal displays

The liquid crystal display (LCD) is one way of showing circuit conditions of digital instruments. Liquid crystal displays are used in clocks, multimeters, timers, surveying equipment, and medical equipment. LCDs often last for more than 50,000 hours. Fig. 8-17 shows a four digit LCD unit.

Light emitting diodes

Many instruments use light emitting diodes (LEDs) for display. Fig. 8-18 shows common LED displays. These have low operating voltages (5 volts) and come in a number of display colors such as green, yellow, and red.

One LED programmable display is shown in Fig. 8-19. It is used with computers and microcomputers having read/write memories.

Fig. 8-18. Various LED displays.

Instruments and Measurements

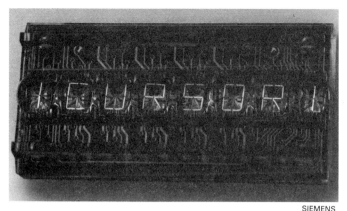

Fig. 8-19. Programmable LED display.

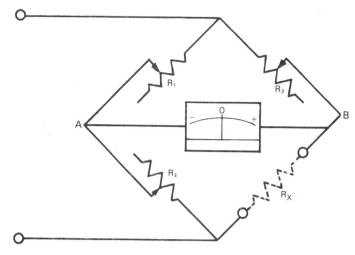

Fig. 8-20. the Wheatstone bridge circuit.

REVIEW QUESTIONS FOR SECTION 8.3

1. A VOM:
   a. Has a meter, power source, variable resistor, and a load.
   b. Comes in series or shunt types.
   c. Is portable and inexpensive.
   d. All of the above.
2. A _____ VOM measures ac and dc voltage, ac and dc current, resistance, and decibel ratings.
3. What is the basic difference between a digital and an analog display?
4. What are some of the uses of liquid crystal displays?

8.4 OTHER METERS

WHEATSTONE BRIDGE

Many types of bridge circuits are used in electrical measurements. One type is the WHEATSTONE BRIDGE. It is used to measure unknown resistance. Fig. 8-20 is the schematic diagram.

R_1, R_2, and R_3 are precision variable resistors. R_X is the unknown. When voltage is applied to the circuit, current will flow through the branch R_1R_2 and parallel branch R_3R_X. Each resistor will cause a voltage drop. The bridge is balanced by placing a meter between points A and B, and adjusting the variable resistors until the meter reads zero. The balanced status shows that:

$$E_{R_1} = E_{R_3}$$

and

$$E_{R_2} = E_{R_X}$$

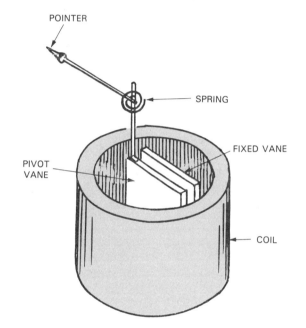

Fig. 8-21. Operating principle of the iron vane meter movement.

Since voltage is proportional to resistance,

$$\frac{R_1}{R_2} = \frac{R_3}{R_X} \text{ and } R_X = \frac{R_2 \times R_3}{R_1}$$

The values of the precision variable resistors are shown on an accurately calibrated dial.

IRON VANE METER MOVEMENT

The operation principle of the iron vane meter movement is shown in Fig. 8-21.

Two pieces of iron are placed in the hollow core of a solenoid, they both become magnetized with the same polarity, when a current passes through the solenoid.

Electricity and Electronics

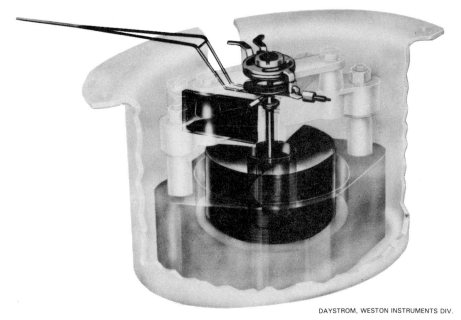

Fig. 8-22. Phantom view of a concentric type iron vane meter movement.

Because like poles repel each other, the two pieces of metal are repelled. One piece of metal is fixed and the other pivoted. The pivoted piece can turn away from the fixed metal. An indicating needle is attached to the moving vane. This also has hair springs so that the vane must move against the spring tension for good readings.

An applied voltage causes current to flow in the solenoid and creates the magnetic field. The moving vane is repelled against the spring according to the strength of the magnetic field. The needle may show either voltage or current. It is calibrated for the magnitude (average size) of the applied voltage or current.

When the iron vane movement is used for a voltmeter, the solenoid is commonly wound with many turns of fine wire. Proper multiplier resistances may be used to increase the range of the meter. Range switches are used.

When used as an ammeter, the solenoid has a few turns of heavy wire. This is because the coil must be connected in series with the circuit and carry the circuit current.

Regardless of the polarity of the applied voltage or current, the iron vane meter movement always deflects in the same direction. (Two north poles repel as well as two south poles.) Either ac or dc may be measured with this instrument. Generally, this type of meter is best for high power circuit measurements. A phantom view of interior construction of the concentric type iron vane meter is shown in Fig. 8-22.

THE WATTMETER

Power in an electric circuit is equal to the product of the voltage and the current. To devise a meter that can read in watts, a movement similar to the D'Arsonval movement can be used. The permanent magnetic field, however, is replaced with coils from an electromagnet. This type of meter is referred to as an electrodynamometer movement.

A circuit diagram of a simple electrodynamometer movement is shown in Fig. 8-23. It is a wattmeter. A moving coil, with the proper multiplier resistance, is connected across the voltage in the circuit. The coils of the electromagnet are connected in series with the circuit under measurement. The action between the two

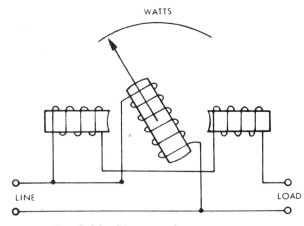

Fig. 8-23. Diagram of a wattmeter.

120

fields is proportional to the product of the voltage and the current. Deflection of the indicating needle is read on a calibrated scale in watts. A commercial meter movement is shown in Fig. 8-24.

A wattmeter measures the instantaneous power used in a circuit. A watt-hour meter measures the amount of power used in a given time. It is installed by a power company on the outside of a home. As a watt-hour is a small unit, standard meters read in kilowatt-hours or 1000 watt-hours (kWh). Electric power consumed is paid for at current rates per kWh.

The watt-hour meter is a complicated type of induction motor. It uses field coils in series with the line current and also field coils connected across the line voltage. An aluminum disc rotates within these fields, at a rate proportional to the power consumed. The disc is geared to an indicating dial. The dial shows the amount of power used.

To read the watt-hour meter, see Fig. 8-25. Dial A, on the right, reads from 0-10 units of kilowatt hours. For each revolution of dial A, dial B reads one. For each revolution of dial B, dial C reads one. For each revolution of dial C, dial D reads one. Mathematically, dial A reads in units of one; dial B reads in units of ten; dial C reads in units of one hundred; dial D reads in units of one thousand.

To read the meter illustrated in Fig. 8-25:
  Dial A points to 5.
  Dial B points between 5 and 6.
  Dial C points between 2 and 3.
  Dial D points between 4 and 5.
Therefore, the correct reading would be 4255 kWh. Notice that when the indicating arrow is between numbers, the lower number or the number just passed should be used.

To practice reading meters, why not read the meter at your home each day for several days and compute the power used? The power consumed may be found by subtracting the previous reading from the present reading to obtain the difference.

AC METERS

Typical dc meter movement, such as the D'Arsonval movement, may also be used to measure alternating currents and voltages with certain changes in the circuit.

Recall that the dc meter must always be connected with the correct polarity. Incorrect polarity will cause a reversal of current through the moving coil. This would cause deflection to the left below zero and possibly damage the meter. If an ac voltage were connected to the meter, the repeated changes in direction of the current would cause the needle to vibrate at zero.

In order to measure this ac voltage, a rectifier must be used. Detailed discussion of rectifiers will be included in Chapter 15. For now, however, a RECTIFIER may be described as a device which permits current to flow in only one direction. When used in a meter, the rectifier allows one half-cycle of the ac current to pass. But it cuts off the current when it reverses itself in the second half-cycle. A simple voltmeter circuit with a rectifier is shown in Fig. 8-26.

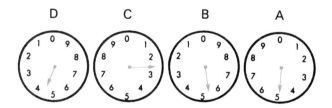

Fig. 8-25. The dials of the watt-hour meter.

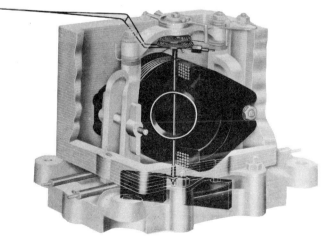

DAYSTROM, WESTON INSTRUMENTS DIV.

Fig. 8-24. Phantom view of the electrodynamometer movement used in a wattmeter.

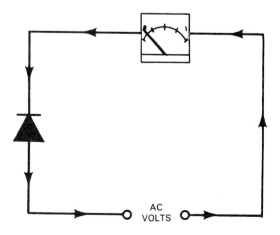

Fig. 8-26. Basic circuit for ac meter, using a single rectifier.

Notice the symbol used for the rectifier. Its output is a pulsating direct current. It will cause the meter to deflect in one direction only. An improved circuit, which serves the same purpose, is the full wave bridge rectifier, Fig. 8-27. By using four rectifiers, both half cycles of an ac current may be applied to the meter.

When ac voltage is applied to terminals AB, and A is positive, electron flow is from B to 1 to 2 to 3 to 4 to A. When B is positive, electron flow is from A to 4 to 2 to 3 to 1 to B. Notice that, regardless of the polarity of the ac voltage, the current flow from point 2 to 3 through the meter is always in the same direction. The deflection of the meter needle usually is calibrated to show effective or rms voltage. Effective value of an ac voltage is equal to .707 times peak value. This is the most useful ac voltage value to measure.

LOADING A CIRCUIT

When a voltmeter is connected across a circuit to measure a potential difference, it is in parallel with the load in the circuit. This may introduce errors in voltage measurement. This is very common when a meter with a low sensitivity is used.

In Fig. 8-28, two 10,000 ohm resistors form a voltage divider across a ten volt source. The voltage drops across both R_1 and R_2 is 5 volts. If a meter with a sensitivity of 1000 ohms/volt on the ten volt range is used to measure the voltage across R_1, the meter resistance will be in parallel with R_1. The combined resistance will be equal to:

$$\frac{10,000 \ \Omega}{2} = 5000 \ \Omega$$

With the meter connected, the total circuit resistance becomes:

$$10,000 \ \Omega + 5000 \ \Omega = 15,000 \ \Omega$$

The current is .00066 amps.

By Ohm's Law, $E_{R_1} = 3.4$ V and $E_{R_2} = 6.6$ V. The meter has caused an error of more than one volt due to its shunting effect. To avoid an excess of errors resulting from this, a more sensitive meter should be used.

In Fig. 8-29, a 5000 ohms/volt meter is used. The combined resistance of meter and R_1 equals 8333 ohms. The total circuit resistance equals 18,300 ohms. By Ohm's Law, I = .00054 amps, $E_{R_1} = 4.6$ volts, and $E_{R_2} = 5.4$ volts. An error of .4 volts still exists, but the increased sensitivity of the meter has reduced this error. More costly meters with a sensitivity of 20,000 ohms/volt would reduce the error to an amount that is barely noticed.

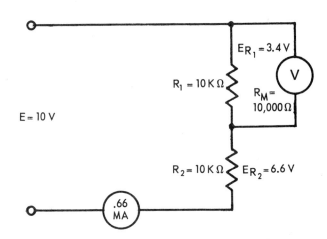

Fig. 8-28. The meter loads the circuit and introduces an error in the voltage reading.

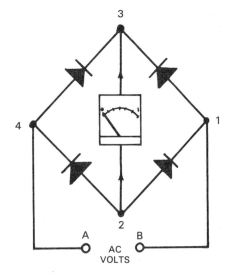

Fig. 8-27. A bridge rectifier circuit for an ac meter.

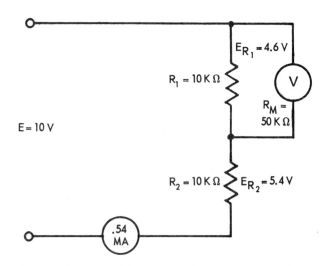

Fig. 8-29. A sensitive meter gives accurate readings.

HOW TO USE A METER

1. A meter is a delicate instrument. Handle it with care and respect. Jarring, pounding, dropping, and otherwise rough treatment may destroy a valuable meter.
2. When measuring voltage, the meter must be connected across the circuit (in parallel or shunt). Start on the HIGHEST range when measuring an unknown voltage and move to lower range for increased accuracy. Observe correct polarity! RED or positive lead to the POSITIVE side of the circuit. BLACK or negative lead to the negative or grounded side of the circuit.
3. A meter has its greatest accuracy at about two-thirds deflection on the scale. Use the range which gives as close to this deflection as possible.
4. When measuring current, the meter must be connected in SERIES with the circuit. A wire must be disconnected or unsoldered to insert the meter. It is wise to make a rough current calculation by Ohm's Law to determine proper current range on the meter. Observe correct polarity. Red or positive to positive side of circuit and black to negative side.
5. When measuring resistance, be certain that no power is applied to the circuit and that capacitors are discharged. It is not necessary to observe polarity. Always zero and adjust meter on proper range before measurements are made. A closed circuit has zero resistance. An open circuit has infinite resistance. On all meter measurements, make a "flash check" before permanently connecting the meter in the circuit. What does this mean? Make your decision on how the meter should be connected in the circuit but leave one meter lead disconnected. Quickly touch and remove this unconnected meter lead, while observing the meter. Does the needle move in the wrong direction? Change polarity. Does the needle move too violently? Change to a higher range. The flash check will save you many dollars in meter repairs and wasted time.
6. High voltages are often measured in electronic circuits. Make sure your meter connections do not short or touch other parts of the circuit. Use a well-insulated test prod to reach blocked test points in the circuit.

THE OSCILLOSCOPE

The oscilloscope permits observation of waveforms in an electronic circuit. It is a priceless service instrument. A detailed description of the oscilloscope is beyond the scope of this text. However, students should know basic principles and operating controls.

Waveforms are displayed on a cathode ray tube (CRT) in a fashion similar to your television. An electron beam from a gun in the CRT sweeps from left to right using saw-tooth voltages produced by the sweep oscillator. This produces a single line (usually green) on the face of the tube. If an ac voltage is applied to the vertical deflection circuit, then a wave is produced on the CRT. This wave represents the instantaneous voltage during the cycles of the ac input.

The horizontal sweep oscillator can be adjusted through a wide range of frequencies. Assume that it is adjusted at 60 Hz. It is then producing 60 straight lines per second across the face of the CRT. The time for one sweep would be 1/60 of a second.

Now, a 60 Hz ac voltage is applied to the vertical input terminals. This wave starts at zero, rises to maximum positive and returns to zero. It then decreases to maximum negative and returns to zero. This all occurs during one cycle. The period of this cycle is 1/60 of a second. Therefore, the horizontal sweep and the input signal voltage are synchronized. One wave appears on the screen. If the horizontal sweep were set at 30 Hz, which is only one half as fast as the 60 Hz input signal, then two complete waves would appear on the screen. The horizontal frequency may be adjusted so that the wave forms appear STATIONARY (not moving).

Oscilloscope familiarization

The function of the basic oscilloscope controls will be described, using Figs. 8-30 and 8-31.

Fig. 8-30. Four channel oscilloscope. (TEKTRONIX)

Electricity and Electronics

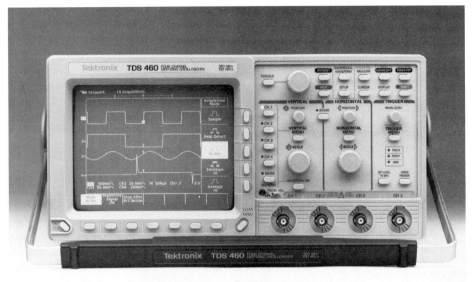

Fig. 8-31. Multiple-use oscilloscope showing a number of waveforms.

INTENSITY: Controls the brightness of the trace on the screen. It is combined with the on-off switch.

FOCUS: Used to adjust the electron beam in order to make a sharp trace on the screen.

V-CENTERING: Moving the trace up or down, such as to locate it in the center on the screen.

H-CENTERING: Moving the beam right or left to center the wave pattern on the screen.

BAND RESPONSE SWITCH: Provides either a wide or narrow band response.

V-POLARITY: Used to invert a pattern.

SWEEP VERNIER: Fine frequency adjustment for horizontal sweep oscillator.

H-GAIN: Increases the length of pattern by expanding it horizontally.

V-RANGE: Selects the voltage range for measurement.

V-CAL: Makes calibration adjustments so that peak voltages may be read directly on the scope. Sets waveform peaks to the required height.

SWEEP: Selects the range of frequencies for the sweep oscillator.

SYNC: Switch for either internal or external synchronization pulses. These pulses are used to lock input signal and horizontal sweep for a stationary pattern.

V-INPUT: Terminal for input to vertical response circuits.

GND: Ground.

Z-AXIS: Terminal used to modulate the intensity of the trace.

PHASE: Used with "line sync" so that sweep oscillator will be synchronized correctly.

SYNC TERMINAL: Terminal to which an external sync pulse may be applied.

H-INPUT: Switch used to change the horizontal sweep from the selective internal oscillator, to 60 Hz line, or to an external oscillator.

CALIBRATION SCALE: Plastic scale which fits over the face of the CRT. The instrument is calibrated so that these scales may be used for direct measurements of the peak value of waveforms.

Always refer to the instruction manual of the oscilloscope which you are using for explanations of specific controls.

REVIEW QUESTIONS FOR SECTION 8.4

1. What is the purpose of a Wheatstone bridge?
2. A wattmeter measures both _____ and _____ in a circuit to give watts.
3. In the dials of the wattmeter shown below, draw arrows on each dial to show a total reading of 2837 kilowatt hours.

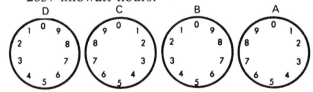

4. A _____ permits current to flow in only one direction.
5. Give four rules to follow when using a meter.
6. The display for an oscilloscope is a _____ _____ _____.
7. What can the oscilloscope do that a VOM cannot?

Instruments and Measurements

SUMMARY

1. The basic meter movement used for many instruments is the moving-coil galvanometer, or the D'Arsonval movement.
2. Analog meters use a scale with continuous variable values. Digital meters give values in discrete amounts in units 0 through 9.
3. It is vital to observe correct polarity in the use of analog meters.
4. Ammeters measure current and are connected in series in the circuit. Shunts are resistors connected in parallel with ammeters to increase the range of the meter.
5. Voltmeters measure voltage and are connected in parallel in the circuit. Multiplier resistors are connected in series with voltmeters to increase the range of the meter.
6. The sensitivity of a meter is an indication of its quality. Sensitivity is measured in an ohms-per-volt rating.
7. There are two basic types of ohmmeter circuits: series and shunt. Use depends on how the unknown resistor is placed in the circuit to be measured.
8. A multimeter is one instrument that will measure a number of values such as current, voltage, and resistance.
9. Digital meters (DVM) have several advantages: they are rugged, small, accurate, and portable.
10. Liquid crystal (LCD) displays are used in many test instruments.
11. The Wheatstone bridge is used to obtain accurate measurements.
12. A wattmeter measures electrical power consumed in wattage.
13. AC meters use rectifiers to convert ac to dc in order that a regular dc meter can be used.
14. Oscilloscopes use waveforms to show what is happening in a circuit.

TEST YOUR KNOWLEDGE, Chapter 8

Please do not write in the text. Place your answers on a separate sheet of paper.

1. The D'Arsonval movement measures _____ and _____.
2. What is the purpose of a dampener?
3. A _____ is an alternate path, supplied for measuring large currents.
4. A meter movement has a moving coil resistance of 50 ohms and requires .001 amperes for full scale deflection. Compute the shunt values for the meter to read in the following ranges:
   a. 0-1 mA.
   b. 0-10 mA.
   c. 0-50 mA.
5. Compute the multipliers for the meter in Question 4 to use on the following ranges:
   a. 0-1 volts.
   b. 01-10 volts.
   c. 0-100 volts.
6. A meter used to measure the value of an unknown resistance is called a(n):
   a. Multimeter.
   b. Ohmmeter.
   c. Multiplier resistor.
   d. None of the above.
7. Make a sketch of the series ohmmeter circuit.
8. The Wheatstone bridge is used to measure an unknown resistance. Solve for the unknown resistance when the following values will balance the bridge:
   $R_1 = 500$ ohms, $R_2 = 100$ ohms, $R_3 = 300$ ohms, $R_X = ?$
9. What is the major difference between an iron vane ammeter and an iron vane voltmeter?
10. The _____ permits observation of waveforms in an electronic circuit.

MATCHING QUESTIONS: Match each of the following terms with their correct definitions.
   a. V-cal.   d. H-gain.
   b. Sync.   e. GND.
   c. Focus.   f. Phase.

11. Ground.
12. Makes calibration adjustments so peak voltages may be read directly on scope.
13. Adjusts electron beam to make sharp trace on screen.
14. Use with line sync for synchronization of sweep oscillator.
15. Increases length of pattern by expanding it horizontally.
16. Switch for internal or external synchronization pulses.

FOR DISCUSSION

1. Discuss the safety rules and precautions to follow when working with meters.
2. Why is a rectifier necessary in measuring an ac current or voltage?
3. What is meant by "loading a circuit" with a meter?
4. How can electronic circuitry be used to test human hearing?
5. Who was Alexander Graham Bell?

Part II Summary
APPLIED ELECTRICITY

IMPORTANT POINTS

1. Magnetism is the source most used for producing electricity.
2. Magnets have two different poles: north and south. Like poles repel and unlike poles attract each other.
3. Flux lines exist around magnetized materials. These lines represent the magnet's field of force.
4. The stronger the magnet, the stronger the flux density.
5. The ability of a material to conduct magnetic lines of force is permeability. The opposition to this is called reluctance.
6. Electromagnets are produced by a current flowing through a coil. Their field strength can be increased and decreased by varying the current. Electromagnets can also be turned on and off.
7. A generator converts mechanical energy to electrical energy.
8. Lenz's Law states that the induced voltage in any circuit is always in such a direction to oppose the effect that produces it. This is sometimes called electrical inertia.
9. Generators are generally very efficient. However, they have some inefficiency in the form of copper losses, hysteresis losses, and eddy currents.
10. Generators can produce ac or dc by the use of either slip rings or commutators on the armature.
11. The usable value for an ac sine wave is the effective value or root mean square (RMS) value.
12. Sine waves can be either in phase or out of phase, depending on the types and values of components in the circuit.
13. Instruments are used to give exact measurement of specific values in a circuit.
14. Ammeters are used in series to measure current. Voltmeters are connected in parallel and measure voltage.
15. Ohmmeters have their own power source and measure resistance.
16. Analog and digital meters are used as basic measurement devices in circuits.
17. An oscilloscope is a test instrument. It uses a cathode ray tube (CRT) which permits observation of the frequency and amplitude of waveforms. Frequency ranges up to 100 MHz are used.

SUMMARY QUESTIONS

1. 1200 V_{eff} = _____ V_{max}.
2. 29.5 V_{max} = _____ V_{eff}.
3. Draw waveforms showing current that lags voltage by 90 degrees.
4. 88 megahertz = _____ hertz.
5. Find the value of the multiplier resistor (R_X) in the circuit shown.

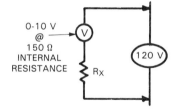

6. Find the value of the shunt (R_S) in the circuit shown.

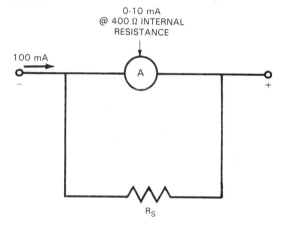

Part III

ALTERNATING CURRENT CIRCUITS

9 Inductance and RL Circuits
10 Capacitance and RC Circuits
11 Tuned Circuits and RCL Networks
12 Electric Motors

Electrical oppositions that are found in alternating current do not exist in direct current. These resistances to change in either current or voltage are the reason we study inductance and capacitance. Inductance and capacitance provide the basis for induction motors, transformers, capacitors, coils, and many other components. A study of these important concepts in Chapters 9 through 12 will give the student an excellent background in alternating current.

Chapter 9, *Inductance and RL Circuits,* presents the theory and operation of inductors and transformers in electrical circuits. Chapter 9 also discusses inductors coupled with resistors to give reactance in ac circuits.

Capacitance is an interesting concept in electricity. It is covered in Chapter 10. Presented are the factors which affect capacitance, the types of capacitors, reactance, and RC circuits.

Chapter 11, *Tuned Circuits and RLC Networks,* deals with ac circuits which have resistors, capacitors, and inductors placed in the circuits. Resonance, and its effect in an ac circuit, is presented. The basis for oscillators, the tank circuit, is explained.

Electric motors are covered in detail in Chapter 12, *Motors.* The principles of motor operation are discussed along with the different types of motors.

A thorough grasp of alternating current theory is important for a solid understanding of electricity and electronics.

Electricity and Electronics

Transformers in the bodies of computers convert 110 Vac into a voltage the computer can use.

What a useful tool is magnetism! Earlier you learned some ways it reacts with electricity. For instance, it allows you to generate electricity. Now you will learn how it transfers electrical energy from one circuit to another. Or it can be used to change voltage as well.

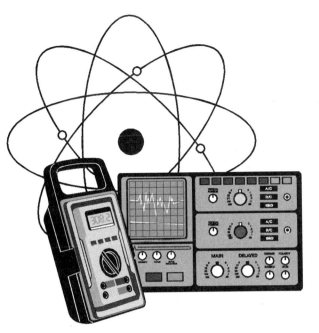

Chapter 9

INDUCTANCE AND RL CIRCUITS

After studying this chapter, you will be able to:
- Define the terms inductor and inductance.
- Explain how inductance affects a current.
- Discuss the relationship between mutual inductance and transformers.
- Outline the operation of transformers.
- Describe the effect of inductance in ac circuits.
- Use various measuring and computing methods to determine the values of currents and voltages in inductive circuits.

The study of electricity and electronics revolves around inductance, capacitance, resistance, and combinations of these in series and parallel circuits. Resistance in a circuit was studied in Chapter 2. This chapter will help you answer these questions:
1. What is an inductor and inductance?
2. What is the effect of inductance in a circuit?
3. What methods are used to measure and compute values of current and voltage in an inductive circuit?

9.1 INDUCTANCE

INDUCTANCE may be defined as the property in an electric circuit which resists a change in current. This resistance to a change of current is the result of the energy stored within the magnetic field of a coil. A coil of wire will have inductance.

Again, set up the experiment in Chapter 7, Fig. 7-1, using the hollow coil A connected to a galvanometer. Move the permanent magnet in and out of the coil, and notice the deflection of the needle in the meter. When the magnet moves in, current will flow in one direction. The current will reverse direction when the magnet is withdrawn. No current will flow unless the magnet is moved.

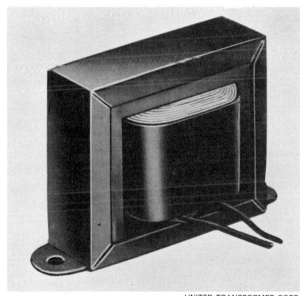

UNITED TRANSFORMER CORP.
Fig. 9-1. This choke is used in the filter section of a power supply.

This is the same principle explained in Chapter 7, on the theory of generators. By moving the magnet, a voltage is induced in the coil which causes current to show on the meter. There is a distinct link between magnet movement and current flow direction. This is an application of LENZ'S LAW: The field created by induced current is of such a polarity that it opposes the field of the permanent magnet.

A coil connected to a source of direct current will build up a magnetic field when the circuit is closed. The expanding magnetic field cutting across the coil windings will induce a counter emf or voltage. This will oppose the source voltage and oppose the rise in current. When the current reaches its maximum value and there is no further change, there is no longer an induced counter emf. The current is then only limited by the ohmic resistance of the wire. However, if the source voltage is disconnected, the current tends to fall to zero. But the collapsing magnetic field again induces a counter emf, which retards the reduction of current.

The inductance of the coil resists any change in current value. The symbol for inductance is L. Inductance is measured in a unit called a HENRY (H). A henry represents the inductance of a coil if one volt of induced emf is produced when the current is changing at the rate of one ampere per second. This may be expressed mathematically as,

$$E = L \frac{\Delta I}{\Delta t}$$

E equals the induced voltage, L equals the inductance in henrys, ΔI equals the change of current in amperes, and Δt equals the change of time in seconds. The symbol Δ means a "change in."

The strength of an induced voltage depends upon the strength of the field. It follows then that a stronger magnetic field would produce higher induced voltage. The magnetic field may be strengthened by inserting a core in the coil. An iron core has higher permeability than air and concentrates the lines of force. This was explained in the study of magnetism.

Large inductors are most often wound on laminated iron cores. Their inductance is measured in henrys. Smaller inductors, used at higher frequencies, may have powdered iron or air cores. They have inductance measured in millihenrys, (1/1000 of a henry) and microhenrys, (1/1,000,000 of a henry). An inductor used in a power supply called a choke is shown in Fig. 9-1. Radio frequency chokes using air cores are shown in Fig. 9-2.

A changing current through an inductor produces an expanding or collapsing magnetic field cutting across the wires of the coil. A counter emf is induced which opposes the change of current. This is called SELF-INDUCTION. The strength of the self-induction depends on the number of turns of wire in the coil, the link between the length of the coil and its diameter, and the permeability of the core.

This effect may be observed by setting up the experiment in Fig. 9-3. First, connect the light directly

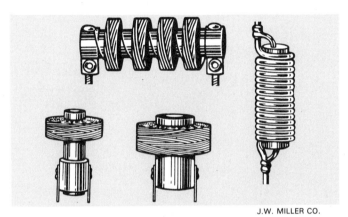

Fig. 9-2. Radio frequency (RF) chokes.

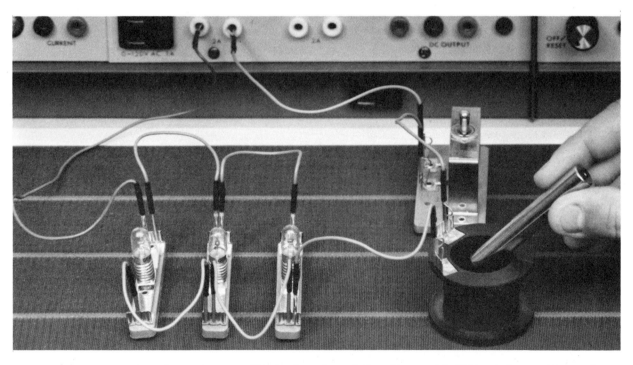

Fig. 9-3. Use an electrodemonstrator to observe self-induction of a coil. Experiment with different sizes and types of cores.

Inductance and RL Circuits

to the 6 V power source. Note that as the switch is closed, the light will burn at full brilliance instantly. Connect coil F in series with the light and close the switch. There will be a slight delay in approaching full brilliance. This is due to the inductance of the coil. Turn the light off and insert the iron core in the coil. Once again close the switch. The light comes to brilliance rather slowly and not to full brilliance. Why does the core increase the delay in brilliance?

Repeat the above experiment, using coil D. Self-inductance is increased in this coil by the number of turns of fine wire in the coil. Move the core in and out of the coil. Note the change in brilliance due to the change in self-inductance. This is used to dim houselights in theaters.

TRANSIENT RESPONSES

The response of the current and voltage in a circuit after an instant change in applied voltage is known as a TRANSIENT RESPONSE.

Refer to the diagram in Fig. 9-4. A coil is connected to a dc voltage source. When the switch is closed, the current will build up gradually. This is due to the self-induction of the coil and the internal resistance of the battery. When the switch is opened, the current will decay in like manner. The rise and decay of the current in this circuit is shown by the graph in Fig. 9-5.

It is important to understand that the opposition to the rise or decay of current occurs only when there is a change in applied voltage. When there is no change, the current remains at its steady state value, depending only on the resistance of the coil.

A resistance is joined in series with an inductor in the RL circuit in Fig. 9-6. The action of the voltage and current should be studied for increases and decreases. For example, when the switch is closed, the counter emf (E_L) of the coil equals the source voltage. Since no current has started to flow, the IR drop across R equals zero. As the current gradually builds up, the voltage E_R increases. And counter emf E_L decreases until a steady state condition exists. All the voltage drop is E_R, and there is no drop across L.

If the RL circuit is shorted by another switch, S_2 in Fig. 9-7, the stored energy in the field of L instantly develops a voltage and the circuit is discharged by current flowing through R. The graphs showing the charge and discharge of the circuit appear in Fig. 9-8.

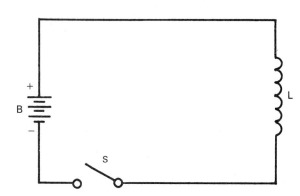

Fig. 9-4. Schematic diagram of an L circuit.

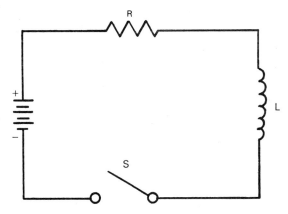

Fig. 9-6. A diagram of the RL circuit.

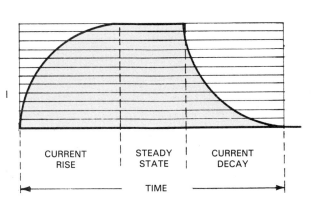

Fig. 9-5. Graph shows the transient response of the L circuit as the switch is closed and opened.

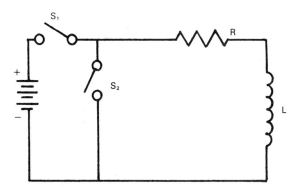

Fig. 9-7. The coil L is shorted through switch S_2. The magnetic field collapses.

The baseline, or X axis, of these graphs represents time. The transient response of an RL circuit does require a definite time, depending upon the values of R and L. This is called the TIME CONSTANT of the circuit. It may be found by the formula,

$$t = \frac{L \text{ (in henrys)}}{R \text{ (in ohms)}}$$

where t equals the time in seconds for the current to increase to 63.2 percent of its maximum value or to decrease to 36.7 percent. In most cases, the circuit is considered charged or discharged after five time constant periods. Fig. 9-9 gives the voltage at the end of each time constant, assuming E source is 100 volts. Refer again to Fig. 9-8 where these points are plotted. A clear picture of the rise and decay of the voltage and current can be seen.

REVIEW QUESTIONS FOR SECTION 9.1

1. Define inductance.
2. What is the difference between inductance and resistance?
3. The unit for inductance is the _____.
4. Find the time constant in the circuit shown.

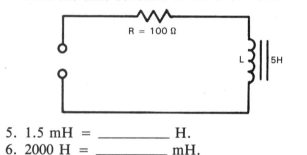

5. 1.5 mH = _____ H.
6. 2000 H = _____ mH.

9.2 MUTUAL INDUCTANCE

In the study of generator regulation, you learned that the magnetic fields created by the windings of the cutouts can be additive or cancelling. When two coils are within magnetic reach of each other so that the flux lines of one coil link with or cut-across the other coil, they have MUTUAL INDUCTANCE. If the coils are close to each other, many magnetic lines in the flux link with the second coil. On the other hand, if the coils were a distance apart, there might be very little linkage.

The mutual inductance of two coils can be increased if a common iron core is used for both coils. The degree to which the lines of force of one coil link with the windings of the second coil is called COUPLING. If all lines of one cut across all the turns of the other it is UNITY COUPLING. A number of percentages of coupling may exist due to the mechanical position of

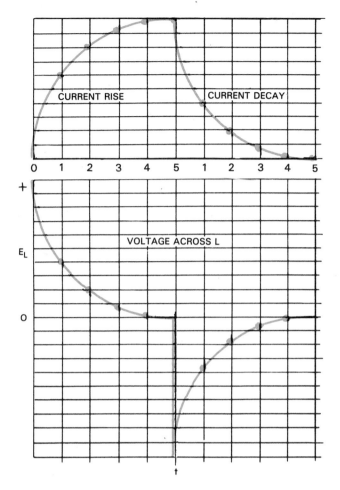

Fig. 9-8. The transient response curves for current and inductive voltage of the RL circuit.

TIME CONSTANT	CHARGING	DISCHARGING
1	63.2 V	36.8 V
2	86.5 V	13.5 V
3	95 V	5 V
4	98 V	2 V
5	99 V	1 V

Fig. 9-9. Assuming the E source is 100 volts, these are the voltages at the end of each time constant.

the coils. The amount of mutual inductance may be found by the formula,

$$M = k\sqrt{L_1 L_2},$$

where M equals mutual inductance in henrys, k equals the coefficient or percentage of coupling, and L_1 and L_2 equal the inductance of respective coils.

The mutual inductance must be considered when two or more inductors are connected in series or parallel in a circuit.

THE TRANSFORMER

A common application of mutual inductance is the TRANSFORMER. A transformer is a device used to transfer energy from one circuit to another by electromagnetic induction. A transformer consists of two or more coils of wire wound around a common laminated iron core. The coupling between the coils approaches UNITY. Transformers have no moving parts and require very little care. They are simple, rugged, efficient devices, Fig. 9-10.

The construction of a simple transformer is shown in Fig. 9-11, along with its schematic symbol. The first winding or input winding is called the PRIMARY. This winding receives the energy from the source. The second winding or output winding is called the SECONDARY. The output load is attached to this winding.

The energy in the secondary is the result of the mutual induction between the secondary and the primary windings. The varying magnetic field of the primary cuts across the windings of the secondary. This induces a voltage in the secondary. Therefore, the transformer is a device which must work on an alternating current or a pulsating direct current. The primary field must be a moving magnetic field in order for induction to take place.

In Fig. 9-12, two types of construction are shown: core construction and shell construction. In the core type, the coils surround the core. In the shell type, the core surrounds the coil.

In either type of construction, a strong coupling must exist between the primary and secondary windings. Then very little flux loss or leakage will occur. In core construction, parts of both windings are wound on each leg of the core. In shell construction, the windings are in alternate layers.

Turns ratio and voltage ratio

A key advantage of transformers is that they can be used to increase or decrease voltage. This is done by increasing or decreasing the number of turns on the secondary winding.

Fig. 9-10. A power transformer. It is used in the power supply for electronic circuits.

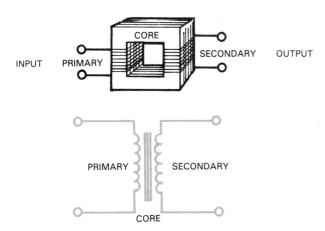

Fig. 9-11. Top. Basic construction of a transformer. Bottom. Schematic symbol of a transformer used in circuit diagrams.

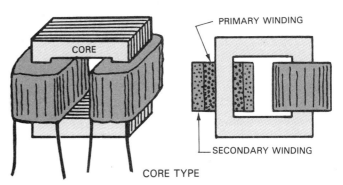

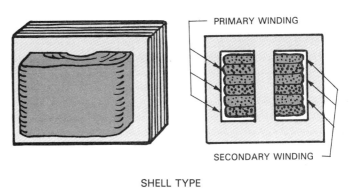

Fig. 9-12. Construction of core type and shell type transformers.

On a secondary having less turns than the primary, voltage is decreased. In this case, voltage STEPS DOWN, Fig. 9-13. Likewise, when the secondary has more turns than the primary, voltage is increased, Fig. 9-14. It STEPS UP.

The ratio between the number of turns in the primary and secondary is called the TURNS RATIO. Referring back to Fig. 9-13, the turns ratio would be:

$$\frac{N_p}{N_s} = \frac{6}{4} \text{ or } 6:4$$

where N equals the number of turns.

The VOLTAGE RATIO is the ratio between the voltage of the primary and secondary. It is in the same proportion as the turns ratio:

$$\frac{E_p}{E_s} = \frac{N_p}{N_s} = \frac{I_s}{I_p}$$

where E equals voltage and I equals amperage.

Refer to Fig. 9-15. The transformer has 200 turns in the primary and 1000 turns in the secondary. If the applied voltage is 117 volts ac, what is the secondary voltage?

$$\frac{117 \text{ V}}{E_s} = \frac{200}{1000}$$

Transposing this equation,

$$E_s = \frac{1000 \times 117}{200} = 585 \text{ V}$$

This is an example of a step-up transformer.

What if this transformer was made with a 10 turn secondary? What then would the secondary voltage be?

$$\frac{117 \text{ V}}{E_s} = \frac{200}{10}$$

$$E_s = \frac{117 \times 10}{200} = 5.85 \text{ V}$$

This is a step-down transformer.

Taps

The power transformer in Fig. 9-16 has a number of taps on the secondary winding. A TAP is a fixed electrical connection made to a winding at a point other than its terminals. Taps provide a number of voltages for use.

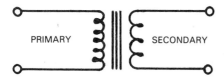

Fig. 9-13. In a step-down transformer, the output voltage is lower than the input voltage.

Fig. 9-14. In a step-up transformer, the output voltage is higher than the input voltage.

UNITED TRANSFORMER CORP.

Fig. 9-16. Typical power supply transformer used in radios, TVs, and transmitters.

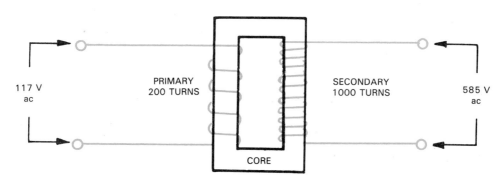

Fig. 9-15. Turns ratio helps determine output voltage.

Inductance and RL Circuits

If a tap is made on the secondary winding as shown in Fig. 9-17, two voltages can be received from the output. These voltages are 180 degrees out of phase with each other. This will be discussed further in Chapter 15.

The primary of a power transformer can also be tapped. This allows use of varying input voltages. This feature is useful when electrical or electronic equipment is moved from one place to another place having a different ac voltage.

Transformer power

The power used in the secondary circuit must be supplied by the primary. Assuming that the transformer is 100 percent efficient, then the power in the secondary, $I_s \times E_s$, must equal the power in the primary, $I_p \times E_p$.

For example, a step-up transformer will produce 300 volts in the secondary when 100 volts ac is applied to the primary. If a 100 ohm load is applied to the secondary, Fig. 9-18, a current of 3 amperes would flow,

$$I = \frac{300 \text{ volts}}{100 \text{ ohms}} = 3 \text{ amps}$$

The power used in the secondary would be $P = I_s \times E_s = 3 \text{ amps} \times 300 \text{ volts} = 900 \text{ watts}$. Since the primary must supply this power,

$$I_p = \frac{P_p}{E_p} = \frac{900 \text{ watts}}{100 \text{ volts}} = 9 \text{ amps, and}$$

$$I_s \times E_s = I_p \times E_p = 900 \text{ watts}$$

The key point of transformer action is that as VOLTAGE INCREASES in the secondary, CURRENT DECREASES in the secondary.

Power companies use that key point as the basis for production of electricity. In the study of conductors, you learned that all wires have some resistance. You will also recall that the power loss in a circuit or conductor is $P = I^2R$. A long length of wire, having a resistance of 2 ohms and carrying a 10 ampere current, will have a power loss of $P = I^2R = 10^2 \times 2 = 200$ watts. If the current is doubled to 20 amperes, then the loss would be $20^2 \times 2 = 800$ watts, or four times as much as at 10 amperes.

Power loss varies as the square of the current. Therefore, it is more economical to raise the voltage, decrease the current, and reduce power loss in the transmission lines. Transformers are used for this purpose, Fig. 9-19.

An electric appliance in your home uses 10.25 amperes of current. How much power is consumed when you turn it on?

$$P = I \times E = 10.25 \times 117 = 1200 \text{ watts}$$

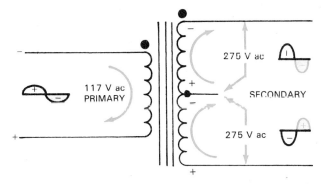

Fig. 9-17. Tap on secondary winding.

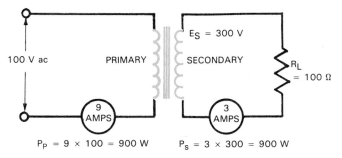

Fig. 9-18. Relationship between voltage, amperage, and power in the primary and secondary of a transformer.

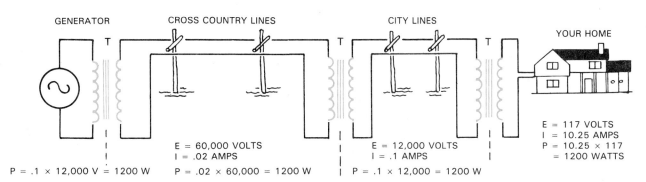

Fig. 9-19. Use of transformers in the transmission of electricity reduces line losses.

In 12,000 volt city lines, a current of .1 amp would flow: .1 amp × 12,000 volts = 1200 watts.

In cross country power lines, a current of .02 amp would flow: .02 amp × 60,000 volts = 1200 watts.

By raising voltage, power companies are able to supply large cities and industries with electricity. They use small wires for transmission lines. And while smaller wires do have greater resistance per foot, the loss from this resistance is only in direct proportion to the current used. Power companies select a wire that has the least resistance, yet is large enough to carry the expected current. The wire must also be strong enough to withstand high winds, ice, and snow.

Autotransformers

The common transformer consists of two windings: primary and secondary. While the two windings are not physically joined, a magnetic coupling exists between them. The electrical separation between the windings is called ISOLATION. Isolation reduces the chance for shock. See Fig. 9-20.

An autotransformer has only one winding. The primary and secondary windings are physically joined at some point. There is no isolation. This increases the risk of electrical shock. However, the autotransformer is most often used in relatively low-voltage applications.

Autotransformers can be either step-up or step-down, Fig. 9-21.

Transformer losses

Three types of losses go along with transformer construction. All losses result in heat.

COPPER LOSSES are the result of resistance of wire used in the transformer windings. These are also called I_2R losses. They vary as the square of the current according to Ohm's Law and the Power Law.

EDDY CURRENT LOSSES are caused by small whirlpools of current induced in the core material. These losses are reduced by using laminated core construction. Each lamination is insulated from its bordering layer by varnish. This cuts the number of paths on which currents can flow.

HYSTERESIS LOSSES, or molecular friction, are the result of magnetic particles changing polarity in step with induced voltage. Special alloys and heat-treating processes are used to make core materials, which reduces hysteresis loss.

Reasons for using a transformer

The basic reasons for using a power, or filament, transformer are to step up or step down voltage or current; to provide two voltages 180 degrees out of phase with each other; or to provide isolation from the primary to the secondary.

Fig. 9-20. Isolated variable alternating current supply.

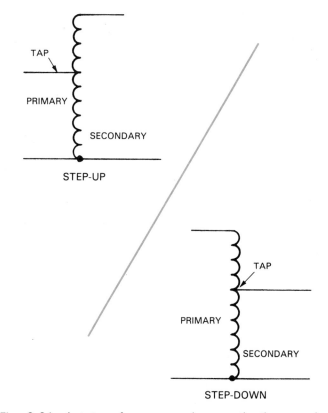

Fig. 9-21. Autotransformers are less costly than regular transformers. Taps may be made variable so that the secondary voltage can be changed.

THE INDUCTION COIL

A transformer requires an alternating current or a pulsating direct current. Transformer action can be observed using an induction coil, Fig. 9-22 and 9-23.

Inductance and RL Circuits

The armature and point action in the buzzer circuit cause the magnetic field of the coil to expand and collapse. It is a constantly moving field. A secondary coil of many fine wire turns is placed over the primary buzzer coil. A high voltage is induced in the secondary.

A capacitor has been added to this circuit to improve its action. Capacitance will be studied in a later chapter.

LESSON IN SAFETY: People vary in their ability to withstand an electrical shock. Be very CAREFUL if you build the circuit shown in Fig. 9-22.

Induction circuit breaker

One common use of the induction coil is in circuit breakers. Circuit breakers, discussed in Chapter 6, use an inductor to trigger a switch to shut off the current in a circuit. Current traveling through the coil of a circuit breaker produces a magnetic field. When a certain level of current is reached, the strength of the magnetic field breaks the circuit. The circuit must be reset manually.

Many circuit breakers are designed to protect equipment or prevent fires. The circuit breakers in Fig. 9-24 are designed to protect people from electric shock. These breakers trigger with a very small fluctuation in the current. The buttons on top allow the breakers to be tested and reset.

PHASE RELATIONSHIP IN TRANSFORMERS

The output voltage of a transformer may be in phase or 180 degrees out of phase with the primary voltage. This depends on the direction of the windings and the method of connection.

Restating Lenz's Law, the polarity of the induced voltage will be opposite to the voltage producing it. A diagram of a transformer and the waveforms of the primary and secondary voltage are shown in Fig. 9-25. The direction of the magnetic flux is shown by arrows

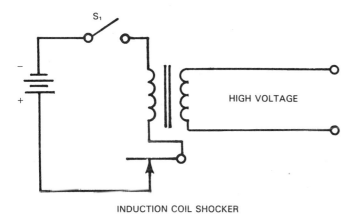

Fig. 9-22. Schematic of an induction coil.

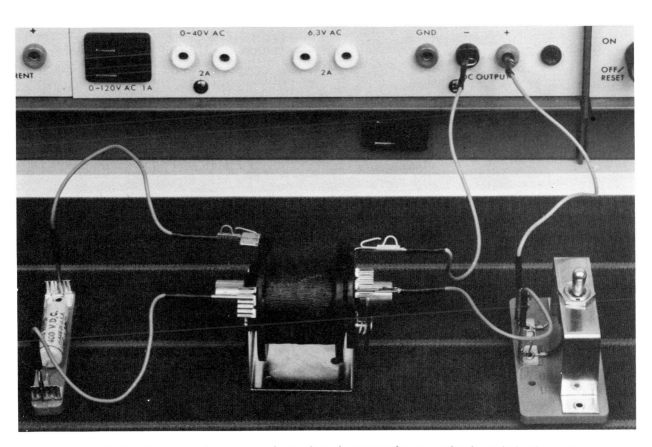

Fig. 9-23. An electrodemonstator is used to show transformer action in an induction coil.

within the core. This type of transformer inverts the alternating voltage wave applied to the primary.

SERIES AND PARALLEL INDUCTANCE

When coils are joined in series so that no mutual inductance exists between them, the total inductance is,

$$L_T = L_1 + L_2 + L_3$$

The individual inductances add like series resistors. However, when there is mutual inductance the formula becomes,

$$L_T = L_1 + L_2 \pm 2M$$

The plus or minus sign ($\pm$) before the M means that plus (+) should be used when the coils aid each other and minus (−) when the coils oppose each other.

When inductors are connected in parallel without mutual inductance,

$$\frac{1}{L_T} = \frac{1}{L_1} + \frac{1}{L_2} + \frac{1}{L_3}$$

or

$$L_T = \frac{L_1 \times L_2}{L_1 + L_2}$$

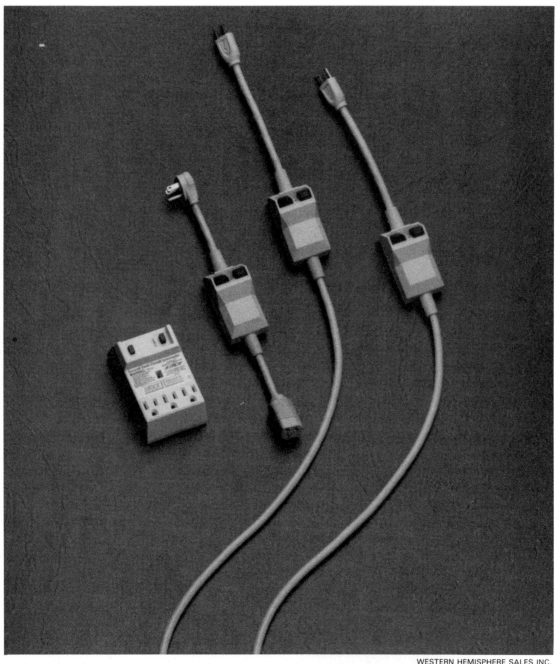

WESTERN HEMISPHERE SALES INC.

Fig. 9-24. Circuit breakers use inductors to protect equipment, property, and people.

Inductance and RL Circuits

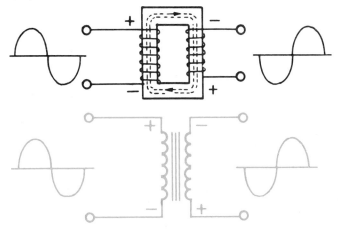

Fig. 9-25. Phase relationship between the primary and secondary of the transformer. The schematic symbol is shown.

Mutual inductance is used widely in electricity and electronics. A skilled technician must have a complete understanding of the characteristics of circuits containing these components.

REVIEW QUESTIONS FOR SECTION 9.2

1. The effect of two or more coils sharing the energy of one coil is called _____ _____.
2. Name two common types of transformer construction.
3. A transformer having more turns in the secondary than the primary is a:
   a. Step-up transformer.
   b. Tap.
   c. Step-down transformer.
   d. None of the above.
4. Solve for the voltage in the secondary of the transformer shown.

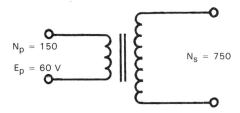

5. Determine the power used in a step-down transformer producing 600 volts in the primary, 250 volts in the secondary, with a 150 ohm load applied to the secondary.
6. Determine the amount of primary current flowing for the transformer in Question 5.
7. List three types of transformer losses.
8. A 1.5 H and a 500 mH coil are connected in series. What is their total inductance in henrys?

9.3 INDUCTANCE IN AC CIRCUITS

In an ac circuit, the applied voltage varies and reverses polarity constantly. Thus, self-inductance develops counter emfs, which oppose the source voltage.

This event may be again observed by setting up Fig. 9-3. Connect a light to a 6 volt dc source. Note its brilliance. Now, connect the light to a 6 volt ac transformer and note the brilliance. Is it brighter or dimmer? The inductance has some indirect effect on the current flow.

Opposition to an alternating current due to inductance or capacitance is called REACTANCE. Its letter symbol is X. Because it is caused by an inductor, its symbol is X_L, called INDUCTIVE REACTANCE. It is measured in ohms.

INDUCED CURRENT AND VOLTAGE

Remember that the induced voltage in a coil is the counter emf. It opposes the source. Therefore, it is 180 degrees out of phase with the source voltage. The greatest counter emf is induced when the current change is at maximum.

Study Fig. 9-26. The greatest rate of change of current is at 90 degrees and 270 degrees. At these points, the applied voltage must be at maximum. The induced voltage is at maximum. At 180 degrees and 360 degrees, the current change is minimum. The current is at its maximum value, ready to start its decline. Note that in a circuit containing pure inductance, the current is 90 degrees out of phase with the applied voltage. The current is lagging behind the voltage. The magnitude of the reactive force opposing the flow of an ac is measured in ohms. It may be expressed mathematically as, $X_L = 2\pi f L$, where X_L equals inductive reactance in ohms, f equals frequency hertz, L equals inductance in henrys, and π equals 3.1416.

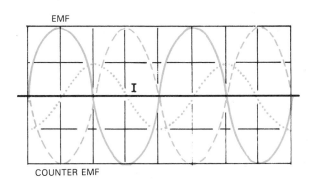

Fig. 9-26. Solid line is emf. Dashed line is counter emf force. Dotted line is current. Greatest counter emf occurs when current is changing at its most rapid rate.

Remember, as frequency or inductance is increased, the inductive reactance increases. They are in direct proportion.

An inductance in a circuit behaves differently as various ac voltages are applied, Fig. 9-27. An 8 henry inductor connected to a dc (f = 0) has a reactance of zero. Using the formula, $X_L = 2\pi fL$, the reactance is computed for frequencies of 50, 100, 500, and 1000 Hz. A partial graph of the results shows the linear increase of reactance as the frequency is increased.

In Fig. 9-28, frequency is held constant at 100 Hz. The reactance is plotted as the inductance is increased from .32 henrys to .8 henrys. Notice the linear increase in reactance as the inductance is increased.

These principles can also be applied to filter and coupling circuits. However, you must first understand the traits of reactance and how they change as a result of inductance and frequency of applied voltage.

POWER IN INDUCTIVE CIRCUITS

Power consumed in a purely resistive circuit is the product of the voltage and the current: $P = I \times E$. This is the actual power used by a circuit. It is called TRUE POWER, given in units of watts. This is not true, however, in an ac circuit containing inductance only. A current will flow, bound by the reactance of the circuit, and the power used to build the magnetic field will be returned to the source when the field collapses. This is REACTIVE POWER. Units of measure are volt-amperes reactive (VARs), Fig. 9-29. Note that true power in a purely inductive circuit is zero.

The combination of true power and reactive power give rise to a measure of power that appears to be delivered to a load. It is called APPARENT POWER. It is equal to the product of the effective voltage and the effective current. Units are volt-amperes. For example, a 100-V_{eff} applied voltage to a certain inductive circuit causes a 10-A_{eff} current. The apparent power will equal:

100 volts × 10 amperes = 1000 volt-amperes

Watts are not used for apparent power.

The ratio of true power to apparent power in an ac circuit is called the POWER FACTOR. It is found using trigonometry. It is equal to the cosine of the phase displacement between current and voltage. Mathematically,

$$\text{Power factor (PF)} = \cos\theta = \frac{\text{true power}}{\text{apparent power}}$$

Note that by rearranging this formula, we arrive at the formula for true power:

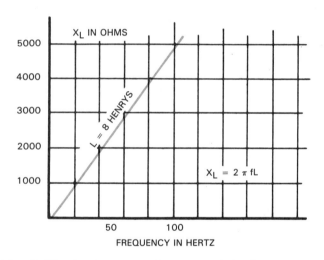

Fig. 9-27. As frequency increases, inductive reactance increases.

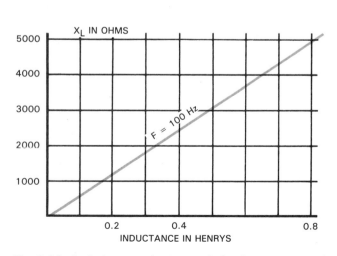

Fig. 9-28. As inductance increases, inductive reactance also increases.

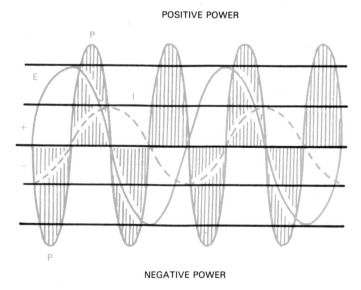

Fig. 9-29. A circuit containing inductance only. The true power is zero and current lags the voltage by 90 degrees.

Inductance and RL Circuits

True power = (apparent power) Cos θ
= $E_{eff} I_{eff}$ Cos θ

As an example, assuming our previous circuit is purely inductive, determine the power factor and phase displacement.

$$PF = \text{Cos } \theta = \frac{0}{1000} = 0$$

The angle whose cosine is 0 is 90 degrees. This tells us that current and voltage in the purely inductive circuit are 90 degrees out of phase.

RESISTANCE AND INDUCTANCE IN AN AC CIRCUIT

When resistance and inductance are in series in a circuit, power is used. If a circuit contains only resistance, the current and voltage are in phase, Fig. 9-30. The power consumed in equal to $I \times E$. Even though the polarity of the voltage changes and the current reverses, positive power is consumed. A resistance consumes power despite the direction in which the current is moving. The power factor in a circuit of this type is cos 0 degrees equals 1. The apparent power is equal to the true power.

The circuit traits change when an inductor is added in series with the resistor. This does not have to be a resistive component. The wire from which the coil is wound will have a certain amount of resistance.

Refer to Fig. 9-31. The series resistance equals 300 ohms. The inductive reactance equals 400 ohms. This reactive component will cause the current to lag by an angle of 90 degrees or less. The forces opposing the current can be thought of as the resistance and reactance. They are 90 degrees out of phase. To find the final opposition, add the vectors of the two forces. Refer to Fig. 9-32.

The vectors can be added using the graph. Place the tail of the X_L vector on the arrowhead of the R vector. Then draw a vector from starting point, θ, to the head of the vector, X_L. This is the magnitude and direction of the combined forces.

The total opposition to an alternating current in a circuit having resistance and reactance is called IMPEDANCE. It is measured in ohms.

Impedance problems are commonly solved using the PYTHAGOREAN THEOREM. This theorem states that the hypotenuse of a right triangle is equal to the square root of the sum of the squares of the two sides.

In the problem X_L equals 400 Ω and R equals 300 Ω, find the impedance.

$$Z = \sqrt{300^2 + 400^2}$$
$$= \sqrt{90{,}000 + 160{,}000}$$
$$= 500 \text{ } \Omega$$

or, $Z = \sqrt{R^2 + X_L^2}$

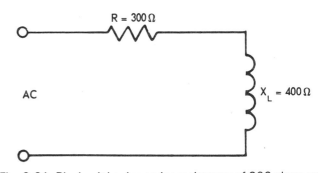

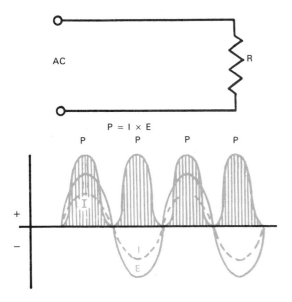

Fig. 9-30. In a purely resistive circuit, true power and apparent power are the same. The current and voltage are in phase.

Fig. 9-31. RL circuit having series resistance of 300 ohms and inductive reactance of 400 ohms.

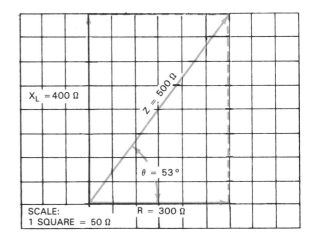

Fig. 9-32. Vector addition of X_L and R, which are 90 degrees out of phase.

The impedance of the circuit is 500 ohms. The angle between vector Z and vector R is called the PHASE ANGLE (θ). It represents the phase displacement between the current and voltage that results from the reactive component. So, Cos θ equals the power factor and also equals $\frac{R}{Z}$ (see Appendix page 398). Therefore,

PF = Cos θ

Cos θ = $\frac{R}{Z}$ or $\frac{300}{500}$ or .6

The angle whose cosine is .6 is 53.10 degrees (approximately). The current lags the voltage by an angle of 53.1 degrees. The true power in this circuit equals the apparent power times the power factor, or cos θ:

True power = apparent power × cos θ

and, Cos θ = $\frac{\text{True power}}{\text{Apparent power}}$

The waveforms for current, voltage, and power are drawn in Fig. 9-33. Assuming an applied ac voltage of 100 V, the current in the circuit will equal:

I = $\frac{E}{Z}$ or I = $\frac{100}{500}$ = .2 amps

The apparent power equals:
Apparent power = E × I
or, .2 × 100 = 20 volt-amperes
The true power equals:
True power = I × E × cos θ
= .2 × 100 × cos 53.1 degrees
= 20 × .6
= 12 watts

These figures can be checked by inserting them into this formula:

PF = $\frac{\text{True power}}{\text{Apparent power}}$

= $\frac{12}{20}$ = $\frac{3}{5}$ = .6

This is not only theory. In practice, the power factor must be considered whenever a power company connects power lines to a manufacturing plant. Industries must keep the power factor of their circuits and machinery within specified limits, or pay the power company a premium.

A reactive power is sometimes called WATTLESS POWER. It is returned to the circuit. In Fig. 9-33, the power in the shaded areas above the zero line is used. The power below the line is wattless power.

OHM'S LAW FOR AC CIRCUITS

In computing circuit values in ac circuits, Ohm's Law is used with one exception. Z is used in place of R. Z represents the total resistive force opposing the current. Therefore,

I = $\frac{E}{Z}$, E = IZ, and Z = $\frac{E}{I}$

A series circuit contains an 8 henry choke and a 4000 ohm resistor. It is connected across a 200 volt, 60 Hz ac source, Fig. 9-34.

First, find the reactance of L:
X_L = 2 π fL
= 2 × 3.14 × 60 × 8
= 3014 Ω (3000 Ω approximately)
Find the impedance of the circuit:
Z = $\sqrt{R^2 + X_L^2}$
= $\sqrt{4000^2 + 3000^2}$
= $\sqrt{16{,}000{,}000 + 9{,}000{,}000}$
= $\sqrt{25{,}000{,}000}$
= 5000 Ω

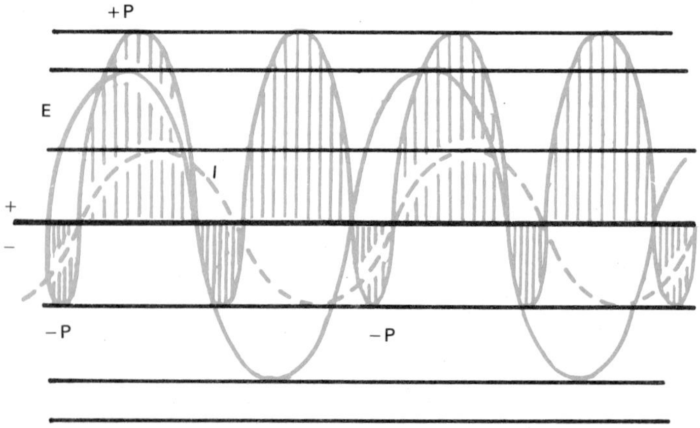

Fig. 9-33. The relationship between voltage, current, and power in the circuit described in the text.

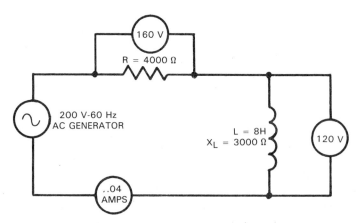

Fig. 9-34. Circuit having an 8 henry choke, 4000 ohm resistor connected across a 200 volt, 60 Hz ac source.

Inductance and RL Circuits

Find the current in the circuit:
$$I = \frac{E}{Z} \text{ or } \frac{200}{5000} = .04 \text{ amps}$$

Find the voltage drop across R and X_L:
$$E_R = I \times R = .04 \times 4000 = 160 \text{ volts}$$
$$E_{X_L} = I \times X_L = .04 \times 3000 = 120 \text{ volts}$$

The sum of the voltage drops does not equal the applied voltage. The reason for this is that the two voltages are 90 degrees out of phase. This requires vector addition:

$$E_S = \sqrt{160^2 + 120^2}$$
$$= \sqrt{25,600 + 14,400}$$
$$= \sqrt{40,000}$$
$$= 200 \text{ volts (approximately)}$$

Find the phase angle θ between I and E:

$$\cos \theta = \frac{R}{Z}$$
$$= \frac{4000 \, \Omega}{5000 \, \Omega}$$
$$= .8$$

.8 is the cosine of angle 36 degrees.

Find the true power and apparent power:
Apparent power = $I \times E$
= $.04 \times 200$
= 8 volt-amperes
True power = $I \times E \times \cos \theta$
= $.04 \times 200 \times .8$
= 6.4 watts

Note that some figures have been approximated to make the problem clearer.

PARALLEL RL CIRCUIT

In a circuit containing a resistance and an inductance in parallel, the voltage of each circuit element will be the same as the source voltage. Further, there will be no phase difference among them. This is because they are all in parallel.

There will be a phase difference among the total and branch currents, however. Current will be in phase with voltage in the resistive branch. It will lag the voltage across the inductor by 90°. Total current will lag the source voltage by some angle between zero and 90°.

In the parallel RL circuit, we do not find impedance through a vector sum. Instead, we apply Ohm's law:

$$Z = \frac{E_S}{I_T}$$

As stated, voltage in the parallel RL circuit is in phase with I_R, and it leads I_L by 90°. Thus, we can say that I_R leads I_L by 90°. The parallel RL circuit, then, has two current components—I_R and I_L. Both of these can be represented by phasors. Since they are out of phase, we cannot simply add the two components together to figure the total circuit current. We must find their *phasor sum*:

$$I_T = \sqrt{I_R^2 + I_L^2}$$

Note that the phasor for I_L is below the horizontal reference. This is because I_L lags voltage—the horizontal reference for the parallel RL circuit. (Since a phasor, or rotating vector, rotates counterclockwise, a *lagging* phasor would be behind, or clockwise to, a *leading* phasor.)

The phase angle, as stated, represents the phase displacement between current and voltage resulting from a reactive circuit element. For the parallel RL circuit, it is found on the *current* phasor diagram. The horizontal reference of this circuit is voltage, since it is common to all circuit elements.

In the current phasor diagram, we do not have a voltage phasor. We use I_R as the horizontal component since it is in phase with voltage. Using this diagram, the phase angle may be found from:

$$\theta = \arctan \frac{I_L}{I_R}$$

Using the circuit with assigned values in Fig. 9-35, determine the phase angle between applied voltage and current. Draw the current phasor diagram and find the circuit impedance.

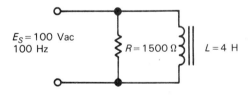

Fig. 9-35. A parallel RL circuit.

Step 1. Compute the value of inductive reactance, X_L.
$$X_L = 2\pi f L = 6.28 \times 100 \text{ Hz} \times 4 \text{ H} = 2512 \, \Omega$$

Step 2. Compute branch currents. Using Ohm's law:

$$I_R = \frac{E_S}{R} = \frac{100 \text{ V}}{1500 \, \Omega} = 0.067 \text{ A}$$

$$I_L = \frac{E_S}{X_L} = \frac{100 \text{ V}}{2512 \, \Omega} = 0.04 \text{ A}$$

Step 3. Determine the phase angle to see by how much the circuit current lags the voltage.

$$\theta = \arctan \frac{I_L}{I_R} = \arctan \frac{0.04 \text{ A}}{0.067 \text{ A}}$$

$$= \arctan 0.597$$

$$= 30.8°$$

Step 4. Draw the current phasor diagram. Use any convenient scale. Remember that I_R is drawn as the horizontal component. I_L is drawn downward at 90° from I_R since it lags voltage—the horizontal *reference*. See Fig. 9-36.

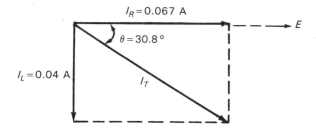

Fig. 9-36. Current phasor diagram.

Step 5. Find the total circuit current.

$$I_T = \sqrt{I_R^2 + I_L^2} = \sqrt{(0.067 \text{ A})^2 + (0.04 \text{ A})^2}$$

$$= 0.078 \text{ A}$$

Step 6. Find the impedance of the circuit. Using Ohm's law:

$$Z = \frac{E_S}{I_T} = \frac{100 \text{ V}}{0.078 \text{ A}} = 1282 \text{ }\Omega$$

Once again, it is important to understand how the various quantities would be affected by a change in frequency. The effects of changing frequency are summarized in Fig. 9-37. It shows how values change in a parallel RL circuit as frequency is changed and inductance is held at a constant value.

REVIEW QUESTIONS FOR SECTION 9.3

1. What is reactance?
2. The actual opposition to a change in current offered by an inductor is _____ _____.
3. Find X_L in the following diagram.

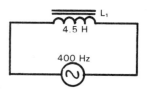

4. The actual power used by a circuit is the _____ power.
5. Find the values in the following figure.
   a. X_L.
   b. Z.
   c. I (lags, leads) E by _____ degrees.

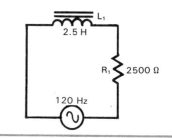

SUMMARY

1. Inductance is that property of a circuit which opposes any change in current flow.
2. Inductors give inductance to a circuit. Other names for them are coils, reactors, and chokes.
3. The unit for inductance is the henry (H).
4. Some factors which affect inductance are:
   a. Number of turns of coil.
   b. Diameter of coil.
   c. Core material of coil.
   d. Type of winding.
   e. Spacing between the windings.

f	X_L	θ	Z	I_R	I_L	I_T
increases ↑	increases ↑	decreases ↓	increases ↑	remains constant	decreases ↓	decreases ↓
decreases ↓	decreases ↓	increases ↑	decreases ↓	remains constant	increases ↑	increases ↑

Fig. 9-37. Table shows the effect on various values in a parallel RL circuit as frequency is changed and the inductance value is held constant.

Inductance and RL Circuits

5. The time constant of a coil is the amount of time it takes for the current to rise from 0 to 63.2 percent of its maximum value. The formula for time constant is:

$$t = \frac{L \text{ (in henrys)}}{R \text{ (in ohms)}}$$

6. Mutual inductance is two or more coils sharing the energy of one. It is the basis for the operation of a transformer.
7. In a transformer, energy is fed in the primary winding (input) and is taken out of the secondary winding (output).
8. A step-up transformer, steps up voltage. A step-down transformer steps down voltage.
9. In a transformer, if the secondary voltage is stepped up a certain amount, the secondary current will be stepped down proportionally.
10. The formula for turns ratio is:

$$\frac{E_{primary}}{E_{secondary}} = \frac{N_{primary}}{N_{secondary}} = \frac{I_{secondary}}{I_{primary}}$$

11. A center tap in a transformer allows the secondary winding to provide two equal voltages. One will be 180 degrees out of phase with the other.
12. The purpose of isolation in a transformer is to electrically separate the secondary winding from the primary winding. Then there is no physical connection between the two windings. This allows magnetic coupling of the field from the primary to the secondary winding.
13. There are three basic types of transformer losses. The three losses are: copper losses, eddy current losses, and hysteresis losses.
14. The following formulas are used to compute inductors in a circuit (all inductors must be the same measuring unit before using the formula):

IN SERIES, $L_T = L_1 + L_2 + L_3 \ldots$

IN PARALLEL, $\frac{1}{L_T} = \frac{1}{L_1} + \frac{1}{L_2} + \frac{1}{L_3} \ldots$

or

$$L_T = \frac{L_1 \times L_2}{L_1 + L_2}$$

15. Inductive reactance (X_L) is the opposition offered to a change in current flow by an inductor. It is measured in ohms.
16. The formula for inductive reactance is $X_L = 2\pi fL$, where π equals 3.14, f equals a frequency in hertz, and L equals inductance in henrys.
17. True power is power actually used in a circuit. Apparent power is the product of the effective voltage times the effective current.
18. Power factor is the relationship between true power and apparent power.
19. The total opposition to a change of current flow in an inductive and resistive circuit is the impedance (Z). It is measured in ohms.
20. There are three methods for computing impedance: Pythagorean theorem, trigonometry, and graph method.

TEST YOUR KNOWLEDGE, Chapter 9

Please do not write in the text. Place your answers on a separate sheet of paper.

1. Draw a graph showing the rise and decay of current in a RL circuit.
2. What is the formula for determining time constant?
3. Draw the symbol for a power transformer.
4. The primary of a transformer has 200 turns of wire and the secondary has 800 turns. 117 volt ac is applied to primary. What is the secondary voltage?
5. A load attached to the secondary of the transformer in Question 8 draws a 100 mA current. What is the primary current?
6. List the three losses associated with transformers. Explain one of these losses.
7. The key point of transformer action is that as _____ increases in the secondary, _____ decreases in the secondary.
8. Reactance is measured in _____.
9. In a series RL circuit, L = 2 H, R = 500 Ω, E_S = 100 V, 60 Hz ac. Find:
   a. X_L.
   b. Z.
   c. I.
   d. E_R.
   e. E_L.
   f. θ.
   g. PF.
   h. True power.
   i. Apparent power.
10. On graph paper, draw vectors and sine curves representing current and voltage of the circuit in Question 13. With a red pencil, draw the power curve.

FOR DISCUSSION

1. Why do power companies use high voltages in cross country transmission lines?
2. What is the difference between apparent power and true power?
3. Discuss the differences between a step-up and step-down transformer.

Electricity and Electronics

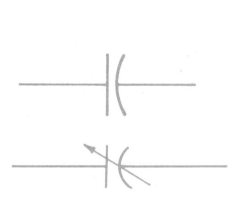

Top. Schematic symbol for fixed capacitor. Bottom. Schematic symbol for variable capacitor.

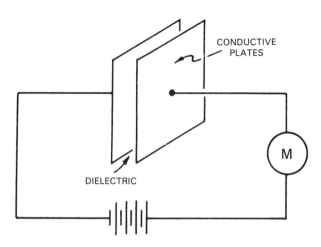

A capacitor is a device that temporarily stores an electric charge. It accepts or returns this charge in order to maintain a constant voltage.

DEVRY INSTITUTES

Some forms of computer memory make use of small capacitors.

Capacitance is useful in electric and electronic circuits. Filters and memory devices depend on capacitors. Capacitors are used to provide radio transmission and reception. In most circuits, 10 to 20 percent of the components are capacitors.

Chapter 10

CAPACITANCE AND RC CIRCUITS

After studying this chapter, you will be able to:
- *Define capacitance and capacitor.*
- *Explain how a capacitor behaves in a circuit.*
- *Discuss the effect of capacitance on an ac circuit.*
- *Describe the results of combining capacitance and resistance in a circuit.*
- *Identify common applications for capacitors in electronic equipment.*

10.1 CAPACITANCE AND THE CAPACITOR

Recall that inductance is the property of a circuit that opposes any change in current. CAPACITANCE, on the other hand, opposes any change in voltage. A CAPACITOR is a device that temporarily stores an electric charge. It accepts or returns this charge in order to maintain a constant voltage.

The capacitor is made of two plates of conductive material, separated by insulation. This insulation is called a DIELECTRIC, Fig. 10-1. The plates are connected to a dc voltage source. The circuit appears to be an open circuit because the plates do not contact each other. However, the meter in the circuit will show some current flow when the switch is closed.

Study Fig. 10-2. As the switch is closed, electrons from the negative terminal of the source flow to one plate of the capacitor. These electrons repel the electrons from the second plate (like charges repel) and will be drawn to the positive terminal of the source. The capacitor is now charged to the same potential as the source and is opposing the source voltage. If the capacitor is removed from the circuit, it will remain charged. The energy is stored within the electric field of the capacitor.

It is important to remember that in the circuit in Fig. 10-2, no electrons flowed through the capacitor. This

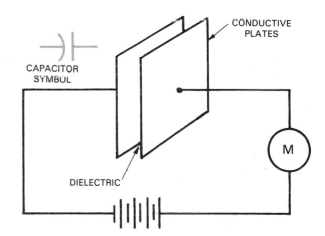

Fig. 10-1. Basic form of a capacitor.

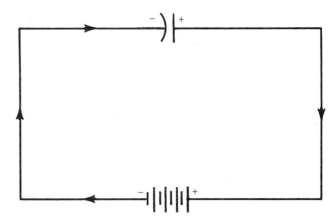

Fig. 10-2. The capacitor charges to the source voltage.

is because a capacitor blocks direct current. However, one plate did become negatively charged and the other positively charged. A strong electric field existed between them.

149

Insulating or dielectric materials vary in their ability to support an electric field. This is known as the DIELECTRIC CONSTANT of the material. The constants of various materials are shown in Fig. 10-3. These numbers are based on comparison with the constant of dry air. The value of dry air is 1.

LESSON IN SAFETY: Many large capacitors in radios, TVs, and other electronic equipment retain their charge even after power is turned off. Discharge these capacitors by shorting terminals to the chassis with an insulated screwdriver. If this is not done, the voltages may destroy test equipment, and those working on the equipment may receive a severe shock!

Capacitance is determined by the number of electrons that may be stored in the capacitor for each volt of applied voltage. Capacitance is measured in FARADS. A farad represents a charge of one coulomb which raises the potential one volt. This equation is,

$$C \text{ (in farads)} = \frac{Q \text{ (in coulombs)}}{E \text{ (in volts)}}$$

Capacitors used in electronic work have capacities measured in microfarads (1/1,000,000 farad) and picofarads (1/1,000,000 of 1/1,000,000 farad). Microfarad is commonly written as μF. It is sometimes written as mfd. Picofarad is written as pF. A conversion chart for these units is shown in Fig. 10-4.

Capacitance is determined by:
1. The material used as a dielectric.
2. The area of the plates. (The larger the plate area, the greater the capacitance.)
3. The distance between the plates. (The greater the distance, the lesser the capacitance.)

These factors are related in the mathematical formulas,

$$C \text{ (in pF)} = .225 \frac{KA(n-1)}{d},$$

where K equals the dielectric constant, A equals the area of one side of one plate in square inches, d equals the distance between plates in inches, and n equals the number of plates.

This formula points out the following facts:
1. Capacity INCREASES as the area of the plates increases, or as the dielectric constant increases.
2. Capacity DECREASES as the distance between plates increases.

TYPES OF CAPACITORS

Capacitors are made in hundreds of sizes and types. Several of these will be discussed in the following section.

MATERIAL	DIELECTRIC CONSTANT
AIR	1
WAX PAPER	3.5
MICA	6
GLASS (WINDOW)	8
PURE WATER	81

Fig. 10-3. Dielectric constants. Larger numbers are better able to support electric fields.

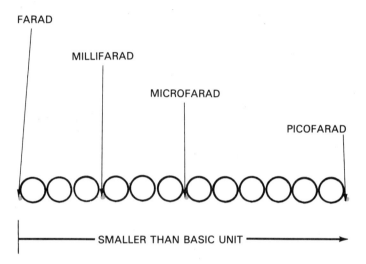

Fig. 10-4. Prefixes used with the farad.

VARIABLE CAPACITORS consist of metal plates that join together as the shaft turns, Fig. 10-5. The stationary plate is called the STATOR. The rotating plate is called the ROTOR.

Variable capacitors are used in radio transmitters and receivers. They are the component that is turned when tuning a radio. This capacitor is at maximum capacity when the plates are fully meshed. The schematic symbol for a variable capacitor is shown in Fig. 10-6.

FIXED PAPER CAPACITORS are made of layers of tinfoil. The dielectric is made of waxed paper. See Figs. 10-7 and 10-8.

The wires extending from the ends connect to the foil plates. the assembly is tightly rolled into a cylinder and sealed with special compounds. Some capacitors are enclosed in plastic for rigidity. These capacitors can withstand severe heat, moisture, and shock.

RECTANGULAR OIL FILLED CAPACITORS, Fig. 10-9, are hermetically sealed in metal cans. They are oil filled and have very high insulation resistance. This type of capacitor is used in power supplies of radio transmitters and other electronic equipment.

Capacitance and RC Circuits

Fig. 10-5. Variable capacitors are made in many types and sizes.

SCHEMATIC SYMBOL FOR VARIABLE CAPACITOR

Fig. 10-6. Schematic symbol for a variable capacitor.

Fig. 10-7. A fixed paper capacitor, enclosed in plastic.

SCHEMATIC SYMBOL FOR FIXED CAPACITOR

Fig. 10-8. The schematic symbol for a fixed capacitor.

Fig. 10-9. Rectangular oil filled capacitor.

CAN TYPE ELECTROLYTIC CAPACITORS use different methods of plate construction, Fig. 10-10.

Some capacitors have aluminum plates and a wet or dry electrolyte of borax or carbonate. A dc voltage is applied during manufacturing. Electrolytic action creates a thin layer of aluminum oxide that deposits on the positive plate. This coating insulates the plate from the electrolyte. The negative plate is connected to the electrolyte. The electrolyte and positive plates form the capacitor. These capacitors are useful when a large amount of capacity is needed in a small space.

151

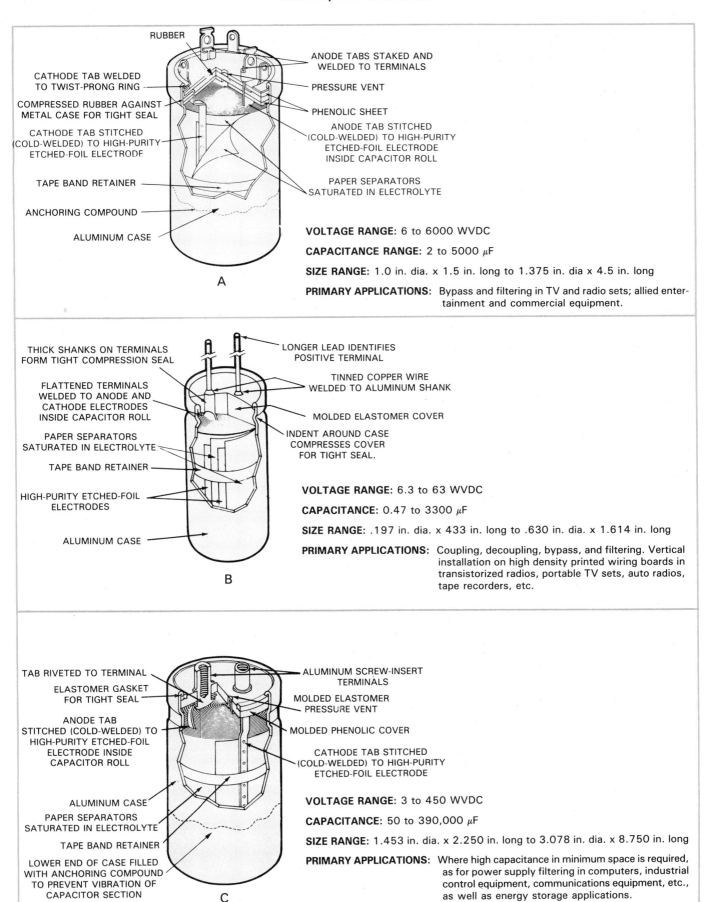

Fig. 10-10. This chart shows a number of can type electrolytic capacitors, along with their voltage ratings and common uses. A—Basic can type. B—Single-ended. C—Cylindrical.

Capacitance and RC Circuits

Polarity of these capacitors is very important. A reverse connection can destroy them. The cans may contain from one to four different capacitors. The terminals are marked by △, ◠, ▢, and ▭ symbols. The metal can is usually the common negative terminal for all the capacitors. A special metal and fiber mounting plate is supplied for easy installation on a chassis.

TUBULAR ELECTROLYTIC CAPACITOR construction is similar to the can type, Fig. 10-11. The main advantage of these tubular capacitors is their smaller size. They have a metal case enclosed in an insulating tube. They are also made with two, three, or four units in one cylinder.

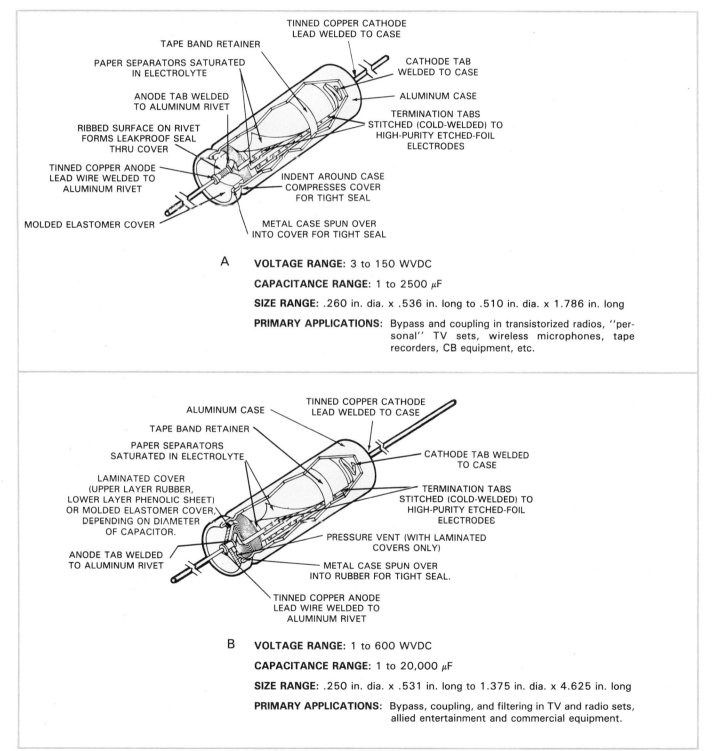

Fig. 10-11. Tubular electrolytic capacitor. A—Standard tubular. B—Economy tubular.

A very popular small capacitor used a great deal in radio and TV work is the CERAMIC CAPACITOR, Fig. 10-12. It is made of a special ceramic dielectric. The silver plates of the capacitor are fixed on the dielectric. The whole component is treated with special insulation that can withstand heat and moisture.

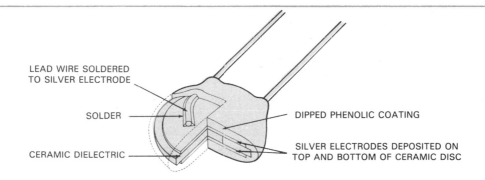

GENERAL APPLICATION

VOLTAGE RANGE: 100 to 7500 WVDC

CAPACITANCE RANGE: 1.5 pF to 0.1 µF

SIZE RANGE: .296 in. dia. x .156 in. thick to .937 in. dia. x .375 in. thick

PRIMARY APPLICATIONS: Entertainment, commercial, and industrial equipment where temperature coefficient is not important, such as bypass and coupling in IF and RF circuits.

TEMPERATURE-COMPENSATING

VOLTAGE RATING: 1000 WVDC

CAPACITANCE RANGE: 1 to 2200 pF

SIZE RANGE: .296 in. dia. x .156 in. thick to .937 in. dia. x .234 in. thick

PRIMARY APPLICATIONS: Standard, and higher quality, equipment where capacitance change with temperature is undesirable: tuned circuits, in TV, FM, VHF, and AM receivers, where oscillator drift must be eliminated.

HIGH-K

VOLTAGE RANGE: 16 to 1000 WVDC

CAPACITANCE RANGE: 0.0001 to .47 µF

SIZE RANGE: .250 in. dia. x .156 in. thick to .875 in. dia. x .187 in. thick.

PRIMARY APPLICATIONS: Bypass and coupling in entertainment, commercial, and industrial equipment, where additional capacitance will not affect circuit operation.

A

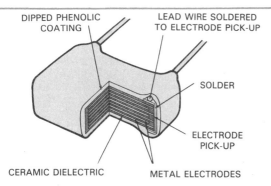

VOLTAGE RANGE: 50 and 100 WVDC

CAPACITANCE RANGE: 10 pF to 4.7 µF

SIZE RANGE: .200 in. x .200 in. x .125 in. to .500 in. x .500 in. x .238 in.

PRIMARY APPLICATIONS: Use in circuitry where capacitors with EIA Characteristics Z5U, X7R, and COG must be selected to meet specific requirements.

B

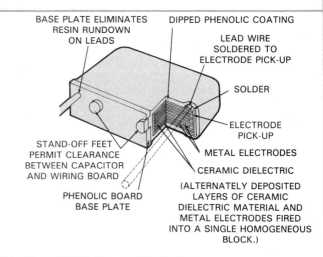

VOLTAGE RATINGS: 50 and 100 WVDC

CAPACITANCE RANGE: 51 pF to 4.7 µF

SIZE RANGE: .225 in. x .250 in. x .125 in. to .500 in. x .525 in. x .260 in.

PRIMARY APPLICATIONS: Similar to standard resin-dipped types, with addition of terminal base and standoff feet for improved performance on printed wiring boards.

C

SPRAGUE PRODUCTS CO.

Fig. 10-12. This chart shows several types of ceramic capacitors. Also listed are their voltage ratings and common uses. A—Disc type. B—Multi-layer resin dipped. C—Multi-layer terminal base.

Capacitance and RC Circuits

Fig. 10-13. Mica capacitors.

Fig. 10-14. Several types of trimmer capacitors.

MICA CAPACITORS, Fig. 10-13, are small capacitors. They are made by stacking tinfoil plates together with thin sheets of mica as the dielectric. The assembly is then molded into a plastic case.

A TRIMMER CAPACITOR, Fig. 10-14, is a type of variable capacitor. The adjustable screw compresses the plates and increases capacitance. Mica is used as a dielectric.

Trimmer capacitors are used where fine adjustments of capacitance are needed. They are used with larger capacitors and are connected in parallel with them.

To adjust trimmer capacitors, turn the screw with a special fiber or plastic screwdriver called an ALIGNMENT TOOL. Do not use a screwdriver. The capacitance effect would cause inaccurate adjustment.

TANTALUM CAPACITORS are a new type of capacitor, similar to aluminum electrolytic capacitors, Fig. 10-15. However, tantalum capacitors use tantalum, not aluminum, for the electrode.

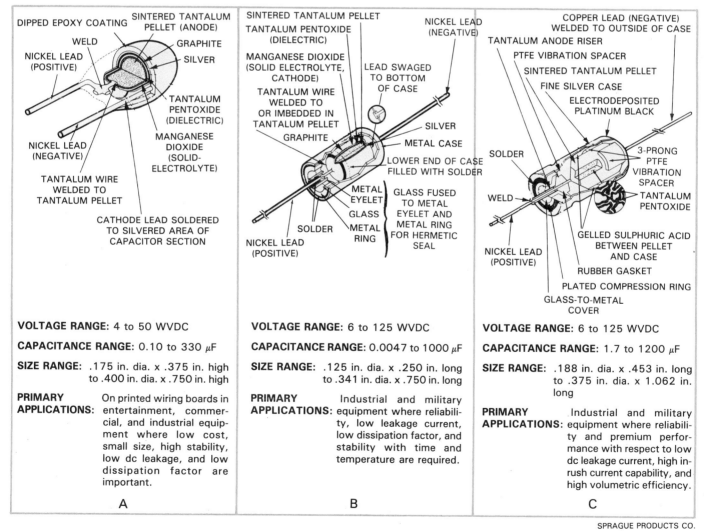

Fig. 10-15. This chart shows three types of tantalum capacitors. Voltage ratings and common uses are also listed.
A—Epoxy dipped solid electrolyte. B—Hermetically sealed solid electrolyte. C—Hermetically sealed sintered-anode.

155

Tantalum capacitors have three distinct advantages that make them quite useful. They have a larger capacitance over a smaller area. This makes them ideal for smaller circuits. Tantalum capacitors have a long shelf life. And, because tantalum resists most acids, tantalum capacitors have less leakage current.

REVIEW QUESTIONS FOR SECTION 10.1

1. A capacitor is made of two _____ of conductive material, separated by an insulator called the _____.
2. The ability of an insulator to support an electric field is known as the _____ _____.
3. Find the missing values for the following circuits:
   a. 20 volts, 1000 μF.
   b. 10 coulombs, 40 volts.
   c. 200 coulombs, 750 volts.
4. Convert the following units:
   a. 6240 pF = _____ μF.
   b. .05 μF = _____ pF.
   c. 150 μF = _____ F.
   d. .005 F = _____ μF.
5. Name the three items that determine capacitance.
6. List six types of capacitors.

10.2 TRANSIENT RESPONSE OF THE CAPACITOR

Recall that the response of current and voltage in a circuit immediately after a change in applied voltage is called TRANSIENT RESPONSE.

Refer to Fig. 10-16. A capacitor and a resistor are connected in series across a voltage source. When the switch is closed, maximum current will flow. Current gradually decreases until the capacitor has reached its full charge, which is the same as the applied voltage.

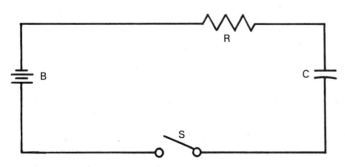

Fig. 10-16. This series RC circuit demonstrates the transient response of a capacitor.

However, the voltage initially is zero across the capacitor. When the switch is closed, this voltage gradually builds up to the value of the source voltage. This is shown in Fig. 10-17. When the switch is opened, the capacitor remains charged. Theoretically, it would remain charged. But there is always some leakage through the dielectric, and during a period of time the capacitor will discharge itself.

In Fig. 10-18, the series combination of charged capacitor and resistor are short-circuited by providing a discharge path. Because there is no opposing voltage, the discharge current will instantly rise to maximum and gradually fall off to zero. The combined graph of the charge and discharge of C is shown in Fig. 10-19.

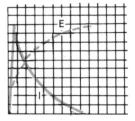

Fig. 10-17. Current and voltage in the series RC circuit.

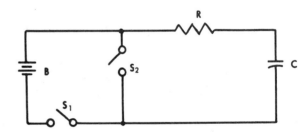

Fig. 10-18. A short circuit occurs in the RC circuit when switch S₂ is closed.

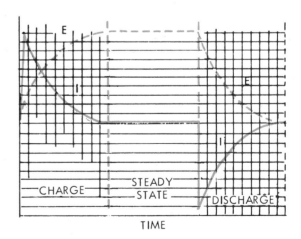

Fig. 10-19. This combination graph shows the rise and decay of current and voltage in the series RC circuit.

Capacitance and RC Circuits

Voltages also appear across R and C in this circuit. The voltage across R is a result of current flow, E = IR. Then the maximum voltage appears across R when maximum current is flowing. This is right after the switch is closed in Fig. 10-16, and after the discharge switch is closed in Fig. 10-18. In either case, the voltage across R drops off or decays as the capacitor approaches charge or discharge. The graph of the voltage across R is drawn in Fig. 10-20.

RC TIME CONSTANT

During the charge and discharge of the series RC network just outlined, a period of time elapsed. This time is indicated along the base or x-axis of the graphs in Fig. 10-19 and 10-20. To review Chapter 6, the amount of time needed for the capacitor to charge or discharge 63.2 percent is known as the TIME CONSTANT of the circuit. The formula to determine the time constant in RC circuits is:

t (in seconds) = R (in ohms) × C (in farads)

For complete charge or discharge, five time constant periods are required. Assuming a source voltage equal to 100 volts, Fig. 10-21 shows the time constant, percentage, and voltage.

For example, a .1 microfarad capacitor is connected in series with a megohm (1,000,000 ohm) resistor across a 100 volt source. How much time will elapse during the charging of C? (Using powers of ten is explained in the Appendix.)

$$t = .1 \times 10^{-6} \times 10^{6} = .1 \text{ second}$$

One tenth of a second is the time constant. C would charge to 63.2 volts during this time. At the end of five time constants, or 5 × .1 = .5 seconds, the capacitor is fully charged.

Set up the circuit in Fig. 10-22. This neon lamp will remain out until a certain voltage is reached. At this voltage, called the IGNITION or FIRING VOLTAGE, the lamp glows and offers little resistance in the circuit.

In this circuit, capacitor C charges through resistance R. When the voltage across C develops to the ignition voltage, the lamp glows. C is quickly discharged. This cycle repeats over and over, causing the neon light to flash. The frequency of the flashes may be changed by varying either the value of C or R. A telephone with a neon accent is shown in Fig. 10-23.

Time constants have many uses in electronic circuits. Timing circuits are used in industry to control the sequence and duration of machine operations. A photographic enlarger uses a time delay circuit to control exposure time.

RC BLINKER PROJECT

You can produce an RC blinker project that has LED's that flash alternately. See the schematic in Fig. 10-24.

The project is easy to build. Using the parts list, Fig. 10-25, assemble the parts according to Fig. 10-24. Install the complete RC circuit and cover the circuit with papier-mâché.

TIME CONSTANT	PERCENT OF VOLTAGE	E CHARGING	E DISCHARGING
1	63.2%	63.2 V	36.8 V
2	86.5%	86.5 V	13.5 V
3	95.0%	95.0 V	5 V
4	98.0%	98.0 V	2 V
5	99+%	99+ V	1 V

Fig. 10-21. A source voltage of 100 volts will create the time constant, percentage, and voltage shown.

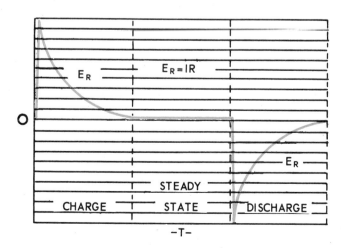

Fig. 10-20. This graph shows the voltage drop across R as the capacitor is charged and discharged.

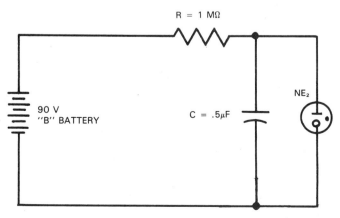

Fig. 10-22. This flashing circuit is called a relaxation oscillator.

Electricity and Electronics

Fig. 10-23. Neon lights are used in many new products. Here, a neon light highlights this designer phone.

```
1—LM3909 INTEGRATED CIRCUIT
1—100 Ω, 1/2 WATT RESISTOR
1—300 Ω, 1/2 WATT RESISTOR
1—510 Ω, 1/2 WATT RESISTOR
1—4,300 Ω, 1/2 WATT RESISTOR
1—400 µF, ELECTROLYTIC CAPACITOR (15 WVDC)
2—LEDs
POWER SOURCE: 10-15 VOLTS DC
PC BOARD
8 PIN IC SOCKET
```

Fig. 10-25. Parts list for RC blinker project.

PARALLEL AND SERIES CIRCUITS

When capacitors are connected in parallel, Fig. 10-26, the total capacitance is equal to the sum of the individual capacitances,

$$C_T = C_1 + C_2 + C_3 \ldots$$

When two capacitors are connected in series, Fig. 10-27, total capacitance is,

$$C_T = \frac{C_1 \times C_2}{C_1 + C_2}$$

When two or more capacitors are connected in series, Fig. 10-28, then,

$$\frac{1}{C_T} = \frac{1}{C_1} + \frac{1}{C_2} + \frac{1}{C_3}$$

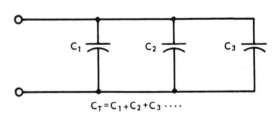

Fig. 10-26. Capacitors in parallel.

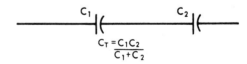

Fig. 10-27. Two capacitors in series.

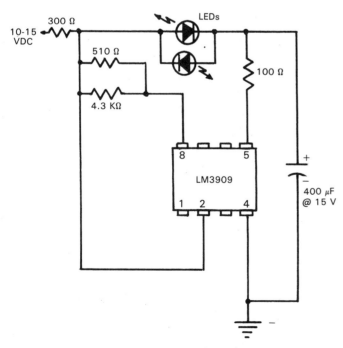

Fig. 10-24. Schematic for RC blinker project.

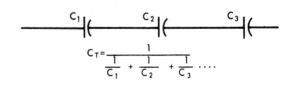

Fig. 10-28. Two or more capacitors in series.

Capacitance and RC Circuits

The dielectrics used for capacitors can only withstand certain voltages. If this voltage is exceeded, the dielectric will break down and arcing will result. This maximum voltage is known as the WORKING VOLTAGE (WV). Exceeding the working voltage can cause a short circuit. This will ruin other parts of the circuit connected to the dielectric.

Increased voltage ratings require special materials and thicker dielectrics. When a capacitor is replaced, check its capacitance value and dc working voltage.

When a capacitor is used in an ac circuit, the working voltage should safely exceed the peak ac voltage. For example, a 120 volt effective ac voltage has a peak voltage of 120 V × 1.414 = 169.7 volts. Any capacitors used must exceed 169.7 volts.

To review, refer to Fig. 10-30. As the ac voltage starts to rise, the current flow is maximum because C is in a discharged state. As C becomes charged to the peak ac voltage, the charging current drops to zero (A). As the voltage begins to drop, the discharging current begins to rise in a negative direction and reaches maximum at the point of zero voltage (B). This phase difference keeps going throughout each cycle. In a purely capacitive circuit, the current leads the voltage by an angle of 90 degrees.

The size of the current in the circuit depends upon the size of the capacitor. Larger capacitors (more capacitance) require a larger current to charge them. Frequency of the ac voltage also affects current flow.

REVIEW QUESTIONS FOR SECTION 10.2

1. Find the time constants for the following circuits:
   a. 500 µF capacitor, 2500 ohms resistance.
   b. 225 µF capacitor, 3200 ohms resistance.
   c. 67 mF capacitor, 180 ohms resistance.
2. Give the total capacitance for the circuits below.

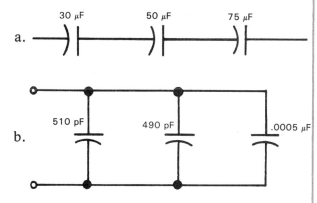

3. Can a capacitor which has a working dc voltage of 25 (WVDC) be used in a circuit which requires 10 WVDC? Explain your answer.

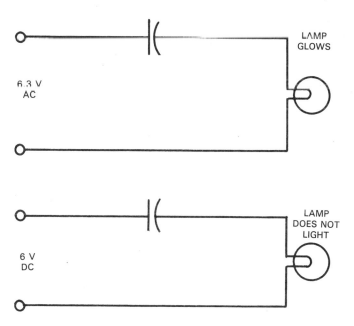

Fig. 10-29. A light will glow when connected to an ac source. The capacitor blocks direct current.

10.3 CAPACITANCE IN AC CIRCUITS

When an ac voltage is applied to a capacitor, the plates charge to one polarity during the first half-cycle. During the next half-cycle, they charge to the opposite polarity. An ac meter in the circuit shows a current flowing at all times.

To prove this, connect a light and capacitor in series to a 6 volt dc source, Fig. 10-29. Does the light glow? Now connect the same circuit to a 6 volt ac source. Notice that the light burns dimly. This shows that alternating current is flowing as a result of the alternate charging of the capacitor.

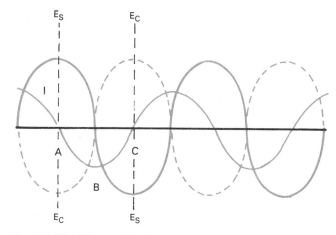

Fig. 10-30. When a circuit is connected to ac, applied voltage, current, and voltage across C appear as shown.

Current flow depends upon the rate of charge and discharge of the capacitor. As the frequency of the ac is increased, current increases. These links are stated in the formula,

$$X_C = \frac{1}{2\pi fC}$$

where X_C equals the capacitive reactance in ohms, f equals the frequency in hertz, and C equals the capacitance in farads.

Like inductive reactance, resistance to the flow of an alternating current resulting from capacitance is called CAPACITIVE REACTANCE. It is measured in ohms, like dc resistance. The formula states that:
1. As the frequency increases, X_C decreases.
2. As capacitance increases, X_C decreases.

What, then, is the reactance of a 10 µF capacitor working in a circuit at a frequency of 120 hertz?

$$X_C = \frac{1}{2\pi \times 120 \times 10 \times 10^{-6}}$$
$$= \frac{1}{6.28 \times 120 \times 10 \times 10^{-6}}$$
$$= \frac{1}{.007536} = 132.7\ \Omega$$

The reactance of a .1 µF capacitor as frequency is varied can be seen in Fig. 10-31. As frequency is changed to 50, 100, 1000, and 10,000 Hz, each reactance is computed using the formula,

$$X_C = \frac{1}{2\pi fC}$$

In Fig. 10-32, the frequency is held constant at 100 Hz. The reactance is plotted for capacitors of .5 µF, .05 µF, and .001 µF. These are common capacitor sizes used in electronic work. They are used in filtering, coupling, and bypassing networks. Fig. 10-33 shows a meter used to test the value of capacitors.

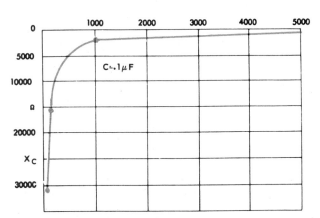

Fig. 10-31. As frequency increases, capacitive reactance decreases.

POWER IN CAPACITIVE CIRCUITS

When a capacitor is discharged, the energy stored in the dielectric is returned to the circuit. This is similar to inductance, which returns the energy stored in the magnetic field to the circuit. In either case, electrical energy is used temporarily by the reactive circuit. This power in a capacitive circuit is also called WATTLESS POWER.

In Fig. 10-34, the voltage and current waveforms are drawn for a circuit containing pure capacitance. The power waveform results from plotting the products of the instantaneous voltage and current at selected points.

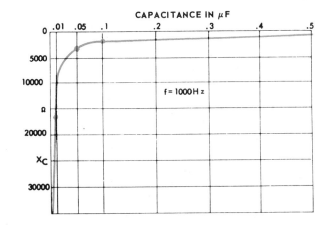

Fig. 10-32. As capacitance increases, reactance decreases.

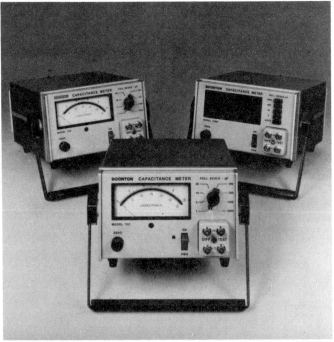

BOONTON ELECTRONICS

Fig. 10-33. Capacitance meters have either analog or digital reading displays.

Capacitance and RC Circuits

The power waveform shows that equal amounts of positive power and negative power are used by the circuit. This results in zero power being used. The TRUE POWER, or actual power used, then, is zero. (Refer to Chapter 9, Section 9.3, POWER IN INDUCTIVE CIRCUITS, and compare these circuits.)

The APPARENT POWER is equal to the product of the effective voltage and the effective current. Look at Fig. 10-35. An applied ac voltage to the capacitive circuit causes a 10 ampere current. The apparent power will equal,

100 V × 10 amps = 1000 volt-amperes

The ratio of true power to apparent power in an ac circuit is called the POWER FACTOR. It is found using trigonometry. It is the cosine of the phase angle between current and voltage. Mathematically,

Power factor (PF) = $\cos \theta$ = $\dfrac{\text{true power}}{\text{apparent power}}$

As an example, assuming our previous circuit is purely capacitive, determine the power factor and phase displacement.

$$PF = \cos \theta = \frac{0}{1000} = 0$$

The angle whose cosine is 0 is 90 degrees. This tells us that current and voltage in the purely capacitive circuit are 90 degrees out of phase.

RESISTANCE AND CAPACITANCE IN AN AC CIRCUIT

When resistance is present in a circuit, power is used. If the circuit contains only resistance, then the voltage and current are in phase. There is no phase angle θ and the power factor is one (Cos 0 degrees = 1). The apparent power equals the true power.

The circuit traits change when capacitance is added in series with the resistor. The capacitive reactance is also a force which resists the flow of an alternating current. Because the capacitive reactance causes a 90 degree phase displacement, the total resistance to an ac current must be the vector sum of the X_C and R. These vectors are drawn in Fig. 10-36. Assuming the circuit has 300 ohms resistance and 400 ohms capacitive reactance, then the resulting force is 500 ohms. This is called the IMPEDANCE, Z, of the circuit.

$$Z = \sqrt{R^2 + X_C^2}$$
$$Z = \sqrt{300^2 + 400^2}$$
$$= \sqrt{90{,}000 + 160{,}000}$$
$$= 500 \ \Omega$$

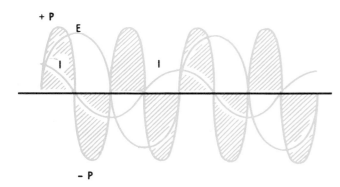

Fig. 10-34. These waveforms show current, voltage, and power in a purely capacitive circuit.

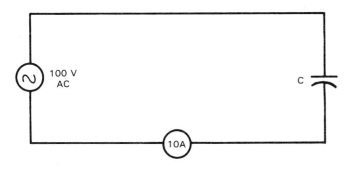

Fig. 10-35. Schematic for the theoretical capacitive circuit.

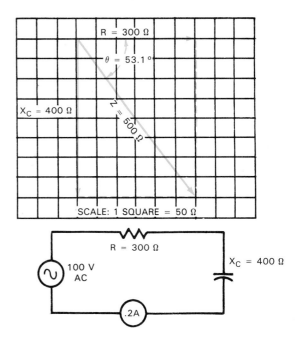

Fig. 10-36. Top. The vector relationship between R and X_C and the resulting vector, Z, for impedance. Bottom. The circuit for the problem in the text.

The angle between vector Z and vector R represents the phase displacement between the current and the voltage as a result of the reactive component. This is angle θ (theta). Since cosine θ is equal to the power factor and to $\frac{R}{Z}$,

$$PF = \cos \theta$$

$$\text{and } \cos \theta = \frac{R}{Z}$$
$$= \frac{300}{500} = .6$$

An angle with a cosine of .6 is 53.1 degrees (approximately). The current then leads the voltage by an angle of 53.1 degrees.

The true power in this circuit equals,

True power = Apparent power × Cos θ
= Apparent power × Power factor

The power factor is also the relationship between the true power and the apparent power. This may be expressed as,

Cos θ = Power factor
= $\frac{\text{True power}}{\text{Apparent power}}$

Using Ohm's Law for ac circuits, the current flowing in Fig. 10-36 (100 volts ac applied) is equal to,

$$I = \frac{E}{Z} \text{ or } I = \frac{100}{500} = .2 \text{ amps}$$

The apparent power is then,

I × E = .2 × 100 = 20 volt-amperes

True power equals,

True power = I × E × Cos θ
= .2 × 100 × Cos 53.1 degrees
= 20 × .6
= 12 watts

These values can be proved correct using this equation:

$$PF = \frac{\text{True power}}{\text{Apparent power}} = \frac{12}{20} = .6$$

All circuits have three electrical properties: R, L, and C. These properties come in many combinations, including R, RL, and RC networks. In Chapter 11, RCL circuits will be studied. Always review previous lessons if the theory covered is not clearly understood. The study of electricity and electronics requires a firm grasp of all previous lessons before progressing to more complex uses.

PARALLEL RC CIRCUIT

In a circuit containing a resistance and a capacitance in parallel, the voltage of each circuit element will be the same as the source voltage. Further, there will be no phase difference among them. This is because they are all in parallel.

There will be a phase difference among the total and branch currents, however. Current will be in phase with voltage in the resistive branch. It will lead the voltage across the capacitor by 90°. Total current will lead the source voltage by some angle between zero and 90°.

In the parallel RC circuit, we do not find impedance through a vector sum. Instead, we apply Ohm's law:

$$Z = \frac{E_S}{I_T}$$

As stated, voltage in the parallel RC circuit is in phase with I_R, and it lags I_C by 90°. Thus, we can say that I_R lags I_C by 90°. The parallel RC circuit, then, has two current components—I_R and I_C. Both of these can be represented by phasors. Since they are out of phase, we cannot simply add the two components together to figure the total circuit current. We must find their phasor sum:

$$I_T = \sqrt{I_R^2 + I_C^2}$$

Note that the phasor for I_C is above the horizontal reference. This is because I_C leads voltage—the horizontal reference for the parallel RC circuit.

For the parallel RC circuit, phase angle is found on the current phasor diagram. The horizontal reference of this circuit is voltage, since it is common to all circuit elements. On the current phasor diagram, the horizontal component is I_R since it is in phase with voltage. The phase angle, then, is the angle between I_R and the total current. This is the phase displacement resulting from the reactive element. In the parallel RC circuit, phase angle is:

$$\theta = \arctan \frac{I_C}{I_R}$$

Using the circuit with assigned values in Fig. 10-37, determine the phase angle between applied voltage and current. Draw the current phasor diagram and find the circuit impedance.

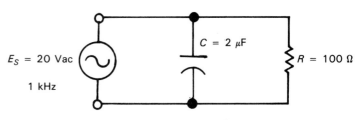

Fig. 10-35. A parallel RC circuit.

Step 1. Compute the value of capacitive reactance, X_C.

$$X_C = \frac{0.159}{fC} = \frac{0.159}{(1 \times 10^3 \text{ Hz})(2 \times 10^{-6} \text{ F})} \cong 80 \text{ }\Omega$$

Step 2. Compute branch currents. Using Ohm's law:

$$I_R = \frac{E_S}{R} = \frac{20 \text{ V}}{100 \text{ }\Omega} = 0.2 \text{ A}$$

$$I_C = \frac{E_S}{X_C} = \frac{20 \text{ V}}{80 \text{ }\Omega} = 0.25 \text{ A}$$

Step 3. Determine the phase angle to see by how much the circuit current leads the voltage.

$$\theta = \arctan \frac{I_C}{I_R} = \arctan \frac{0.25 \text{ A}}{0.2 \text{ A}} = \arctan 1.25 = 51.3°$$

Step 4. Draw the current phasor diagram. Use any convenient scale. Remember that I_R is drawn as the horizontal component. I_C is drawn upward at 90° from I_R since it leads voltage — the horizontal reference. See Fig. 10-38.

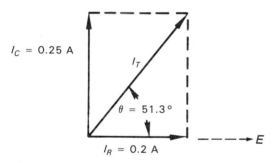

Fig. 10-38. Current phasor diagram.

Step 5. Find the total circuit current.

$$I_T = \sqrt{I_R{}^2 + I_C{}^2} = \sqrt{(0.2 \text{ A})^2 + (0.25 \text{ A})^2} = 0.32 \text{ A}$$

Step 6. Find the impedance of the circuit. Using Ohm's law:

$$Z = \frac{E_S}{I_T} = \frac{20 \text{ V}}{0.32 \text{ A}} = 62.5 \text{ }\Omega$$

REVIEW QUESTIONS FOR SECTION 10.3

1. In the equation $X_C = \frac{1}{2\pi fC}$, _____ equals the capacitive reactance in _____, _____ equals capacitance in _____, and _____ equals the frequency in _____.
2. Define capacitive reactance.

---

3. Find X_C in the circuit shown.

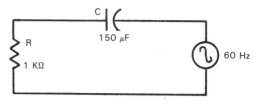

4. Find the following values using the circuit shown.
   a. Z.
   b. Cos θ.
   c. I _____ (leads, lags) E by _____°.

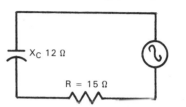

SUMMARY

1. Capacitance is that property of a circuit which opposes any change in voltage.
2. A capacitor is a device that temporarily stores an electric charge. It is made up of two plates of conductive material separated by insulation, called the dielectric.
3. Capacitance is measured in farads. Commonly used units are the microfarad (μF) and picofarad (pF).
4. Factors affecting capacitance are:
   a. Distance between plates.
   b. Plate area.
   c. Dielectric material.
5. There are many types of fixed and variable capacitors.
6. RC time constant can be found using the equation:
   t (in seconds) = R (in ohms) × C (in farads)
7. Formulas for capacitors in series and parallel are:
   Series: $\frac{1}{C_T} = \frac{1}{C_1} + \frac{1}{C_2} + \frac{1}{C_3}$
   and $C_T = \frac{C_1 \times C_2}{C_1 + C_1}$
   Parallel: $C_T = C_1 + C_2 + C_3 \ldots$
8. Working voltage is the maximum voltage that can be steadily applied to a capacitor without creating an arc.
9. The formula to find capacitive reactance is:

$$X_C = \frac{1}{2\pi fC}$$

10. Capacitive reactance (X_C) is opposition to the flow of an ac resulting from capacitance.

11. True power = I × E × Cos θ.
12. Apparent power = I × E.
13. Power factor = $\frac{\text{True power}}{\text{Apparent power}}$.
14. Z = $\sqrt{R^2 + X_C^2}$.
15. Cos θ = $\frac{R}{Z}$.

TEST YOUR KNOWLEDGE, Chapter 10

Please do not write in the text. Place your answers on a separate sheet of paper.
1. Define capacitance.
2. What safety precautions should be followed when working on radio and TV capacitors?
3. With respect to dielectric constants, the larger the number the _____ (better, worse) it supports an electric field.
4. Capacitance is determined by:
   a. Dielectric material.
   b. Plate area.
   c. Distance between plates.
   d. All of the above.
5. The time constant of an RC circuit is equal to _____ × _____.
6. Determine total capacitance for:
   a. Two capacitors connected in series with values of 680 μF and 200 μF.
   b. Four capacitors connected in parallel with values of 25 μF, 60 μF, 2 μF, and 4 μF.
   c. Five capacitors connected in parallel with values of 1150 pF, 97 pF, 130 pF, 1240 pF, and 50 pF.
7. The symbol for capacitive reactance is _____. It is measured in _____ and can be found using the formula _____.
8. A capacitor has a value of .1 μF. Find its reactance value at:
   a. 50 Hz.
   b. 100 Hz.
   c. 1000 Hz.
   d. 10 Hz.
9. As capacitance increases, X_C _____. As frequency increases, X_C _____.

MATCHING QUESTIONS: Match each of the following terms with their correct definitions.
   a. Apparent power.
   b. Power factor.
   c. True power.
   d. Cos θ.
   e. Impedance.
10. $\frac{R}{Z}$.
11. I × E.
12. I × E × Cos θ.
13. $\sqrt{R^2 + X_C^2}$.
14. $\frac{\text{True power}}{\text{Apparent power}}$.
15. A series circuit has 300 ohms resistance and .1 μF capacitance. It is connected to a 50 volt 400 Hz source. Find:
   a. X_C.
   b. Z.
   c. E_{X_C}.
   d. E_R.
   e. I.
   f. θ.
16. The frequency of the current in a circuit is 1000 Hz. What are the reactance values of the following capacitors?
   a. .5 μF.
   b. .1 μF.
   c. .05 μF.
   d. .01 μF.
   e. .001 μF.

FOR DISCUSSION

1. Choose one type of capacitor and research it. Make a report to the class, addressing construction of the capacitor and its various applications.
2. Discuss the effect of using a common screwdriver to adjust a trimmer capacitor. Why is it necessary to use a plastic or fiber screwdriver?
3. Research the reason why a capacitor blocks dc, but passes ac.
4. Why does a theoretical circuit containing only capacitance consume no power?
5. Should a capacitor with a working voltage of 150 volts be used in circuit with 117 volts ac supplied from your house circuit? Explain your answer.

Chapter 11

TUNED CIRCUITS AND RCL NETWORKS

After studying this chapter, you will be able to:
- *Explain resonant frequency and how it affects various RCL circuits.*
- *Discuss the characteristics of a series RCL circuit at its resonant frequency.*
- *Discuss the characteristics of a parallel RCL circuit at its resonant frequency.*
- *Describe filtering action.*
- *List four types of filters and name various applications of each.*

11.1 RCL NETWORKS

The principles of resistance, capacitance, and inductance should be reviewed before studying combination circuits.

1. In an ac circuit containing resistance only, the applied voltage and current are in phase. There is no reactive power. The power consumed by the circuit is equal to the product of volts times amperes.
2. In an ac circuit containing inductance only, the current lags the voltage by an angle of 90 degrees. They are not in phase. The power consumed by the circuit is zero.
3. In an ac circuit containing resistance and inductance, the current will lag the voltage by a phase angle of less than 90 degrees. The total resistive force is the vector sum of the resistance and the inductive reactance. This is the impedance of the circuit.
4. In an ac circuit containing capacitance only, the current leads the voltage by an angle of 90 degrees. The power consumed is zero.
5. In an ac circuit containing resistance and capacitance, the current will lead the voltage by an angle of less than 90 degrees. The impedance is equal to the vector sum of the resistance and the capacitive reactance.
6. In an ac circuit containing resistance, inductance, and capacitance (RCL circuit) in series, the resulting impedance is equal to the vector addition of R in ohms, X_L in ohms, and X_C in ohms.

RESONANCE

A special condition exists in an RCL circuit when it is energized at a frequency at which the inductive reactance is equal to the capacitive reactance ($X_L = X_C$). Since X_L increases as frequency increases, and X_C decreases as frequency increases, there is one frequency at which they are both equal. This is the RESONANT FREQUENCY of the circuit, or f_o. A series or parallel RCL circuit at resonant frequency is known as a TUNED CIRCUIT.

In the vector diagram, Fig. 11-1, X_L equals 100, X_C equals 100, and R equals 50. X_L and X_C are opposing each other because they are 180 degrees out of phase.

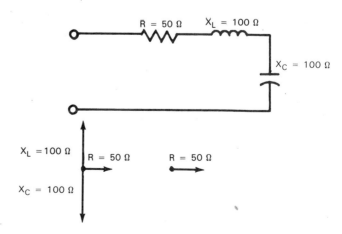

Fig. 11-1. Resonance exists when $X_L = X_C$ and the circuit appears as a resistive circuit.

The algebraic sum of these vectors is zero, so only resistance of 50 ohms remains. The current and voltage are in phase. This particular circuit frequency may be figured using the resonant frequency formula:

$$f_o = \frac{1}{2\pi\sqrt{LC}},$$

where f_o equals resonant frequency in hertz, L equals inductance in henrys and C equals capacitance in farads.

This formula is arrived at using the following steps:

At resonance, $X_L = X_C$ or,

$$2\pi fL = \frac{1}{2\pi fC}$$

Move $2\pi L$ to the right side of the equation by dividing both sides by $2\pi L$,

$$f = \frac{1}{(2\pi)^2 fLC}$$

Move f to the left side by multiplying both sides by f,

$$f^2 = \frac{1}{(2\pi)^2 LC}$$

Take square root of both sides of the equation,

$$f = \frac{1}{2\pi\sqrt{LC}}$$

Simplify the equation further,

$$\frac{1}{2\pi} = \frac{1}{6.28} = .159$$

Therefore, $f_o = \frac{.159}{\sqrt{LC}}$ (approximately)

What is the resonant frequency of a circuit which has 200 μH inductance and 200 pF capacitance?

$$f_o = \frac{.159}{\sqrt{200 \times 10^{-6} \times 200 \times 10^{-12}}}$$

$$= \frac{.159}{\sqrt{40,000 \times 10^{-18}}}$$

$$= \frac{.159}{\sqrt{4 \times 10^4 \times 10^{-18}}} = \frac{.159}{\sqrt{4 \times 10^{-14}}} = \frac{.159}{2 \times 10^{-7}}$$

$$= .08 \times 10^7 \times 800,000 \text{ Hz}$$

$$= 800 \text{ KHz or } .8 \text{ MHz}$$

THE ACCEPTOR CIRCUIT

The action of a series RCL circuit at its resonant frequency is drawn in Fig. 11-2. At resonance $X_L = X_C$, so the impedance of the circuit equals,

$$Z = \sqrt{R^2 + (X_L - X_C)^2}$$

$$= \sqrt{R^2} \text{ or } R$$

Only the ohmic resistance impedes the current flow in the circuit. In a series resonant circuit the impedance is minimum. At frequencies above or below the resonant frequency, X_L is not equal to X_C and the reactive component INCREASES the impedance of the circuit.

The response of a series tuned circuit appears as a bell-shaped curve, Fig. 11-3. Notice that the impedance of the circuit is minimum at resonance. Note also that maximum current flows at resonance and rapidly falls off on either side of resonance. This is due to the increased impedance, Fig. 11-4. This is called an ACCEPTOR CIRCUIT. It provides maximum response to currents at its resonant frequency.

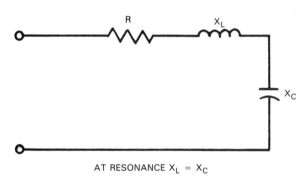

Fig. 11-2. The series resonant circuit is an acceptor circuit.

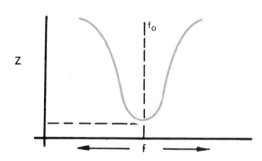

Fig. 11-3. The curve shows the increase in impedance as the frequency is varied above or below resonance.

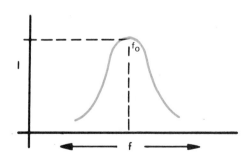

Fig. 11-4. The response curve showing the current flow in the circuit above or below resonance.

Refer to Fig. 11-5. This circuit has 10 ohms of resistance, 200 µH of inductance, and 200 pF of capacitance. These components are connected across a 500 µV RF generator at 800 kilohertz. The resonant frequency of this RCL circuit is 800 kilohertz, so X_L equals X_C at f_o. Z equals 10 ohms and,

$$I = \frac{E}{Z} \text{ or } \frac{500 \times 10^{-6}}{10}$$
$$= 50 \, \mu A$$

Also,

$$X_L = 2\pi \times 800 \times 10^3 \times 200 \times 10^{-6}$$
$$= 1000 \, \Omega \text{ approx.}$$

and,

$$X_C = \frac{1}{2\pi \times 800 \times 10^3 \times 200 \times 10^{-12}}$$
$$= 1000 \, \Omega \text{ approx.}$$

The voltage drops around the circuit equal,

$E_R = 50 \, \mu A \times 10 = 500 \, \mu V$
$E_{X_L} = 50 \, \mu A \times 1000 = 50,000 \, \mu V$
$E_{X_C} = 50 \, \mu A \times 1000 = 50,000 \, \mu V$

It appears that the sum of these voltages would be 100,500 µV. However, E_{X_L} and E_{X_C} are 180 degrees out of phase. Therefore, a vector addition must be made,

$$E_{source} = \sqrt{(E_R)^2 + (E_{X_L} - E_{X_C})^2}$$
$$= 500 \, \mu V$$

Note that, at resonance, the voltage drops across X_L and X_C are equal. Also, the voltage drop across R equals the source voltage.

In a resonant circuit, an interchange of energy between the inductance and the capacitance builds up voltage which exceeds the supply voltage. In the first half-cycle, the magnetic field of the inductance stores the energy of the discharging capacitor. In the next half-cycle, the stored energy in the magnetic field of the inductance charges the capacitor. This action occurs back and forth, limited only by the series resistance between the two components. At resonance, the charging times for the inductance and capacitance must be the same, or they would have a cancelling effect.

Summarizing the series tuned circuit:
1. At resonance, the impedance is minimum and the line current is maximum.
2. At resonance, the voltage drop, E_{X_L}, is equal to E_{X_C}, but is 180 degrees out of phase.
3. The vector sum of all voltage drops equals the applied voltage. Fig. 11-6 shows how values change as circuit frequency changes.

THE TANK CIRCUIT

Study Fig. 11-7. It shows a parallel tuned circuit. This type of tuned circuit behaves differently than a series tuned acceptor circuit.

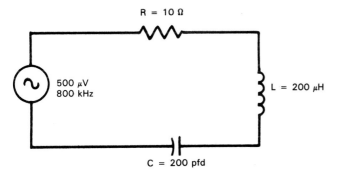

Fig. 11-5. A series tuned acceptor circuit.

f KHz	X_L OHMS	X_C OHMS	Z OHMS	I µA	E_{X_L} µV	E_{X_C} µV
200	250	4000	3750	.13	33	520
400	500	2000	1500	.33	165	660
600	750	1500	750	.66	495	990
800	1000	1000	10	5.0	50,000	50,000
1000	1250	800	450	1.1	1375	880
1200	1500	666	833	.6	900	400
1600	2000	500	1500	.33	660	165

Fig. 11-6. As circuit frequency changes, other values also change. The figures used are rounded to nearest whole number to ease understanding. Resistance of 10 ohms is insignificant. It is not included in computations, except at resonance.

Electricity and Electronics

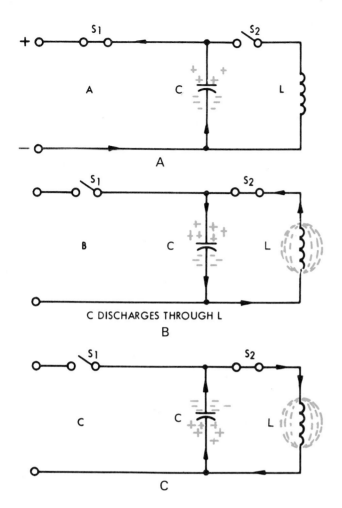

Fig. 11-7. The charge and discharge of C through inductance L is similar to flywheel action. It is called a tank circuit. A—Capacitor charges. B—Capacitor discharges. C—Capacitor discharges in reverse direction.

Switch S_1 is closed and capacitor C is charged to the supply voltage. When S_1 is opened, C remains charged. When S_2 is closed, the capacitor discharges through L in the direction of the arrows. As the current flows through L, a magnetic field builds around L. This field remains as long as the current flows. When the charges on the plates of C become equalized, current ceases to flow. The magnetic field around L then collapses. The energy stored in this field is returned to the circuit. In Chapter 9, we learned that the induced emf opposes the current change.

This is also true in the parallel circuit. When the current drops to zero as a result of the discharged capacitor, a current is induced by the collapsing magnetic field. This then drives a charge onto the capacitor, but opposite to the original polarity. Later the capacitor discharges in the opposite direction. Once again the same cycle of events occurs. The capacitor again becomes charged as in its original state. This discharge-charge cycle repeats over and over. The current periodically changes direction in the circuit. It is OSCILLATING.

The periodic current changes in the circuit may be described as FLYWHEEL ACTION. The circuit is called a TANK CIRCUIT. If no energy were used during the cycles of oscillation, the circuit might oscillate indefinitely. But there is always some resistance due to coil windings and circuit connections. This resistance uses up the energy stored in the circuit and dampens out the oscillation. The AMPLITUDE of each successive oscillation decreases due to the resistance.

Compare the oscillators of a tank circuit to a child on a swing. The child may be swinging back and forth, or oscillating, but if no one adds a little push to the swing, the amplitude of the swing will decrease until it comes to rest. If not for friction or resistance, the swing might continue forever. Now if the swing was pushed every time it reached its maximum backward position, the added energy would replace the energy lost by friction. The full swinging action would continue.

The tank circuit is similar. If pulses of energy are added to the oscillating tank circuit at the correct frequency, it will continue to oscillate.

What is the meaning of correct frequency? During the discharge and charge of the capacitor C in Fig. 11-7, a set amount of time must pass. In other words, during one cycle of oscillation a set interval of time must pass. The number of cycles that occur in one second would be called the FREQUENCY of oscillation. It is measured in CYCLES PER SECOND (cps) or HERTZ (Hz).

Close study of this circuit would reveal that if C or L were made larger so that they required a longer time to charge, the frequency of the circuit would decrease. This relationship may be shown using the resonance formula:

$$f_o = \frac{1}{2\pi \sqrt{LC}},$$

where f_o equals resonant frequency in hertz, L equals inductance in henrys, and C equals capacitance in farads.

In electronic work, L usually has values in microhenrys (μH) and C in picofarads (pF). If this is the case, the equation may be simplified to,

$$f_o = \frac{.159}{\sqrt{LC}},$$

where f_o equals resonant frequency in megahertz, L equals inductance in microhenrys, and C equals capacitance in picofarads.

THE REJECT CIRCUIT

When a parallel tuned circuit is connected across a variable frequency generator, minimum line current flows at the resonant frequency of the tuned circuit, Fig. 11-8. The resonant frequency may be learned by observing the minimum value or "dip" in the line current. This is measured by a current meter in the line. Radio transmitter operators always "dip the final." This means that they tune the final tank circuit to resonance which is indicated by a dip in current flow in the final circuit.

Since line current in the circuit at resonance is minimum, a parallel tuned circuit will have maximum line impedance (Z). At frequencies other than resonance, the impedance is much less. So, the circuit rejects signals at or near its resonant frequency, and allows signals of frequencies other than resonance to pass. The characteristics are shown in the response curves in Fig. 11-9.

Why does a parallel tuned tank present maximum impedance at resonance? At resonance, $X_L = X_C$, and both have reactive values in parallel across the generator source. It would appear that these two reactive branches would combine to form a low reactive path for the line current. However, the current flowing in the X_L branch is lagging the applied voltage. And the current in the X_C branch is leading the applied voltage. The currents, therefore, are 180 degrees out of phase and cancel each other. The total line current is the sum of the branch currents. It is zero, except for a small amount of current which flows due to the resistance of the wire in the coil.

An example may make clear these circuit characteristics. Refer to Fig. 11-10. A 200 μH inductor and a 200 pF capacitor are connected in parallel across a generator source of 500 μV. The resistance, 10 ohms, represents the lumped resistance of the wire of the inductor. These same parts were used in the study of the series resonant circuit. Compare the two.

The resonant frequency of this tuned circuit is:

$$f_o = \frac{.159}{\sqrt{LC}} \text{ or } \frac{.159}{\sqrt{200 \times 200}}$$

$$= \frac{.159}{200} = .8 \text{ MHz (approx.)}$$

(Note use of convenient formula when L is in μH, C is in pF, and f_o is in megahertz.)

At resonance:

$$X_L = 2\pi fL \text{ or } 1000 \text{ }\Omega$$

$$X_C = \frac{1}{2\pi fC} \text{ or } 1000 \text{ }\Omega$$

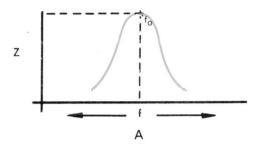

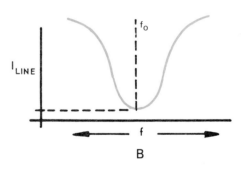

Fig. 11-9. A—Curve shows maximum impedance at resonant frequency. B—Circuit response shows minimum line current at resonance.

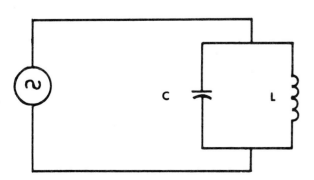

Fig. 11-8. A parallel LC circuit connected to a variable frequency generator.

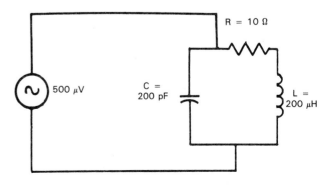

Fig. 11-10. This parallel resonant circuit has a 200 μH inductor and a 200 pF capacitor. They are connected in parallel across a 500 μV generator source.

The voltage across both branches of the parallel circuit is the same as the applied voltage, or 500 μV. The current, therefore, in the X_L branch is:

$$I_{X_L} = \frac{500 \ \mu V}{1000 \ \Omega} \text{ or } \frac{500 \times 10^{-6}}{10^3}$$

$$= 500 \times 10^{-9} \text{ amps} = .5 \ \mu \text{ amps}$$

The current in the X_C branch is:

$$I_{X_C} = \frac{500 \ \mu V}{1000} = .5 \ \mu A$$

Since these two currents are 180 degrees out of phase:

$$I_{X_L} - I_{X_C} = .5 \ \mu A - .5 \ \mu A = 0$$

In this exercise, R has been overlooked because of its small value. However, some current will flow as a result of R. For frequency above and below resonance, the action of the circuit may be seen in Fig. 11-11.

Note that the line current is the difference between the branch currents. Fig. 11-11 could be carried a step further to show the decreasing impedance due to frequencies other than resonance for,

$$Z = \frac{E}{I}$$

As I increases, Z must decrease.

Q OF TUNED CIRCUITS

The Q OF THE CIRCUIT is the link between inductive reactance and resistance in the circuit, or

$$Q = \frac{X_L}{R}$$

In a series circuit, for example, if the inductive reactance is 1000 ohms at resonance, and the resistance of the wire of the coil is 10 ohms, the FIGURE OF MERIT Q would be:

$$Q = \frac{1000}{10} = 100$$

The Q of a circuit indicates the sharpness of the reject or accept characteristics of the circuit. It is a quality factor. In the series acceptor circuit, an increase in resistance reduces the maximum current at resonance, Fig. 11-12. The Q may also be used to determine the rise in voltage across L or C at resonance:

$$E_{X_L} = E_{X_C} = Q \times E_S$$

Refer back to Fig. 11-5. The supply voltage is 500 μV and the Q is 100. Therefore, the voltage rise across X_L or X_C at resonance is equal to:

$$E_{X_L} = E_{X_C} = 100 \times 500 \ \mu V = 50,000 \ \mu V$$

If the circuit had a lower Q, the magnified voltage would be much less. For example, increase R to 20 ohms. Then the Q would equal 50 and the voltage rise would equal:

$$E_{X_L} = E_{X_C} = 50 \times 500 \ \mu V = 25,000 \ \mu V$$

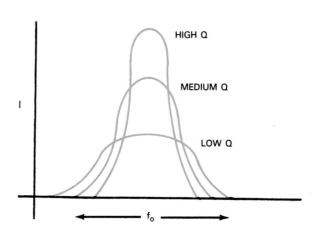

Fig. 11-12. As the Q of a circuit is lowered, the curve flattens out. Its selectivity decreases and its band pass increases.

FREQUENCY KHz V	X_L (Ω)	X_C (Ω)	I_{X_L} (μA)	I_{X_C} (μA)	I_{Line} (μA)
200	250	4000	2.0	.125	1.875
400	500	2000	1.0	.25	1.75
600	750	1500	.66	.33	.33
800	1000	1000	.5	.5	0
1000	1250	800	.4	.625	.225
1200	1500	666	.33	.75	.42
1600	2000	500	.25	1.00	.75

Fig. 11-11. Circuit performance for frequency above and below resonance.

Tuned Circuits and RCL Networks

High Q circuits are very useful in selective electronic circuits. Typical Qs range from 50 to 250. The higher the value of Q, the greater response of the circuit at resonance. Also, a high Q circuit will have increased selectivity. Selectivity is set by bandwidth. Bandwidth is the band of frequencies above and below resonance in which the circuit response does not fall below 70.7 percent (half power point) of the response at resonance.

The bandwidth of the tuned circuit may be found by the formula:

$$Bw = \frac{f_o}{Q}$$

Continuing with the previous problem, the bandwidth equals:

$$Bw = \frac{800,000 \text{ Hz}}{100} = 8000 \text{ Hz}$$

Since 800 kHz (the resonant frequency) is at the maximum response point, the bandwidth will extend 4000 Hz below resonance and 4000 Hz above resonance. The circuit may be considered, then, as passing all frequencies between 796 kHz and 804 kHz. Beyond either of these limits, the response will fall below the 70.7 percent value.

Q may be computed for parallel tuned circuits using X_L at resonance and R, which is the resistance of the coil L. The Q of a tank circuit may be used to learn the maximum impedance of the circuit at resonance:

$$Z = Q \times X_L,$$

where Z equals impedance at resonance, Q equals $\frac{X_L}{R}$ at resonance, and X_L equals reactance at resonance.

Referring back to Fig. 11-10, the impedance at resonance would be:

$$Z = 100 \times 1000 \text{ ohms} = 100,000 \text{ ohms}$$

One final example, a resistance shunt across or in parallel with a tank circuit, is shown in Fig. 11-13. R_S is called a damping resistor. It broadens the frequency response of the circuit, because it carries a part of the line current which cannot be cancelled at resonance. Shunt damping lowers the Q of the circuit and makes it less selective.

LOADING THE TANK CIRCUIT

The parallel tuned circuit is used to couple energy from one circuit to another. Coupling transformers in radio and television sets use the tuned circuit to transfer signals from one stage to another.

A radio transmitter uses a coupling device attached to a tank circuit to feed energy to the antenna system. If another coil is inductively coupled to the coil in the tank circuit, the varying magnetic field of the oscillating tank inductance will induce a current to flow in the coil.

Fig. 11-14 shows the circuit of an intermediate frequency transformer used in a radio receiver.

Both the primary and secondary of this transformer are tuned circuits. Maximum radio frequency energy transfer may be achieved by tuning the circuits to resonance. The intermediate frequency used in most radios is 455 kHz. (In television, 10.7 MHz IF is used.) The IF transformer is tuned for maximum response at this frequency. A band pass of 5 kHz above and below the center frequency is maintained. Signals between 450 kHz and 460 kHz may pass without attenuation (decrease in amplitude or intensity). The process of transformer adjustment is called ALIGNMENT.

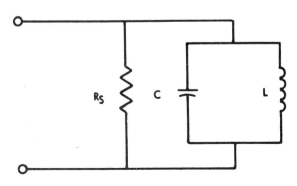

Fig. 11-13. The resistor, R_S, is a damping resistor. It broadens the circuit response.

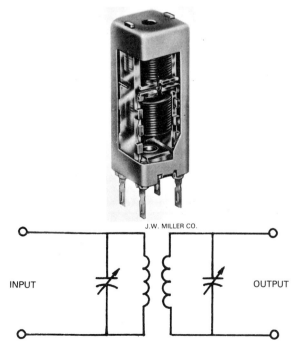

Fig. 11-14. Top. An intermediate frequency transformer used to couple energy from one tube to another in a radio receiver. Bottom. Schematic for the transformer.

REVIEW QUESTIONS FOR SECTION 11.1

1. Resonant frequency is a special condition of RCL circuits in which _____ _____ _____.
2. Find the resonant frequency using the values given in the following circuit.

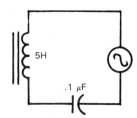

3. In a series resonant circuit, the impedance is _____.
4. Periodic current changes in a tank circuit are described as _____ _____.
5. In a parallel tuned circuit, at resonance the circuit _____ currents at the resonant frequency.
6. The relationship between X_L and R in a circuit is stated by the _____ of the circuit.

11.2 FILTERING CIRCUITS

Inductance and capacitance are quite useful in electronic circuitry. One example of this usefulness is the filter. A FILTER is a circuit that separates specific frequencies.

There are many filter designs. For example, they may be designed to pass low frequencies and reject high frequencies; to reject low frequencies and pass high frequencies; to either pass or reject specified frequency bands.

Each filter type is named according to its function. There are four types of filters: low-pass, high-pass, band-pass, and band-reject.

There is also a type of filter used to adjust a pulsating current coming from a rectifier. A RECTIFIER is a device that converts ac into a pulsating dc. Because most electronic circuitry needs a pure direct current, the dc must be adjusted. This filter circuit adjusts the current. It will be discussed in Chapter 15, POWER SUPPLIES.

FILTERING ACTION

In the study of filtering action, these points should be reviewed:
1. A capacitor will block a direct current, but will pass an alternating current.
2. A conductor may carry a current which has both a dc component and an ac component.

The graph in Fig. 11-15A shows the flow of a steady 10 volt direct current. Fig. 11-15B shows an ac voltage whose peak value is 10 volts. Fig. 11-15C shows these two voltages combined. This wave depicts the sum of the two waves.

Note the new axis of the ac voltage. It now varies above and below the 10 volt dc level. Because the axis has been raised by the dc voltage, the ac voltage no longer reverses polarity. The current is a varying direct current.

Its amplitude varies between 0 and 20 volts. When this varying direct current is connected to a circuit, Fig. 11-16, the capacitor C immediately charges up to the average dc level of voltage. In this case, that would equal

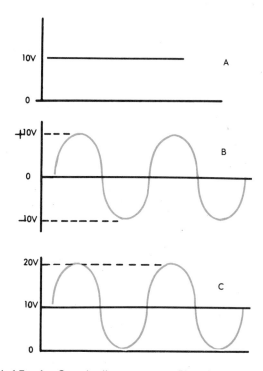

Fig. 11-15. A—Steady direct current. B—10 volt peak alternating current. C—Combined dc and ac.

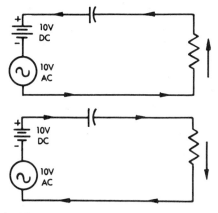

Fig. 11-16. The ac generator and dc source are connected in series to the RC circuit.

10 volts. When the incoming voltage rises to 20 volts, current flows through R to charge capacitor C. In the next half-cycle, the incoming voltage drops to zero, and capacitor C discharges to zero through R.

IMPORTANT: A voltage appears across R due to the charge and discharge of C. The output of this circuit, then, taken from across R, represents only the ac component of the incoming signal voltage. See Fig. 11-17. The dc component is blocked by capacitor C. There is little phase shift between the input and output waves. This is because the value of R is chosen as ten times or more the value of the reactance of C at the input voltage frequency. When this ratio of resistance to reactance is maintained, the phase shift does not affect the waves.

EXPERIMENT: Select a 1 μF capacitor from your stock of parts. Connect the circuit as shown in Fig. 11-18. 60 Hz ac will be used in this experiment. Compute the reactance of C:

$$X_C = \frac{.16}{fC} \text{ or } \frac{.16}{60 \times 1 \times 10^{-6}}$$
$$= .0026 \times 10^6$$
$$= 2600 \text{ ohms}$$

R should equal 10 times X_C or 26,000 ohms. A 25,000 ohm resistor is suitable for this experiment. Connect the 6.3 volts (rms) terminals of a power supply in series with the 10 volt dc supply. Connect to the circuit. Measurements across C with a dc voltmeter should read 10 volts. An ac meter across the same points should read zero. An ac meter across R should read 6.3 volts and a dc meter, zero.

CONCLUSIONS: The dc component of the incoming signal appears across C. The ac component appears across R. This important principle will be discussed in more detail in later chapters. It is called RC coupling.

BYPASSING

At times it is necessary to create a voltage drop across a resistor resulting from only the dc component of a signal voltage. This is done by bypassing an ac signal or voltage around a resistance. Circuits that do this task are shown in Figs. 11-19A and 11-19B.

Fig. 11-19A shows two 1500 ohm resistors joined to a source of 10 volts dc. Total resistance equals $R_1 + R_2$ or 3000 ohms. The current in the circuit is:

$$I = \frac{E}{R} \text{ or } \frac{10 \text{ V}}{3000 \text{ Ω}} = .0033 \text{ amps or 3.3 mA}$$

The voltage drop across each resistor is:

$$E_{R_1} = I \times R \text{ or } .0033 \times 1500 = 5 \text{ V}$$
$$E_{R_2} = I \times R \text{ or } .0033 \times 1500 = 5 \text{ V}$$

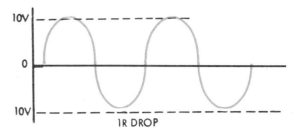

Fig. 11-17. The voltage across R is the result of the charge-discharge current and represents the ac component of the signal.

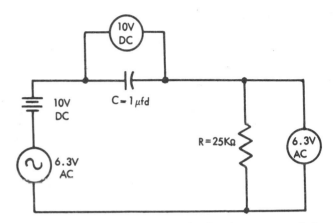

Fig. 11-18. This circuit shows a dc signal component being blocked and an ac component being passed.

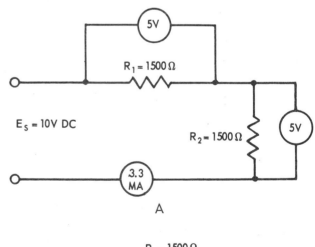

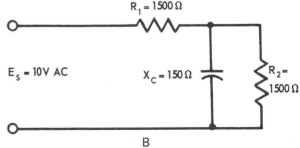

Fig. 11-19. A—A dc voltage produces equal drops across R_1 and R_2. B—The voltage drop across R_2 is held constant by bypassing the ac component.

Connect a 10 volt ac source now and an ac voltage will appear across BOTH resistors. But we wish to bypass this ac component around R_2. To do this, connect a capacitor that will have low reactance to the ac voltage in parallel with R_2. If the reactance of C is one-tenth of resistance R, the greater part of the varying current will flow through C and not through R.

Now use the values given in Fig. 11-19B. C has a 150 ohm reactance to the ac frequency. This is one-tenth the value of R_2. For the most part, the impedance of $R_2 \| X_C$ (R_2 and X_C in parallel) can be given a value of 150 ohms. Voltage distribution around the circuit may be measured or computed.

Total resistance equals $R_1 + R_2 \| X_C$ or 1650 ohms, and $R_2 \| X_C$ represents one-eleventh of the total resistance to alternating current. Because voltage drop is a function of resistance, ten-elevenths of the voltage appears across R_1 and only one-eleventh across $R_2 \| X_C$. The applied ac voltage is 10 volts, so:

$$E_{R_1} = 9.1 \text{ V and } E_{R_2} = .9 \text{ V}$$

If a 10 V dc voltage is connected in series with the 10 V ac voltage for the combined input voltage, the dc voltage divides equally between R_1 and R_2 with 5 volts each. C is an open circuit for a direct current.

The ac voltage divides in the ratio mentioned above. Most of this voltage appears across R_1. The voltage across R_2 remains fairly constant due to the bypass capacitor action.

To summarize: choose a capacitor which will form a low reactance path around a resistor or to ground for currents of chosen frequencies.

LOW-PASS FILTERS

At times, a filter is needed which will pass low frequencies, yet decrease the high frequency currents. This filter is called a LOW-PASS FILTER. The circuit always has a resistance or an inductor in series with the incoming signal voltage. And it also has a capacitor in shunt or across the line, Fig. 11-20.

As the frequency is increased, the reactance of L increases so that a larger amount of the voltage appears across L. Also as the frequency increases, the reactance of C decreases. This provides a bypass for the higher frequency currents around the load resistance R.

In both respects, the higher frequencies appear in small amounts across the load. Low frequencies develop higher voltages across the load. Low frequencies are passed, high frequencies are rejected.

HIGH-PASS FILTER

The opposite to the low-pass filter is the HIGH-PASS FILTER. It passes chosen high frequency current and rejects low frequency currents. The circuit includes a capacitor in series with the incoming signal voltage and an inductance shunt across the line, Fig. 11-21. As the frequency increases, X_L increases. A higher voltage is developed across L and R in parallel. As frequency increases, X_C decreases, providing a low reactance path for high frequency signals. Low frequencies are shunted or bypassed around the load R by the low reactance of L at low frequencies.

To improve the filtering action of both low- and high-pass filters, two or more sections are often joined. These sections are named according to the circuit make-up. In Fig. 11-22, the schematic drawings and the names of several types of filters are given.

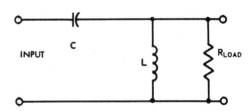

Fig. 11-21. A high-pass filter circuit.

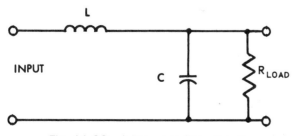

Fig. 11-20. A low-pass filter circuit.

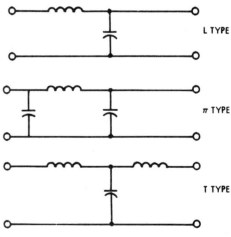

Fig. 11-22. Filter circuits may be named according to circuit configuration.

TUNED CIRCUIT FILTERS

Earlier in this chapter, series and parallel resonant circuits were said to be accept or reject circuits. These tuned circuits may be used as filters because they are able to give maximum response at their tuned resonant frequency.

In Fig. 11-23, a series resonant circuit is used as a band-pass filter. It accepts only currents near its resonant frequency. In Fig. 11-24, the series resonant circuit is shunted across, or parallel to, the load. This provides a low impedance path around or bypassing the load for currents at resonant frequencies. In this case the filter would respond as a band-reject filter.

The circuits in Fig. 11-25 use a parallel tuned circuit. Its effect is the opposite of the series tuned circuit. When the tank circuit is in series with the incoming currents, it provides maximum impedance. It rejects the frequencies at its resonant frequency. When the tank circuit is shunt across the load, it causes the maximum response across the load at resonance. Frequencies other than resonance are bypassed because of the decreased impedance of the shunt tuned tank circuit.

Combinations of both series and parallel tuned circuits may be used to provide sharper cutoff and greater attenuation. A high-pass filter attenuates all signals below Channel 2, yet all TV channels are passed through to the receiver. This eliminates noise interference.

THE NOMOGRAPH

The nomograph, or alignment chart, shown in Fig. 11-26, provides a rapid and handy method for solving problems involving X_L, X_C, and resonance. These examples will aid you in using and understanding this chart.

1. What is the reactance of a .01 µF capacitor at 10 kHz? Use upper chart. Place a ruler or straight edge at 10 kHz on the frequency scale and .01 on the capacitance scale. The rule crosses the reactance scale at 1590 ohms.

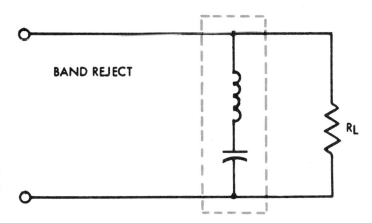

Fig. 11-24. A series tuned circuit arranged as a band-reject filter.

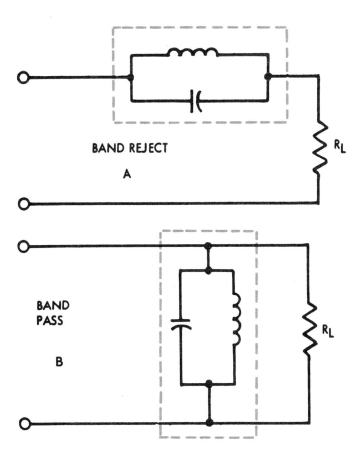

Fig. 11-25. A—The parallel tuned circuit as a band-reject filter. B—The parallel tuned circuit as a band-pass filter.

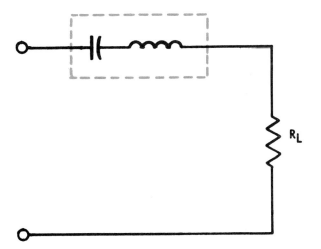

Fig. 11-23. The series tuned circuit used as a band-pass filter.

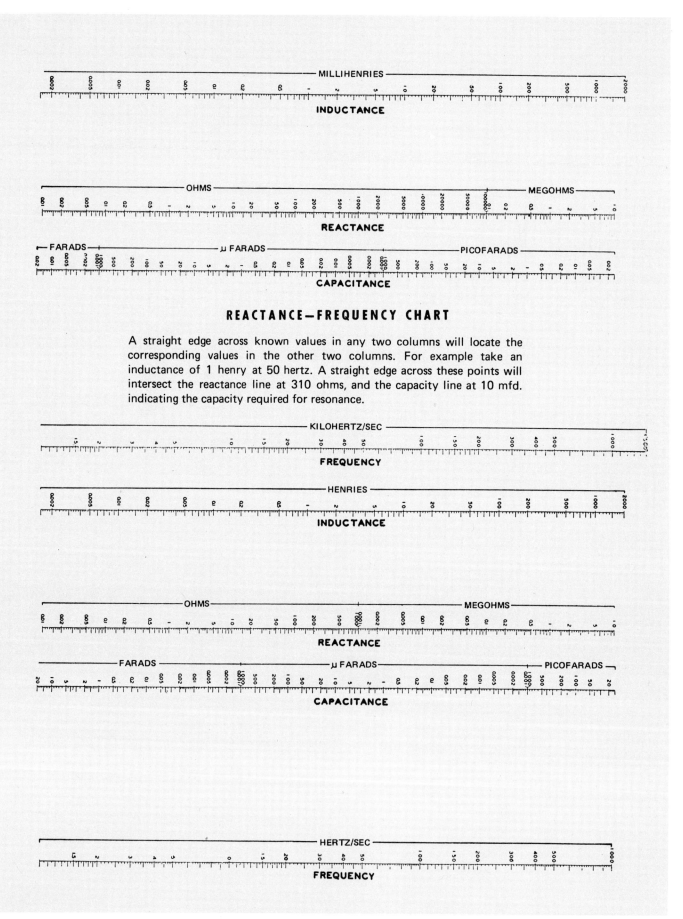

Fig. 11-26. Nomograph or alignment chart.

2. What is the reactance of an 8 henry choke coil at 60 Hz? Use lower chart. Place rule on 60 Hz and 8 H and read 3014 ohms.
3. What is the resonant frequency of a tuned circuit when C equals 250 pF and L equals 200 mH? Use upper chart. Place rule on 200 mH and 250 pF and read 22.5 kHz on frequency scale.

REVIEW QUESTIONS FOR SECTION 11.2

1. Circuits that separate ac variations from dc are _____.
2. A _____ will pass ac and block dc.
3. In a low-pass filter, the _____ _____ are rejected.
4. When a series resonant circuit is used as a band-pass filter, it accepts only current near its _____ _____.
5. What is a nomograph?

SUMMARY

1. Networks are ac circuits which have resistors, capacitors, and inductors placed in the circuit to pass, reject, or control current.
2. Resonant frequency of a circuit occurs when the inductive reactance (X_L) is equal to the capacitive reactance (X_C).
3. The formula for the frequency of resonance is:
$$f_o = \frac{1}{2\pi \sqrt{LC}}$$
It can be simplified to:
$$f_o = \frac{.159}{\sqrt{LC}}$$
4. In a series resonant circuit, the current is maximum and the impedance is minimum at the resonant frequency.
5. Tank circuits oscillate, providing an ac signal at a desired frequency. A tank circuit schematic is shown below.

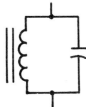

6. In a parallel resonant circuit, the current is minimum and the impedance is maximum at the resonant frequency.
7. Parallel resonant circuits reject currents at resonant frequency.
8. Series resonant circuits pass currents at resonant frequency.
9. The Q of a circuit describes the relationship between X_L and R. The formula is:
$$Q = \frac{X_L}{R}$$
10. Four types of filtering circuits are:
    a. Low-pass filter.
    b. High-pass filter.
    c. Band-pass filter.
    d. Band-reject filter.
11. A nomograph is a chart that can be used to solve problems involving X_L, X_C, and resonance.

TEST YOUR KNOWLEDGE, Chapter 11

Please do not write in the text. Place your answers on a separate sheet of paper.
1. What is the formula for resonance?
2. Compute the resonant frequency of the following circuits:
   a. L = 100 μH, C = 250 pF.
   b. L = 200 μH, C = 130 pF.
   c. L = 8 μH, C = 1 μF.
3. A series resonant circuit is sometimes called an _____ circuit.
4. Explain flywheel action of a tank circuit.
5. A parallel tuned circuit is sometimes called a _____ circuit.
6. At resonance in a parallel tuned circuit, line current is _____.
7. State the formula for determining the Q of a circuit.
8. A filter is a circuit that separates specific _____.
9. Name four general types of filters.
10. Distinguish between a high-pass and a low-pass filter.

FOR DISCUSSION

1. How can a series resonant circuit develop voltages higher than the applied voltage?
2. Explain flywheel action and relate it to frequency.
3. A simplified formula for resonant frequency is $f_o = \frac{.159}{\sqrt{LC}}$, where f_o is in MHz, L is in μH, and C is in pF. Explain mathematically the derivation of this formula from $f_o = \frac{1}{2\pi \sqrt{LC}}$.
4. Why is resonance indicated by a dip in the line current of a parallel RCL circuit?
5. What is the relationship between the Q and the selectivity of a tuned circuit?

Electricity and Electronics

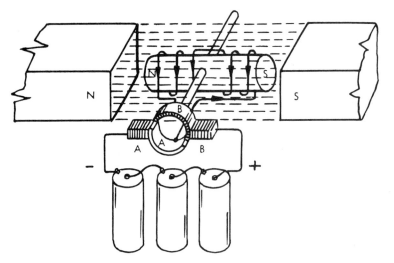

Schematic shows the theory of dc motor operation. Current can be traced from external power source, through brushes and commutators to armature.

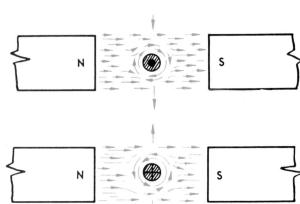

Current flow through conductor in a magnetic field determines the relative motion of the conductor.

DEVRY INSTITUTES

This student is using an electric motor to drive a robotic arm.

Electric motors are important to most of the technology humans use. Motors depend on the principles of magnetism, force, power, and electricity. Motors are similar to generators.

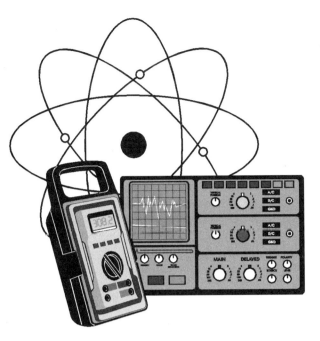

Chapter 12

ELECTRIC MOTORS

After studying this chapter, you will be able to:
Explain the operating principles of dc motors.
Identify various dc motors.
Discuss the purpose for and operation of motor starting circuits.
Identify and explain the operation of various induction motors.

One of the most important developments in the field of electricity is the ELECTRIC MOTOR. It is a device that converts electrical power to mechanical power. Motors are used for refrigeration and air conditioning. They are also used to mix foods and clean carpets. Motors are used to operate grinders, mixers, pumps, and other machinery. Motors also power bench saws, lathes, and various wood and metal machines.

12.1 MOTOR OPERATION PRINCIPLES

Motor operation depends on the interaction of magnetic fields. The Laws of Magnetism state that:

LIKE POLES REPEL EACH OTHER
UNLIKE POLES ATTRACT EACH OTHER

or

A NORTH POLE REPELS A NORTH POLE
A SOUTH POLE REPELS A SOUTH POLE

but

A NORTH POLE ATTRACTS A SOUTH POLE

The theory of the simple dc motor is shown in Figs. 12-1 through 12-9.
To make the motor more powerful, permanent field magnets may be replaced by electromagnets, called

Fig. 12-1. A magnetic field exists between the north and south poles of a permanent magnet.

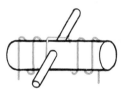

Fig. 12-2. An electromagnet is wound on an iron core and the core is placed on a shaft so it can rotate. This assembly is called the armature.

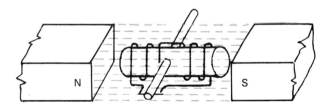

Fig. 12-3. The armature is placed in the permanent magnetic field.

FIELD WINDINGS. These windings may have an independent source of voltage connected to them. Or, they may be connected in series or parallel with the armature windings, to a single voltage source. All three types are shown in Fig. 12-10.

Electricity and Electronics

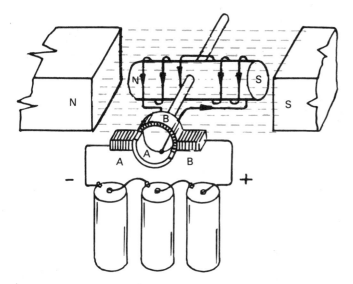

Fig. 12-4. The ends of the armature coil are connected to semicircular sections of metal called commutators. Brushes contact the rotating commutator sections and energize the armature coil from an external power source. (Recall that the polarity of the armature electromagnets depends on the direction of current flow through the coil.) A battery is connected to the brushes. Current flows into brush A to commutator section A, through the coil to section B, and back to the battery through brush B, completing the circuit. The armature coil is magnetized as indicated in the sketch.

Fig. 12-5. The north pole of the armature is repelled by the north pole of the field magnet. The south pole of the armature is repelled by the south pole of the field magnet. The armature turns one quarter revolution, or 90 degrees.

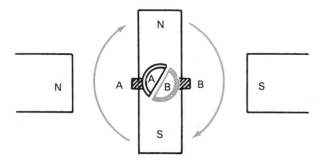

Fig. 12-6. The north pole of the armature is attracted by the south pole of the field magnet. The south pole of the armature is attracted by the north pole of the field. The armature turns another quarter turn. It has now turned one-half revolution.

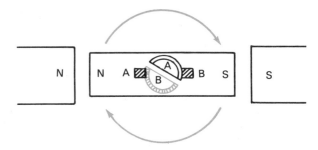

Fig. 12-7. As the commutator sections turn with the armature, section B contacts brush A and section A contacts brush B. The current now flows into section B and out section A. The current has been reversed in the armature due to commutator switching action. This current reversal changes the polarity of the armature, so that unlike poles are next to each other.

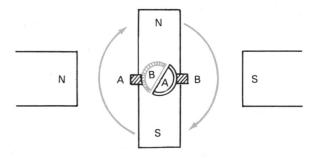

Fig. 12-8. Like poles repel each other and the armature turns another quarter turn.

Fig. 12-9. Unlike poles attract each other and the armature turns the last quarter turn, completing one revolution. The commutator and brushes are now lined up in their original positions, which causes the current to reverse in the armature again. The armature continues to rotate by repulsion and attraction. The current is reversed at each one-half revolution by the commutator.

Construct a trial motor, Fig. 12-11. Connect the motor first in series, then in parallel as a shunt motor. Compare the speed and power of the two motors.

In industry, motors are made in a slightly different manner than discussed. Rotational force comes from the interaction between the magnetic field found around a current carrying conductor and a fixed

Electric Motors

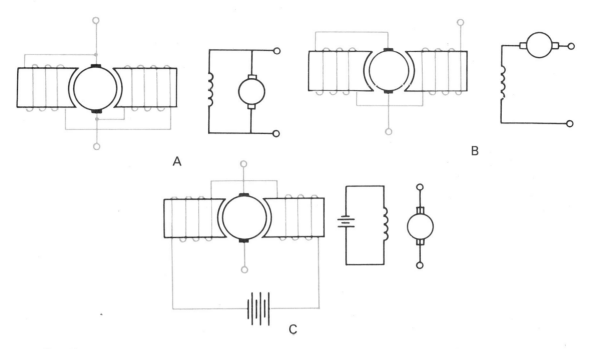

Fig. 12-10. Sketches and schematic diagrams of field winding connections. A—Shunt wound motor is connected in parallel. B—Series wound motor. C—Independently excited field motor.

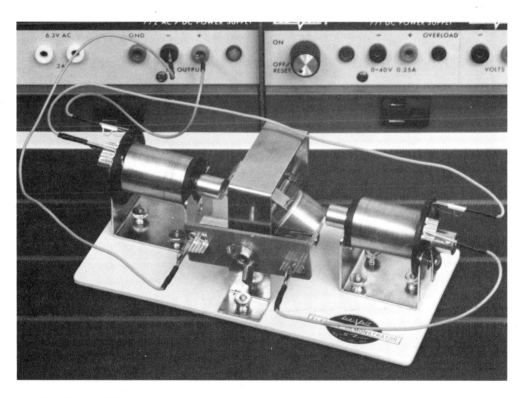

Fig. 12-11. The trial motor may be set up to operate as a series or shunt motor.

magnetic field. A conductor carrying a current has a magnetic field around it. The direction of the field depends on the direction of the current. When this conductor is placed in a fixed magnetic field, the interaction between the two fields causes motion. See Figs. 12-12 through 12-16.

Armature coils on industry motors are connected to commutator sections, as in the trial motor. The theory of operation is similar. A practical motor has several armature coils wound in separate slots around the core. Each coil has a commutator section. Increasing the number of field poles gives the motor greater power.

181

Fig. 12-12. A magnetic field exists between the poles of a permanent magnet. The arrows indicate the direction of field.

Fig. 12-13. A current carrying conductor has a magnetic field; its direction depends on direction of current. Use the left hand rule to determine direction.

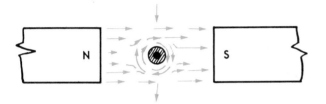

Fig. 12-14. The field around the conductor flows with the permanent field above the conductor, but opposes the permanent field below the conductor. The conductor will move toward the weakened field.

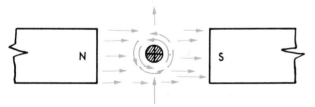

Fig. 12-15. The current has been reversed in the conductor, causing the conductor field to reverse. Now the field is reinforced below the conductor and weakened above the conductor. The conductor will move up.

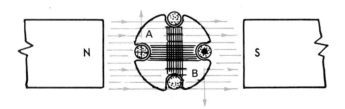

Fig. 12-16. The single conductor is replaced by a coil of conductors wound in the slots of an armature core. Notice how the interaction of the two fields will produce rotation. Coil side A moves up and coil side B moves down. The rotation is clockwise.

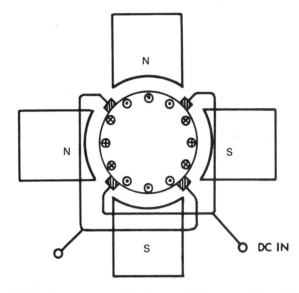

Fig. 12-17. The torque of the motor is increased by adding armature coils and field coils.

A four-pole motor is sketched in Fig. 12-17. The current divides into four parts. The current flowing in windings under each field pole produces rotation. This then increases the turning power, or TORQUE, of the motor.

COUNTER EMF

In Chapter 7, you learned that when a conductor cuts through a magnetic field, voltage is induced in the moving conductor. And while a motor is meant to convert electrical energy into mechanical energy, when the armature begins to rotate, the motor also becomes a generator.

The induced electromotive force opposes the applied emf. It is called COUNTER EMF. It is a result of the generator action of the motor. The counter emf magnitude increases as the rotation speed and field strength increase. Therefore,

$$\text{Counter emf} = \text{Speed} \times \text{Field strength} \times K$$

K equals other constants in any motor, such as the number of windings. The actual effective voltage then applied to the windings in the armature must equal,

$$E_{source} - E_{counter} = E_{armature}$$

The current flowing in the armature windings at any given instant may be found using Ohm's Law when the ohmic resistance of the windings is known:

$$I_{armature} = \frac{E_{armature}}{R_{armature}}$$

COMMUTATION AND INTERPOLES

As the motor armature rotates, the current in the armature windings routinely reverses. This is caused by commutator action. Due to the self-inductance of the windings, the current does not instantly reverse. This results in sparking at the commutator brushes.

There are a number of methods for preventing sparks. In one method, the brushes are moved slightly against the direction of rotation and the counter emf is used to induce the previous pole. The counter emf opposes the self-induction caused by the decreasing current in the coil. Sparking is eliminated.

However, this is not a practical method for preventing sparks. That is because as the load varies on the motor, the brush position must be changed by hand. Instead, larger motors use interpoles to reduce the sparking. An INTERPOLE is a smaller field pole placed midway between main field poles. The interpole has the same polarity as the main field poles and follows the main pole in direction of rotation. Interpoles are also called COMMUTATING POLES.

A counter emf is developed as the armature passes the interpole. It overcomes the emf caused by self-induction in the armature windings. The windings of the interpole are connected in series with the armature and carry the armature current. Thus, interpole field strength varies as the load varies. And it provides automatic control of commutator sparking, Fig. 12-18.

SPEED REGULATION

Many motors are designed for special purposes. Some develop full power under load. Others must be brought up to speed before the load is applied. When the speed for a motor is determined on the job, the motor should maintain that speed under varying load conditions. A ratio of the speed under no-load to the speed under full-load can be expressed as a percentage of the full-load speed. This is called the PERCENT OF SPEED REGULATION and it equals,

$$\frac{\text{Speed no-load} - \text{Speed full-load}}{\text{Speed full-load}} \times 100\%$$

A low speed regulation percentage means that the motor operates at a somewhat constant speed, regardless of load applied.

REVIEW QUESTIONS FOR SECTION 12.1

1. The motor is used to convert _____ energy into _____ energy.
2. Outline how a simple dc motor operates.
3. The turning power of a motor is called the _____.
4. An _____ is a smaller field pole placed midway between main field poles.
5. Give the formula to find percent of speed regulation.

12.2 DC MOTORS

THE SHUNT DC MOTOR

The shunt motor has the field windings shunt across or in parallel to the armature, Fig. 12-19. The shunt motor is commonly called a CONSTANT SPEED MOTOR. It is used in driving machine tools and other machines needing relatively constant speed under varying loads.

In the shunt motor, both the field and the armature are connected across the power line. Under no-load conditions, the counter emf is almost equal to the line voltage. Very little armature current flows and very little torque is developed. When a load is applied and

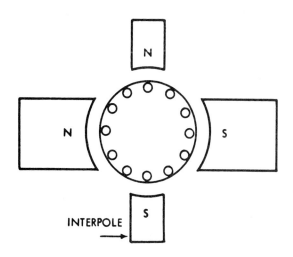

Fig. 12-18. Interpoles reduce sparking at the commutator.

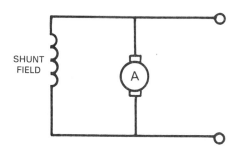

Fig. 12-19. Schematic of a shunt motor.

the armature decreases its speed, the counter emf also decreases. This increases the armature current and the torque. When the torque matches the load, the motor remains at constant speed, Fig. 12-20.

The total current used by this motor is the sum of the field and armature currents. The input power may be computed using Watt's Law:

Power = Applied E × Total Current

The output power will be different because the motor is not one hundred percent efficient.

THE SERIES DC MOTOR

In the series wound motor, the field windings are connected in series with the armature windings, Fig. 12-21. All the line current must flow through both the field and armature windings. Under loaded conditions, the counter emf opposes the line voltage and keeps the current at a safe level.

If the load were suddenly removed, the armature would speed up and develop a higher counter emf (cemf). This would reduce the current flowing through the field and reduce the field strength. In turn, the motor would increase its speed because:

$$\text{Speed} = \frac{\text{cemf}}{\text{Field Strength} \times K}$$

NO-LOAD	HIGH COUNTER EMF LOW ARMATURE CURRENT LOW TORQUE
FULL-LOAD	DECREASED COUNTER EMF INCREASED ARMATURE CURRENT INCREASED TORQUE

Fig. 12-20. Shunt motor load conditions.

This action builds on itself. The motor would reach a speed where the armature would fly apart, due to centrifugal force. Thus, a series motor is never operated without a load. Furthermore, the series motor should be connected directly to a machine or through gears. It is not safe to use a belt from motor to machine. If the belt should break or slip off, the motor would "run wild" and may destroy itself.

A key advantage of the series motor is its ability to develop a high torque under load. Under load conditions, the armature speed is low and the cemf is low. This results in a high armature current and increased torque. Series motors have heavy armature windings to carry these high currents. As the motor increases in speed, the cemf builds up, the line current decreases, and the torque decreases. Series motors are used on electric trains, cranes and hoists, and other traction equipment.

COMPOUND DC MOTORS

The compound motor has both the series and the shunt field. It combines the good points of each type of motor, Fig. 12-22.

The series windings also carry the armature current. It consists of a number of heavy turns of wire. The shunt field winding consists of many turns of finer wire. Both windings are wound on the same field poles.

There are two methods used to connect these windings. If the magnetic field on the series winding aids or reinforces the magnetic field of the shunt winding, the motor is said to be a CUMULATIVE COMPOUND MOTOR. If the two windings are connected to oppose each other magnetically, the motor is a DIFFERENTIAL COMPOUND MOTOR. A detailed study of compound motors is beyond the scope of this text. However, the characteristics of different types should be noted.

The differential compound motor behaves much like the shunt motor. The starting torque is low. It has good speed regulation if loads do not vary greatly. However, it is not widely accepted.

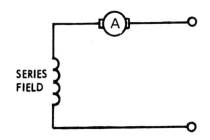

Fig. 12-21. Schematic of a series motor.

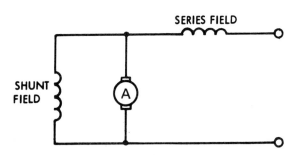

Fig. 12-22. Schematic of a compound motor.

Electric Motors

The cumulative compound motor develops high starting torque. It is used where heavy loads are applied and some variance in speeds can be tolerated. The load may be safely removed from this motor.

MOTOR STARTING CIRCUITS

Armature resistance in most motors is quite low. When the motor is stopped or at low speed, very little cemf is developed to oppose the line voltage. Therefore, dangerously high currents will flow. To protect the motor until it builds up its speed and cemf, a current limiting resistor may be placed in series with the armature.

A simplified motor starting circuit is drawn in Fig. 12-23. The resistance is variable. Starting from a dead stop, maximum resistance is inserted in the armature circuit. As the speed builds up, the resistance is slowly decreased until the motor reaches full speed. The resistance may then be removed.

Using Ohm's Law, assume the armature resistance is .1 ohm and the applied voltage 100 volts. Then the initial armature current would be:

$$I_{armature} = \frac{100 \text{ V}}{.1} \text{ or } 1000 \text{ amps}$$

This current would burn the insulation off the wires and the motor would be destroyed. When the motor is up to speed and cemf has developed, then the voltage across the armature is equal to,

$$E_{line} - cemf = E_{armature}$$

If the motor develops a cemf of 95 volts as it approaches full speed, the armature current would be:

$$I_{armature} = \frac{100 \text{ V} - 95 \text{ V}}{.1} \text{ or } 50 \text{ amps}$$

This is a safe current for this motor armature to carry. Therefore, start the motor with the full series resistance in the armature circuit. Then gradually decrease the resistance until the motor can limit the current by it own cemf.

Starting resistance may be adjusted by hand using a lever. The lever decreases the resistance step-by-step, until the motor reaches full speed, Fig. 12-24.

In Step 1, the maximum resistance is in series. With each following step, the resistance decreases. At Step 4, the lever arm is held magnetically by the holding coil, which is also in series with the circuit. If the line voltage should fail, the lever arm would snap back to the off position and the motor would have to be restarted. This is a protective device.

Operators must use good judgment when starting a motor. They must not move the lever to the next lower resistance position until the motor has gained sufficient speed.

Automatic starters remove the possibility of human error and also permit remote starting of the motor. Several devices are used.

Fig. 12-25 shows one type of starting switch. When

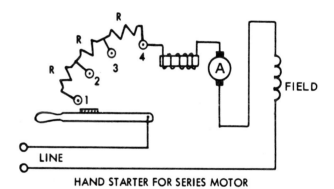

HAND STARTER FOR SERIES MOTOR

Fig. 12-24. Typical step motor starter circuit.

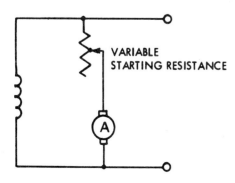

Fig. 12-23. Simplified motor starting circuit.

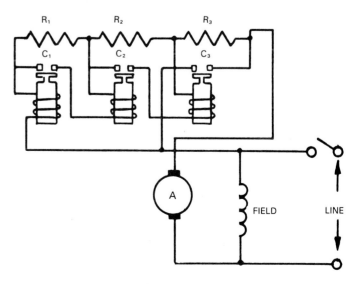

Fig. 12-25. Circuit for an automatic motor starter.

line switch S is closed, voltage is applied to the field across the line. It is also applied to the armature through R_1, R_2, and R_3 in series with the armature CONTACTORS C_1, C_2, and C_3. The contactors will not close until the current has decreased to a preset value.

The first surge of current in the armature circuit flows through series coil of C_1 which holds it open. As the motor increases speed and the armature current decreases, C_1 contactor closes and cuts R_1 out of the circuit. Similar action occurs in C_2 and C_3. When the motor reaches full speed, all contactors will be closed. The protective resistance is no longer needed.

One more type of motor starter is the PUSH-BUTTON STARTER. You may have seen motors equipped with starting boxes containing red and black push buttons marked stop and start, Fig. 12-26. The circuit diagram is shown in Fig. 12-27.

This starter is a relay switch which operates on only a momentary pulse of current. Follow the action. When the black button is pushed, the contacts close and energize the relay coil, which closes contact S. The push button does not have to be held down because once the relay switch is closed, it remains closed or LOCKED OUT. The closed switch S completes the circuit to the motor or device. The contacts of the stop or red button are normally closed. To stop the motor, a brief push on the red button opens the circuit to the relay. This allows contact S to open. For larger motors, the push button starter may be used to trigger other starting devices.

THYRISTOR MOTOR CONTROLS

The thyristor is a family of semiconductor devices that includes the silicon controlled rectifier (SCR) and the triac. These devices are used in many circuits, including motor controls, to control the speed of a motor, Fig. 12-28.

A half-wave SCR motor control circuit is shown in Fig. 12-29. This circuit controls up to seven amps in a universal motor. These are often found in hand drills and hand saber saws.

The triac controls the full ac wave, while the SCR only controls half the wave. A full-wave triac circuit is shown in Fig. 12-30.

Fig. 12-31 shows a number of SCR packages. The pencil points to an unencapsulated pellet. The two types in the foreground can handle less than an amp of current. The larger types can control up to 2000 amps.

Fig. 12-26. Push button starter.

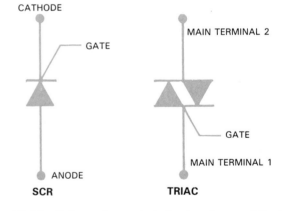

Fig. 12-28. Schematic symbols for two types of thyristors.

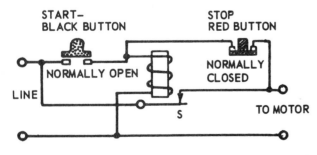

Fig. 12-27. Circuit of push-button starter showing how relay is "locked out."

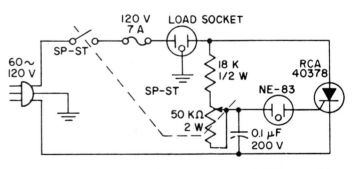

Fig. 12-29. Half-wave SCR motor control (RCA circuit).

Electric Motors

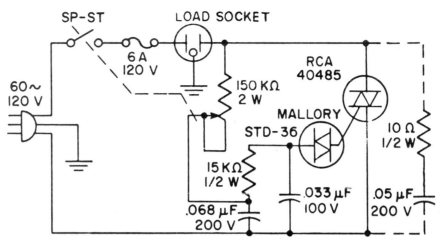

Fig. 12-30. Triac motor control (RCA circuit).

Fig. 12-31. A variety of SCR packages.

THE UNIVERSAL MOTOR

The Laws of Magnetism were used to explain the operation of the dc motor. But, will a dc motor operate on alternating current? Yes, to a limited extent.

With an alternating current the poles of both the field and armature windings will periodically reverse. However, since two north poles repel each other, as do two south poles, motor action continues in the dc motor when ac is applied. For best results a series motor should be used. When the shunt motor is connected to ac, the inductance of the field windings causes a phase displacement. Motor action will be impaired. Conduct an experiment. Connect the dc motor made in the laboratory as a series motor. Apply about 6 to 8 volts ac to its terminals.

When a universal type motor is used in industry, the series wound type is preferred. These motors are not used for heavy-duty purposes because of the large amount of sparking at the brushes. Commercial types are used for small fans, drills, and grinders.

REVIEW QUESTIONS FOR SECTION 12.2

1. Name the type of motor shown in the following schematic.

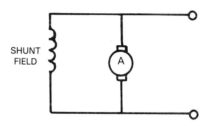

2. What is a key advantage of a series motor?
3. There are two types of compound motors. Name them.
4. _____ are a type of starting switch that will not close until the current has decreased to a preset value.
5. A _____ controls a half-wave of ac.
   a. SCR.
   b. Triac.
   c. Universal motor.
   d. None of the above.

12.3 INDUCTION MOTORS

The induction motor is a special class of motor. It is designed to operate on alternating current. It is made with the usual field poles and windings. The field poles and windings remain stationary and are called the STATOR, Fig. 12-32. An alternating current is connected to these field coils. The polarity of the coils changes from time to time, based on the frequency of the current.

Electricity and Electronics

Fig. 12-32. A fractional horsepower induction motor. Notice the stator coils and rotor.

Think of these windings as the primary of a transformer. Another coil of wire closely coupled to this primary will induce a current to flow in it. According to Lenz's Law, then, the induced current will produce a field polarity opposite to the field which causes it. In this case that is the primary or stator windings. Therefore, there is an attractive force between the two coils.

In the induction motor these secondary coils are mounted in slots in a laminated armature. The armature is called the ROTOR. See Fig. 12-33. This motor will not start rotation, it will only hum. If started by hand, it will keep rotating at a speed of 3600 rpm, at 60 Hz ac. The 3600 is derived from $60 \times 60 = 3600$, or the number of cycles per minute. The alternations of the ac are functioning like the commutator in a dc motor. The current and polarity in both reverse routinely.

Fig. 12-33. The rotor and centrifugal switch assembly in an induction motor.

Electric Motors

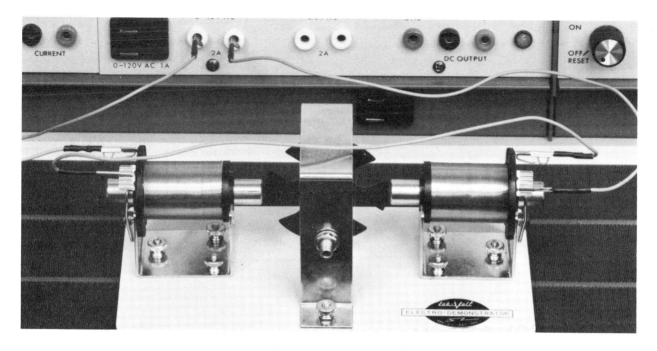

Fig. 12-34. The electrodemonstrator set up as an ac motor.

Replace the armature on the laboratory made motor with the ac rotor, Fig. 12-34. Connect the field windings to a 6 volt ac source. The motor will not start by itself. But twirl the rotor with your fingers, and the motor continues to rotate. Some method is needed to start the induction motor.

THREE-PHASE INDUCTION MOTOR

A three-phase alternating current has three separate currents. They are 120 degrees out of phase with each other. This type of ac can produce a rotating magnetic field in the stator. If the field rotates, then the rotor will follow the field around.

Part of a three pole stator is drawn in Fig. 12-35. Phase 1 current is connected to Pole A. Phase 2 is connected to Pole B. Phase 3 is connected to Pole C. When Phase 1 rises to maximum, the rotor is attracted to Pole A. As Phase 1 starts to decrease, Phase 2 rises to maximum. It attracts the rotor to Pole B. As Phase 2 starts to decrease, Phase 3 rises to maximum. It attracts the rotor to Pole C. Starting again with Pole A and Phase 1, the field rotates around the stator and attracts the rotor with it. The phase relationship between currents is shown using sine waves in Fig. 12-36.

The rotor consists of a laminated iron cylinder with heavy closed loops of copper wire. These are imbedded in slots around the surface. Because the rotor looks like a cage, it is called a SQUIRREL CAGE rotor.

A polyphase motor is shown in Fig. 12-37. The rotor cannot follow the rotating field exactly. If both were

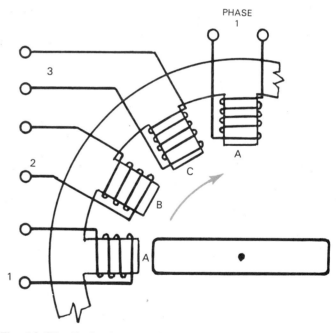

Fig. 12-35. Each phase of the three-phase current is connected in sequence to stator coils.

in step, the rotor would not cut across any magnetic lines of force. No current would be induced in the rotor coils. A certain amount of SLIP exists between the rotating field and the rotor. Generally, this is about 4 to 5 percent. Therefore the rotor turns at 1720 rpm when the field rotates at 1800 rpm. As the motor is loaded, slippage increases and heavier currents are induced in the rotor.

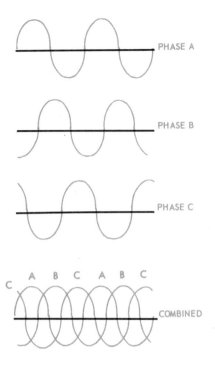

Fig. 12-36. The phase relationship between currents in a three-phase power source.

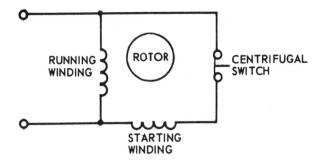

Fig. 12-38. Schematic of a split-phase motor, showing starting windings.

DELCO—REMY DIV., GENERAL MOTORS CORP.

Fig. 12-37. Cutaway view of a polyphase motor.

SINGLE-PHASE INDUCTION MOTORS

In industry, three-phase electricity is used to start large induction motors. At home, such currents are not readily available. So, other means must be provided for starting an induction motor. If a single-phase current is divided into a polyphase current, the starting problem would be solved. This is called SPLITTING PHASES.

Two methods are used to produce a SPLIT-PHASE. You learned that an inductance causes current to lag in a circuit. If the single-phase current flows through two parallel paths that contain unequal amounts of inductance, one current will lag behind the other. A phase displacement will exist between them. A two-phase current has been made of a single-phase current.

The many turns of wire that comprise the field windings of the induction motor are excellent inductances. They can be used for the purpose of phase splitting. A schematic diagram of a split-phase motor is shown in Fig. 12-38.

The STARTING windings consist of many turns of relatively fine wire. The RUNNING windings have many turns of heavier wire. The windings are placed in slots around the inside of the stator.

When electricity is applied to the motor terminals, current flows in both windings because they are in parallel. The main winding has many turns of wire and high inductance. This causes the current to lag almost 90 degrees. The starter windings of fine wire have much less inductance. Current does not lag by as great an angle. The phase displacement creates a rotating field similar to the three-phase motor and the rotor starts to run. When the motor reaches speed, a centrifugal switch open the starter winding circuit. Then the motor runs as a single-phase inductor motor. The fine wire in the starting winding will not stand constant use. It is cut out of the circuit after it has performed its function.

The second method of phase splitting involves the use of a capacitor. Recall that capacitance causes current to lead applied voltage. If a capacitor is connected in series with the starting winding, a much larger phase displacement can be created. The capacitor is switched out of the circuit when the motor approaches full speed, Fig. 12-39. This motor is called a CAPACITOR-START induction motor. These motors are used for many jobs around the laboratory and home. They maintain a fairly constant speed under varying loads.

Electric Motors

They do not develop a strong starting torque compared to the three-phase squirrel cage motor. A capacitor-start motor is shown in Fig. 12-32. Note the starting capacitor mounted on top.

REPULSION INDUCTION MOTOR

A repulsion induction motor resembles a dc motor because it also has a commutator and brushes. However, in the repulsion induction motor, the brushes are connected to each other, rather than to the source of power. The brushes are arranged so that only chosen armature coils are closed or complete circuits at any time. The currents induced by transformer action in the rotor windings are displaced by the shorting brushes. Poles are created which oppose the stator poles.

Rotation is started by repulsion. When the motor reaches about 75 percent of its speed, a centrifugal shorting ring "shorts out" all the commutator sections. The motor runs as an induction motor. The repulsion induction motor has high starting torque, but does not quickly come up to speed under load. It is being replaced by the less costly capacitor-start motor.

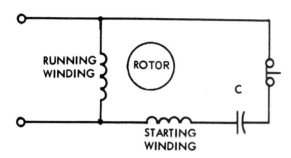

Fig. 12-39. Diagram of a capacitor-start, split-phase motor.

SHADED POLE MOTOR

A two-pole motor having an uncommon method for starting rotation is shown in Fig. 12-40. The rotor is a squirrel cage type. The stator resembles the common dc motor, except that a slot is cut in the face of each pole and a single turn of heavy wire is wound in the slot. This is called SHADING THE POLE. The single turn of wire is the SHADING COIL.

The action of the shading coil can be seen in Fig. 12-41. As the current rises in the first quarter cycle of the ac wave, a magnetic field is made in the field winding. However, the expanding magnetic field cuts across the shading coil. This induces a current and polarity which opposes that of the field coil. This tends to decrease and weaken the total field on the side of the shading coil.

At the top of the ac wave very little current change occurs. The total magnetic field becomes equalized across the face of the pole. When the current starts to

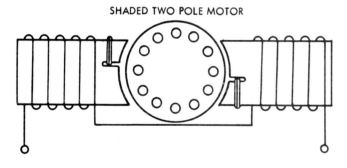

Fig. 12-40. Sketch of a shaded pole motor.

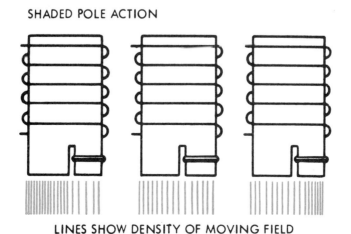

Fig. 12-41. The magnetic field moves from left to right as the result of shading.

decrease in the next quarter cycle, the polarity of the induced current in the shading coil resists the decreasing current. So the pole has a strong magnetic field on the right side. Notice that during this half-cycle the magnetic field has moved from left to right. This movement of the field causes the rotor to start rotation.

The starting torque of this motor is quite small. Therefore, it has many uses, such as in small fans and electric clocks.

REVIEW QUESTIONS FOR SECTION 12.3

1. What is an induction motor?
2. The stationary field poles and windings in an induction motor are called a _____.
3. A three-phase alternating current has three separate currents that are _____ out of phase with each other.
   a. 90 degrees.
   b. 45 degrees.
   c. 180 degrees.
   d. 120 degrees.
4. What is meant by splitting phases?
5. The _____ windings have many turns of fine wire, while _____ windings are made with many turns of heavier wire.
6. How does a repulsion induction motor differ from a standard dc motor?

SUMMARY

1. A motor is a device for changing electrical energy to mechanical energy.
2. Like poles repel each other; unlike poles attract each other.
3. Counter emf, or cemf, is the induced voltage that opposes the applied voltage.
4. The rotation of the motor produces turning or twisting power called torque.
5. Percent of Speed Regulation =
   $$\frac{\text{Speed no-load} - \text{Speed full-load}}{\text{Speed full-load}} \times 100$$
6. The series motor has the field windings connected in series.
7. Shunt motors have their field windings connected in parallel.
8. Compound motors have their field windings connected in series and parallel.
9. There are many types of motor starting circuits. Those discussed in the chapter include both manual and automatic types.
10. A silicon controlled rectifier is a semiconductor device that controls half of the sine wave while a triac controls the full wave.
11. In an induction motor, the field poles and windings remain stationary.
12. Induction motors need starting circuits. They have starting and running windings.

TEST YOUR KNOWLEDGE, Chapter 12

Please do not write in the text. Place your answers on a separate sheet of paper.

1. What is the purpose of field windings in the operation of a dc motor?
2. Explain generator action in a motor.
3. Interpoles are used to prevent _____ at the commutator brushes.
4. Draw diagrams of a series motor and of a shunt motor.
5. Motor starting circuits are intended to protect motors until they build up _____ and _____.
6. What are two advantages of automatic starting circuits?
7. The _____ is a family of semiconductor devices.
8. Explain the operation of a three-phase induction motor.
9. A _____ _____ is a heavy wire wound into a slot cut on the face of a pole.

FOR DISCUSSION

1. Why does a motor increase in speed when its field strength is decreased?
2. Why does a motor get hot when overloaded?
3. If a load is removed from a series motor, it will destroy itself by centrifugal force. Explain this occurrence.
4. Why are motor starters necessary on heavy-duty motors?
5. Will your experimental dc motor run on 6 volts ac? Explain your answer.
6. Compare an induction motor to a transformer.

Electric Motors

Small electric motors are used to move sheets of paper through laser printers.

Part III Summary

ALTERNATING CURRENT

IMPORTANT POINTS

1. Inductance is that property of a circuit which opposes any change in current. Capacitance is that property of a circuit which opposes any change in voltage.
2. Components which give inductance to a circuit are called inductors, chokes, coils, or reactors.
3. Time constant (t) in an ac circuit is the amount of time it takes for the current in an inductive or capacitive circuit to rise from 0 to 63.2% of its maximum value. There are two formulas for time constant. The formula for inductance is:

$$t = \frac{L \text{ (in henrys)}}{R \text{ (in ohms)}}$$

The formula for capacitance is:

$$t = R \text{ (in ohms)} \times C \text{ (in farads)}$$

4. Mutual inductance, which is the basis for transformer action, occurs when two or more coils share the energy of one coil.
5. When a tap is made on the secondary, two voltages are produced, one being 180 degrees out of phase with the other.
6. The basic reasons for using a transformer are:
   a. To step up or step down voltage or current.
   b. To provide two voltages 180 degrees out of phase with each other.
   c. To provide isolation from the primary to the secondary.
7. Transformer losses can be copper losses, hysteresis losses, or eddy current losses.
8. Formulas for resistors, inductors, and capacitors in series or parallel are shown below. All units used in the formulas must have the same prefix or basic value.

COMPONENT	IN SERIES	IN PARALLEL
RESISTORS	$R_T = R_1 + R_2 + R_3$	$\frac{1}{R_T} = \frac{1}{R_1} + \frac{1}{R_2} + \frac{1}{R_3}$
INDUCTORS	$L_T = L_1 + L_2 + L_3$	$\frac{1}{L_T} = \frac{1}{L_1} + \frac{1}{L_2} + \frac{1}{L_3}$
CAPACITORS	$\frac{1}{C_T} = \frac{1}{C_1} + \frac{1}{C_2} + \frac{1}{C_3}$	$C_T = C_1 + C_2 + C_3$

SUMMARY REVIEW POINT 8

9. Inductive reactance (X_L) is the actual opposition to change in current offered by an inductor. Capacitive reactance (X_C) is the actual opposition to a change in voltage offered by a capacitor. Both are measured in ohms. The formula for inductive reactance is:

$$X_L = 2\pi fL$$

The formula for capacitive reactance is:

$$X_C = \frac{1}{2\pi fC}$$

10. The total opposition to a change in either current or voltage in a circuit, by either an inductor or a capacitor, is impedance (Z). Impedance is measured in ohms.
11. Factors which affect resistance, inductance, and capacitance in a circuit are shown on the following page.

ELECTRICAL OPPOSITION	SYMBOLS	FACTORS THAT AFFECT ELECTRICAL OPPOSITION
RESISTANCE (R) (measured in ohms, Ω)	—⋀⋀⋀— —⋀⋀⋀— —⋀⋀⋀—	1. Length of conductor. 2. Surface area of conductor. 3. Temperature of conductor. 4. Material of conductor.
INDUCTANCE (L) (measured in henrys, H)	(coil symbols)	1. Number of windings. 2. Types of winding. 3. Type and make-up of core. 4. Diameter of coil winding. 5. Spacing between windings.
CAPACITANCE (C) (measured in farads, F)	(capacitor symbols)	1. Distance between plates. 2. Plate area. 3. Dielectric material.

SUMMARY REVIEW POINT 11

12. Resonance occurs when $X_C = X_L$.
13. In a series resonant circuit, the current is maximum and the impedance is minimum at the resonant frequency.
14. In a parallel resonant circuit, the current is minimum and the impedance is maximum at the resonant frequency.
15. The Q of a circuit describes the relationship between X_L and R:

$$Q = \frac{X_L}{R}$$

16. A motor is a device which converts electrical energy to mechanical energy.
17. There are three basic types of field winding connections in motors:
    a. Series.
    b. Parallel.
    c. Compound.

SUMMARY QUESTIONS

1. Convert the following units:
   a. 6.4 mH = _____ μH.
   b. 2200 μF = _____ F.
   c. 5 H = _____ mH.
   d. 16,000 pF = _____ μF.
2. Solve for the unknown quantities.
   I_S = _____ A.
   E_P = _____ V.

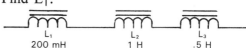

$N_P = 800$ $N_S = 3200$
$I_P = 4\ A$ $E_S = 60\ V$

3. Find L_T.

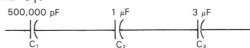

L_1 200 mH L_2 1 H L_3 .5 H

4. Find C_T.

500,000 pF   1 μF   3 μF
C_1 C_2 C_3

5. Find X_L in the circuit shown.

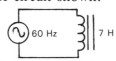

60 Hz   7 H

6. Find the missing values in the circuit shown.
   Z = _____
   Phase L = _____.
   I _____ (leads, lags) E by _____.

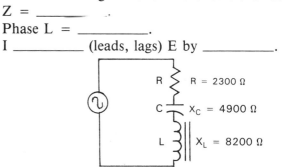

R = 2300 Ω
X_C = 4900 Ω
X_L = 8200 Ω

7. Find X_C.

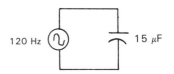

120 Hz   15 μF

Part IV

ELECTRONIC DEVICES AND APPLICATIONS

13 Basic Electronic Devices
14 Integrated Circuits

In the history of electronics, there are three generations of devices: vacuum tubes, transistors, and integrated circuits. The invention, development, and use of each of these devices has had an impact on the electronics field. The amount of time from invention to development to use has shortened with each of these three generations. And each new device is more reliable, less expensive, and more capable than its older counterpart. In Part IV, you will learn more about some of these devices.

Chapter 13 is titled *Basic Electronic Devices*. The primary devices presented are semiconductors such as diodes and transistors. Because they are no longer in wide use, vacuum tubes will be touched on briefly.

Integrated circuits are covered in detail in Chapter 14. Their production is explained, and several types of ICs are discussed.

Electricity and Electronics

Various electronic devices and components are assembled into modern system used in such applications as desktop publishing. Shown here are a Touchscreen computer and a LaserJet printer.

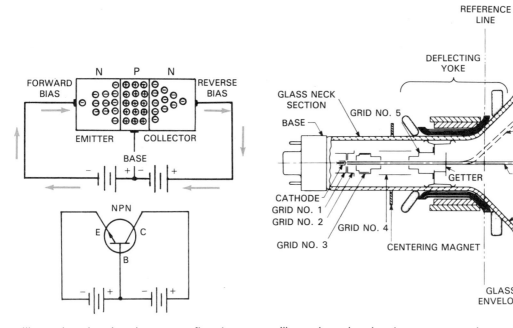

Illustration showing the current flow in a NPN transistor.

Illustrations showing the components in a typical CRT used in a television.

Electronics is so much a part of our lives. We ought to become familiar with the devices and terms. We may take computers for granted, but none of them would work without semiconductor materials, diodes, and transistors.

Chapter 13

BASIC ELECTRONIC DEVICES

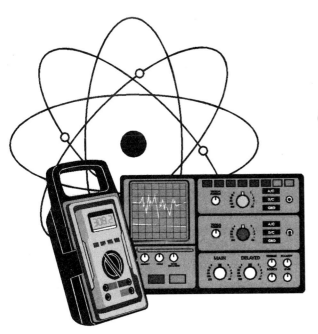

After studying this chapter, you will be able to:
- *Define electronics.*
- *Explain the doping process.*
- *Discuss various types of semiconductor diodes.*
- *Summarize the operating theories of bipolar and field effect transistors.*
- *Distinguish between diodes, triodes, tetrodes, and pentodes.*
- *Illustrate the operation of a cathode ray tube.*

ELECTRONICS may be defined as the study of the flow of electrons in active devices such as integrated circuits, transistors, and vacuum tubes. A time line showing the development of these three generations of electronic devices is shown in Fig. 13-1. During experiments in 1883 with the incandescent lamp, Thomas Edison noticed that electrons could be driven off the hot filament of a lamp. He called this the "Edison Effect." In 1904, J.A. Fleming built a diode (two element vacuum tube) that changed ac to dc (rectification). That same year, Lee DeForest invented the first vacuum tube amplifier. It was called a triode. This material will be discussed in detail later in this chapter.

The first solid-state amplifier device was called the transistor. It was invented at Bell Laboratories in 1948. The technology used in transistors was key to the development of the third generation device, the integrated circuit.

13.1 SEMICONDUCTORS

On July 1, 1948, the New York Herald Tribune announced:

"A tiny device that serves nearly all of the functions of a conventional vacuum tube, and holds wide promise for radio, telephone and electronics was demonstrated yesterday by the Bell Telephone Laboratories scientists who developed it. Known as the transistor . . ."

The scientists, John Bardeen and Walter Brattain invented the point contact transistor. It had two wires carefully fused on a crystal of germanium. William Shockly followed by creating the bipolar, or junction, transistor. These inventions were also the start of microelectronics.

The transistor provided instant circuit operation. No warm-up time was needed as with the vacuum tube circuit. Large amounts of power were not needed. The transistor was, and is, known for its small size, long life, and light weight.

The study of transistors or semiconductors requires understanding of new concepts in electronic theory. Briefly, what have we learned about conductors? We have learned that a copper wire is a good conductor. We have also learned that certain materials, such as glass and rubber, are insulators. Between insulators and conductors are other materials. These are known as semiconductors. The transistor is a semiconductor. It is neither a good conductor nor a good insulator.

ATOMIC CHARACTERISTICS

To review Chapter 1, diagrams of the atomic structures of germanium (Ge) and silicon (Si) are shown in Fig. 13-2. Silicon is in great supply. It makes up 28 percent of the world's surface. The atomic number of Ge is 32 and of Si, 14. Recall that the atomic number is the number of electrons in orbit outside the nucleus of an atom. Therefore, germanium has 32 electrons in its orbit. Silicon has 14 electrons in its orbit.

Note that both germanium and silicon have four electrons in their outer rings. These electrons on the outer

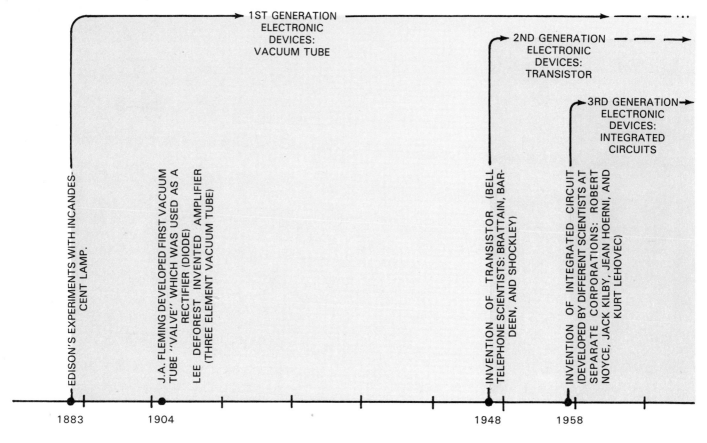

Fig. 13-1. Time line of notable electronic devices.

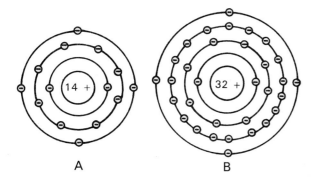

Fig. 13-2. Atomic structures. A—Silicon has an atomic number of 14. There are 14 protons within the nucleus and 14 electrons orbiting the nucleus. B—Germanium has an atomic number of 32. There are 32 protons in the nucleus and 32 electrons orbiting the nucleus.

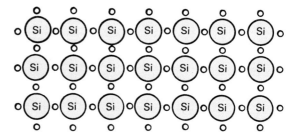

Fig. 13-3. The crystalline structure of silicon. The electrons in the outer ring only are shown.

Fig. 13-3 shows a silicon structure. It is a pure insulator. It will not conduct electricity.

CONDUCTION OF ELECTRICITY

Current flow in a conductor results from energy transfer by electrons. An example of this is shown in Fig. 13-4. An electron is added to one end of the conductor. Another electron leaves the opposite end. This chain reaction is the conduction of electricity by ELECTRONS.

rings are called VALENCE ELECTRONS. In some atoms, valence electrons are able to bond with the valence electrons of another atom. This is called a COVALENT BOND. The atoms are each bonded to their own nucleus and to each other. This bond creates a LATTICE CRYSTALLINE structure. Germanium and silicon both lend themselves to this type of bond.

Basic Electronic Devices

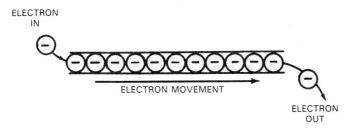

Fig. 13-4. Electricity conducted by the movement of electrons.

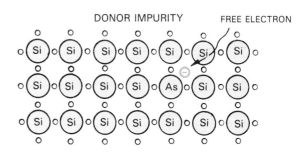

Fig. 13-7. Arsenic (As) is a pentavalent impurity. It leaves free electrons and is called a donor impurity.

DOPING

Recall that PURE silicon is a good insulator. Adding an impurity to the silicon will change the conduction traits of the material. The process of adding impurities to pure semiconductor material is called DOPING. The result of doping is an EXTRINSIC SEMICONDUCTOR. This means the material is not in its pure form. A pure semiconductor is INTRINSIC.

Trivalent and pentavalent impurities are most often used for doping in transistors. TRIVALENT impurities have three valence electrons. Pentavalent impurities have five valence electrons. See Fig. 13-6.

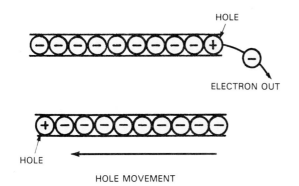

Fig. 13-5. Electricity conducted by the movement of holes.

In Fig. 13-5, an electron is removed from one end of the conductor. This leaves a hole where there should be an electron, but there is not. This hole is strongly attracted to another electron. It is positive. The next electron fills up the hole and leaves a positive hole in its place. Each positive hole is filled in turn by the next electron. Then the vacant hole appears at the other end of the conductor. This chain reaction is the conduction of electricity by HOLES. The study of transistors is concerned with conduction by electrons and by holes.

Adding doping elements to semiconductor material results in either an excess or shortage of electrons in the covalent bond. For example, arsenic, a pentavalent, is added to silicon. Four valence electrons of arsenic will form a covalent bond with the four valence electrons of silicon. One electron from the arsenic will remain free, Fig. 13-7. Arsenic donated a free electron to the structure. All pentavalent impurities donate free electrons. They are called DONOR IMPURITIES. In this structure, electricity is conducted by (negative)

	ELEMENT	SYMBOL	ATOMIC NUMBER	VALENCE ELECTRONS
TRIVALENT ACCEPTOR ELEMENT PRODUCES P TYPE MATERIALS	BORON	B	5	3
	ALUMINUM	Al	13	3
	GALLIUM	Ga	31	3
	INDIUM	In	49	3
QUADVALENT PURE SEMI-CONDUCTOR MATERIAL	SILICON	Si	14	4
	GERMANIUM	Ge	32	4
PENTAVALENT DONOR ELEMENT PRODUCES N TYPE MATERIALS	PHOSPHORUS	P	15	5
	ARSENIC	As	33	5
	ANTIMONY	Sb	51	5

Fig. 13-6. Comparison of trivalent, quadvalent, and pentavalent elements. Quadvalent elements are pure semiconductors.

electrons. Therefore, the crystal becomes an N TYPE crystal.

What occurs when a trivalent impurity is added to a pure semiconductor material? It has only three electrons to join in covalent bond. One more electron is needed to complete the lattice structure. A hole remains in the place of the missing electron, Fig. 13-8. This hole will accept an electron. A trivalent impurity, then, is called an ACCEPTOR IMPURITY. In this structure, electricity is conducted by (positive) holes. The crystal become a P TYPE crystal.

Fig. 13-9 shows conduction through an N type crystal. Fig. 13-10 shows conduction through a P type crystal. Notice that the hole moves in reverse of the electron flow.

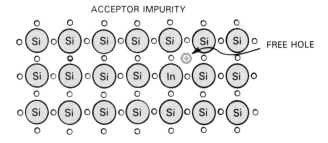

Fig. 13-8. Indium (In) is a trivalent impurity. It leaves positive holes and is called an acceptor impurity.

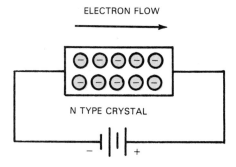

Fig. 13-9. Conduction through an N type crystal by electrons.

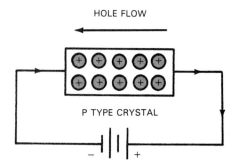

Fig. 13-10. Conduction through a P type crystal by holes.

REVIEW QUESTIONS FOR SECTION 13.1

1. Define electronics.
2. Electrons on the outer ring of an atom are called _____ _____.
3. What is a covalent bond?
4. Doping:
   a. Is the process of adding impurities to pure semiconductor material.
   b. Results in an extrinsic semiconductor.
   c. Changes the conduction traits of a pure material.
   d. All of the above.
5. Pentavalent impurities are known as _____ impurities and they produce _____ type crystals. Trivalent impurities are known as _____ impurities and they produce _____ type crystals.

13.2 SEMICONDUCTOR DIODES

A DIODE is a device designed to permit electron flow in one direction and block flow from the other direction. A diode consists of two electrodes: a cathode and an anode. A CATHODE is an electrode that emits (gives off) electrons. An ANODE collects the electrons and puts them to use.

A semiconductor diode is the result of fusion between a small N type crystal and a P type crystal, Fig. 13-11. At the junction of the two crystals, the carriers (electrons and holes) tend to diffuse. Some electrons move across the barrier to join holes. Some holes move across the barrier to join electrons. Remember that unlike charges attract each other.

Due to diffusion, a small voltage or potential exists between the regions near the junction. How does this happen? The region of the P crystal near the junction becomes negative. It has taken electrons from the N crystal. The region of the N crystal near the junction becomes positive. It has lost some electrons, but gained

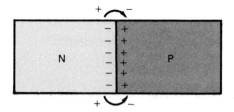

Fig. 13-11. A semiconductor diode made of N and P crystals.

holes. This voltage or potential is called a POTENTIAL HILL or POTENTIAL BARRIER. The barrier prevents the other electrons and holes in the crystal from joining.

The symbol for a semiconductor diode is shown in Fig. 13-12. Using the electron theory of current flow, the arrow side of the symbol denotes the anode portion of the diode. The anode contains the P type material. The bar side of the symbol denotes the cathode portion of the diode. The cathode contains the N type material.

A voltage (potential) is connected across a diode in Fig. 13-13. The positive terminal of the source is connected to the P crystal. The negative source is connected to the N crystal.

The negative electrons in the N crystal move toward the barrier. The positive holes in the P crystal move toward the barrier. The source voltage opposes the potential hill and reduces its barrier effect. This allows the electrons and holes to join at the barrier. Therefore, current flows in the circuit. It flows in the P crystal by holes. It flows in the N crystal by electrons. The diode is BIASED in a FORWARD direction.

Fig. 13-14 shows the same junction diode BIASED in a REVERSE direction. The positive source is connected to the N crystal and the negative source to the P crystal.

In reverse bias, the source voltage aids the potential hill. Electron/hole combinations are limited at the junction. The electrons in the N crystal are attracted to the positive source terminal. Very little current will flow in the circuit. Reverse voltage can be increased to a point where the diode will break down.

POINT CONTACT DIODES

The point contact diode is used for detection and rectification. It consists of a small piece of N type germanium crystal. Against this is pressed a fine phosphor bronze wire. While the diode is being made, a high current is run through the diode from wire to crystal. This forms a P type region around the contact point in the germanium crystal. The point contact diode, therefore, has both the P and N type crystals. Its operation is like that of the junction diode. A crystal diode used in a radio detector is shown in Fig. 13-15, with its schematic symbol. This diode will be used in a number of projects found in the text.

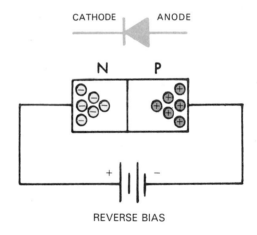

Fig. 13-14. Conduction through a junction diode biased in a reverse direction.

Fig. 13-12. Symbol for a semiconductor diode.

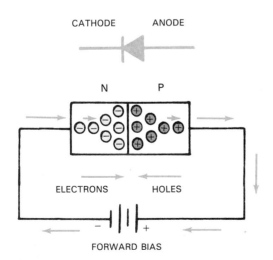

Fig. 13-13. Conduction through a junction diode biased in the forward direction.

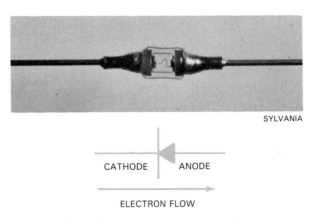

Fig. 13-15. A typical crystal diode.

SILICON RECTIFIERS

Diodes conduct current more easily in one direction than in the other. This process is called RECTIFICATION. A key semiconductor diode is the SILICON RECTIFIER. This device is used in the power rectification circuits that will be discussed in Chapter 15.

Silicon rectifiers come in a variety of shapes and sizes. Fig. 13-16 shows some common diode outlines.

Silicon rectifiers have high forward-to-reverse current ratios. They can achieve rectification efficiencies of greater than 99 percent. These rectifiers are very small and light. They are built to last a long time. They can be made resistant to shock and mishandling. They do not decay with age like a vacuum tube rectifier does.

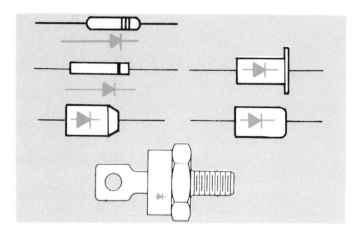

Fig. 13-16. Some widely used silicon rectifier outlines.

DIODE CHARACTERISTICS AND RATINGS

Ratings of diodes and rectifiers are commonly based on current and voltage capabilities and PEAK INVERSE VOLTAGE (PIV). The PIV rating is used by manufacturers. It defines the greatest reverse voltage that can be applied across a rectifier without damaging it. An example of this is a 1 amp at 50 PIV for a 1N 4001 silicon rectifier.

Many diodes are identified by an "1N." Examples of this are 1N 4001 or 1N 5400. Some manufacturers use their own labels: HEP 320, SK 3051, or 276-1102.

SERIES AND PARALLEL RECTIFIER ARRANGEMENTS

Diodes or rectifiers can be connected in series or parallel. This is done to improve the voltage or current capabilities of a single rectifier. Connecting diodes in series increases voltage ratings over the value of a single rectifier. Diodes may be connected in parallel to improve the current handling ability of the combination over that of only one diode.

TESTING DIODES

You have learned that diodes with a forward bias direction have a low resistance to current flow. A reverse bias direction has a high resistance. These diodes may be tested using an ohmmeter.

To test a diode in the forward bias direction, connect the diode across the ohmmeter. Set the range switch on the low resistance setting (X1 or X10). The reading should be low. A low reading is from 50 to 1000 ohms. See Fig. 13-17A.

To test a diode in the reverse bias direction, connect the diode across the ohmmeter leads. Set the range

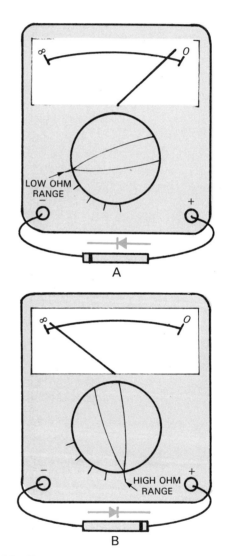

Fig. 13-17. Testing a diode using an ohmmeter. A—Testing forward bias. B—Testing reverse bias.

switch on a high resistance setting, such as X10 k, X100 k, or X1 Meg. A good diode will show a high resistance. See Fig. 13-17B.

Basic Electronic Devices

Many new digital meters have a diode test function built in. One model has a diode symbol on the range switch scale, Fig. 13-18. A properly connected diode can be tested by turning the switch to that function.

LIGHT EMITTING DIODES

Light emitting diodes (LEDs) are special function diodes. When connected in the forward bias direction, they emit (give off) light, Fig. 13-19. Light emitting diodes are made from semiconductor compounds such as gallium arsenide, gallium arsenide phosphide, and gallium phosphide.

An LED receives energy from the dc power source, or battery. This causes the electrons and holes to combine in the PN junction region of the diode. This creates PHOTONS. Each photon is from one hole (+ charge) and one electron (− charge), and is a "particle" of light that can be seen. Different LED colors use different materials in their manufacture. Fig. 13-20 shows the operating principle of an LED.

Basing diagrams for some common LEDs are shown in Fig. 13-21. Note that the cathode lead is the shortest lead. It is also closest to the notch on certain diodes. LEDs have many uses. For example, they are used on the instrument panel of cars, clocks, watches, etc.

> **REVIEW QUESTIONS FOR SECTION 13.2**
> 1. A _____ is a device that will pass current flow in one direction and block it in the other direction.
> 2. What occurrence prevents electrons and holes in a crystal from joining?
> 3. Draw and label the symbol for a semiconductor diode.
> 4. In _____ bias, the source voltage aids the potential hill. In _____ bias, the source voltage opposes the potential hill.
> 5. Look up a 1N 4003 diode in a parts catalog and determine the current rating and PIV rating.
> 6. Explain how a light emitting diode works.

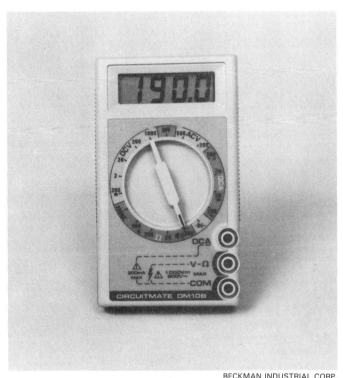

BECKMAN INDUSTRIAL CORP.

Fig. 13-18. This digital meter can also be used to test diodes.

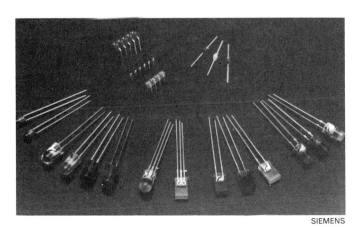

SIEMENS

Fig. 13-19. Various types of LEDs.

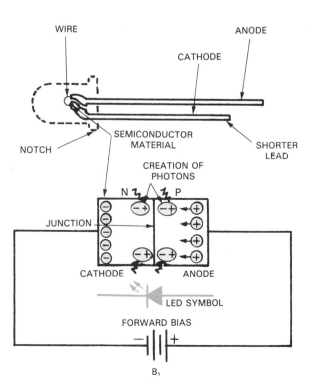

Fig. 13-20. LED operation.

Electricity and Electronics

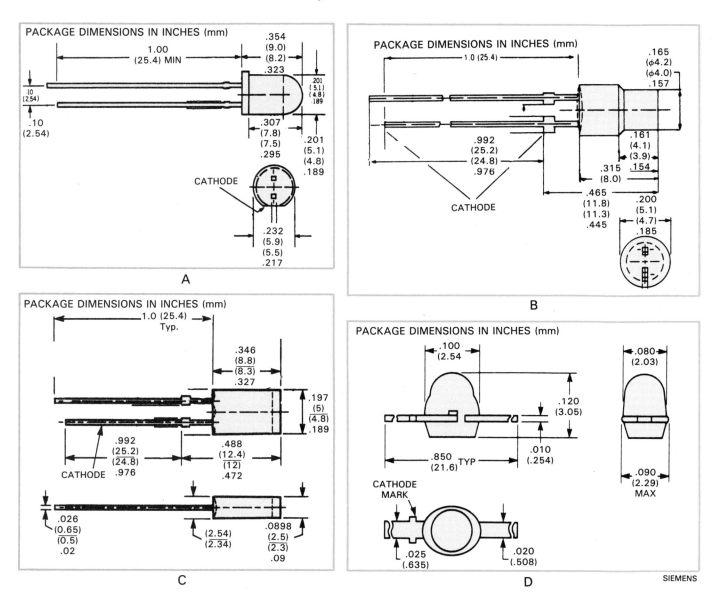

Fig. 13-21. LED basing diagrams A—T 1 3/4 LED lamp. B—Cylindrical LED lamp. C—Rectangular LED lamp. D—Miniature axial lead LED lamp.

13.3 TRANSISTORS

Transistors are key devices in electronics for several reasons. They are able to amplify current. They can create ac signals at desired frequencies (oscillation). And they can be used as switching devices. This makes them important in computer circuits.

The bipolar transistor consists of three layers of impure semiconductor crystals. This transistor has two junctions. There are two types of bipolar transistors, PNP and NPN. Blocks and schematic symbols for these are shown in Fig. 13-22.

A PNP bipolar transistor has a thin layer of N type crystals placed between two P type crystals. The NPN bipolar transistor has a thin layer of P type crystal placed between two N type crystals. In both types, the first crystal is called the EMITTER. The center section if called the BASE. The third crystal is called the COLLECTOR.

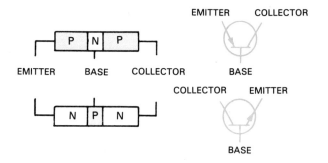

Fig. 13-22. Block diagrams and symbols for NPN and PNP transistors.

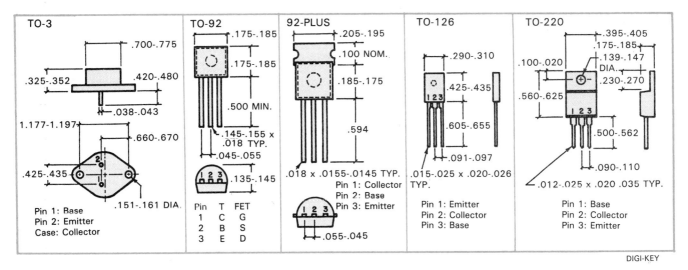

Fig. 13-23. Basing diagrams for five transistors.

In the schematic symbols, notice the direction of the arrow. This indicates whether it is a PNP or NPN transistor. Fig. 13-23 shows basing diagrams for five transistors.

The theory of NPN transistor operation is shown in Fig. 13-24. Two batteries are used for the sake of understanding. The negative terminal of the battery is connected to the N emitter. The positive terminal of the same battery is connected to the P type base. Therefore, the emitter-base circuit is forward biased.

In the collector circuit, the N collector is connected to the positive battery terminal. The P base is connected to the negative terminal. The collector-base circuit is reverse biased.

Electrons enter the emitter from the negative battery source and flow toward the junction. The forward bias has reduced the potential hill of the first junction. The electrons then combine with the hole carriers in the base to complete the emitter-base circuit. However, the base is a very thin section, about .001 in. Most of the electrons flow on through to the collector. This electron flow is aided by the low potential hill of the second PN junction.

Approximately 95 to 98 percent of the current through the transistor is from emitter to collector. About 2 to 5 percent of the current moves between emitter and base. A small change in emitter bias voltage causes a somewhat large change in emitter-collector current. The emitter-base current change is quite small, however.

A PNP transistor has a P type material for the emitter, an N type material for the base, and a P type material for the collector. See Fig. 13-25. The power supply or battery must be connected the opposite way of a NPN transistor. Like the NPN transistor, the emitter-to-base circuit has forward bias. And the collector-to-base circuit has reverse bias. In a PNP transistor, most carriers in the emitter-to-collector are holes.

FIELD EFFECT TRANSISTOR (FET)

The major difference between a transistor and a vacuum tube is that a vacuum tube is controlled by voltage and a transistor is controlled by current. All transistors are current amplifiers except one, the FIELD EFFECT TRANSISTOR. This special component is a voltage controlled amplifier. The construction of the device is shown in Fig. 13-26.

An N type material is diffused into a P type substrate material to form a thin channel. A gate of P material

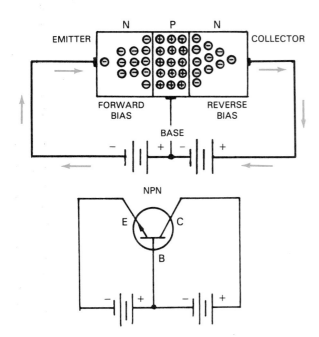

Fig. 13-24. The current flow in a NPN transistor.

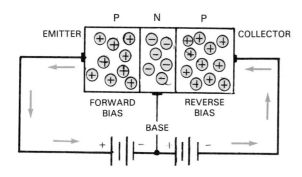

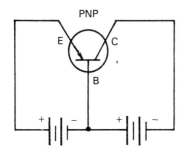

Fig. 13-25. The current flow in a PNP transistor.

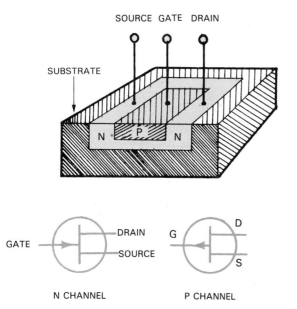

Fig. 13-26. Physical construction of an FET and its schematic symbol.

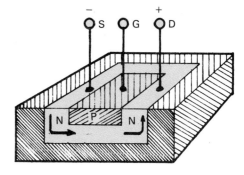

Fig. 13-27. Maximum current flows from source to drain. The channel is unrestricted.

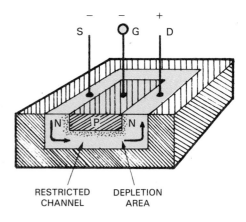

Fig. 13-28. Increased depletion area due to reverse bias of drain-gate junction will restrict the channel and reduce current flow.

is placed between the source and drain. Majority carriers perform conduction in an FET. In this case, electrons are the major current carriers between the source and drain. This current is shown in Fig. 13-27.

A DEPLETION AREA is void of current carriers. This can occur in a PN junction of a semiconductor diode. It is the result of electrons and holes diffusing across the junction. When the PN junction diode is reverse biased, the depletion area becomes wider.

This same effect can occur in an FET. It happens when the gate junction is reverse biased in respect to the source and drain, Fig. 13-28. The conduction channel between the source and drain is now restricted. The drain current is greatly reduced. A negative voltage can be applied to the gate which will cut off the current between source and drain. This is called PINCH OFF voltage. In normal operations, the gate is not forward biased.

Because a somewhat small voltage change at the gate produces a large change in drain current, the device is able to amplify.

The major advantage of the FET over the typical transistor is its high input impedance.

REVIEW QUESTIONS FOR SECTION 13.3

1. Name the three parts of a bipolar transistor.
2. Draw the symbols for an NPN transistor and a PNP transistor. Label the parts.
3. Name the three parts of a field effect transistor.
4. What is a depletion area?

13.4 VACUUM TUBES

Vacuum tube use has been replaced in large part by the use of transistors and integrated circuits. The vacuum tube still finds some special uses, such as in cathode rate tubes (CRTs). For the most part, however, vacuum tubes use more power, take up more space, and produce more heat then transistors and integrated circuits.

The vacuum tube is a direct result of Thomas Edison's experiment with the incandescent lamp. In this experiment, a metal plate was placed in the vacuum bulb containing the light. A battery and series meter were connected between the light filament and the plate. At this point, Edison discovered that electric current would flow through the light, if the positive terminal of the battery was connected to the plate. If the negative battery terminal was connected to the plate, however, no current would flow. See Fig. 13-29. Because the electron theory was not yet discovered, the flow of current through the light was a mystery.

Today, however, we understand electron theory. Certain metals and metal oxides give up free electrons when heated. The heat supplies enough energy to the electrons that they break away from the forces holding them in orbit. They become free electrons. This is known as THERMIONIC EMISSION.

In Edison's experiment, a cloud of free electrons was emitted from the filament of the light bulb. When a positive plate was placed in the bulb, the free electrons were attracted to the plate. (Unlike charges attract.) A flow of electrons means a current is present. The meter in the plate circuit meant that electrons were passing from filament to plate. The vacuum tube uses the principles discovered in the laboratory of Thomas Edison.

THERMIONIC EMITTERS

Better emitting materials have been developed in the years since Edison's carbon filament. Many materials, when heated to the point of emission, will melt. Tungsten seemed to be the best material for many years. It required a great deal of heat for proper emission, yet it was strong and durable. Tungsten is still used in large, high-power vacuum tubes.

To lower working temperatures and power usage, the thoriated-tungsten emitter was developed. It consisted of a thin layer of thorium placed on the tungsten emitter. Thorium is one of the heaviest metallic elements. Its symbol is TH, its atomic weight, 232.2. It is mined in India. This type of emitter produces the proper emission at much lower temperatures than pure tungsten.

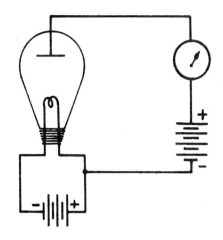

Fig. 13-29. Circuit diagram of the Edison discovery.

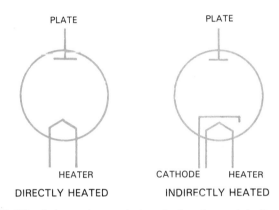

Fig. 13-30. Schematic symbols of directly and indirectly heated cathodes.

The most efficient emitter is the oxide-coated type. The emitter is metal, such as nickel. On this is formed a thin layer of barium or strontium oxide. Because of its low power usage and high emission, this type is widely used in vacuum tubes for radios, televisions, and other electronic devices.

CATHODES

The emitter in the vacuum tube is called the CATHODE. Heat may be supplied either directly or indirectly. Both methods have advantages. Schematic diagrams of both types are shown in Fig. 13-30.

Portable equipment that uses batteries as a source of power works best with directly heated cathodes. These is less heat loss. The filament can be designed so that only a small amount of power is consumed during use.

When an ac source is handy, the indirectly heated cathode is more useful. There is little power loss. And the heater voltage source and cathode can be separated. This ends any humming in the circuit.

DIODES

Recall that a diode consists of two electrodes. In the case of a vacuum tube diode, the two electrodes are the cathode and plate. Again, the cathode may be heated directly or indirectly. The plate is a round piece of metal that surrounds all elements in the tube. The plate acts as the collector of electrons emitted from the cathode. See Fig. 13-31.

The heater connections are H_1 and H_2. They may be joined to an ac source. The first number in a tube name is the approximate voltage that should be applied to the heaters. For example, a 6H6 tube needs 6.3 volts; a 12AX7 needs 12.6 volts; a 25Z6 needs 25 volts.

When the tube in Fig. 13-31 is turned on, the heaters indirectly heat the cathode. This causes thermionic emission of electrons. If the diode plate is joined to the positive terminal of the battery in the circuit, electrons will flow in the circuit from cathode to plate. If the connections are reversed, no electrons will flow. This electron tube acts as a one-way valve. It permits electron flow in one direction only.

At a certain temperature, the cathode emits the largest number of electrons. These form a space charge around the cathode. The cathode has been made slightly positive because of emitting electrons. Some of these electrons are attracted back to the cathode. When the plate is made positive, many electrons are attracted to it. This is electron flow. The plate can be made MORE positive by applying a higher voltage. A greater number of electrons are attracted. Then there is an increase in current.

Fig. 13-32 shows the increase in current as a result of an increase in plate voltage. At some point, as voltage is increased, all the emitted electrons will be attracted to the plate. Any further increase in voltage will not increase the current. This is called the SATURATION POINT of the electron tube.

TRIODES

A TRIODE is a three element tube made of a cathode, plate, and GRID. It was developed by Dr. Lee DeForest. In his experiments, DeForest inserted a fine wire mesh between the cathode and the plate of the tube. In doing this, he was able to control electron flow through the tube.

The grid in the triode most often has a cylindrical shape. It surrounds the cathode. Space between grid wires allows electrons room to pass through to the plate. The grid controls electron flow. It is commonly called the CONTROL GRID.

Electron flow is controlled by changes in plate voltage. In the triode, the grid also affects electron

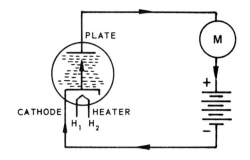

Fig. 13-31. Diode circuit shows direction of electron flow.

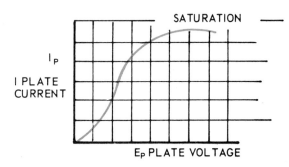

Fig. 13-32. As plate voltage increases, plate current increases.

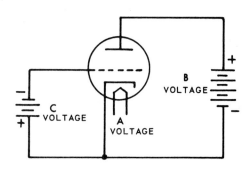

Fig. 13-33. This triode circuit shows the connections for plate voltage and grid bias voltage.

flow. For example, a negative grid will repel many electrodes back to the cathode. This limits the number of electrons passing on to the plate. As the grid is made more and more negative, a point is reached where no electrons flow to the plate. This is the CUT-OFF point of the tube. It is the negative voltage amount applied to the control grid that stops electron flow. The voltage applied to the control grid is the BIAS VOLTAGE. At cut off, it is called the CUT-OFF BIAS.

A triode with both plate and grid voltage is shown in Fig. 13-33. Notice that the grid bias battery has its negative terminal connected to the grid. In electronic work, these voltages have specific names. A voltage is for the heaters in the tube. B voltage is for the plate of the tube. C voltage is for the grid of the tube.

Portable radios and equipment use dry batteries, called A and B batteries, for voltage sources. In a radio working on ac power, these voltages come from the power supply. Earlier portable radios also used C batteries. C bias voltage, however, may now be obtained by other methods.

Current flow through an electron tube as the grid bias is changed is shown in Fig. 13-34. The plate voltage is held at a constant value. The curve in this graph is plotted by measuring the value of current at each change of grid voltage. At a grid bias of negative 2 volts, the current is 8 mA. At negative 6 volts, the current drops to 3 mA.

TETRODES

Without neutralization circuits, the triode is limited as an amplifier. This is due to the shunting effect of electrode capacitance at high frequencies, Fig. 13-35. To overcome this drawback, another grid is inserted in the triode. It is called the SCREEN GRID. It is placed between the control grid and the plate. This four element tube (cathode, control grid, screen grid, and plate) is called a TETRODE, Fig. 13-36.

The screen grid is bypassed to ground, externally, through a capacitor. The grid is a good screen between the control grid and plate. And it stops the grid-plate capacitance (C_{gp}). A dc voltage, slightly less in value than the plate voltage, is applied to the screen grid. The screen grid increases the speed of the electrons passing between cathode and plate. Some electrons are attached to the screen grid. This will cause a current to flow in the screen circuit. However, most electrons pass through the screen to the plate. While a tetrode has four elements, a pentode has five elements.

High amplification is possible in the tetrode because the control grid is placed quite close to the cathode. A decrease or increase in plate voltage has little effect on the plate current. This is due to the isolation effect of the screen grid. Tetrode transconductance is somewhat high.

PENTODES

In the tetrode, the electrons speed up to strike the plate with force. When this occurs, loosely held electrons on the plate are knocked off into free space. They form a space charge around the plate. This is called SECONDARY EMISSION.

Some of these electrons are drawn to the screen. They detract from useful plate current through the tube. This effect is more pronounced when plate voltage is below screen voltage.

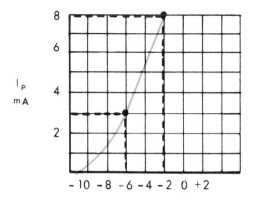

Fig. 13-34. The change in plate current, I_p, as a result of a change in grid voltage, E_g. The plate voltage is held at a constant value.

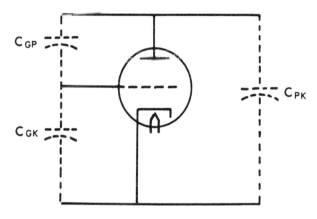

Fig. 13-35. Dotted lines show capacitance between the elements in a tube.

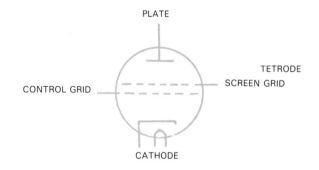

Fig. 13-36. Symbol for a tetrode.

To overcome the drawbacks of secondary emission, a third grid is placed in the tube between the screen grid and plate. This grid is called the SUPPRESSOR GRID. The tube now has five elements (cathode, control grid, screen grid, suppressor grid, and plate). It is called PENTODE.

Electricity and Electronics

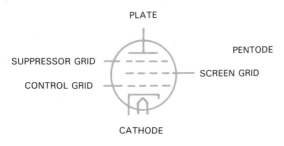

Fig. 13-37. Symbol for a pentode.

The suppressor grid is connected internally, to the cathode. It is at cathode potential. This grid repels free electrons resulting from secondary emission. It drives them back to the plate, Fig. 13-37.

These tubes have high amplification factors, high plate resistance, and high transconductance. Interelectrode capacitance is at a minimum. Pentodes are used as rf amplifiers and as audio power amplifiers. See Fig. 13-38.

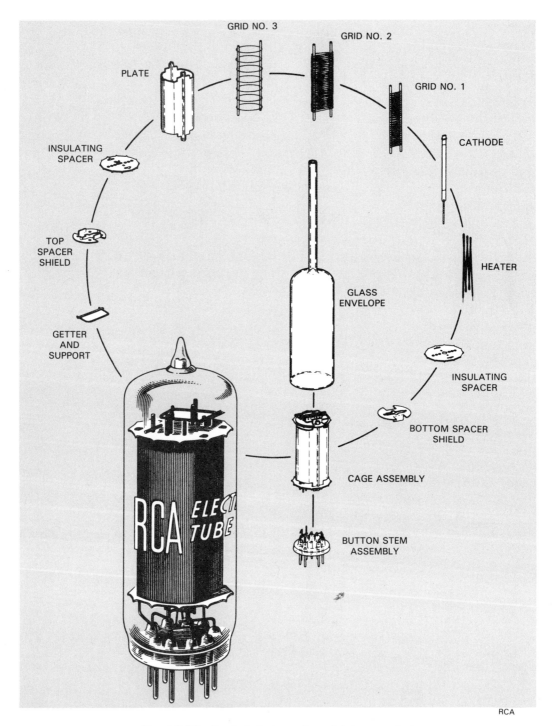

Fig. 13-38. Exploded view of a miniature pentode.

212

CATHODE RAY TUBES

Cathode ray tubes (CRTs) are also vacuum tubes. They work in the same manner. Common examples of CRTs include oscilloscope displays and computer monitors, Fig. 13-39.

The television tube is also a type of CRT. Refer to Fig. 13-40. On the left is the cathode and heater. The cathode emits a space cloud of electrons. Around the cathode is a metal cylinder. It is closed at one end except for a small hole (aperture). This controls the flow of electrons into the CRT. It is like the control grid. It also causes emitted electrons to move through the small hole in the shape of a beam.

Grid 2 works at a higher positive potential. It is strongly attracted to the negative electrons. This causes the electrons to speed up. Grids 3, 4, and 5 further accelerate and focus the electrons, resulting in a narrow beam of electrons.

Magnetic deflection coils around the neck of the tube move the electron beam. The beam is moved from left to right, top to bottom.

The inside of the picture tube face is treated with phosphor, a special type of luminescent material. The luminescence is a combination of fluorescence, the emission of light during the stimulus, and phosphorescence, the continuation of light after the stimulus has ceased. As the electrons strike the screen, the chemicals glow. The degree of brilliance is related to the strength of the beam. The picture is observed through the glass front of the tube.

The inside of the tube bulb contains a conductive coating. External connections to this coating are made from a high voltage power supply. When the electron beam strikes the CRT screen, electrons are knocked off. These are called secondary emission. These electrons are collected by the inner coating, completing the circuit.

LESSON IN SAFETY: Very high, dangerous voltages are present at the external connections to this anode (usually from 8000 to over 20,000 volts). Use caution when working close to these connections. Work with only one hand, to reduce the chance of death from accidental shock. Keep one hand in your pocket!

Fig. 13-39. Cathode ray tubes are used as display screens in computer monitors.

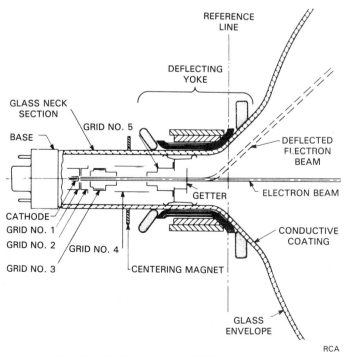

Fig. 13-40. Sketch of a CRT in a television.

REVIEW QUESTIONS FOR SECTION 13.4

1. Define thermionic emission.
2. The most efficient emitter is the _____-coated type.
3. The maximum limit of electron flow a vacuum tube will give off, regardless of increases in plate voltage, is the _____ _____.
4. The voltage applied to the control grid is the _____ voltage.
5. High amplification is possible with the:
   a. Diode.
   b. Triode.
   c. Tetrode.
   d. None of the above.
6. Secondary emission is a problem connected with what tube?

SUMMARY

1. The three generations of electronic devices are:
   a. The vacuum tube; developed in the late 19th century and discovered in 1904 as a rectifier device.
   b. The transistor; invented in 1948.
   c. The integrated circuit; invented in 1958.
2. Semiconductors are materials that fall in between conductors and insulators in terms of resistance. Common quadvalent elements used for basic semiconductor devices are silicon and germanium.
3. Trivalent and pentavalent elements are added to pure silicon or germanium to make N type or P type semiconductors. This process is called doping.
4. Rectification is the process of changing ac to dc. Diodes are PN semiconductor devices which can be used as rectifiers.
5. Reverse bias provides high resistance in a diode. Forwad bias provides low resistance in a diode.
6. Diodes are labeled by a 1N prefix. They are rated by a peak inverse voltage (PIV) valve and a maximum current rating.
7. Diodes can be tested with an ohmmeter or a semiconductor tester.
8. Light emitting diodes (LEDs) are special function diodes that, when connected in a forward bias direction, give off light. Light consists of photons. Photons are created by the joining of electrons and holes in the PN junction of the diode.
9. The transistor is a semiconductor device capable of amplification, oscillation, and switching.
10. The three basic leads of a transistor are the emitter, base, and collector.
11. Vacuum tubes are not used much now in electronic circuits, with the exception of the cathode ray tube. Transistors and integrated circuits are used more often in devices such as amplifiers, oscillators, and electronic switches.
12. Cathode ray tubes are special vacuum tubes used in television, oscilloscope, and computer screens.

TEST YOUR KNOWLEDGE, Chapter 13

Please do not write in the text. Place your answers on a separate sheet of paper.

1. A semiconductor is neither a good _____ nor a good _____.
2. Define atomic number.
3. Electricity can be conducted in two ways. Name these ways.
4. Explain the difference between donor impurities and acceptor impurities.
5. A diode consists of two electrodes. Name these electrodes and state their functions.
6. A potential hill or barrier prevents electrons and holes from _____.
7. In forward bias, the barrier effect is _____. In reverse bias, the barrier effect is _____.
8. A point contact crystal has:
   a. P crystals.
   b. N crystals.
   c. Both P and N crystals.
   d. None of the above.
9. What is PIV?
10. A field effect transistor is a _____ controlled device.
11. Name three disadvantages vacuum tubes have over transistors and integrated circuits.
12. Draw and label the symbols for a diode, triode, tetrode, and pentode.
13. Determine the approximate voltage needed by the following tubes.
    a. 5Y3.
    b. 12AX7.
    c. 35W4.
14. What is the purpose of the grid in a triode?
15. In what tube is the suppressor grid found?
16. The arrow of an NPN transistor points in. True or False?

FOR DISCUSSION

1. Why is it important for electronic circuits to be made smaller and smaller?
2. Why has the vacuum tube lost out to the transistor and IC in the field of electronics?
3. What is meant by solid-state?
4. How is the cathode ray tube used for both television screen and oscilloscope displays? What are some likely replacements in the future for cathode ray tubes?

Basic Electronic Devices

FISHER

Many basic electronic devices studied in this chapter are combined in this television. For example, a Cathode Ray Tube is utilized as a picture tube, while many other electronic devices are used to produce a sharp picture and clear sound for entertainment.

Electricity and Electronics

GOULD AMI SEMICONDUCTORS

Integrated circuits in dual in-line packages (DIPs). Notice the gull-wing-shaped leads on the smaller IC. They allow for surface mounting of the device.

INTEGRATED CIRCUITS

ADVANTAGES	• Low cost • Higher switching speed • Low power consumption • Small size • High component density
DISADVANTAGES	• Some components cannot be fabricated • Large amounts of current and voltage cannot be handled

Integrated circuits compared to transistors.

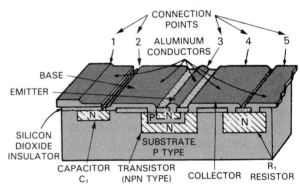

Illustration showing how various components are integrated into a circuit.

The integrated circuit (IC), or chip, has become the basis for most of our modern microelectronics. Chips, while very small in size, can perform the functions of transistors, diodes, resistors, and capacitors.

216

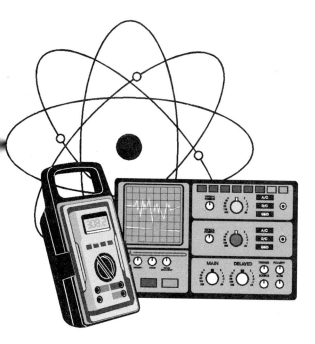

Chapter 14

INTEGRATED CIRCUITS

After studying this chapter, you will be able to:
- Define integrated circuit (IC).
- Give a brief history of the IC.
- Discuss the construction of an IC.
- Illustrate the steps in construction of an IC.

An INTEGRATED CIRCUIT, or IC, is a complete electronic circuit contained in one package, Fig. 14-1. This package often includes transistors, diodes, resistors, and capacitors along with the connecting wiring and terminals. An IC is also called a "chip."

As discussed in Chapter 13, the transistor was invented in 1947 by Brattain, Bardeen, and Shockley of Bell Laboratories. The transistor did what Lee DeForest's triode amplifier did. And it did not need heat to operate. Also, it was solid (solid-state) and much smaller. Transistors were first used in hearing aids and in small transistor radios. Small size and efficient operation made transistors useful in defense items.

BELL LABORATORIES

Fig. 14-1. A single silicon chip. It contains 150,000 transistors.

Transistors were also used in the newly developed electronic computer circuits of the 1950s and early 1960s. Computers used thousands of switching circuits. Transistors were able to quickly perform this switching function. But as computer circuits became larger, electronic circuits needed to become smaller. Because components of circuits had to be wired together, making smaller circuits was a complex task. Printed circuits helped, but wiring was still bulky. This problem was solved by integrating all these components into one solid piece of material. See Fig. 14-2.

14.1 HISTORY OF THE INTEGRATED CIRCUIT

In 1952, G.W.A. Dummer of the Royal Radar Establishment (Great Britain) had the idea for an integrated circuit. However, his ideas were not put into work at that time. In 1957, a new process for planar transistors was developed at Fairchild Semiconductors. This allowed semiconductor emitters, bases, and other parts to be made on the surface of a silicon wafer.

In early 1958, Jack Kilby of the Texas Instruments corporation was working on making micromodules. These were to be made by printing the components on a ceramic wafer. He realized that semiconductors and other components could be made on the same surface by a manufacturing process. The first commercially-produced integrated circuit resulted from this work. It was made on a thin wafer of germanium. It still had wire connections; a major problem of wiring together large amounts of transistors and other conductors.

About the same time, another process for making ICs was being studied at Fairchild Semiconductors. Using the principles of planar transistor manufacturing, Robert Noyce used silicon dioxide dopants to protect PN junctions.

The integrated circuit has changed the electronics field. In 1965, about 30 components could be put on a silicon chip 5 millimeters (3/16 in.) square. By 1982, that number had increased to 1,000,000, Fig. 14-3.

┌─ REVIEW QUESTIONS FOR SECTION 14.1 ─┐

1. What is an integrated circuit?
2. Who invented the transistor? The IC?
3. List the advantages and disadvantages of ICs when compared with transistors.

14.2 IC CONSTRUCTION

An integrated circuit is made on a thin piece of silicon with a one to two inch diameter. Each piece of silicon can contain from 100 to 1000 (and more) circuits. Mak-

INTEGRATED CIRCUITS

ADVANTAGES	• Low cost • Higher switching speed • Low power consumption • Small size • High component density
DISADVANTAGES	• Some components cannot be fabricated • Large amounts of current and voltage cannot be handled

Fig. 14-2. The advantages and disadvantages of ICs when compared with transistors.

DATE	TYPE OF COMPONENT INTEGRATION	LEVEL OF COMPONENT INTEGRATION
1964	SMALL SCALE INTEGRATION	UP TO 10 COMPONENTS OR GATES.
1968–1969	MEDIUM SCALE INTEGRATION (MSI)	UP TO 100 COMPONENTS OR GATES.
1970	LARGE SCALE INTEGRATION (LSI)	UP TO 1000 COMPONENTS OR GATES.
Early 1980s	VERY LARGE SCALE INTEGRATION (VLSI)	1000 OR MORE COMPONENTS OR GATES.
Late 1980s	MEGA INTEGRATION	1 MILLION OR MORE COMPONENTS PER CHIP.

Fig. 14-3. Levels of integration.

ing an integrated circuit is a detailed process. Many steps are involved in creating this tiny device.

The CIRCUIT DESIGNER begins the production process by designing the complete integrated circuit. One factor affecting design is the intended use of the IC. With this in mind, the designer plans the best IC for use. He or she submits the completed design in the form of a schematic diagram.

From this schematic diagram, the LAYOUT DESIGNER creates a detailed technical drawing. The circuit is drawn much larger than the final product, Fig. 14-4. This is done so when the drawing is reduced there will be enough space between parts. It is important that none of the lines touch each other. If they do, when the circuit is tested it will short out.

Next, each circuit layout is PHOTOGRAPHICALLY REDUCED. It is not unusual for the layout to be

Integrated Circuits

Fig. 14-4. This circuit diagram will be reduced hundreds of times before it is used to make circuits.

Fig. 14-5. Sliced and cut silicon wafers.

reduced 400 times. A reduced layout allows for hundreds of circuits to be put on one wafer.

Working plates are made from the reduced layouts. These plates are called PHOTOMASKS. Each photomask goes with a certain step in the production process. Each mask contains a large number of actual sized, identical parts. The photomasks are now ready for use. Production of the IC itself can now begin.

Crystals are needed to build ICs. That is because the structure of an IC is a pure silicon crystal. To make the crystal, liquid silicon is purified. A solid silicon particle, or SEED, is dipped into the melted silicon. It is slowly withdrawn and placed in a cool area. This solid then grows into a long, thin crystal. Small portions of impurities are then added. The impurities give the silicon its electrical traits. The grown crystal is sliced into wafers about .5 mm thick, Fig. 14-5. The wafers are then polished. This rids the wafer of surface scratches and contaminants.

On the thin wafers of doped silicon, the basic building process begins. The circuit is built, layer by layer, on the silicon wafer, or SUBSTRATE. Each layer receives a pattern from the photomask.

The first layer on top of the silicon is a layer of N type material. It is grown on the wafer. It is called the EPITAXIAL layer, Fig. 14-6. Epitaxy is a growth of one crystal on the surface of another crystal. This is the collector for a transistor or an element of a diode.

Next, a thin coat of silicon dioxide is grown over the N type material, Fig. 14-7. This is done by exposing the wafer to an oxygen atmosphere at about 1000 degrees Celsius.

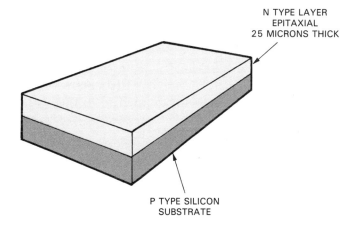

Fig. 14-6. Growing N type material on P type substrate.

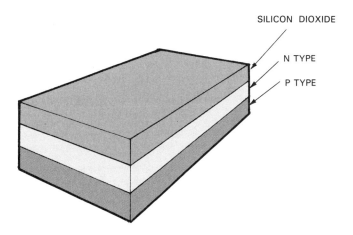

Fig. 14-7. A silicon dioxide layer is placed on top of the N type layer.

Next, a thin coat of a light-sensitive emulsion is placed over the N type layer. The emulsion is called PHOTORESIST. In a process called PHOTOENGRAVING, the photomask is placed over the N type layer. Then the entire wafer is exposed to ultraviolet light, Fig. 14-8. The light causes the image of the photomask to transfer to the wafer.

The exposed photoresist hardens. The areas covered by the mask remain soft. Acids or solvents are used to etch away the unexposed area of the photoresist. This leaves the layer of the N type silicon exposed, Fig. 14-9.

The exposed N type layer is further etched away by very hot gases. A chemical washes away any remaining hardened photoresist to expose all N type silicon dioxide.

As parts of the IC are constructed, they must be isolated from each other. This is done by diffusion. DIFFUSION is a process in which impurities are doped into the silicon wafer to form the needed junctions. Diffusion forms islands of N type materials backing P type materials.

The wafer is diffused using boron. The boron cuts into and forms a P type material on all areas not protected by the silicon dioxide. The wafer has isolated islands of N type material, Fig. 14-10. NP junctions form around each island. There are back-to-back diodes between each N type island.

During diffusion, a new layer of silicon dioxide forms over the diffused P type areas, as well as on top of the islands, Fig. 14-11.

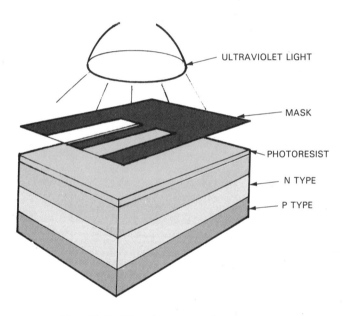

Fig. 14-8. The photoengraving process.

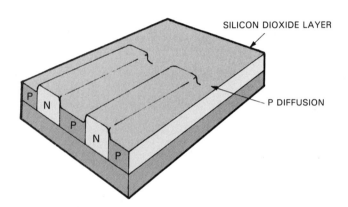

Fig. 14-10. N type material remains after P diffusion.

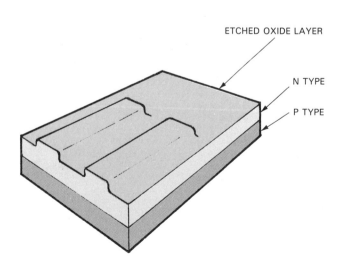

Fig. 14-9. The first masking and etching isolates components.

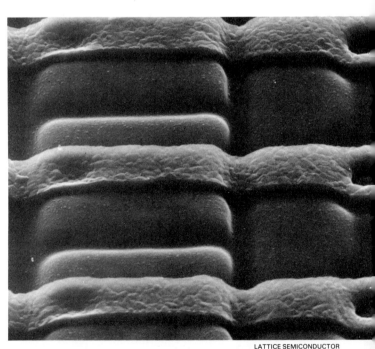

Fig. 14-11. P type diffusant on N type silicon dioxide.

Integrated Circuits

The wafer is again coated with photoresist and exposed under a photomask. Areas in the N type islands are etched away. Once again, the wafer is subjected to a P type diffusant. This forms areas for transistor base regions, resistors, or elements of diodes or capacitors. The wafer is then reoxidized, Fig. 14-12.

The wafer is again masked and exposed to open windows in the P type regions. A phosphorus diffusant is used to produce N type regions for diodes and capacitors. Small windows are also etched through to the N layer for electrical connections, Fig. 14-13. The total wafer is again given an oxide coating.

The monolithic circuit is now complete except for the ALUMINUM INTERCONNECTIONS. The aluminum interconnections join islands. They also join the circuit to other circuits and other devices.

A thin coat of aluminum is vacuum-deposited over the entire circuit. Then, the aluminum coating is sensitized and exposed through another special mask. After etching only the interconnecting aluminum remains. It forms a pattern between transistors, diodes, and resistors, Fig. 14-14. The completed circuits are then tested, Fig. 14-15. In one test the circuits are used to perform a series of electrical tasks, Fig. 14-16.

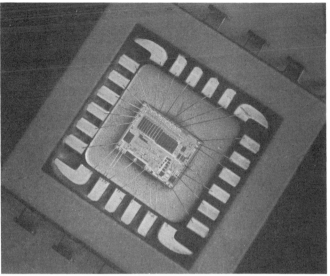

Fig. 14-14. Aluminum interconnections.

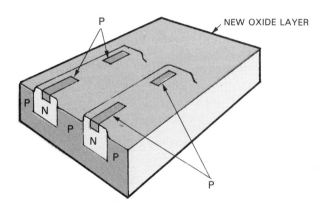

Fig. 14-12. P type regions are diffused in the N type islands.

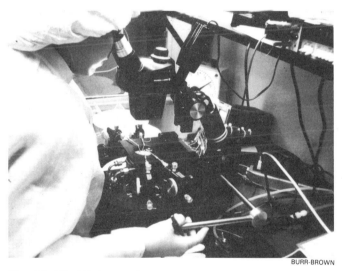

Fig. 14-15. Wafer testing must take place in a sterile area. Even a small piece of dust can ruin a circuit.

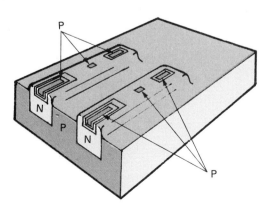

Fig. 14-13. Windows are opened in the P type regions.

Fig. 14-16. IC test clip.

After testing, the wafers are separated into individual chips. This is usually done by scribing with a diamond-tipped tool. The chips are then mounted onto a small can or flat package, Fig. 14-17. Leads are bonded. The ICs are washed. The cavities that hold the ICs are sealed. Finally, the ICs are shipped to a distributor.

RESISTORS

The process just discussed is used to make semiconductor materials on ICs. This process can be used to make resistors, capacitors, and diodes.

Recall that N type or P type materials have certain resistances. Resistance depends on the physical size of the material (length or surface area) and the amount of dopants in the material. Semiconductors are made with very pure silicon. Through the doping process, impure trivalent or pentavalent atoms are added to produce the N type or P type substrate material. For example, a P type silicon material is used as the substrate. An N type material is diffused into the surface of the chip, Fig. 14-18. Then another P type material is added to the N type material. Metal leads are fastened to the end of this P type material. The P type material and its two connections are used for a resistor.

CAPACITORS

Like resistors, capacitors can be made in an integrated circuit. Values for these capacitors are very small. However, they are still able to perform functions of coupling and storage. Fig. 14-19 shows how a capacitor can be made in an integrated circuit.

PUTTING IT TOGETHER

An example of how a transistor, resistor, and capacitor can be integrated into one circuit is shown in Fig. 14-20. Keep in mind that thousands of these circuits can be placed onto an area the size of the head of a pin.

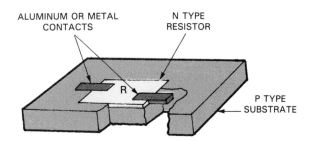

Fig. 14-18. Resistors made on an integrated circuit.

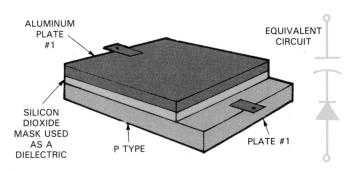

Fig. 14-19. Capacitors made on an IC.

Fig. 14-17. ICs can be mounted in several ways.

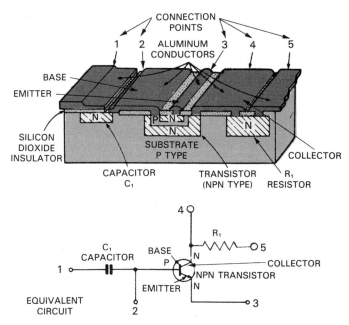

Fig. 14-20. Various components integrated into a small circuit.

Integrated Circuits

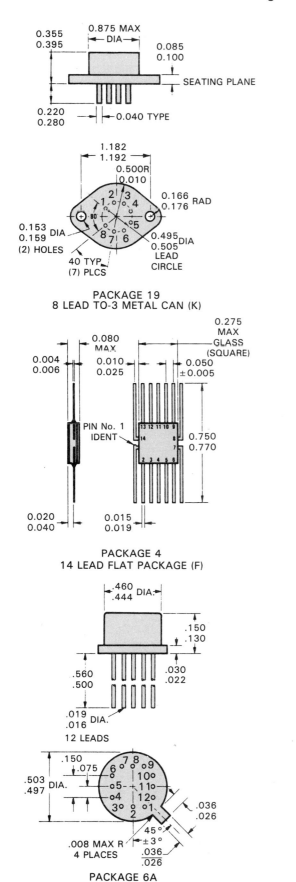

Fig. 14-21. Outlines for several types of ICs. Note the manner in which the pins are numbered.

COMMON TYPES OF ICs

Integrated circuits come in two basic types. The type depends on their function. These types are LINEAR and DIGITAL.

Linear ICs have variable outputs, controlled by variable inputs. These ICs are also called ANALOG devices or circuits. Linear ICs are components in linear amplifiers, operational amplifiers, voltage regulators/buffers, voltage comparators, analog switches, and audio amplifier circuits. Linear ICs will be discussed in more detail in Chapter 16.

Digital integrated circuits are used as switches. Their output operates in either on or off conditions. They are found in many logic and gate circuits in computers. These will be discussed in Chapter 17.

Several IC designs, including pin numbering systems and dimensions, are shown in Fig. 14-21.

REVIEW QUESTIONS FOR SECTION 14.2

1. What is the purpose of photomask?
2. The silicon wafer on which an IC is built layer by layer is called the _____.
3. A light-sensitive emulsion that accepts the transferred image of the photomask is called:
   a. Photoengraver.
   b. Film.
   c. Photoresist.
   d. None of the above.
4. What is diffusion?
5. What is the purpose of aluminum interconnections?
6. Give the pin number (in blue) for the IC below. (Refer to an IC data handbook.)

7. Name the two common types of ICs.

SUMMARY

1. An integrated circuit, or IC, is a complete electronic circuit in a small package. It contains many transistors, diodes, resistors, and capacitors.
2. ICs are also called chips.
3. Advantages of the IC are its low cost, high component density, high switching speed, low power consumption, and small size. The disadvantages are only certain parts can be built into an IC and it is limited in voltage and current amounts it can handle.

4. The IC production process is a detailed, involved one, in which one circuit is built on another circuit.
5. ICs come in many styles, sizes, and housing outlines.
6. The two basic types of ICs are linear and digital.

TEST YOUR KNOWLEDGE, Chapter 14

Please do not write in the text. Place your answers on a separate sheet of paper.

1. An IC:
   a. Is a complete electronic circuit.
   b. Often contains transistors, diodes, resistors, and capacitors.
   c. Is also called a chip.
   d. All of the above.
2. The _____ _____ begins the IC production process by illustrating the complete IC.
3. Why is the initial drawing of an IC done on such a large scale?
4. Why are silicon wafers polished?
5. _____ is the growth of one crystal on the surface of another crystal.
6. During photoengraving, the covered areas remain _____.
7. Label the layers in the following sketch.

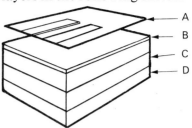

8. During diffusion, _____ are added to the silicon wafer to form needed junctions.
9. Why are completed ICs tested?
10. Linear ICs are also called _____ devices or circuits.
11. Digital integrated circuits are used as _____.

FOR DISCUSSION

1. What are the two primary functions of an electronic amplifier device?
2. Explain how resistors are made on an IC chip.
3. Explain the difference between linear and digital ICs.
4. What has caused the drop in IC cost in the past 10 years?

Integrated Circuits

SOLAREX CORP.

Photovoltaic module for powering a 12 V refrigeration circulator pump. Unit has film of amorphous (uncrystallized) silicon as the absorber. The amorphorous construction saves wafer cutting time. Some scientists believe that integrated circuit chips will one day be made of amorphous silicon.

Part IV Summary

ELECTRONIC DEVICES AND APPLICATIONS

IMPORTANT POINTS

1. Electronics is the study of the flow of electrons in active devices such as transistors, integrated circuits, and vacuum tubes.
2. The transistor was invented in 1948.
3. Semiconductors are made of materials that are somewhere in between conductors and insulators in terms of resistance.
4. Pentavalent elements have five electrons in their outermost shell. Trivalent elements have three electrons in their outermost shell.
5. Diodes are rectifiers that are able to pass current easily in one direction and block it in the other direction.
6. Diodes have high resistance in the reverse bias direction and low resistance in the forward bias direction.
7. Transistors are semiconductor amplifying devices. They are either PNP and NPN.
8. Vacuum tubes were the first electronic amplifier devices. However, they have been largely replaced by transistors and ICs.
9. Cathode ray tubes are vacuum tubes. Some uses include TV video display units, oscilloscopes, and computer monitors.
10. Integrated circuits (ICs) are complete electronic circuits contained in one package. ICs are also called chips.
11. ICs are either linear or digital.

SUMMARY QUESTIONS

1. Look up a 1N 4004 diode and find out the following specifications:
   a. PIV.
   b. Reverse amperage.
2. What type of device is a 2N 5089? Give its specifications and case type.
3. Refer to an IC data handbook. What is a 555 type IC? What is its function? How does it differ from a 556? What is the case outline for the 555?
4. What is the purpose of an IC tester?

Part V

BASIC ELECTRONIC CIRCUITS

15 Power Supplies
16 Amplifiers and Linear Integrated Circuits
17 Digital Circuits
18 Oscillators

Basic electronic circuits are very important to the operation of all electronic products. While the layout of each circuit may differ, their purpose remains the same.

Chapter 15 is *Power Supplies*. The purpose of any power supply is to provide the voltage and current necessary to properly operate an electronic device. Chapter 15 discusses power supply designs and major circuit block functions. Three excellent projects are provided for students to construct.

The topic of Chapter 16 is *Amplifiers and Linear Integrated Circuits*. Amplification and amplifier opertion are major topics of discussion in this chapter. Linear ICs function and operation are also discussed.

Digital Electronic Circuits is the title of Chapter 17. The binary numbering system is explained, along with switching circuits. Types of logic gates are discussed. Digital electronics applications are also highlighted. An excellent project, Binary Bingo, is found at the end of Chapter 17.

Chapter 18, *Oscillators,* covers the generation of signals or waveforms in electronics. A number of oscillators are discussed.

Electricity and Electronics

KEPCO, INC.

One type of power supply. Name several uses for this power supply.

Power supplies operated from 120 V outlets are often more convenient than battery, generator, alternator, or solar power supplies. Voltage regulation is easy to obtain with power supplies operated from 120 V. Most power supplies for transistor circuits are adjustable from 0 V to 50 V. The adjustable power supply is very useful for getting data about circuit operation and for troubleshooting.

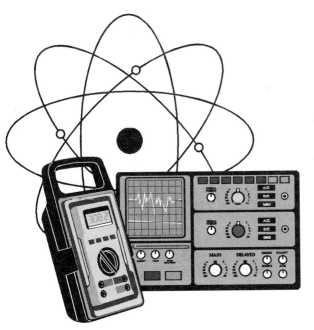

Chapter 15

POWER SUPPLIES

After studying this chapter, you will be able to:
- Identify the power supply functions.
- Explain the differences between half-wave and full-wave rectification.
- Outline various methods for regulating voltage.
- Discuss methods for raising voltages.
- Construct simple power supplies.

Proper operation of electronic equipment requires a number of source voltages. Low dc voltages are needed to operate ICs and transistors. High voltages are needed to operate CRTs and other devices. Batteries can provide all of these voltages. However, the voltages are commonly supplied by the local power company at 115 volt ac and a frequency of 60 hertz. Different voltages are needed to operate some equipment.

A POWER SUPPLY is an electronic circuit designed to provide various ac and dc voltages for equipment operation.

15.1 POWER SUPPLY FUNCTIONS

The complete power supply circuit can perform these functions:
1. Step up or step down, by transformer action, ac line voltage to required voltages.
2. Change ac voltage to pulsating dc voltage by either half-wave or full-wave rectification.
3. Filter pulsating dc voltage to a pure dc voltage for equipment use.
4. Provide some method of voltage division to meet equipment needs.
5. Regulate power supply output in proportion to the applied load.

See Figs. 15-1 and 15-2 for common commercial power supplies.

Fig. 15-1. Typical bench style power supply. It is to provide dc voltages for equipment.

Fig. 15-2. Low voltage dc power supply.

231

Electricity and Electronics

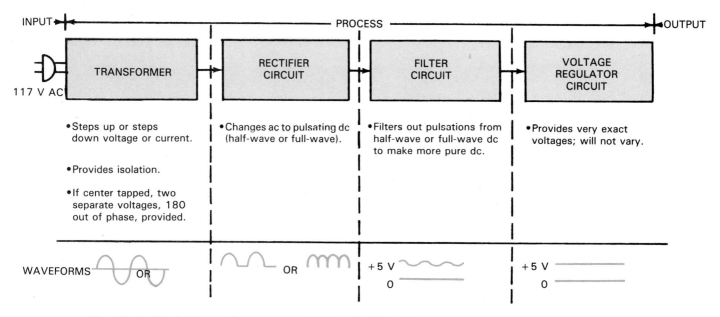

Fig. 15-3. Block diagram for power supply. Input is 117 volts ac. Processes used in typical power supply are shown below the blocks. Output of power supply can be dc or ac. Output of this supply is 5 volts dc.

A block diagram illustrating these functions is shown in Fig. 15-3. Note that certain functions are not found in every power supply.

THE POWER TRANSFORMER

The first device in a power supply is the transformer. Its purpose is to step up or step down alternating source voltage to values needed for radio, TV, or electronic circuit use. Review transformer action, Chapter 9.

Most transformers do not have any electrical connection between the secondary and primary windings. This is called ISOLATION. This is a safety feature. It helps prevent shocks in the secondary. Body or hands must be joined across both of the secondary connections in order to receive a shock.

This is not true in the primary with commercial ac provided by the power company. One connection is HOT. The other is GROUNDED or NEUTRAL. Standing on the ground, touching the hot connection will result in a shock. Touching the ground connection will not result in a shock. See Fig. 15-4.

Secondary windings can be tapped to provide different voltages. A tap placed midway between the two ends of a secondary winding is called a CENTER TAP. Many power supplies use a center tap secondary transformer winding. The tapped voltages, Fig. 15-5, are 180 degrees out of phase with respect to the center tap.

A variety of transformers can be found in nearly all electronic devices. Students should know the basic theory and purpose of the transformer.

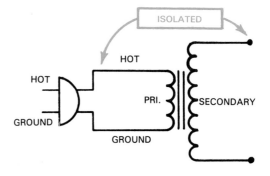

Fig. 15-4. Isolation in a transformer.

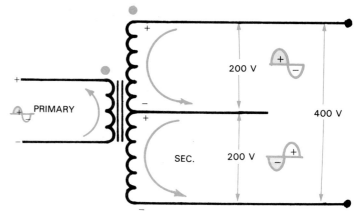

Fig. 15-5. A center tap transformer.

LESSON IN SAFETY: Transformers produce high voltages that can be very dangerous. Proper respect and extreme caution must be used at all times when working with, or measuring, high voltages.

232

HALF-WAVE RECTIFICATION

The process of changing an alternating current to a pulsating direct current is called RECTIFICATION. In Fig. 15-6, the output of a transformer is connected to a diode and a load resistor, in series. The input voltage to the transformer appears as a sine wave. From time to time polarity reverses at the frequency of the applied voltage. The output voltage of the transformer secondary also appears as a sine wave. Magnitude depends on the turns ratio. It is 180 degrees out of phase with the primary. The top of the transformer (point A) is joined to the diode anode. During the first half-cycle, point A is positive. The diode conducts, producing a voltage drop across R equal to IR. During the second half-cycle, point A is negative. The diode anode is also negative. No conduction takes place. No IR drop appears across R.

An oscilloscope connected across R produces the wave form shown in Fig. 15-7. Note that the B side of the transformer is connected to ground. The output of this circuit is pulses of current flowing in only ONE direction and at the SAME frequency as the input voltage. The output is a pulsating direct current.

Only one half of the ac input wave is used to produce the output voltage. This is called a HALF-WAVE RECTIFIER. Look at the polarity of the output voltage. One end of the resistor R is connected to ground. The electron or current flow is from ground to the cathode. This makes the cathode end of R positive as shown.

A negative rectifier can be made by reversing the diode in the circuit, Fig. 15-8. The diode conducts when the cathode becomes negative. This is the same as making the anode positive. The current through R would be from anode to ground. This would make the anode end of R negative and the ground end of R more positive. Voltages taken from across R, the output, would be negative in respect to ground. This circuit is called an INVERTED DIODE. It is used when a negative supply voltage is required.

It is possible to have a power supply that provides half-wave rectification without the use of a transformer. This circuit is not isolated. There is no step up or step down of voltages of currents. But it is a simpler, less costly design. And since there is no transformer, it can be used in smaller spaces, Fig. 15-9.

FULL-WAVE RECTIFICATION

The pulsating direct voltage output of a half-wave rectifier is hard to filter to a pure dc voltage. The half-wave rectifier uses only one half of the input ac wave. By using two diodes, both half-cycles of the input wave can be rectified. Refer to Figs. 15-10 and 15-11. To do this, a center tap is made on the secondary winding. The secondary winding is attached to ground.

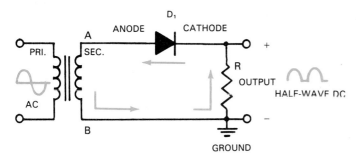

Fig. 15-6. Basic diode rectifier schematic.

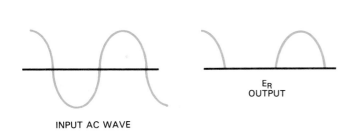

Fig. 15-7. Input and output wave forms of a diode rectifier.

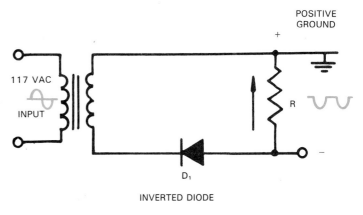

Fig. 15-8. An inverted diode produces a negative voltage.

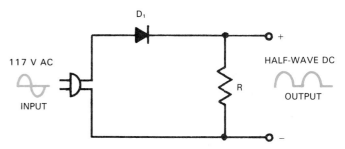

Fig. 15-9. Half-wave rectification without a transformer.

Electricity and Electronics

In Fig. 15-10, point A is positive and diode anode D_1 is positive. Current flows as shown by the arrows. During the second half of the input cycle, point B is positive, diode anode D_2 is positive, and current flows as shown in Fig. 15-11. No matter which diode is conducting, the current through load resistor R is always in the same direction. Both positive and negative half-cycles of the input voltage cause current through R in the same direction.

Output voltage of this FULL-WAVE RECTIFIER is taken from across R. It consists of direct current pulses at twice the frequency of input voltage, Fig. 15-12. To produce full-wave rectification in this circuit, secondary voltage was cut in half by the center tap.

The diodes, D_1 and D_2, used in Figs. 15-10 and 15-11, are packaged both individually and in pairs. Fig. 15-13 shows two rectifiers. The center tap, or center lead, is used as the connection for the cathodes. The cathodes are wired together.

The bridge rectifier

It is not always necessary to use a center tapped transformer for full-wave rectification. Full secondary voltage can be rectified by using four diodes in a BRIDGE RECTIFIER CIRCUIT, Fig. 15-14 and 15-15. Two circuits are shown so that current flow can be observed on each half cycle.

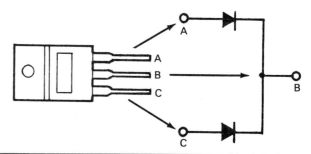

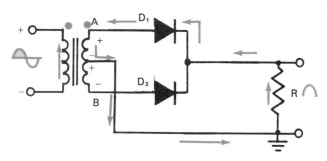

Fig. 15-10. Arrows show current in full-wave rectifier during first half-cycle.

INTERNATIONAL RECTIFIER
Fig. 15-13. Dual diodes with center tap.

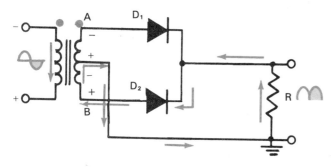

Fig. 15-11. Direction of current during second half-cycle.

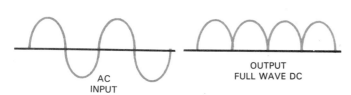

Fig. 15-12. The wave forms of input and output of full-wave diode rectifier.

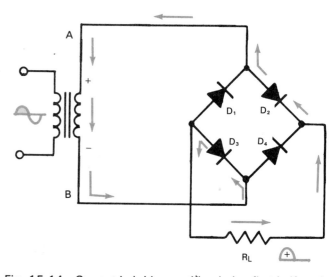

Fig. 15-14. Current in bridge rectifier during first half-cycle.

Power Supplies

In Fig. 15-14, point A of transformer secondary is positive. Current flows in the direction of the arrows. When point B is positive, current flows as in Fig. 15-15. Again, notice that the current through R is always in one direction. Both halves of the input voltage are rectified and the full voltage of the transformer is used.

Bridge rectifiers can be used in circuits not having transformers. Without transformers, voltages or current will not be stepped up or down. There will be no isolation. These circuits are also called LINE-OPERATED BRIDGE CIRCUITS, Fig. 15-16.

FILTERS

The output of either the half-wave or full-wave rectifier is a pulsating voltage. Before it can be applied to the other circuits, the pulsations must be reduced. A more pure dc is needed. This is obtained using a FILTER NETWORK, Fig. 15-17. The line, E_{av}, shows the average voltage of the wave. It is equal to .637 × peak voltage. The shaded portion of the wave above the average line is equal in area to the shaded portion below the line. Movement above and below the average voltage is called the ac RIPPLE. It is the ripple that requires filtering. The PERCENTAGE of ripple to the output voltage must be kept to a small value. The ripple percentage may be found using the formula:

$$\% \text{ ripple} = \frac{E_{RMS} \text{ of ripple voltage}}{E_{AV} \text{ of total output voltage}} \times 100\%$$

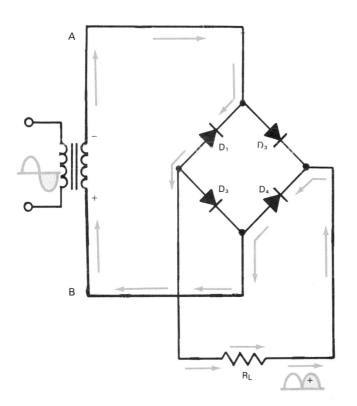

Fig. 15-15. Current in bridge rectifier during second half-cycle.

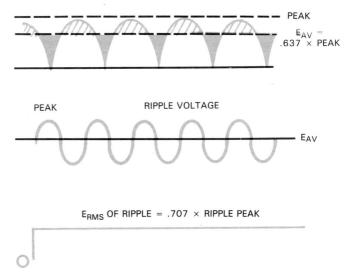

Fig. 15-17. Average value of full-wave rectifier output.

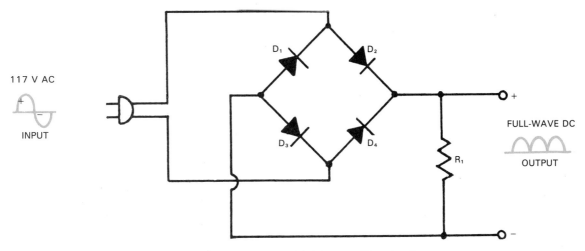

Fig. 15-16. Line-operated bridge rectifier circuit.

A capacitor connected across the rectifier output provides some filtering action, Fig. 15-18. The capacitor is able to store electrons. When the diode or rectifier is conducting, the capacitor charges rapidly to about the peak voltage of the wave. It is limited only by the resistance of the rectifier and the reactance of the transformer windings. Between the pulsations in the wave, voltage from the rectifier drops. The capacitor then discharges through the resistance of the load. The capacitor, in effect, is a storage chamber for electrons. It stores electrons at peak voltage. It supplies electrons to the load when rectifier output is low. See Fig. 15-19.

Capacitors used for this purpose are electrolytic types. This is because large capacitances are needed in a limited space. Common values range from 4 to 2000 microfarads. Working voltages are in excess of the peak voltage from the rectifier.

The filtering action may be improved by adding a choke in series with the load. This filter circuit appears in Fig. 15-20. The filter choke consists of many turns of wire wound on a laminated iron core.

Recall that inductance was that property of a circuit which resisted a change in current. A rise in current induced a counter emf which opposed the rise. A decrease in current induced a counter emf which opposed the decrease. As a result, the choke constantly opposes any change in current. Yet it offers very little opposition to a direct current.

Chokes used in radios have values from 8 to 30 henrys. Current ratings range from 50 to 200 milliamperes. Larger chokes may be used in transmitters and other electronic devices. Filtering action as a result of the filter choke is shown in Fig. 15-21.

A second capacitor can be used in the filter section after the choke, to provide more filter action. See Fig. 15-22. Action of this capacitor is similar to the first capacitor. The circuit configuration appears as the Greek letter π. The filter is called a PI (π) SECTION filter.

The first filtering component is a capacitor. It is called a CAPACITOR INPUT FILTER. In Fig. 15-23

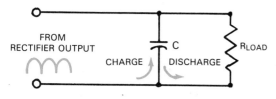

Fig. 15-18. Filtering action of a capacitor.

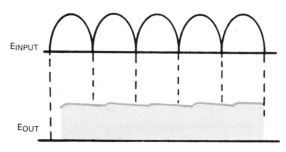

Fig. 15-21. Waveforms show filtering action of capacitor and choke together.

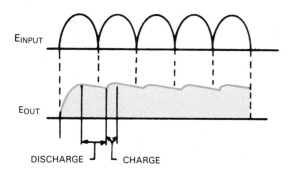

Fig. 15-19. Input and output of capacitor filter showing change in wave form.

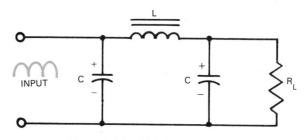

Fig. 15-22. Pi (π) section filter.

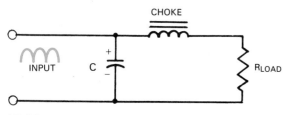

Fig. 15-20. Further filtering is produced by the choke in series with the load.

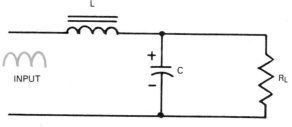

Fig. 15-23. Choke input L filter.

Power Supplies

the choke is the first filtering component. It is called a CHOKE INPUT FILTER. The circuit configuration now appears as an inverted L. It is called an L SECTION FILTER. Several of these filter sections may be used in series to provide added filtering.

In the capacitor input filter, the capacitor charges to the peak voltage of the rectified wave. In the choke input, the charging current for the capacitor is limited by the choke. The capacitor does not charge to the peak voltage. As a result, the output voltage of the power supply using the capacitor input filter is higher than one using the choke input filter.

REVIEW QUESTIONS FOR SECTION 15.1

1. Define power supply.
2. List five functions provided by a complete power supply.
3. The first device in a power supply is the _____.
4. A circuit that changes alternating current to a pulsating direct current is a _____.
5. Name two types of full-wave rectifier circuits.
6. What is a ripple?

15.2 VOLTAGE REGULATION

The output voltage of a power supply will usually decrease when a load is applied. This is not good. It should be stopped. The voltage decrease under load compared to the power supply voltage with no load is called the PERCENTAGE OF VOLTAGE REGULATION. It is one factor used to determine the quality of a power supply. Expressed mathematically,

$$\% \text{ Voltage Regulation} = \frac{E_{nl} - E_{fl}}{E_{fl}} \times 100$$

where E_{nl} equals voltage with no load and E_{fl} equals voltage with full load

For example, a power supply has a no load voltage of 30 volts. This voltage drops to 25 volts when a load is applied. What is its percentage of regulation?

$$\% \text{ Regulation} = \frac{30 \text{ V} - 25 \text{ V}}{25 \text{ V}} \times 100$$

$$= \frac{5}{25} \times 100 = 20\%$$

LOAD RESISTOR

To complete the basic power supply circuit, a load resistor is connected across the supply, Fig. 15-24. This resistor serves:

1. As a BLEEDER. During operation of the power supply, peak voltages are stored in the capacitors of the filter sections. These capacitors remain charged after the equipment is turned off, and can be dangerous if accidentally touched by the technician. The load resistor serves as a "bleeder." It allows these capacitors to discharge when not in use. The wise technician will always take the added precaution and short-out capacitors to ground with an insulated screwdriver.
2. To improve regulation. The load resistor acts as a preload on the power supply. It causes a voltage drop. When equipment is attached to the supply, the added drop is fairly small and the regulation is improved. For example, assume the terminal voltage of a power supply is 30 volts with no load resistor. No equipment is connected to it. When equipment is connected and turned on, the voltage drops to 25 volts. Regulation is 20 percent. (See previous example under voltage regulation.) If the resistor across the power supply produces an initial drop to 26 volts, then output voltage is considered 26 volts. If the equipment now connected to the supply causes the voltage to drop to 25 volts, then the power supply regulation is,

$$\% \text{ Regulation} = \frac{26 \text{ V} - 25 \text{ V}}{25 \text{ V}} \times 100$$

$$= \frac{1 \text{ V}}{25 \text{ V}} \times 100 = 4\%$$

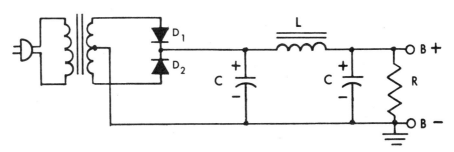

Fig. 15-24. Complete power supply circuit with load resistor.

Electricity and Electronics

The usable voltage of the supply has only varied four percent. A further advantage of preloading the supply is an increase in the choke filtering action. The resistor allows current to flow in the supply at all times. A choke has better filtering action under this current condition than when current varies between a low value and zero.

3. As a voltage divider. The load resistor provides a way of obtaining several voltages from the supply. Replacing the single load resistor with separate resistors in series provides several fixed dc voltages, Fig. 15-25. A sliding tap resistor provides for voltage adjustments. This is called a VOLTAGE DIVIDER. It takes advantage of Ohm's Law which states that the voltage drop across a resistor equals current times resistance, ($E = I \times R$). In Fig. 15-25, the three resistors are equal. The voltage drop divides equally between them. And their sum equals the applied voltage. Other values of resistors may be used, if other voltages are required.

Do not neglect the effect of connecting a load to one of the taps on a voltage divider. This load is in parallel to the voltage divider resistor. It decreases total resistance and, therefore, changes the voltage at that tap.

In Fig. 15-26, the divider consists of three 5 kilohm (kΩ) resistors. The supply of 30 volts divides to 10 volts, 20 volts, and 30 volts at terminals C, B, and A respectively. If a load of 5 kilohm is now connected to terminal C as shown, it is parallel with R_3 and resistance becomes:

$$R_T = \frac{R_3 R_L}{R_3 + R_L} = \frac{5000 \times 5000}{5000 + 5000} = 2500 \ \Omega$$

The total series resistance across the power supply is 5 kΩ + 5 kΩ + 2500 Ω = 12,500 Ω. The current through the divider is:

$$I = \frac{E}{R} = \frac{30}{12,500 \ \Omega} = .0024 \text{ amps}$$

The voltage at point C is:

$$E = I \times R = .0024 \times 2500 \ \Omega = 6 \text{ V}$$

The voltage at point B is 18 volts. If another load were connected to point B, a further change of voltage division would result.

ZENER DIODE

An electronic device that may be used as a voltage regulator is the ZENER DIODE, Fig. 15-27.

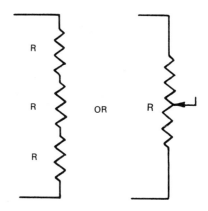

Fig. 15-25. Voltage divider across power supply output.

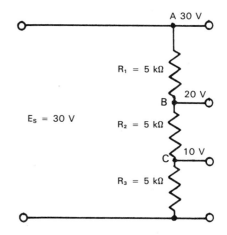

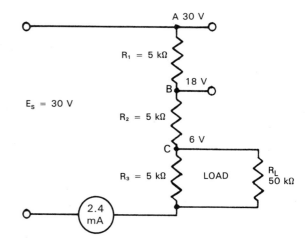

Fig. 15-26. Diagrams show change in resistance in a voltage divider when a load is attached.

Fig. 15-27. Zener diode symbol.

Power Supplies

Fig. 15-28 shows a characteristic curve for a zener diode. When the diode is forward biased, it acts like a diode or a closed switch. When applied voltage is increased, the forward current increases. When the diode is reverse biased, a small reverse current flows to a point where the diode reaches the zener breakdown region, V_z. At this point the zener diode is able to maintain a fairly constant voltage as the current varies over a certain range. Because of this, the diode provides excellent voltage regulation.

Fig. 15-29 shows a zener diode being used as a simple shunt regulator.

VOLTAGE REGULATOR CIRCUIT

Some method for providing a constant voltage output at the power supply under varying load conditions is needed. This method would take into consideration the fact that a voltage drop across a resistor is equal to the product of current and resistance.

This method comes in the form of a circuit and is shown in Fig. 15-30. The total input of the power supply filter is applied to terminals A and B. Regulated output is across points C and B. The voltage regulator used in Figs. 15-30 is often called a three terminal fixed voltage regulator. Common output regulated voltages can be 5, 6, 8, 12, 15, 18, 24 volts, etc. (Various current ratings are available from manufacturers.)

Regulators also come in a number of common transistor package designs (TO-3, TO-39, TO-202, TO-220, etc.). These solid-state regulators are basically blowout proof. They require that a heat sink be used to remove excess heat from the device.

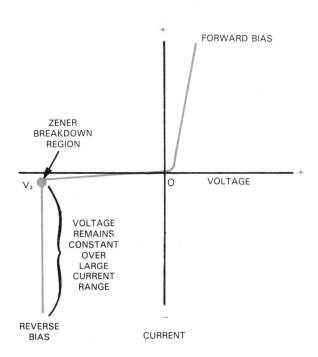

Fig. 15-28. Zener diode characteristics.

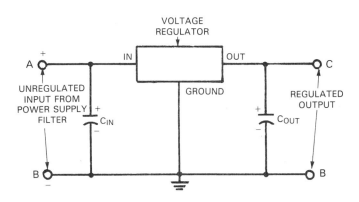

Fig. 15-30. Basic voltage regulator circuit.

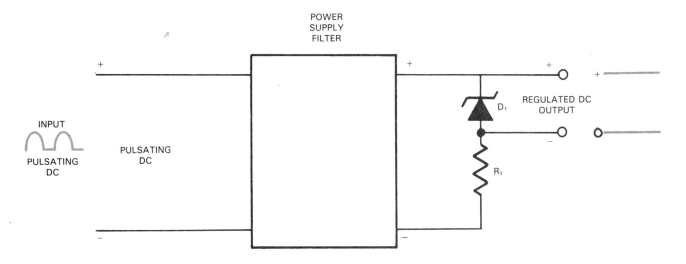

Fig. 15-29. Zener diode as shunt voltage regulator.

239

Electricity and Electronics

The internal circuits used for these voltage regulators are quite complex. They have a number of transistors, diodes, zener diodes, and resistors built into one small package. Fig. 15-31 shows two voltage regulator schematics and their package designs.

A circuit for an adjustable voltage regulator is shown in Fig. 15-32. Values are given for the components. Use a proper heat sink on the voltage regulator case to stop overheating problems. A voltage regulator, on a car, controls the voltage that reaches the alternator's rotor.

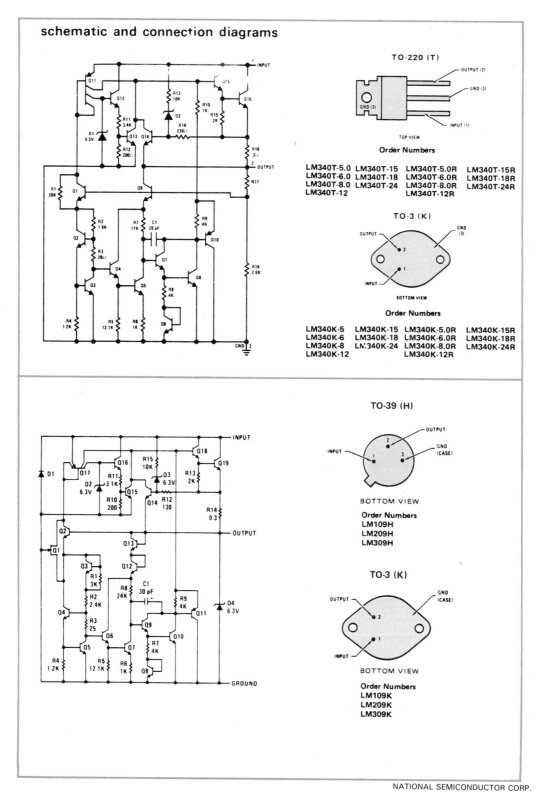

NATIONAL SEMICONDUCTOR CORP.

Fig. 15-31. Schematic and connection diagrams for voltage regulators.

Power Supplies

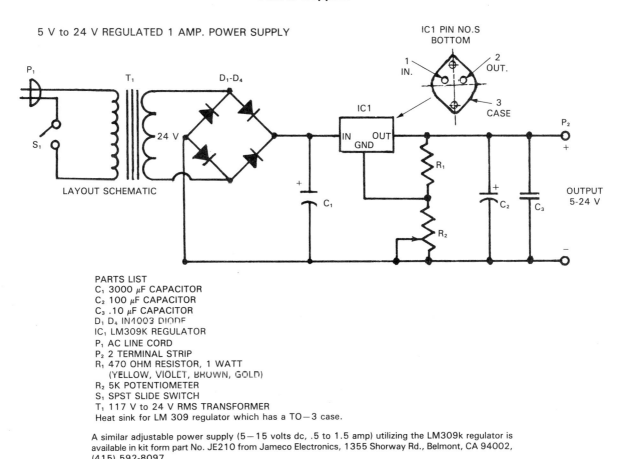

Fig. 15-32. 5 to 24 volts regulated 1 amp power supply.

REVIEW QUESTIONS FOR SECTION 15.2

1. What is the purpose of a voltage regulator in a power supply?
2. A power supply has a no load voltage of 65 volts. Voltage drops to 45 volts when a load is applied. What is the percentage of voltage regulation?
3. List three purposes for using a load resistor in a power supply.
4. A zener diode is an electronic device that acts as a _____ _____.
5. What is the symbol for a zener diode?

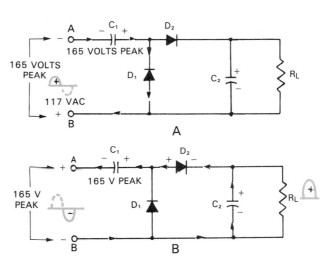

Fig. 15-33. A—During first half-cycle, C_1 charges through conduction of rectifier D_1. B—During second half-cycle, applied line voltage is in series with charge on C_1. Current flows through D_2. C_2 gets the sum of line voltage and that from C_1.

15.3 VOLTAGE DOUBLERS

Up to this point, it has been assumed that the source of power was 117 volt ac found in homes and schools. The transformer was used to step up or step down the voltages required for the electronic circuits. Because transformers are heavy and costly, voltage multiplying circuits have been devised to raise voltages without the use of transformers.

Study Fig. 15-33. It shows the action in a half-wave voltage doubler circuit.

In Fig. 15-33A, the input ac voltage is on the negative half-cycle. As a result, point A is negative. Current flows from point A, through the rectifier D_1, and charges capacitor C_1 to the polarity shown.

During the positive half-cycle, point A is positive. The applied peak voltage of 165 volts is in series with the charged capacitor C_1. In the series connection the voltages add together. So the output from the doubler is the applied voltage plus the voltage of C_1. Current cannot flow through the rectifier D_1 due to its one-sided conduction. The output wave form shows half-wave rectification with an amplitude of about twice the input voltage. Rectifier D_2 permits current to flow in only one direction to the load.

A full-wave voltage doubler is drawn in Fig. 15-34. During the positive peak of the ac input, point A is positive. Current flows from point B, charging C_1 in the polarity shown, through D_1 to point A.

During the negative cycle of the input, point A is negative. Current flows through D_2 to C_2 and charges it to the noted polarity, to point B.

Notice that during one cycle of ac input, capacitors C_1 and C_2 have been charged so that the voltages across C_1 and C_2 are in series. The output is taken from across these capacitors in series. The output voltage is the sum of both voltages or twice the input voltage.

Voltage doubler circuits provide useful high voltages for circuits needing low current. Because output voltage depends on charged capacitors, voltage regulation is poor. Conventional filter circuits are added to smooth out the voltage as in transformer rectifier circuits.

AC-DC SUPPLY

In Fig. 15-35, a common ac-dc power supply is shown. It is used in many small radios and calculators. When used with ac, the output is pulsating dc. This must be filtered. The capacitor input filter is used to secure the higher voltage. Usually, in a radio using this supply, the negative of B− is grounded to the chassis. If the ac plug is inserted the wrong way in the wall socket, severe shock is possible, if ground and chassis is touched.

When this supply is used on dc, the positive side of the dc line must be connected to the anode of the diode. The set will not operate otherwise. Current will only flow from cathode to plate.

LESSON IN SAFETY: The ac-dc radio described in Fig. 15-35 is dangerous if the ac plug is connected the wrong way. One wire of your house wiring system is grounded to a water pipe, which will ground your radio only if the plug is inserted correctly. If the plug is reversed, an accidental contact between the radio and a water faucet would give a severe shock across the body. If the radio came in contact with any grounded table or appliance, sparks will fly.

FLOATING GROUND

Some hazards of the ac-dc supply can be avoided. Isolate the radio chassis from ground by a buffer capacitor. The chassis is not used as a common ground. The negative terminals are wired together in a FLOATING GROUND system. See Fig. 15-36.

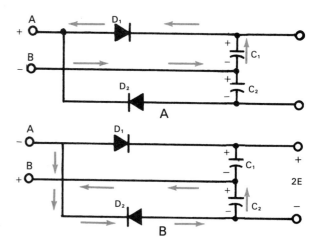

Fig. 15-34. A—C_1 charges during first half-cycle. B—C_1 + C_2 in series.

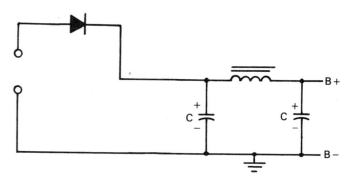

Fig. 15-35. An ac-dc power supply circuit.

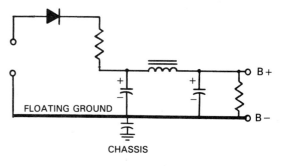

Fig. 15-36. An ac-dc supply with floating ground.

REVIEW QUESTIONS FOR SECTION 15.3

1. What type of circuit is shown below? How does it operate?

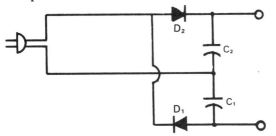

2. A line voltage of 117 volts ac produces a peak voltage of _____ volts when using a voltage doubler.
3. Why are voltage doubler circuits important for electronic equipment?

15.4 POWER SUPPLY CONSTRUCTION

12 VOLT AUTOMOBILE BATTERY TRICKLE CHARGER

A battery charger for an automobile is shown in Fig. 15-37. It is called a TRICKLE charger because it provides a low level of charging current compared to service station chargers. This battery charger is excellent for recharging an automobile battery overnight, or for 12 to 24 hours.

The ammeter shows output current while charging. A minimum amount of output current means that the battery is reaching a maximum charge. Fig. 15-38 shows a power supply schematic that can be used as a battery charger. Wiring is simple. This project is fun to construct. The parts list is shown in Fig. 15-39.

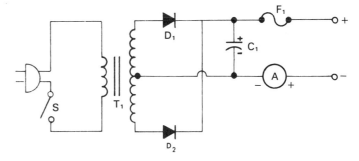

Fig. 15-38. 12 volt power supply.

S	SPST switch
T_1	Step-down transformer, 117 V primary, 12 V CT secondary at 3A
D_1, D_2	Diodes, 50 PIV at 5A (RCA SK 3586)
*C_1	1000-5000 μF, 50 volt capacitor (electrolytic)
F_1	3 amp fuse with holder
A	0-3 amp, ammeter

Also needed: line cord, chassis output leads with alligator clips.

*C_1 is only needed when operating 12 volt automotive equipment such as a radio or a tape player. It is not needed when using this power supply as a battery charger.

Fig. 15-39. Parts list for 12 volt power supply.

This power supply can also be used to operate 12 volt automotive tape players and radios by adding C_1, shown in the schematic.

0-15 VOLT DC LOW AMPERAGE POWER SUPPLY

A 0-15 V dc power can be used to power transistor radios. Or it can provide a low voltage substitute for many dc operated electronic devices. The maximum output current is 100 mA. See Fig. 15-40. See Figs. 15-41 and 15-42 for the schematic and parts list.

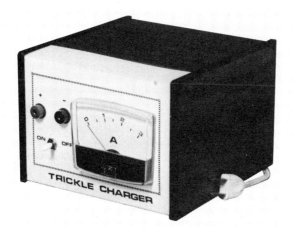

Fig. 15-37. Trickle charger used for recharging batteries.

GRAYMARK ENTERPRISES, INC.

Fig. 15-40. Solid-state, low voltage power supply.

Electricity and Electronics

SCHEMATIC DIAGRAM

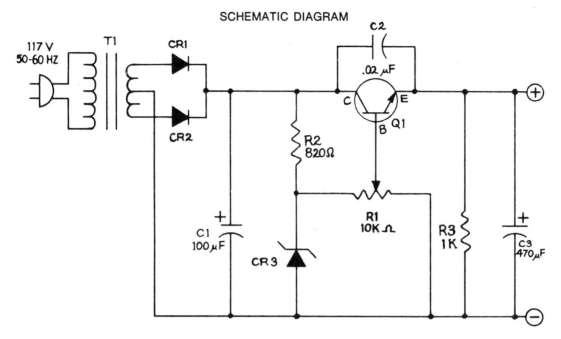

Fig. 15-41. Schematic diagram for 0-15 volt dc power supply.

R_1 —	(1) Potentiometer, 10 k ohms, 2 W (62110)	CR_1, CR_2 —	(2) Diode, silicon, power, 50 PIV, 1 A (62102)
R_2 —	(1) Resistor, solid, 820 ohms, 1/2 W, 10% (61821)	CR_3 —	(1) Diode, zener, 16 V, .5 W RCA SK 3142 (62109)
R_3 —	(1) Resistor, solid, 1 k ohms, 1/2 W, 10% (61377)	Q_1 —	(1) Transistor, power, with 2-insulative washers* (62103)
C_1 —	(1) Capacitor, electrolytic, 100 μF, 35 WV (62100)	T_1 —	(1) Transformer, power, P = 117 V, S = 40 V, at 135 mA* (62104)
C_2 —	(1) Capacitor, ceramic disc., .02 μF, 50 V (61231)		
C_3 —	(1) Capacitor, electrolytic, 470 μF, 16 WV (62101)		

*Attached to panel

This power supply may be ordered as Model 803 from Graymark Enterprises, Inc., P.O. Box 54343, Los Angeles, California 90054. Individual parts may be ordered by quoting the part number to Graymark.

Fig. 15-42. Parts list for Model 803 power supply.

SUMMARY

1. Power supplies are electronic circuits that provide various dc and ac voltages and currents for equipment.
2. The basic circuits in many power supplies include transformers, rectifiers, filters, and voltage regulators. Transformers and voltage regulators may not be found in all power supplies.
3. The purpose of a transformer in a power supply can be to (1) increase voltage and decrease current, (2) increase current and decrease voltage, (3) provide isolation and (4) provide two equal voltages, one 180 degrees out of phase (with center tap).
4. A rectifier changes ac to pulsating dc.
5. Some basic line and transformer operated rectifier circuits are shown in Fig. 15-43.
6. A power supply filter uses capacitors, inductors, and/or resistors to change pulsating dc to pure dc.
7. Voltage regulation can be achieved with a zener diode or a complete voltage regulation circuit.
8. A bleeder resistor is a resistor placed in parallel (shunt) across the output of a power supply. It provides a load; a discharge path for charged current to flow and a limited amount of regulation of voltage.
9. There are two basic types of voltage regulators: half-wave and full-wave. Both operate from the principle that capacitors, which always charge up to a peak value of a wave form, can be connected in series with each other (full-wave) or with a line voltage (half-wave) to provide an added voltage.
10. A floating ground is a common connection electrically separated from the chassis.

Power Supplies

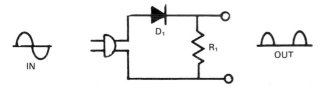
SINGLE DIODE, HALF-WAVE OUTPUT

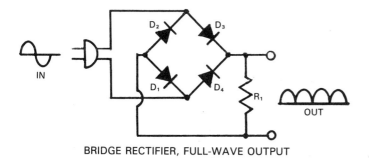
BRIDGE RECTIFIER, FULL-WAVE OUTPUT

A

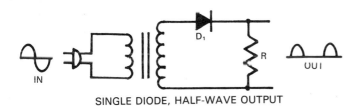
SINGLE DIODE, HALF-WAVE OUTPUT

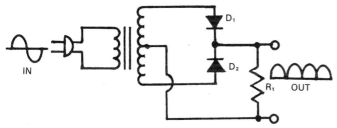
CENTER TAPPED TRANSFORMER, FULL-WAVE OUTPUT

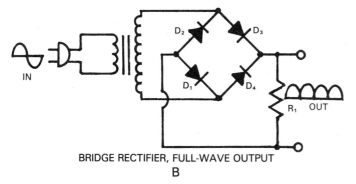
BRIDGE RECTIFIER, FULL-WAVE OUTPUT

B

Fig. 15-43. Rectifier designs. A—Line-operated rectifier circuits. B—Transformer operated rectifier circuits.

TEST YOUR KNOWLEDGE, Chapter 15

Please do not write in the text. Place your answers on a separate sheet of paper.

1. A power supply:
   a. Provides various ac and dc voltages for equipment operation.
   b. Can change ac voltage to pulsating dc voltage.
   c. Can filter pulsating dc voltage to a pure dc voltage.
   d. All of the above.
2. List, in order, the devices a current goes through in a power supply.
3. A lack of electrical connection between secondary and primary windings is known as _____.
4. In order to rectify both half-cycles of the input wave, a _____ _____ is made on the secondary winding.
5. What is the purpose of a filter network?
6. State the equation for finding percentage of ripple.
7. Name two devices that can be used to improve filtering action.
8. Why is the output from a capacitor input filter higher than that of a choke input filter?
9. The terminal voltage of a power supply is 30 volts. When a load is applied, voltage drops to 28 volts. What is the percentage of regulation?
10. An electronic device that acts as a voltage regulator is a:
    a. Diode.
    b. Zener diode.
    c. Bleeder.
    d. None of the above.
11. What is the purpose of a voltage regulator circuit?

FOR DISCUSSION

1. What is the purpose of a power supply?
2. Do all power supplies need filters? Explain.
3. How does a transformer provide isolation?
4. What is the purpose of a center tap on the secondary winding of a transformer?
5. How does a zener diode provide voltage regulation?
6. Explain how a full-wave voltage doubler operates.

ZIRCON CORP.

Some tools, such as this electronic level, must work with analog circuits. A light emitting diode and an electronic sound indicate when the device is level.

Amplifiers and linear integrated circuits are analog devices. Analog means that the output of a circuit changes smoothly and continuously from one value to another. This is in contrast to a digital circuit, which changes abruptly from 0 to 1 or from 1 to 0. The output level of an analog circuit depends on the power supply voltage. If the supply voltage goes up, the circuit output goes up. For this reason, power supplies for amplifiers and linear integrated circuits must be well regulated.

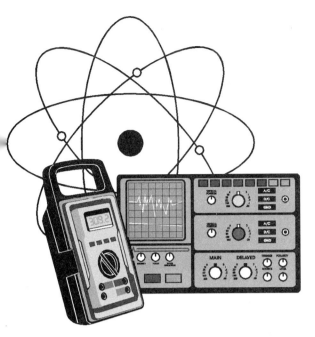

Chapter 16

AMPLIFIERS AND LINEAR INTEGRATED CIRCUITS

After studying this chapter, you will be able to:
- List the components of amplifier circuits and give the function for each component.
- Explain amplifier operation.
- Identify various transistor circuit configurations.
- Discuss the advantages and disadvantages of various methods of amplifier coupling.
- Explain the uses of linear integrated circuits as amplifiers.
- Construct several amplifier projects.

An AMPLIFIER is an electronic circuit that uses a small input signal to control a larger output signal. In physics, weight can be amplified using levers. A small lever is used to lift a larger weight. In electronics, amplifiers have been used since the early twentieth century. Amplification is done using vacuum tubes or semiconductors, such as transistors or integrated circuits.

The amount of amplification in a circuit is known as GAIN. This is the ratio between the output power (current or voltage) and the input power (current or voltage), Fig. 16-1.

16.1 AMPLIFIER CIRCUITS

Amplifier circuits are CONTROL circuits. A small amount of current or voltage can control a larger amount of voltage or current. These circuits produce outputs that vary, or are LINEAR.

Amplifier devices such as transistors or ICs can also be used to SWITCH current on or off. This depends on how these devices are biased in the circuit. Switching circuits will be discussed in Chapter 17.

NPN AND PNP TRANSISTORS

Recall that there are two types of bipolar transistors. They are NPN and PNP. These transistors have three parts: emitter, base, and collector. In the NPN transistor, the emitter is an N type semiconductor material. The base is a P type materials. The collector is an N type material. To remember the direction in which the emitter arrow points for an NPN transistor, recite "**n**ever **p**oints **in**." See Fig. 16-2.

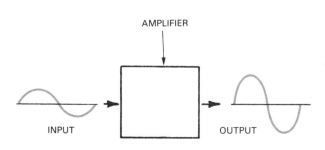

Fig. 16-1. Block diagram of an amplifier.

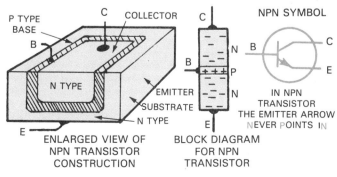

Fig. 16-2. NPN transistor. B—Base. C—Collector. E—Emitter.

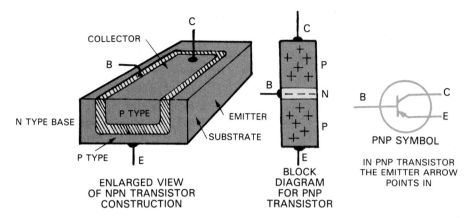

Fig. 16-3. PNP transistor.

In a PNP transistor, the emitter is a P type semiconductor material. The base is an N type material. The collector is a P type material. The arrow on the emitter symbol points in toward the base, Fig. 16-3.

BIASING

For amplifiers to operate properly, they must be correctly biased. BIASING is setting up the correct dc operating voltages between input leads of a transistor.

In a transistor there are two junctions. One is between the emitter and the base. It is called the EMITTER JUNCTION. The other is between the collector and the base. This is commonly referred to as the COLLECTOR JUNCTION. Energy from the internal power source (battery or power supply) is needed to overcome these junction resistances for proper transistor operation.

In an NPN transistor, a BIAS voltage must exist between emitter and base, Fig. 16-4. Voltage applied to these elements with the correct polarity will create current flow. This is known as FORWARD BIAS.

REVERSE BIAS is needed in the collector junction of an NPN transistor, Fig. 16-5.

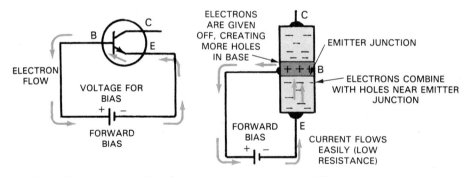

Fig. 16-4. Forward bias for emitter junction in an NPN transistor.

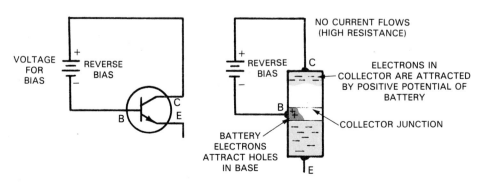

Fig. 16-5. Reverse bias for collector junction in an NPN transistor.

Amplifiers and Linear Integrated Circuits

Both types of bias are needed for the operation of a transistor amplifier. Fig. 16-6 shows a complete NPN transistor circuit. Notice the forward bias in the emitter junction and reverse bias in the collector junction.

The same type of bias is needed for operation of a PNP transistor amplifier, Fig. 16-7. The currents for each circuit are labeled. I_E is for emitter current, I_C for collector current, and I_B for base current. The emitter junction bias is provided by battery B_1. The collector junction bias is provided by battery B_2.

SINGLE BATTERY CIRCUIT

Two voltage sources have been used in all circuits discussed thus far. One source is used for the forward biasing of the emitter junction. The other is used for the reverse biasing of the collector junction. There is no need for two batteries. The amplifier in Fig. 16-8 uses only one battery.

There is no question about the reverse bias of the collector junction. C is connected through R_C to the most negative point of the circuit. That is the negative supply terminal. The most positive point in the circuit is the ground. It is connected directly to the positive terminal of V_{CC}.

R_F and R_B form a resistance voltage divider connected directly across V_{CC}. The voltage at B is less negative than the negative terminal of V_{CC} by the amount of the voltage drop across R_F. It is negative in respect to E. E is at ground; the most positive point in the circuit.

Using the proper values for R_F and R_B, the desired forward bias voltage and current can be established in the emitter junction. The series of R_F and R_B must be large enough so that current drain from the supply battery will be small. This assures long life.

METHODS OF BIAS

The FIXED BIAS method is shown in Fig. 16-9. Notice that R_B has been omitted. This circuit sets a constant base current. It can sense slight changes in the circuit. Proper selection of R sets up the required forward bias voltages and base current.

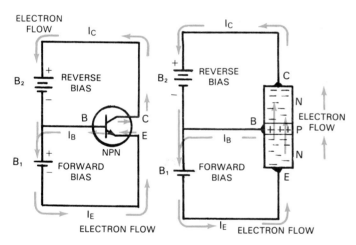

Fig. 16-6. Forward and reverse bias in an NPN transistor amplifier circuit.

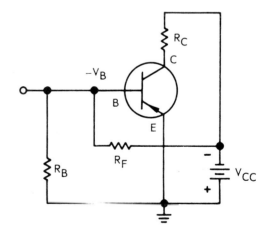

Fig. 16-8. This circuit uses a single power source.

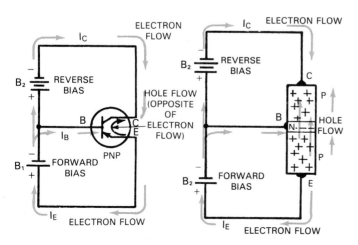

Fig. 16-7. Forward and reverse bias in a PNP transistor amplifier circuit.

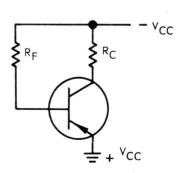

Fig. 16-9. Fixed bias method for connecting a transistor.

The SINGLE BATTERY BIAS scheme is a common method for biasing transistors, Fig. 16-10.

EMITTER BIASING is a third method for setting forward bias of the emitter junction, Fig. 16-11. In this case, the forward bias V_{EE} will set a constant emitter current, I_E. This will produce a voltage drop across R_E. R_E is chosen to provide the proper forward bias. R_B is the return to complete the emitter circuit. A signal applied to the amplifier will produce an ac component in the collector current. This does not upset the emitter bias because a low reactance path around R_E is provided by bypass capacitor C. This circuit ensures a stable operating point for the transistor. A disadvantage, however, is the need for two power supplies or batteries.

A SELF BIAS schematic is shown in Fig. 16-12. This circuit differs from the fixed bias method in that bias resistor R_F is connected to the collector rather than V_{CC}. This method provides a more stable operating point than fixed biasing. Only one power source is needed.

If a fixed collector current is assumed at some operating point, the voltage V_C will be constant. But it will be lower in value than V_{CC}, due to the drop across R_C. Any change in I_C will then change the value of V_C. Since the base is connected to V_C through R_F, a certain amount of degeneration will result.

What is meant by degeneration? A positive signal at the input of the PNP amplifier makes the base more positive, decreases its forward bias, and decreases I_C. A reduction in I_C means a lesser voltage drop across R_C. Collector voltage V_C becomes more negative as it reaches the value of negative V_{CC}. This more negative voltage to the base through R_F tends to increase forward bias. It opposes the increase caused by the input signal.

A signal from the output of a device fed back to the input in a phase relationship that opposes the input is called DEGENERATION. The major disadvantage of self bias is the loss of amplifier gain due to degeneration.

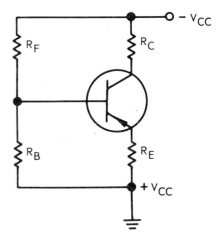

Fig. 16-10. Schematic of a single battery bias.

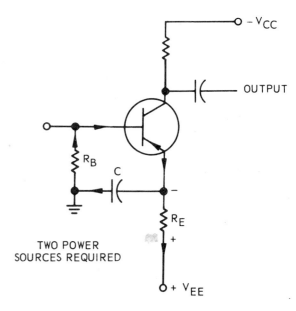

Fig. 16-11. Schematic circuit for emitter biasing.

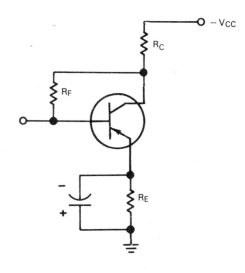

Fig. 16-12. Schematic of self bias circuit.

REVIEW QUESTIONS FOR SECTION 16.1

1. What is an amplifier?
2. Amplifiers can perform two functions. Name them.
3. The process of setting up the correct dc operating voltages between input leads of a transistor is called _____.
4. In a transistor there are two junctions. Name them.
5. Define degeneration.

Amplifiers and Linear Integrated Circuits

16.2 AMPLIFIER OPERATION

As was already stated, an amplifier is an electronic device that uses a small input signal (voltage or current) to control a larger output signal. This works best when the output load, or impedance, of the circuit is greater than the input load, Fig. 16-13. The input is a microphone in the emitter and base control circuit. The output is a loudspeaker connected to the collector and emitter circuit. The load of the loudspeaker is much greater than that of the microphone. The output current in the circuit is about the same as the input current. From what you learned about Ohm's Law, if the current is about the same and the resistance (load or impedance) is 50 times greater in the output than in the input, the output voltage change will be 50 times as great as the input signal. This is how the signal is amplified.

Refer to Fig. 16-14. It shows an NPN amplifier circuit with actual component values and specific voltages. The emitter is common to both the input and output circuits. The ac input signal is impressed across the emitter and base of the transistor Q_1. The output signal is taken across the collector and emitter of Q_1. A capacitor, C_1, is placed in the input signal. This will couple (or pass) the ac input signal to the base for control. A ground is added at a common connection in the emitter, B_1, and B_2 circuit.

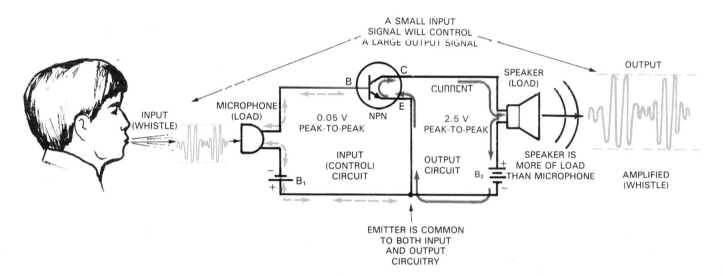

Fig. 16-13. Input circuit controlling output circuit (signal) in an NPN amplifier.

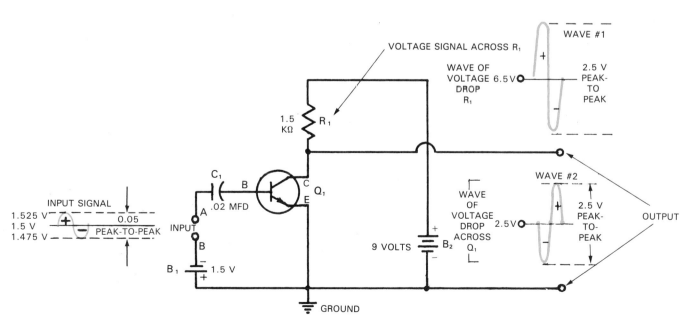

Fig. 16-14. Common emitter NPN amplifier circuit.

A .05 volts peak-to-peak ac input signal is applied across the emitter to base at points A and B. As the signal swings positive, the base-to-emitter forward bias voltage (signal plus bias) will increase (get larger). As a result of this increase, the base current will also increase. This will result in a smaller resistance across the emitter-to-collector of Q_1. This will cause the current in the circuit made of the battery B_2, R_1, and emitter/collector Q_1 to increase. This increase in current flow through a fixed value resistor R_1, will cause a greater voltage drop across the resistor. (See Wave #1). The increase in voltage drop across R_1 will cause a decrease in voltage across Q_1. This is because they are both in series with the 9 volt battery, B_2. (See Wave #2). Keep in mind what happens to the voltage in a series circuit. The total voltage, E_{B2}, is equal to the sum of the voltage drops in that series circuit. This includes the transistor, Q_1, and the resistor, R_1.

During the negative half of the input signal, the bias voltage on the base is decreased or lowered. As a result, the resistance of the emitter-to-collector will increase causing current to be reduced. If the current flowing in the emitter-to-collector circuit is decreased, it will also decrease in R_1, since it is in series with the collector of Q_1. If the current through a fixed value resistor, R_1, is decreased, the voltage drop across it will also decrease. (See negative half of Wave #1). This decrease in voltage across R_1 will cause an increase in voltage drop across the emitter-to-collector of Q_1, as the sine wave swings positive in Wave #2. In any instant, both Wave #1 and Wave #2 must add up to 9 volts, Fig. 16-15. Note that Wave #1 and Wave #2 are 180 degrees out of phase.

A .05 volt signal in the input circuit (emitter-to-base in Q_1) has controlled a 2.5 volt signal in the output (emitter-to-collector in Q_1). This is amplification.

COMPUTING GAIN

There are a number of formulas that can be used to compute types of amplification in transistor amplifier circuits. When the formulas show a triangle or delta (Δ) symbol, this means "the change in" a certain value or waveform.

VOLTAGE GAIN. The voltage gain (A_V) is figured by dividing the change in the output voltage by the change in the input voltage.

$$A_V = \frac{\Delta E_{out}}{\Delta E_{in}} = \frac{E_{out_{p-p}}}{E_{in_{p-p}}}$$

CURRENT GAIN. The current gain (A_I) is figured by dividing the change in the output current by the change in the input current.

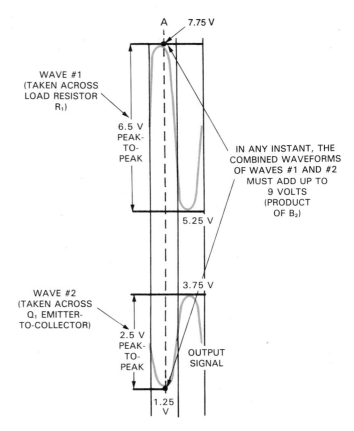

Fig. 16-15. Relationship of output signal and ac signal across R_1.

$$A_I = \frac{\Delta I_{out}}{\Delta I_{in}}$$

POWER GAIN. The power gain (A_P) is figured like Watt's Law. The voltage gain (A_V) is multiplied by the current gain (A_I).

$$A_P = A_V \times A_I$$

Gain is a ratio and, therefore, a dimensionless number. Gain simply tells how much a signal has been amplified. The value is a number, like 100, meaning the voltage gain is 100 times larger in the output than in the input. (Gain is sometimes given in units of decibels, but these are dimensionless. They give relative power levels.)

REVIEW QUESTIONS FOR SECTION 16.2

1. If the change in the output voltage is 200 mV and the change in the input ac signal is 10 mV, what is the voltage gain?
2. Output ac current change is 750μA and ac input current change is 150μA. Find the current gain.
3. Compute the power gain when the output voltage gain is 1.75 V and the current gain is 500 mA.

16.3 TRANSISTOR CIRCUIT CONFIGURATIONS

The transistor can be connected in three circuit configurations. Which configuration is used depends on which element is common to both input and output circuits. Usually the common element is at ground potential.

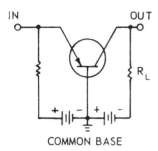

Fig. 16-16. The common base (CB) or grounded base circuit.

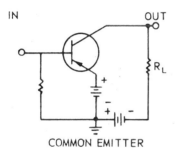

Fig. 16-17. The common emitter (CE) circuit.

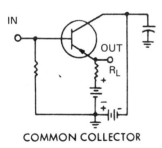

Fig. 16-18. The common collector (CC) circuit.

In Fig. 16-16, a COMMON BASE (CB) amplifier circuit is shown. The ac input signal between emitter and base will vary the forward bias. This is done by alternately adding to and subtracting from the fixed bias. The increase and decrease in current in the emitter-collector circuit produces the amplified voltage across R_L. The input signal and the output signal are in phase. There is no signal inversion.

In Fig. 16-17, a COMMON EMITTER (CE) circuit is drawn. This is the most common transistor circuit. Notice that this circuit resembles the conventional vacuum tube amplifier. In the common emitter circuit, a positive input signal would make the base more positive or less negative with respect to the emitter. This would reduce the forward bias of the circuit. A negative input signal would make the base more negative and increase the forward bias. As a result, the voltage across R_l in the output circuit is 180 degrees out of phase. The signal has been inverted in the same manner that signals are inverted in a vacuum tube circuit.

In the common emitter circuit the gain is the ratio between the change in collector current and the change in base current. The current gain is represented by the Greek letter β (beta).

$$\beta = \frac{\Delta I_c}{\Delta I_b}, V_c \text{ Constant}$$

In this equation, β (beta) equals current gain in common emitter circuit, ΔI_c equals the change in collector current, and ΔI_b equals the change in base current. V_c (collector voltage) is held constant.

Note: In transistor circuits, voltage is represented by a capital V instead of E as used elsewhere.

Fig. 16-18 shows a COMMON COLLECTOR circuit. The signal is applied between the base and collector. The output is taken from the collector-emitter circuit. It will have a very high input impedance and a low output impedance. It is useful as an impedance matching circuit. The input and output signals are in phase. The voltage gain in this circuit must always be less than one.

Fig. 16-19 shows common input and output impedance values for common base (CB), common

TYPE OF CIRCUIT	INPUT IMPEDANCE	OUTPUT IMPEDANCE	A_V	A_I	A_P
COMMON BASE (CB)	LOW 50-150 Ω	HIGH 300 KΩ-500 kΩ	HIGH 500-1500	LESS THAN ONE	MEDIUM 20-30 db
COMMON EMITTER (CE)	MEDIUM 500 Ω-1.5 kΩ	MEDIUM 30 KΩ-50 kΩ	MEDIUM 300-1000	MEDIUM 25-50	HIGH 25-40 db
COMMON COLLECTOR (CC)	HIGH 20 KΩ-500 kΩ	LOW 50 Ω-1 kΩ	LESS THAN ONE	MEDIUM 25-50	MEDIUM 10-20 db

Fig. 16-19. Common input and output impedance values of CB, CE, and CC circuits.

Electricity and Electronics

emitter (CE), and common collector (CC) amplifier circuits. Fig. 16-20 shows the phase relationships between input signals and output signals in CB, CE, and CC amplifiers.

CLASSES OF AMPLIFIERS

Amplifiers are classified according to their bias current. This determines the output waveform compared to the input waveform.

CLASS A. The class A amplifier is biased so that an output signal is produced for 360 degrees of the input signal cycle. See Fig. 16-21. Current I_c flows at all times. The output of a class A amplifier is an amplified copy of the input signal. Its efficiency is low. This class is widely used in high fidelity sound systems.

CLASS B. A class B amplifier is biased so that an output signal is reproduced during 180 degrees of input signal. The class B amplifier is biased close to cutoff, Fig. 16-22. Since collector current flows only one-half of the time, this amplifier has medium efficiency. It is similar to a half-wave rectifier. Its output is distorted.

Two amplifiers operated in the push-pull configuration can restore both halves of the signal in the output. The push-pull circuit has increased efficiency and greater power output. It will be used in many audio amplifiers as the final output stage.

CLASS C. A class C amplifier is biased so that less than 180 degrees of output signal is produced during 360 degrees of input signal. This class finds limited use in transistor circuits.

THERMAL CONSIDERATIONS

Transistors are easy to overheat. There are a few methods that prevent thermal damage to a transistor.

One way to reduce overheating is to use special circuits. As operating temperature increases, there is greater thermal activity in the crystalline structure. And there is a decrease in resistance.

TYPE OF AMPLIFIER	INPUT WAVEFORM	OUTPUT WAVEFORM	STATUS
COMMON BASE (CB)	∿	∿	IN PHASE
COMMON EMITTER (CE)	∿	∿	180° OUT OF PHASE
COMMON COLLECTOR (CC)	∿	∿	IN PHASE

Fig. 16-20. Phase relationships of waveforms in CB, CE, and CC transistor amplifiers.

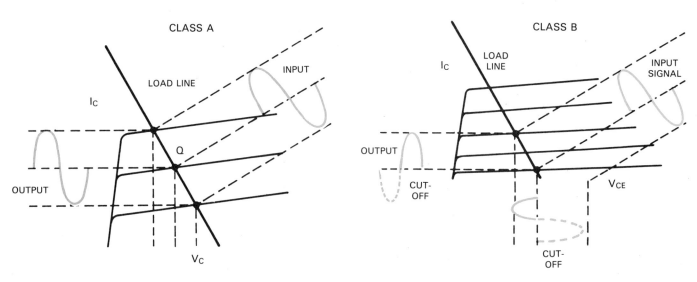

Fig. 16-21. Class A amplifier. It has a 360 degree signal output for a 360 degree input signal.

Fig. 16-22. Class B amplifier. It has a 180 degree signal output for a 360 degree input signal.

Repeated resistance decrease and current increase may cause destruction of the transistor. This is called THERMAL RUNAWAY. In Fig. 16-23, the common emitter amplifier has a stabilizing resistor R_3. An increase in current makes the emitter end of R_3 more negative in respect to ground. The forward bias of the emitter-base circuit decreases. This acts as a limiter to current increase resulting from thermal runaway. C_3 is a bypass capacitor. It prevents degeneration and loss of gain.

Another way to prevent thermal overheating is by using HEAT SINKS. Heat sinks are large pieces of metal, usually with fins, that absorb and disperse heat generated in a transistor. See Fig. 16-24.

TRANSISTOR PRECAUTIONS

1. Do not remove or replace transistors in circuits when the power is on. Surge currents may destroy the transistor.
2. Be very careful with close connections such as those found in miniature transistor circuits. A brief short circuit can burn out a transistor.
3. When measuring resistances in a transistor circuit, remember that your ohmmeter contains a battery. If it is improperly applied, it will burn out a transistor. It is also important to observe the correct polarity when measuring ohms for accurate readings.
4. Soldering the leads of transistors is a skill which you must develop. Heat will destroy the transistor. When soldering a transistor lead, grasp the lead between the transistor and the connection to be soldered with a pair of long nose pliers. These pliers will act as a heat sink and absorb the heat from soldering. Use a pencil type soldering iron of no more than 25 watts.

REVIEW QUESTIONS FOR SECTION 16.3

1. Name three types of transistor connections.
2. This class of amplifier has an output signal that is an amplified copy of the input signal.
   a. Class A.
   b. Class B.
   c. Class C.
   d. None of the above.
3. This class of amplifier has an output wave 180 degrees out of phase with the input signal.
   a. Class A.
   b. Class B.
   c. Class C.
   d. None of the above.
4. The total effect of uncontrolled decreasing resistance and increasing current flow is called _____ _____.
5. What is a heat sink?

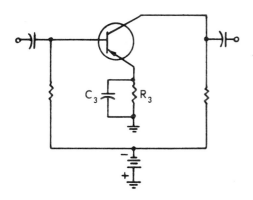

Fig. 16-23. Common emitter stage with stabilizing resistor and bypass capacitor.

Fig. 16-24. Heat sink designs.

16.4 COUPLING AMPLIFIERS

Transistor stages can be connected in series to produce the desired amplification. This is called CASCADING. The key challenge of cascading is matching the high impedance output of one transistor to the low input impedance of the next transistor without severe loss in gain.

Transistor stages may be connected using one of the three primary methods:
1. Transformer coupling.
2. RC coupling.
3. Direct coupling.
4. Push-pull coupling.

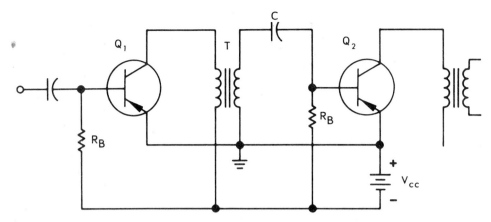

Fig. 16-25. Transformer coupled amplifier circuit.

TRANSFORMER COUPLING

The circuit in Fig. 16-25 shows two stages of transistor amplifier stages coupled with a transformer. Transistor Q_1 has an output impedance of 20 kΩ and Q_2 has in input impedance of 1 kΩ. A severe mismatch and loss of gain will result.

The transformer can match these impedances. A step-down transformer is required. A low secondary voltage means a higher secondary current. This is fine for transistors because they are current operated. Many special sub-ouncer and sub-sub-ouncer transformers have been developed for this purpose, Fig. 16-26.

The purpose of C in the transformer coupling circuit of Fig. 16-25 is to block the dc bias voltage of the transistor from ground. Notice that if C is left out, the transistor base will be grounded directly through the transformer secondary.

The major drawback of transformer coupling, besides cost, is poor frequency response of transformers. They tend to saturate at high audio frequencies. At radio frequencies, inductance and winding capacitance will present problems.

A type of transformer coupling that avoids this problem uses a tapped transformer. These taps can be at medium, low, and high impedance points. Good impedance matching can be attained, as well as good coupling and gain. Study the circuit in Fig. 16-27. For radio frequency amplifier circuits, both the transformer primary and secondary windings can be tuned by C for frequency choice.

There is one other key point about the transformer. The primary impedance of the transformer acts as a collector load for the transistor. This impedance only appears under signal conditions. The load is X_L of primary. From the dc point of view, the only load is the ohmic resistance of the wire used to wind the transformer primary. This information will be helpful when designing a power amplifier.

Fig. 16-26. These sub-ouncer and sub-sub-ouncer transformers are used in transistor circuitry.

RC COUPLING

A simple and less costly method of coupling transistor amplifer stages is to use resistor-capacitor (RC) coupling. See Fig. 16-28.

In Fig. 16-28A, capacitor C will charge V to 10 volts. Only during the charging of C will a current cause a voltage to appear cross R. After C is charged, V_C will equal 10 volts and V_R will equal ϕ.

Amplifiers and Linear Integrated Circuits

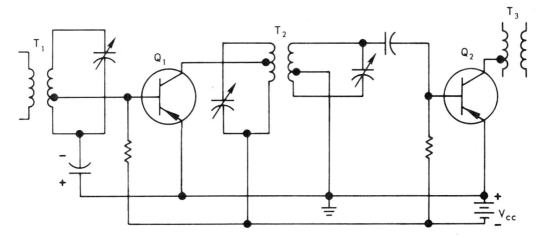

Fig. 16-27. The taps on the transformer primary and secondary windings provide an ideal matching point. The transformer can be designed for good overall gain.

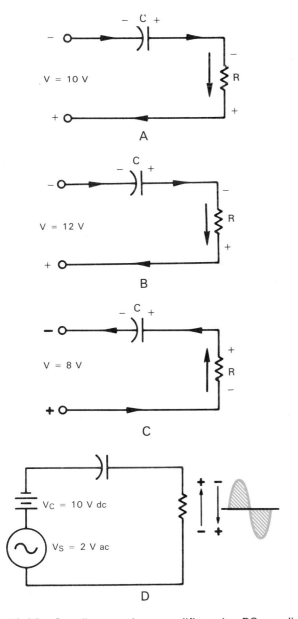

Fig. 16-28. Coupling transistor amplifier using RC coupling.

If source voltage is changed to 12 volts, C will increase its charge to 12 volts also, Fig. 16-28B. The charging current will produce a brief two volt voltage pulse across R in the polarity shown.

If source voltage is changed to 8 volts, C will discharge to 8 volts, Fig. 16-28C. The discharge current will produce a brief voltage pulse across R in the polarity shown.

In Fig. 16-28D, both dc and ac voltage are connected to the RC circuit. The ac signal causes the total voltage to vary between 8 and 12 volts. Therefore, C will charge and discharge at the ac generator frequency.

The voltage across R will rise and fall at the same frequency as the generator voltage. Look at Fig. 16-29. The input signal varies around a dc level of 10 volts. But, the output signal varies around the zero level. The dc component has been removed. A capacitor blocks dc. In this instance, it is called a BLOCKING CAPACITOR.

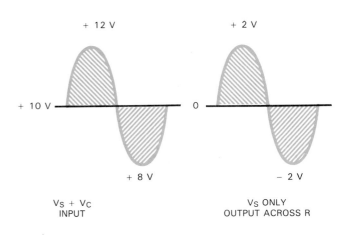

Fig. 16-29. The capacitor blocks the dc voltage, but permits the ac signal to pass.

From a mathematical viewpoint, C and R in series form an ac voltage divider, Fig. 16-30. Voltage division depends on the reactance of X_C at the frequency of the signal. Voltage output across R is a key consideration when the greatest amount of ac voltage must appear across R. With a 1000 Hz frequency and a C value of .01 μF:

$$X_C = \frac{1}{2\pi fC} = \frac{1}{6.28 \times 10^3 \times 1 \times 10^{-8}}$$

$$\cong .16 \times 10^5 = 16{,}000\ \Omega$$

If the value of R is ten or more times greater than X_C, then most of the voltage will appear across R. Assuming R equals 160 kΩ, then:

$$V_R = \frac{R}{R + X_C} \times V$$

$$= \frac{1.6 \times 10^5}{(1.6 \times 10^5) + (1.6 \times 10^4)} \times V$$

$$= \frac{1.6 \times 10^5}{17.6 \times 10^4} \times V = .91 \times 10\ V = 9.1\ V$$

It seems that almost all of the voltage does appear across R. Less than a volt has been lost as signal output. If R were made larger, then even more of the total output would be developed across R. But look at the two stage amplifier using transistors in Fig. 16-31.

Consider the signal voltage at the collector of Q_1. It has a choice of paths to go and will take the easiest path. It can go through R_C or through the coupling network which is in parallel. If the network impedance is higher than R_C, signal currents will go through R_C instead of to the next stage. Therefore, the coupling capacitors used in transistor circuits must have values in the area of 8 to 10 μF.

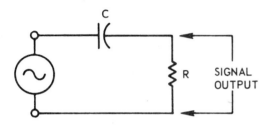

Fig. 16-30. R and C form a voltage divider for the ac signal.

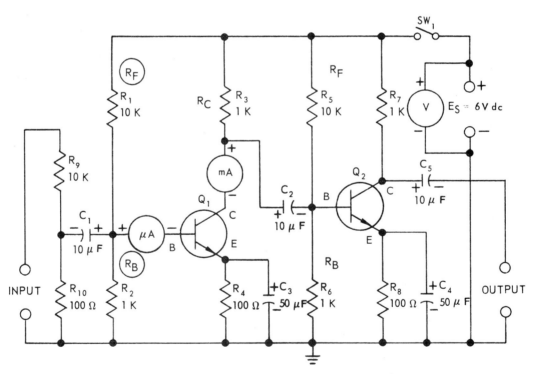

Fig. 16-31. Schematic and parts list for RC coupling.

Amplifiers and Linear Integrated Circuits

To prove the point, assume Q_2 input impedance is 500 Ω. X_C would need to be 50 Ω or less. At 1000 Hz frequency:

$$C = \frac{1}{2\pi f X_C} = \frac{1}{6.28 \times 10^3 \times 5 \times 10} \cong 3.2 \ \mu F$$

At lower amplified frequencies, X_C would be higher. A capacitor of 8 to 10 μF would be required to prevent loss of amplifier gain.

Note also in Fig. 16-31 that R_F and R_B are in parallel with the emitter-base circuit. These resistor values must be high enough so that the signal will not be bypassed around the emitter junction.

The values of R_B and R_F were based on the required bias and stability of the circuit. Current drain from the source was also considered.

Transistor circuit design can become quite complex. It is often a matter of give and take. Compare the output and input impedances of the transistors in the CE configuration. The input can be in the 500 to 1.5 kΩ range. The output impedance in the 30 kΩ to 50 kΩ range. This is a severe mismatch. With the RC coupling, the power gain loss caused in the mismatch must be accepted. When cost is a factor, it may be cheaper to add another transistor stage. This will offset the loss due to mismatch. And it is less costly than purchasing a transformer for interstage matching.

DIRECT COUPLING

In many circuits, very low frequency signals must be amplified or, perhaps, the dc value of a signal must be retained. Amplifier circuits using RC or transformer coupling will block out the dc component. Direct coupling solves this problem, Fig. 16-32.

In this circuit, the collector of Q_1 is joined directly to the base of Q_2. The collector load resistor R_C also acts as a bias resistor for Q_2. Any change of bias current is amplified by the directly coupled circuit. Therefore, it is very sensitive to temperature changes. This can be overcome with stabilizing circuits. Another drawback is that each stage needs a different bias voltage for proper operation. See Figs. 16-33 and 16-34.

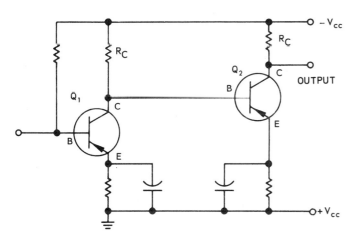

Fig. 16-32. Directly coupled amplifier.

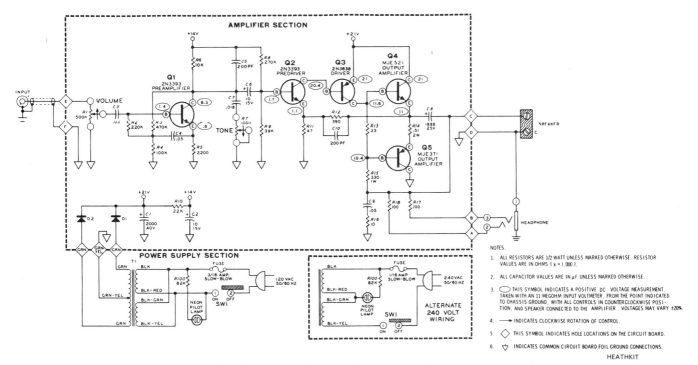

Fig. 16-33. Schematic of a one channel (mono) solid-state amplifier.

259

Electricity and Electronics

Fig. 16-34. One hundred disc player with stereo amplifier.

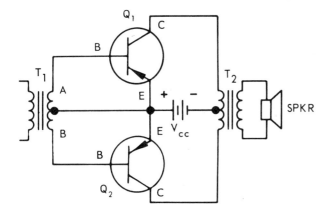

Fig. 16-35. Push-pull amplifier circuit.

PUSH-PULL COUPLING

Two transistors joined in parallel in a push-pull power amplifier circuit will achieve peak power output and efficiency from a power amplifier, Fig. 16-35.

This basic circuit works at zero bias when no signal is applied. No voltage is applied across the EB junction, so the current I_B is zero. When point A of the input transformer become negative (result of a signal), then B of Q_1 is more negative than E. This is a forward bias and Q_1 conducts. At the same time, the base of Q_2 is being driven positive. This increases the reverse bias on Q_2. Q_2 does not conduct.

On the second half of the input signal, the reverse is true. Q_2 is driven into conduction when a negative signal is applied to its base. Q_1 is cut off as a reverse bias is applied by a positive signal to its base. So, first Q_1 conducts, then Q_2 conducts.

On one half-cycle, current I_C flows in Q_1. On the second half-cycle, current I_C flows in Q_2. These transistors are operating Class B. One-half of the time each transistor is resting and cooling. There is no question that this circuit has increased efficiency. In fact, the maximum possible efficiency approaches 78.5 percent.

The two half signals, or waves of current, from the transistors are restored to their original input form by transformer action in the output transformer.

In Fig. 16-36, when Q_1 conducts, current flows as shown by arrows. It creates a magnetic field of one polarity. This denotes a half-wave form in the secondary of the transformer. When Q_2 conducts, the

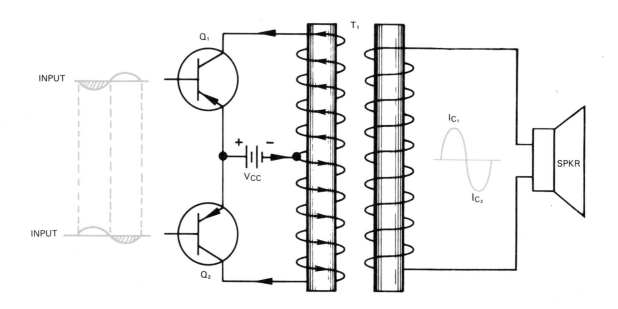

Fig. 16-36. Each half of the signal is joined in the output transformer to restore the original wave.

polarity of the primary reverses. It induces the wave of opposite polarity in the secondary. Thus, the complete wave is restored.

> **REVIEW QUESTIONS FOR SECTION 16.4**
>
> 1. Connecting amplifier stages in series is called _____.
> 2. A _____ can be used to couple amplifier stages together, as well as match input and output impedances.
> 3. Name two disadvantages of transformer coupling.
> 4. Refer to Fig. 16-33. Q_1 is connected to Q_2 by the _____ coupling method.
> 5. If the collector of one amplifier is hooked to the base of the next amplifier, with no other components in between, the coupling method is called _____ _____.
> 6. Connecting two amplifier stages in parallel is referred to as _____-_____ coupling.

16.5 LINEAR INTEGRATED CIRCUITS

Linear integrated circuits have variable outputs controlled by variable inputs. They are like conventional transistor amplifiers except that the amplifier device is housed in an integrated circuit. The primary purpose of a linear IC is to increase or amplify currents, voltages, or wattages.

Linear integrated circuits are called ANALOG DEVICES or CIRCUITS. In this case, linear can mean CONTROLLING or REGULATING since the input signal controls the output signal in a linear IC. The output signal changes are smooth and constant, like the input signal. See Fig. 16-37.

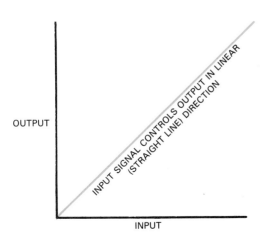

Fig. 16-37. Input versus output in linear ICs.

Digital ICs are used as switches. Their output is always one of two conditions: on or off. These integrated circuits will be discussed in Chapter 17.

The symbol for an amplifier in a circuit is a triangle. This is also used as the IC symbol, Fig. 16-38.

VOLTAGE REGULATORS AND REFERENCES

Voltage regulators are circuits providing exact dc voltages in the output of power supplies. LINEAR IC VOLTAGE REGULATORS are often made on a silicon chip. They come in a number of packages, such as TO-3, TO-5, TO-202, TO-220, and others. Fig. 16-39 shows a common schematic and application for a LM-309 (National Semiconductor) voltage regulator. Many linear IC voltage regulators are immune to blowout.

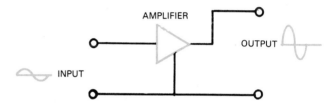

Fig. 16-38. The triangle is used as the symbol for an amplifier in a circuit and an IC.

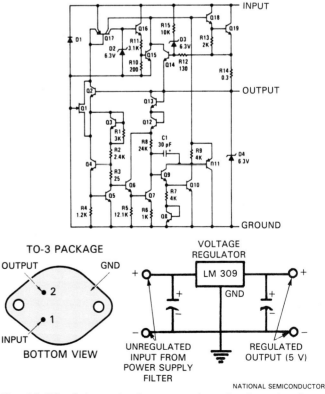

Fig. 16-39. Schematic diagram and application of voltage regulator in a circuit.

Electricity and Electronics

Voltage references are precision, temperature stabilized IC circuits. They are made up of zener diodes, transistors, resistors, and capacitors. They provide exact output regulated voltages. They perform better than common zener diodes.

OPERATIONAL AMPLIFIERS

Operational amplifiers or OP AMPS are one type of linear integrated circuits. They have very high voltage amplification. They are used as basic amplifier circuits in cable TV system video line amplifiers (drivers), two-way intercoms, instrumentation amplifiers, function generators, and servo motor amplifiers. Fig. 16-40 shows an op amp in an instrumentation amplifier.

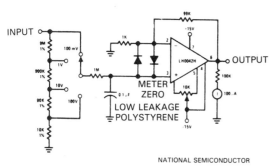

Fig. 16-40. Sensitive, low cost, multirange voltmeter that uses an op amp.

AUTOMOTIVE AND INDUSTRIAL

Current automobiles are full of integrated circuits. Examples include fluid detectors in radiators, tachometers, amplifiers or drivers, spark timing controls, factory testing systems, warning light drivers, microprocessors, computerized suspension systems, and anti-skid devices, Fig. 16-41.

Industrial uses of linear ICs include timing circuits, sensing amplifiers, temperature controllers, flashers, ground fault interrupter circuits, temperature transducers, and fluid detectors. See Fig. 16-42. These are just a few examples.

RADIO AND TV

Linear ICs have made portable miniature tape players with AM-FM radios possible, Fig. 16-43. In radio circuits, linear ICs provide integrated rf/if amplifier stages in one package, FM demodulators, stages stereo audio amplifiers, FM detector/limiter stages, and other circuits.

In TV circuits, linear ICs can be used for complete sound systems, automatic fine tuning circuits, IF amplifier systems, etc.

Other uses of linear ICs include voltage comparators, analog switches, sample and hold circuits, and A to D, D to A circuits. Many manufacturers and suppliers have data books containing application information.

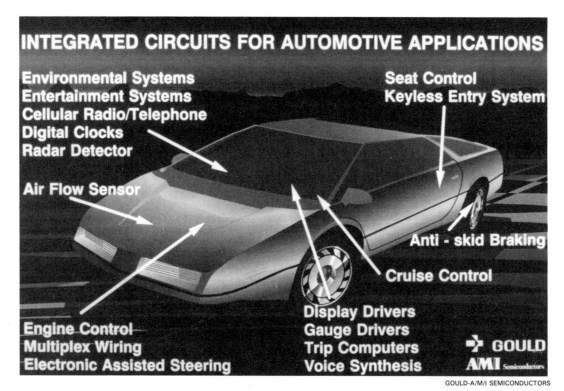

Fig. 16-41. Linear IC use in automobiles.

Amplifiers and Linear Integrated Circuits

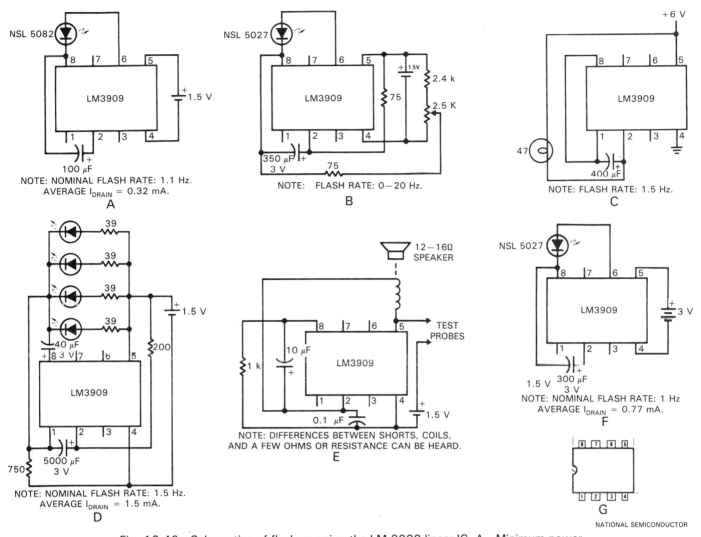

Fig. 16-42. Schematics of flashers using the LM 3909 linear IC. A—Minimum power, 1.5 V. B—Variable flasher. C—Incandescent bulb flasher. D—High efficiency parallel circuit. E—Buzz box continuity and coil checker. F—3 V flasher. G—Pin identification.

Fig. 16-43. Linear ICs made the technology for portable stereos possible.

REVIEW QUESTIONS FOR SECTION 16.5

1. The primary purpose of a linear IC is to _____ currents, voltages, and wattages.
2. _____ ICs are used as switches.
3. What are voltage regulators?
4. _____ _____ are precision, temperature stabilized IC circuits.
5. Draw the op amp symbol.

16.6 LINEAR IC CONSTRUCTION PROJECTS

PORTABLE STEREO AMPLIFIER*

Portable cassette stereos were designed for "take along" music. But with a few connections, these stereos can be used to give full stereo sound.

Electricity and Electronics

This project will outline the steps for making a portable stereo amplifier. And to further increase usefulness, two preamplifiers are included. These will pick up low level input signals such as microphones, guitars, and turntables. With a few small changes to the basic circuit, a PA amplifier can be made. The schematic for the portable stereo amplifier is shown in Fig. 16-44. The parts list is shown in Fig. 16-45.

The Output Stage

The LM 380 is a small audio amplifier. Internal gain is fixed at around 34dB (50 times), Fig. 16-44. A unique input stage allows signals to be referenced to ground. The output is centered somewhere near half the supply voltage. The output is also short circuit proof with internal thermal limiting.

With the supply voltage restraints given and a minimum 8 ohm load, a heat sink on the design shown is not required. A small amount of heat sinking under the board is provided by using the copper tracks as thermal fins. This is not normally enough sinking for the chip to be stretched to its peak capability. In this design, however, and with limited parameters, it should satisfy thermal conditions. With a maximum supply of 15 volts and an 8 ohm load, the output is around 1.5 watts per channel. The input stage is usable with signals from 50 mV to 500 mV RMS.

*The portable stereo amplifier, portable stereo speakers, and megaphone amplifier projects used with permission from Dick Smith's "Fun Way into Electronics, Volume 3," by Dick Smith (B2610). The parts are available from Dick Smith Electronics, Inc., P.O. Box 8021, Redwood City CA 94063, (415) 368-8844. Kit of all components, including PC board, but excluding speakers and power supply (K-2667). A separate 12 V, 500 mA power supply (M-9555) is also available.

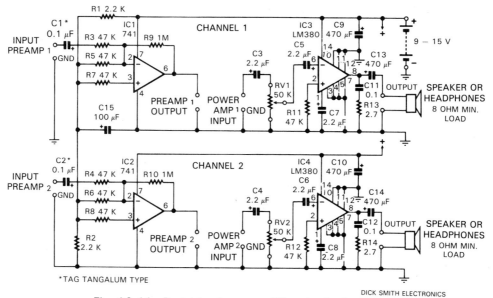

Fig. 16-44. Portable stereo amplifier circuit diagram.

Parts Description		Total Quantity
Resistors:		
2.7 ohm	R_{13}, R_{14}	2
2.2 K	R_1, R_2	2
47 K	$R_3, R_4, R_5, R_6, R_7, R_8$	
	R_{11}, R_{12}	8
1 meg	R_9, R_{10}	2
50 K	Log Potentiometers	
	VR_1, VR_2	2
Capacitors:		
0.1 μF	Ceramic	
	C_{11}, C_{12}	2
0.1 μF	Tag Tantalum	
	C_1, C_2	2
2.2 μF	RB Vertical Electrolytics	
	$C_3, C_4, C_5, C_6, C_7, C_8$	6
100 μF	RB Vertical Electrolytic	
	C_{15}	1
470 μF	RB Vertical Electrolytics	
	$C_9, C_{10}, C_{13}, C_{14}$	4

Parts Description		Total Quantity
Semiconductors:		
741 Op-Amps		
	IC_1, IC_2	2
LM 380	Audio Amp.	
	IC_3, IC_4	2
Miscellaneous:		
PC Board	ZA-1467	1
Hookup Wire	500 mm	1
Optional Components:		
(Not included in basic kit.)		
8 ohm speaker systems		2
Power Supply, 12 V, not less than 200 mA		
eg. M-9530		1
33 ohm ½ Watt resistors		2
47 K ¼ Watt resistors		2
3.5 mm Mini Stereo Plug		
P-1140		1
Hookup Wire—assorted colors		1

Fig. 16-45. Parts list for the portable stereo amplifier.

Amplifiers and Linear Integrated Circuits

The Input Pre-Amplifier Stage

Two noncomitted 741 operational amplifiers have been configured as input amplifiers. Their input stages have been referenced to a common point, that is half the supply voltage, Fig. 16-45. This is put in effect using two 2.2 k resistors, R_1 and R_2 to split the supply. C_{15} is the bypass capacitor at this point. The gain of each of these amplifiers has been fixed at 21 by the input resistors R_3 and R_4 and the feedback resistors, R_9 and R_{10}.

Input capacitors, C_1 and C_2 isolate any DC component from the signal. The output point has been moved to a connection pad on the board. This will allow for peak use. In most cases, this would be used to drive the final stages via capacitors, C_3 and C_4 and volume controls, R_{V_1} and R_{V_2}. If required, three stages could each be used alone (maybe as experimental bench amplifiers). In one use shown later, they are configured slightly different to use the board as a PA amplifier.

With a power supply of 12 volts, the resting current drawn by the total system is 30 to 35 mA. Under driven conditions, the drain could increase to 300 mA or more.

Assembly

1. Check the components against the parts list, shown in Fig. 16-45.
2. Insert all the resistors, cut pigtails, and solder, as shown in Fig. 16-46.
3. Next insert the ICs and solder. This may be an approach you have not used. But because of the size of the capacitors that flank these ICs, it is easier to load in this sequence.
4. Now, insert, trim, and solder the capacitors. Note carefully the polarity of electrolytics.
5. The only connections left are the volume controls, R_{V_1} and R_{V_2}, the input wiring, and the power wires. The input wiring will be inserted based on how you use the system. If the amplifiers will provide "big sound" for your portable stereo, only the output stages will be used. Where a low level signal, like a microphone input, is to be used, the preamplifiers would have to be included.
6. When all connections are checked a final inspection of the soldered joints under the printed circuit boards should be made. Then power can be applied. For a power supply, you will need at least 9 volts at 200 mA.

PORTABLE STEREO SPEAKERS

Using the portable stereo amplifier as an output stage, you can get sound out of typical stereo speakers.

Follow the diagram in Fig. 16-47. Any type of 8 ohm speakers will work. The power supply is not crucial but should be within the 9 to 15 volt limits. A value of 12 volts is ideal.

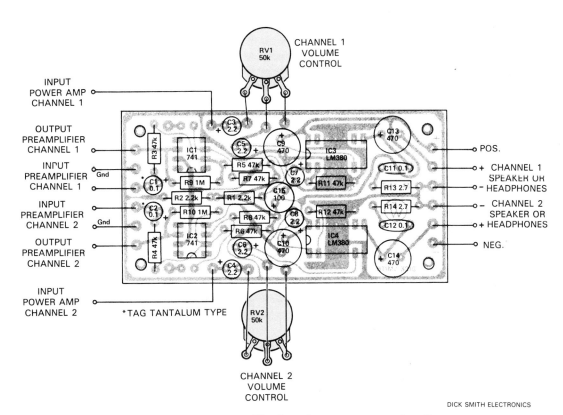

DICK SMITH ELECTRONICS

Fig. 16-46. Portable stereo amplifier board component overlay diagram.

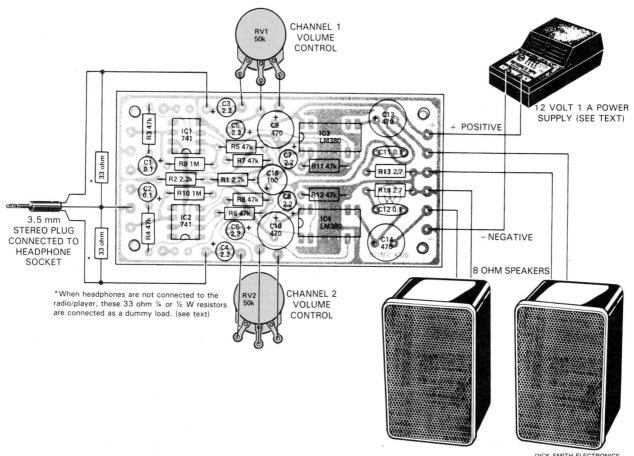

Fig. 16-47. Wiring diagram of a typical system using the amplifier as an output stage to a portable stereo.

The input to the amplifier is connected to the output section only. The front end 741 preamps are not used in this case. The headphone connector to these portable type stereo units is commonly a 3.5 mm plug (for example, the Dick Smith P-1140). Solder one of these plugs to the three wires as shown; one for each channel, with a common ground. This is the body or the long stem of the plug.

The 33 ohm resistor joined across each channel to ground is a dummy load for the radio player when headphones are not connected. If the stereo has two headphone outlets, leave one set of headphones connected. Then the two resistors are not needed. In some stereo types, the headphone plug is the on-off switch.

MEGAPHONE AMPLIFIER

The output of the preamp is connected to the input of the output stage as shown in Fig. 16-48. In this configuration, the system could be used as an experimental amplifier where low level signals are present. It makes a handy bench or workshop amplifier. No special frequency response tailoring has been included around the 741 stages. But it could be used as a microphone amplifier using dynamic or electret types.

These stages are usable with input signals from 3.5 mV to 100 mV RMS.

By using the twin output stages in a "bridge" mode, the output power can be almost doubled.

As Fig. 16-49 shows, the speaker is connected across the active output points of each amplifier. The common in this case is not connected.

Preamplifier 1 is used as an input stage with a gain of 21. It is useful with signals from 3.5 mV to 100 mF. The second preamp has a different function in this case. The gain has been reduced to unity by changing the feedback resistor R_{10} to 47 K. Now this stage becomes an inverter of the signal to the second output stage. This fills the requirements of the bridge output that the input to the output stages are 180 degrees out of phase. This provides twice the voltage swing across the 8 ohms load for a given supply. This increases the output power over a single stage by four. However, the package power dissipation will be the first parameter limiting power delivered into the load. In this case, the power supply is limited to a maximum of 12 volts. The final result is an output power capability of around double that of a single amplifier. The configuration shown is suitable for input signal levels from 3.5 mV to 100 mV RMS. Both the volume controls R_{V1} and

Amplifiers and Linear Integrated Circuits

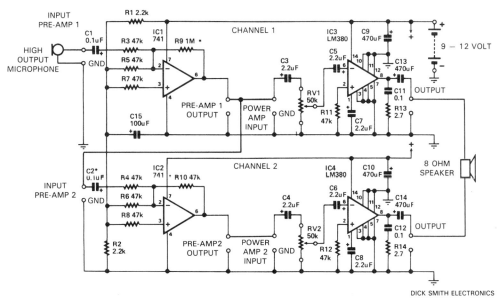

Fig. 16-48. Circuit diagram of megaphone amplifier.

R_{V_2} have to be turned up and down by the same amount to control the output.

For convenience, R_{V_1} and R_{V_2} can be replaced by a dual gang pot.

Where high gain is not required, the circuit has to be changed. As with the modified stage of preamp 2, the feedback resistor R_9 of preamp 1 has to be changed to a value of 47 K so that it only has unity gain. In this way, higher input signals from 50 mV to 500 mV can be handled. This makes the system ideal as a "big sound" amplifier for a portable stereo. The connections can be made to the speaker or earphone sockets as described for the portable stereo amplifier.

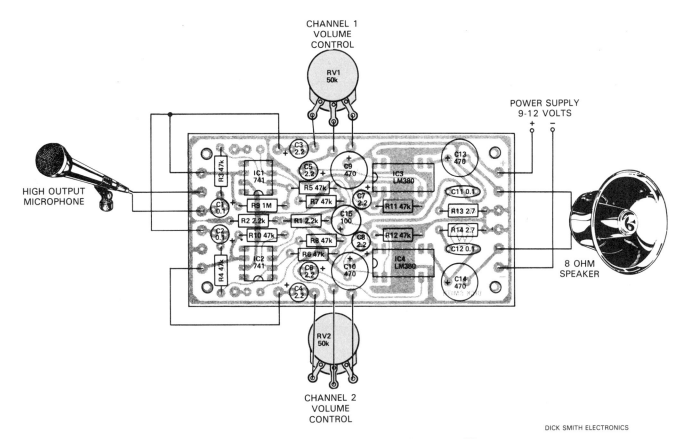

Fig. 16-49. Wiring of a small megaphone amplifier.

TRANSISTORIZED TELEPHONE AMPLIFIER

How many times have you wished that everyone in the room might hear the voice of friends or family over a long distance telephone call? The amplifier circuit shown in Fig. 16-50 can be attached to your telephone. A flick of the switch activates the transistorized amplifier. You may then have a family telephone visit.

The telephone pickup T_4, Fig. 16-50, is connected to two transformer coupled, common emitter stages. These drive a push-pull power amplifier. Potentiometer R_7 is a gain control to adjust the volume of the amplifier. In the layout of the parts, keep transformers T_1 and T_2 as far from the input as possible. This will prevent possible feedback. Keep leads short. C_2, C_3, and C_5 are bypasses that prevent oscillation. If the circuit should oscillate, reverse the connections to the secondary of T_1.

The parts can be mounted on a small sheet of printed circuit board. The amplifier may be placed in a case of your own design.

PAGING AMPLIFIER

A small audio amplifier connected to a microphone is called a paging or announcing amplifier. It will boost a voice so that it may be heard for some distance. The circuit, Fig. 16-51, has enough power to drive the small speaker. A gain control is provided by R_3.

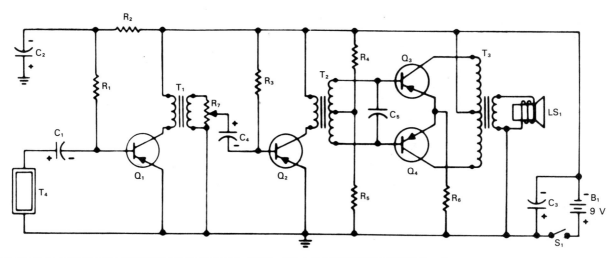

PARTS LIST, TELEPHONE AMPLIFIER	
B_1 — 9 V battery	R_3 — 270 kΩ, 1/2 watt
C_1, C_4 — 2 μF, electrolytic, 15 V	R_4 — 4.7 kΩ, 1/2 watt
C_2, C_3 — 8 μF, electrolytic, 15 V	R_5 — 68 Ω, 1/2 watt
C_5 — 0.01 μF ceramic disc or mica, 100 V	R_6 — 12 Ω, 1/2 watt
LS_1 — 3-6 ohm loudspeaker	R_7 — 100 kΩ, 1/2 watt potentiometer
Q_1, Q_2, Q_3, Q_4 — Sylvania 2N1265 Transistor (or RCA SK3003)	T_1 — Stancor TA-27 or equivalent
R_1 — 470 kΩ, 1/2 watt	T_2 — Stancor TA-35 or equivalent
R_2 — 150 kΩ, 1/2 watt	T_3 — Stancor TA-21 or equivalent
S_1 — SPST switch	T_4 — Telephone pickup coil, Radio Shack 44-533, Shield M-133, or equivalent.

SYLVANIA ELECTRIC CORP.

Fig. 16-50. Wiring diagram and parts list for telephone amplifier.

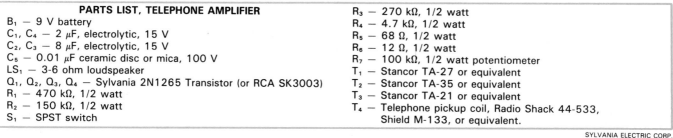

SYLVANIA ELECTRIC CORP.

Fig. 16-51. Wiring diagram for a paging amplifier.

In the paging amplifier, three grounded emitter stages are connected in cascade. The last stage is a push-pull power amplifier. In the layout of parts, it is wise to keep R_3 some distance from transformer T_2 to prevent feedback.

These parts can be mounted on a printed circuit, Fig. 16-52. Construct a new circuit in breadboard style. Test it before final assembly and soldering. Then try designing a case for it.

SUMMARY

1. Amplifiers are electronic circuits which control output signals with input signals.
2. Transistors, integrated circuits, and vacuum tubes are all amplifier devices.
3. Bias is the voltage difference between the two input elements of an amplifier device.
4. In the normal operation of a transistor amplifier, the input elements are usually forward biased while the output elements are reverse biased.
5. Formulas to compute gain are shown in Fig. 16-53.
6. Three transistor circuit configurations are the common base amplifier (CB), common emitter amplifier (CE) and common collector amplifier (CC).
7. Amplifiers may be classified according to their bias current. Some common classes are A, B, and C.
8. Transistors and other semiconductors are sensitive to heat. Heat sinks can be used to absorb and disperse unwanted heat. The heat sink protects the transistor from the heat.
9. Some important methods of coupling amplifier stages are:
   a. Transformer coupling.
   b. RC coupling.
   c. Direct coupling.
   d. Push-pull coupling.
10. Linear ICs are used in amplifier circuits needing constant or variable output voltage or current values.
11. Some applications of linear ICs are voltage regulators, voltage references, operational amplifiers (op amps), voltage comparators, and analog switches.

```
B₁ — 9 V battery
C₁, C₂ — 10 μF electrolytic, 15 V

LS₁ — any 4, 8, or 16 ohm loudspeaker
Q₁ — Sylvania 2N1265 transistor (or RCA SK 3003)

Q₂, Q₃, Q₄ — Sylvania 2N1266 transistor (or RCA SK3005)
R₁ — 68 kΩ, 1/2 watt

R₄ — 120 kΩ, 1/2 watt
R₂, R₆, R₈ — 560 Ω 1/2 watt

R₃ — 5 kΩ, 1/2 watt potentiometer
R₅ — 6.8 kΩ, 1/2 watt

R₇ — 100 Ω, 1/2 watt
R₉ — 10 kΩ, 1/2 watt

T₁ — Stancor TA-5 or equivalent
T₂ — Stancor TA-10 or equivalent
```

Fig. 16-52. Parts list for paging amplifier.

Voltage Gain	$A_V = \dfrac{\Delta E_{out}}{\Delta E_{in}}$
Current Gain	$A_I = \dfrac{\Delta I_{out}}{\Delta I_{in}}$
Power Gain	$A_P = A_V \times A_I$

Fig. 16-53. Formulas for computing gain in transistor amplifiers.

TEST YOUR KNOWLEDGE, Chapter 16

Please do not write in the text. Place your answers on a separate sheet of paper.

1. An amplifier:
   a. Is a control circuit.
   b. Uses a small input signal to control a larger output signal.
   c. Produces outputs that are linear.
   d. All of the above.
2. What is gain?
3. The emitter junction is located between:
   a. Collector and emitter.
   b. Emitter and base.
   c. Collector and base.
   d. None of the above.
4. The collector junction is located between:
   a. Collector and emitter.
   b. Emitter and base.
   c. Collector and base.
   d. None of the above.
5. In an _____ transistor, a bias voltage must exist between emitter and base.
6. Name three methods of bias.
7. Give the formula for computing voltage gain.
8. Give the formula for computing current gain.
9. Give the formula for computing power gain.
10. Draw the common base amplifier circuit.
11. How is the gain in the common emitter circuit computed?
12. The input and output wave forms of a common collector amplifier are:
    a. In phase.
    b. 180 degrees out of phase.
    c. All of the above.
    d. None of the above.
13. Draw a diagram of a directly coupled amplifier.
14. Draw a graph showing the input signal versus the output signal of a linear IC.

FOR DISCUSSION

1. What are some of the characteristics of a good stereo sound system?
2. Discuss the relative merits of RC coupling versus transformer coupling.
3. Discuss the several kinds of distortion possible in an amplifier.
4. Why should transistors be temperature controlled by heat sinks and ventilation?
5. Discuss precautions in the use of and servicing of transistor circuits.

Amplifiers and Linear Integrated Circuits

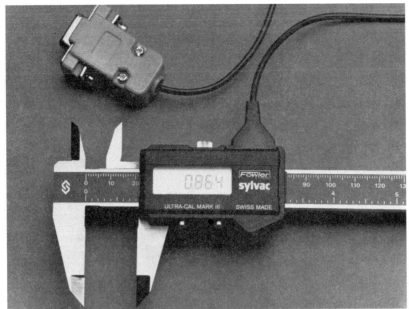

Linear integrated circuits have led to creations such as this electronic caliper. This caliper can be connected directly to any computer or printer.

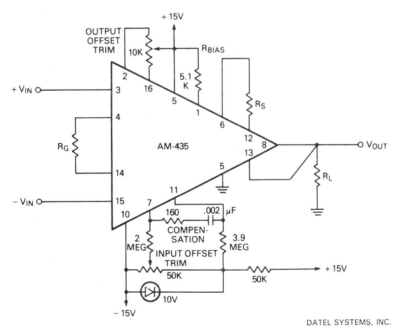

An instrumentation amplifier helps scientists get data from experiments. Inputs can be of the differential type or of the single level type. The sensitivity is adjustable. It is controlled by the ratio of R_S and R_G.

Electricity and Electronics

TEKTRONIC

Computers and testing equipment are based on digital circuit design.

Digital circuits perform the logic operations of computers. Their input and output levels are either 1 or 0. The only numbers used in the binary number system are 1 and 0. Circuits are easy to design, build, and troubleshoot. The supply voltage can vary by plus or minus 3 percent without causing trouble. Boolean logic helps describe and predict the operation of digital circuits. Boolean logic is a condensed form of mathematics similar to algebra. People interested in the field of digital electronics are encouraged to study Boolean logic.

Chapter 17

DIGITAL CIRCUITS

After studying this chapter, you will be able to:
- *Convert decimal numbers to their binary equivalents and binary numbers to their decimal equivalents.*
- *Name seven types of logic gates.*
- *Explain the operation of various types of logic gates.*
- *Use truth tables to determine the output of a logic gate.*
- *Discuss four types of logic families.*
- *Construct and operate the project, Binary Bingo.*

To review, an IC is a complete electronic circuit contained in one package. Usually this package includes transistors, diodes, resistors, and capacitors, along with the connecting wiring and terminals.

A major advantage of the IC is its size. It is possible to wire each component found in an IC. However, this device would require much more space. IC size makes possible high capacity computers that can fit in a briefcase.

ICs are also very stable compared to individual components. And they are less costly to build and operate.

Fig. 17-1 shows an enlarged IC. Note the thousands of components that make up the IC. This IC is used in computers.

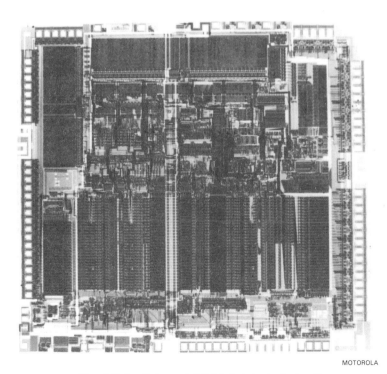

MOTOROLA

Fig. 17-1. Enlarged 32-bit integrated circuit.

There are two types of integrated circuits: LINEAR and DIGITAL. Linear ICs were discussed in Chapter 16. They are used as amplifiers and have variable outputs. Digital ICs are used as switches. They work in either the on or off state.

Digital electronics have many uses in our lives. They create the time display on a wristwatch and the display on a portable CD player, Fig. 17-2. Fig. 17-3 shows a small camcorder. These popular devices record and display both visual and audio information.

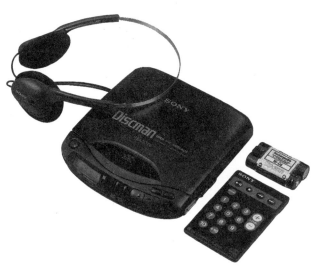

SONY ELECTRONICS CORP.

Fig. 17-2. Digital electronics make this portable CD player and its liquid crystal display possible.

SONY ELECTRONICS CORP.

Fig. 17-3. Camcorders record information in a digital format.

17.1 DIGITAL FUNDAMENTALS

Digital integrated circuits handle digital information using switching circuits. They work as a result of gating circuits and flip-flop circuits being combined. Let us look at gates and how they operate using binary logic.

BINARY NUMBERING SYSTEM

Electronic circuits can be made to act in only two states: ON and OFF. This two state system is called BINARY. It can be compared to a single-pole, single-throw (SPST) switch, Fig. 17-4. A switch that is off represents a 0 in the binary numbering system. Likewise, when a switch is on, this represents a 1. There are only two conditions to the circuit: on or off.

It is not practical to build large electronic logic circuits using manual switches. Manual switches do however, provide a good basis for understanding other switchable electronic components.

The most common electronic component that can be used as a switch is a transistor. It can allow current to flow through it. This would be an on or 1 state. Or it can stop current flow. This would be the off or 0 state. See Fig. 17-5.

In Chapter 16, Class B amplifiers were discussed. These amplifiers are biased at the midpoint of the curve made by the emitter voltage and collector current. Without any input signal (A to B) the amplifiers are turned off. If the input signal changes to a negative level (B to C), the emitter-to-base bias will be decreased. This will cause the output signal across the lamp to swing positive (input signal is 180 degrees out of phase with output signal in a CE circuit).

Simple circuits made up of diodes, transistors, and resistors can perform the basic logic functions, Fig. 17-6. Since the development of the IC, thousands of circuits can be designed and built into a single chip.

With only one SPST switch, you can count to a maximum of 1. Suppose you wanted to count higher. How do you count to a number such as 45 or 79?

There is a basic rule for counting in any system.

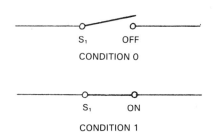

Fig. 17-4. A single-pole, single-throw switch as a binary device.

Digital Circuits

Digits must be recorded one after each other for each count, until the count exceeds the total number of inputs available. Then a second column is started and counting continues. The binary system works as shown in Fig. 17-7.

The digit 0 in the decimal system is 0 in the binary system. Likewise the digit 1 in the decimal system is 1 in the binary system. However, the number 3 in the decimal system is the number 11 in the binary system. The binary number for 6 is 110, Fig. 17-8.

Fig. 17-9 shows the decimal to binary conversion for the numbers 0 through 26.

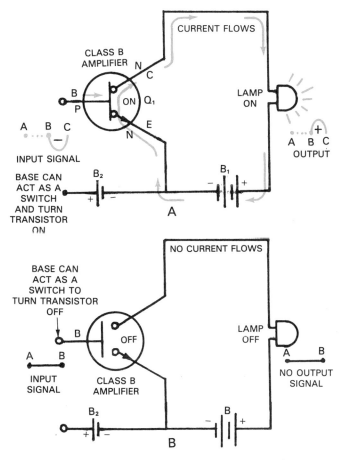

2^7	2^6	2^5	2^4	2^3	2^2	2^1	2^0
128s	64s	32s	16s	8s	4s	2s	1s
8TH NUMBER	7TH NUMBER	6TH NUMBER	5TH NUMBER	4TH NUMBER	3RD NUMBER	2ND NUMBER	1ST NUMBER

Fig. 17-7. The binary numbering system.

2^2	2^1	2^0
4s	2s	1s
1	1	0

Fig. 17-8. Binary number for decimal number 6.

Fig. 17-5. Diagram of transistor acting as a switch. A—On. B—Off.

COMPONENT	ON OR 1 STATE	OFF OR 0 STATE
SPST SWITCH	—o o—	—o o—
MAGNETIC CORE		
DIGITAL PULSE	+5V / 0V	0V
RELAY		

Fig. 17-6. Binary states of components.

DECIMAL			BINARY NUMBER				
HUNDREDS	TENS	ONES	2^4	2^3	2^2	2^1	2^0
			16s	8s	4s	2s	1s
		0					0
		1					1
		2				1	0
		3				1	1
		4			1	0	0
		5			1	0	1
		6			1	1	0
		7			1	1	1
		8		1	0	0	0
		9		1	0	0	1
	1	0		1	0	1	0
	1	1		1	0	1	1
	1	2		1	1	0	0
	1	3		1	1	0	1
	1	4		1	1	1	0
	1	5		1	1	1	1
	1	6	1	0	0	0	0
	1	7	1	0	0	0	1
	1	8	1	0	0	1	0
	1	9	1	0	0	1	1
	2	0	1	0	1	0	0
	2	1	1	0	1	0	1
	2	2	1	0	1	1	0
	2	3	1	0	1	1	1
	2	4	1	1	0	0	0
	2	5	1	1	0	0	1
	2	6	1	1	0	1	0

Fig. 17-9. Decimal to binary conversion table.

Electricity and Electronics

Try this activity. It may help you understand the binary numbering concept. Using five small pieces of tape, number the tape as shown in Fig. 17-10. Place one piece of tape on each finger of your left hand with palm up (in order shown). Fingers pointing up will represent 1s. Fingers folded down will represent 0s. Position your hand as shown in Fig. 17-11. What is the binary number? What is its decimal equivalent?

A larger decimal number, 79, is shown in binary form in Fig. 17-12.

Binary addition is a simple process, Fig. 17-13. Adding 0 and 0, the sum is still 0. Adding 1 and 0, the sum is 1. But adding 1 and 1, the sum is 10. The 0 is placed in the 1s column. The 1 is carried to the 2s column. To add two larger binary numbers, the same steps are followed. Fig. 17-14 shows the addition of 57 (00111001) and 24 (00011000).

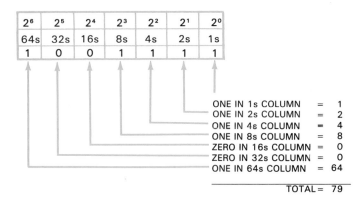

Fig. 17-12. Binary number for decimal number 79.

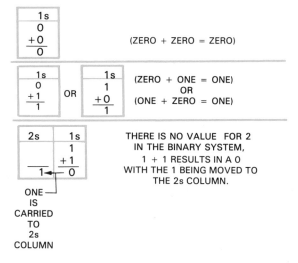

Fig. 17-13. Binary addition fundamentals.

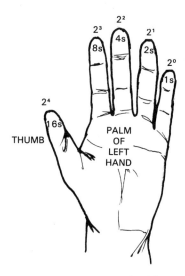

Fig. 17-10. Learning the binary system.

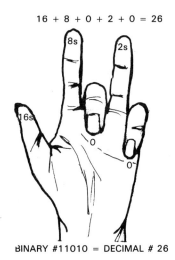

Fig. 17-11. Converting 26 to 11011.

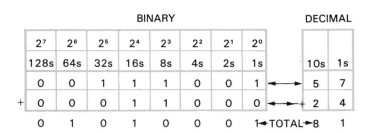

Fig. 17-14. Binary addition of 57 and 24.

VOLTAGE LOGIC LEVELS IN DIGITAL CIRCUITS

We know that digital circuits have only two states of 0 and 1. What operating voltages are needed in a circuit for these two values? That will depend on what type of logic circuitry or "family" is used. Whatever logic family is used, in positive logic, the high value of 1 is called the VALID LOGIC HIGH range. It most often varies from 2 to 3.5 volts to 5 volts. The low,

Digital Circuits

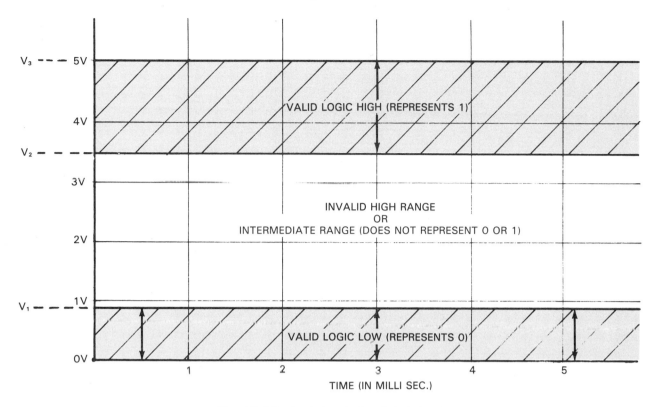

Fig. 17-15. Logic levels for digital circuits.

or 0, value varies from 0 volts to .8 to 1.5 volts. It is shown as the 0 to V_1 range in Fig. 17-15. This is called the VALID LOGIC LOW.

The area in between these two values acts as a buffer range. Any voltage in this range applied to the digital circuit causes confusion in the IC. The IC will not know whether to produce a 0 or a 1. This area is called the INVALID VALUE OR INTERMEDIATE RANGE. In Fig. 17-15 it is shown between V_1 and V_2.

BITS, NIBBLES, AND BYTES

In binary code, the smallest unit of information is the BIT. The bit comes from joining the two words, BInary digiT. A bit can be either 0 or 1. It is only one column or digit in a binary numbering system, Fig. 17-16. Four bits of information make up a NIBBLE. Two nibbles or eight bits make up a BYTE. A byte is a single unit of memory in a computer. For example, a computer with a 256 byte storage can hold 2048 bits of information. This is a small number, however. Most computer memory is given using the terms K (KILOBYTES) or Meg (MEGABYTES). These represent roughly 1000 and 1,000,000 bytes respectively.

Computer storage abilities continue to grow. Many computers can now hold billions of bytes, or GIGABYTES, of storage.

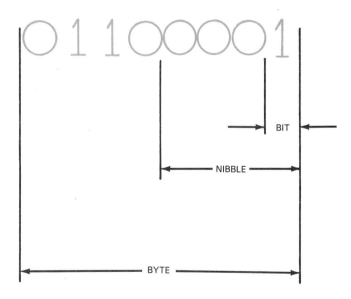

Fig. 17-16. Bits versus nibbles versus bytes.

REVIEW QUESTIONS FOR SECTION 17.1

1. Describe how a transistor biased as a Class B amplifier can act as a switch.
2. What is the binary number for 114?
3. What is the decimal number for 101101?
4. What is the sum of 01001010 and 01011001? What is the decimal number of the sum?

17.2 LOGIC GATES

Electronic switching circuits that decide whether inputs will pass to output or be stopped are called LOGIC GATES. The logic gates that we will discuss are the building blocks for other logic gates. The basic logic gates are:
1. AND GATES.
2. OR GATES.
3. NOT GATES.
4. NAND GATES.
5. NOR GATES.
6. XOR GATES.
7. XNOR GATES.

AND GATES

The AND gate accepts YES and NO inputs. Based on these inputs, the gate decides on the output. It also will be YES or NO. The AND gate symbol is shown in Fig. 17-17.

The AND gate will produce an output of 1 (YES) if ALL inputs are 1 (YES). Logic 1 or YES also equals ON. A simple circuit with switches that functions like an AND gate is shown in Fig. 17-18. Both switches are OFF (0 or NO). The lamp does not burn. When both switches are on, the lamp will burn, Fig. 17-19. An actual AND gate used in a computer is very complex. This circuit using SPST switches was only illustrating AND gate operation.

A binary table that explains the operation of the AND gate is shown in Fig. 17-20. This is called a TRUTH TABLE. The AND gate is used to detect the presence of YES signals or 1s on both inputs, A and B. If this occurs, the output signal will be 1. However, if even one input signal is 0, output will be 0.

The output signal as affected by input signals in the AND gate is shown in Fig. 17-21.

OR GATES

The OR gate provides an output signal of 1 (YES) when either one or the other of its inputs is 1. And if all of the inputs are 0 in an OR gate, the output is a 0 (NO). The OR circuit detects the presence of any input. Its symbol for an OR gate is shown in Fig. 17-22.

The truth table for an OR gate is shown in Fig. 17-23. The schematic for a simulated OR gate is shown in Fig. 17-24. The output signal as affected by input signals in the OR gate is shown in Fig. 17-25.

NOT GATES

The NOT gate is often called an INVERTER because that is what it does in a circuit. The NOT gate is put into a circuit to invert the polarity of the input signal. If the input signal is 1, the output signal will be 0. Likewise, if the input signal is 0, the output will be 1.

The symbol for the NOT gate is shown in Fig. 17-26. Note that the NOT gate symbol has only one input lead. There is also a small circle at the end of the triangle in the symbol.

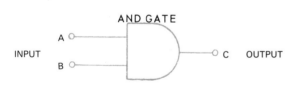

Fig. 17-17. AND gate symbol.

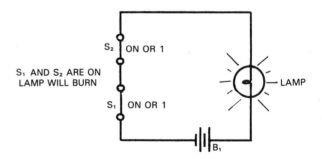

Fig. 17-19. This is the same circuit as in Fig. 17-18. However, the lamp will burn because both switches are on.

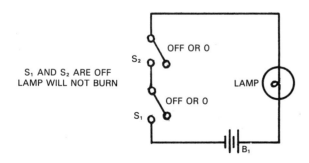

Fig. 17-18. This simple switching circuit simulates the operation of an AND gate. The lamp will not burn because both switches are off.

SIGNAL AT INPUTS		SIGNAL AT OUTPUT
A	B	C
0	0	0
0	1	0
1	0	0
1	1	1

Fig. 17-20. AND gate truth table. Note that only one combination in this table will produce a 1 or YES output.

Digital Circuits

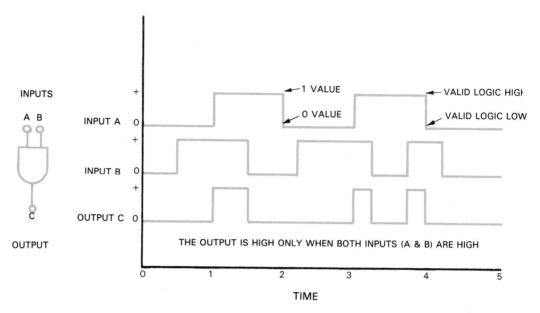

Fig. 17-21. Resulting valid logic highs and lows in an AND gate.

Fig. 17-22. OR gate symbol.

SIGNAL AT INPUTS		SIGNAL AT OUTPUT
A	B	C
0	0	0
0	1	1
1	0	1
1	1	1

Fig. 17-23. OR gate truth table.

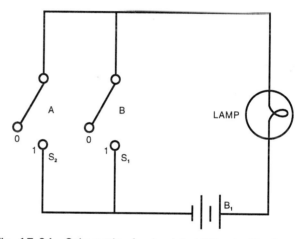

Fig. 17-24. Schematic of a simulated OR gate. The lamp will not burn in this instance.

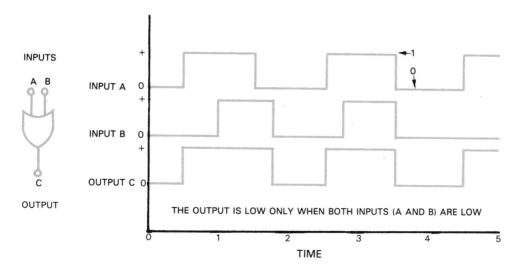

Fig. 17-25. Valid logic highs and lows for OR gate.

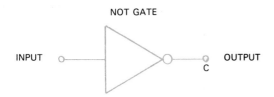

Fig. 17-26. NOT gate symbol.

The schematic for a simulated NOT gate circuit is shown in Fig. 17-27. The NOT gate truth table is shown in Fig. 17-28. And Fig. 17-29 shows the output signal in a NOT gate as affected by the input signals.

NAND GATE

All logic gates are combinations of the basic gates: AND, OR, and NOT. The NAND is a NEGATIVE AND gate. It is made up of an AND gate and a NOT gate that inverts the output. It is also called a NOT AND gate.

The NAND gate symbol is like the AND gate symbol with a circle at the end, Fig. 17-30. The schematic for a simulated NAND gate circuit is shown in Fig. 17-31. The NAND truth table is shown in Fig. 17-32.

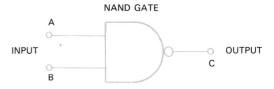

Fig. 17-30. NAND gate symbol.

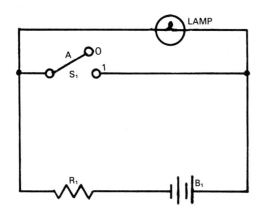

Fig. 17-27. Schematic of simulated NOT gate. The input is 0, so the output is 1.

INPUT	OUTPUT
A	C
0	1
1	0

Fig. 17-28. NOT gate truth table.

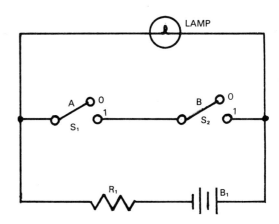

Fig. 17-31. Schematic of a simulated NAND gate. The inputs are 0 and 0, so the output is 1.

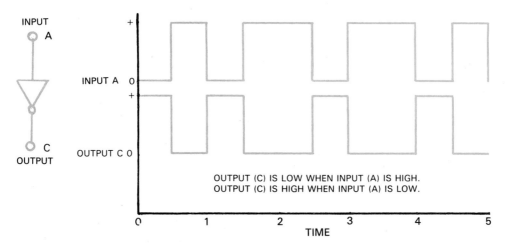

Fig. 17-29. Valid logic highs and lows for NOT gate.

Digital Circuits

It is the reverse of the AND truth table. Fig. 17-33 shows the waveforms in a NAND gate citcuit.

NOR GATE

The NOR gate gives the opposite (or negative) results of the OR gate. It is made up of an OR gate and a NOT gate (inverter). The symbol for a NOR gate is shown in Fig. 17-34. Fig. 17-35 shows the schematic for a simulated NOR circuit.

The NOR circuit is used to test for any kind of input. If there is no input, output will be 1. Likewise, if there is an input, output will be 0. This is shown in the truth table in Fig. 17-36. The waveforms of a NOR gate circuit are shown in Fig. 17-37.

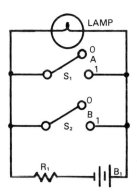

Fig. 17-35. Schematic of simulated NOR gate. The inputs are 0 and 0, so the output is 1.

SIGNAL AT INPUTS		SIGNAL AT OUTPUT
A	B	C
0	0	1
0	1	0
1	0	0
1	1	0

Fig. 17-36. NOR gate truth table.

SIGNAL AT INPUTS		SIGNAL AT OUTPUT
A	B	C
0	0	1
0	1	1
1	0	1
1	1	0

Fig. 17-32. NAND gate truth table.

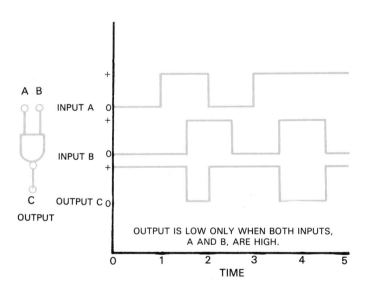

Fig. 17-33. Valid logic highs and lows for NAND gate.

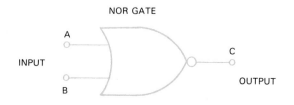

Fig. 17-34. NOR gate symbol.

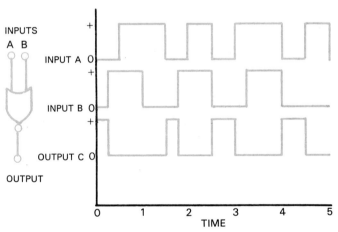

Fig. 17-37. Valid logic highs and lows for NOR gate.

XOR GATE

There is a special type of gate that provides output whenever ANY, but NOT ALL, inputs are logic high. It is called the EXCLUSIVE OR gate or the XOR gate. Recall that in contrast, the OR gate provides output whenever ANY or ALL inputs are logic high.

The symbol for the XOR gate is shown in Fig. 17-38. The circuit that simulates the XOR gate is shown in Fig. 17-39. The XOR gate truth table is shown in Fig. 17-40. The waveforms of a XOR gate circuit are shown in Fig. 17-41.

Electricity and Electronics

Fig. 17-38. XOR gate symbol.

XNOR GATE

A gate similar to the XOR gate is the EXCLUSIVE NOT (XNOR) gate. It is the XOR gate with the output inverted. There is output only if all inputs are logic high or logic low. The symbol for the XNOR gate is shown in Fig. 17-42. The truth table is shown in Fig. 17-43. Notice that if there is only a one in column A or column B, there is a zero in column C. If column A and column B both have a zero or a one, then column C has a one.

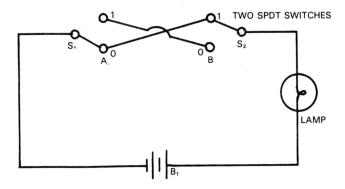

Fig. 17-39. Schematic of simulated XOR gate. Note that it has 2 SPDT switches. One input is 1 and the other is 0. The output will be 1.

Fig. 17-42. XNOR gate symbol.

SIGNALS AT INPUT		SIGNAL AT OUTPUT
A	B	C
0	0	0
0	1	1
1	0	1
1	1	0

Fig. 17-40. XOR gate truth table.

SIGNALS AT INPUT		SIGNALS AT OUTPUT
A	B	C
0	0	1
0	1	0
1	0	0
1	1	1

Fig. 17-43. XNOR gate truth table.

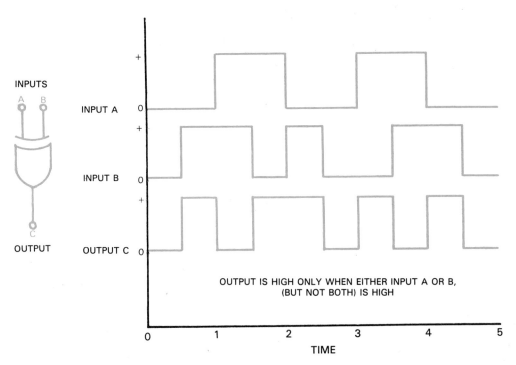

Fig. 17-41. Valid logic highs and lows for XOR gate.

Digital Circuits

> **REVIEW QUESTIONS FOR SECTION 17.2**
>
> 1. Draw the symbols for three logic gates. Label the drawings.
> 2. In an AND gate, input A is 1 and input B is 0. What is output C?
> 3. The OR gate will provide an output signal except:
>   a. When both inputs are 1.
>   b. When both inputs are 0.
>   c. When either one of the inputs is 1.
>   d. None of the above.
> 4. The _____ gate is often called the inverter.
> 5. What gate provides an output when any but not all of the inputs are 1?
> 6. List the truth table for the XNOR gate.

17.3 LOGIC FAMILIES

Manufacturing techniques have a major impact on the arrangement of digital circuits into groups or families. It is crucial that the traits of one logic family match the traits of another family when a number of digital ICs are used in one device. In the time that has passed in the digital electronics field, certain logic families have evolved.

RESISTOR-TRANSISTOR LOGIC

A logic family that is no longer used is the resistor-transistor logic (RTL) family. It used a mixture of resistors and transistors in the gate circuits to perform logic functions. RTL circuits needed somewhat low voltage for proper operation (3.6 to 4.0 volts).

DIODE-TRANSISTOR LOGIC

The diode-transistor logic (DTL) family combined diodes, transistors, and resistors to produce logic gates. This family of circuits was one of the first to be made in digital ICs. They are not used any longer. They have been replaced by more up-to-date logic families.

CMOS LOGIC

The complementary metal-oxide semiconductor logic (CMOS) family uses field effect transistors. CMOS integrated circuits have good resistance to noise. They require only small amounts of power. Voltage and current needs are not crucial. CMOS ICs do have one major problem. They can be damaged by static electricity. To avoid this problem, the worker and work surface should be grounded through a high resistance resistor (3-10 megohms at 5 watts or more).

CMOS ICs contain a manufacturer's identification number. One company uses a CD prefix. Another firm uses a C within the number, such as 74C30. High speed CMOS devices contain an H in the identification number, such as 74HC00 or 74HC190.

TRANSISTOR-TRANSISTOR LOGIC

The transistor-transistor logic (TTL) family is used widely in today's digital electronics. They work quickly. They perform logic functions from about 20 megahertz to 60 megahertz. TTL ICs work faster than CMOS ICs.

TTL ICs can be identified by their number. Their numbers start out with 74: 7400, 7402, 7404, 7408, 7432, and 7486. Logic high (1) for TTLs is 2 volts to 5 volts. Logic low is 0 volts to .8 volts. TTL requires high power dissipation and high current.

High power TTL ICs contain the part number 74H. They consume about twice the power of a regular TTL IC. Low power TTL ICs contain the part number 74L. One-tenth the power of a regular TTL IC is consumed.

A number of gates can be placed in one TTL IC. Fig. 17-44 shows some TTL ICs that contain AND gates, NOT gates (inverters), NAND gates, NOR gates, and XOR gates.

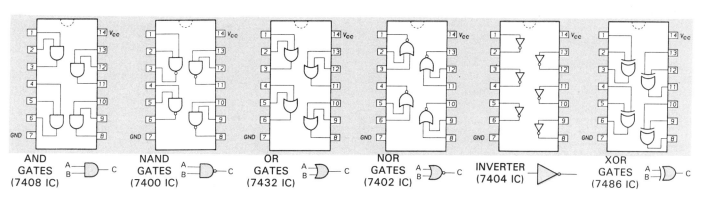

Fig. 17-44. Logic gates in transistor-transistor logic ICs. Note the basing diagram for the ICs.

Electricity and Electronics

OTHER DIGITAL CIRCUITS

Digital circuits are used for many reasons. For example, digital circuits convert ac sine waves to digital pulses, Fig. 17-45. The ac signal is sampled. Various points are given digital values. These are converted to a binary value later.

Fig. 17-46 shows a digital disk that has binary information recorded on the surface. A laser "reads" the information on the recording. The ICs in Fig. 17-46 are also used in digital circuits.

Fig. 17-47. Universal Product Code symbols.

Another example of digital electronics use is the Universal Product Code (UPC). It is found on most grocery items, Fig. 17-47. The UPC is read by a laser (light). The laser improves inventory (item movement) control. Some laser scanning systems contain a computer generated voice that tells what has been purchased and the price of the item.

REVIEW QUESTIONS FOR SECTION 17.3

1. What two logic families are used widely today?
2. A 7408 IC is found in the _____ logic family.
3. List the advantages and major disadvantages of the CMOS ICs.
4. What type of transistor does the CMOS IC use?
5. What number is used to signify a TTL IC?

17.4 DIGITAL IC CONSTRUCTION PROJECTS

BINARY BINGO*

You have learned about binary logic and the binary numbering system. This project uses these principles. Logic design and display functions are the center of this project. When completed, it becomes a game of skill and reaction time. The rules for play are quite simple. However, playing can be frustrating and demanding. The decimal number that is displayed must be matched with a series of four push buttons. These buttons represent the binary equivalent of the decimal number. Skill level can be changed.

The complete schematic for Binary Bingo is shown in Fig. 17-48. It uses a variety of logic gates. The parts list is shown in Fig. 17-49.

Assembly

1. Check your part kit against the parts list.
2. Insert all low profile components. These include links L_1 to L_9. L_{10} is a length of hook-up wire. Also

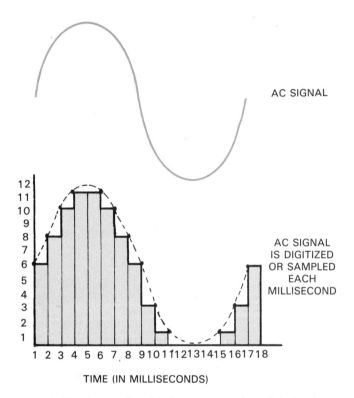

Fig. 17-45. An ac signal being converted to digital pulses.

Fig. 17-46. Digital disc with chips used in circuits.

---

*The Binary Bingo project is used with permission from *Dick Smith's "Fun Way into Electronics, Volume 3,"* by Dick Smith (B2610). The parts are available from Dick Smith Electronics, Inc., P.O. Box 8021, Redwood City, CA 94063, (415) 368-8844. A polybag kit of all components, including PC board but excluding Zippy Box case and AA cells, is available from the catalog, number 2668.

Digital Circuits

insert resistors and diodes. Make sure of the polarity of the diodes. Cut excess lengths and pigtails. Solder as necessary, Fig. 17-50.
3. Now insert all capacitors. Cut and solder pigtails. Note polarity of the electrolytics. The larger 470 μF

capacitor C_6 can be laid on its side to reduce the height. Then the board will fit in the Zippy Box.
4. Carefully insert the ICs and solder. The battery lead with the snap attached can also be soldered. (Red is positive.)

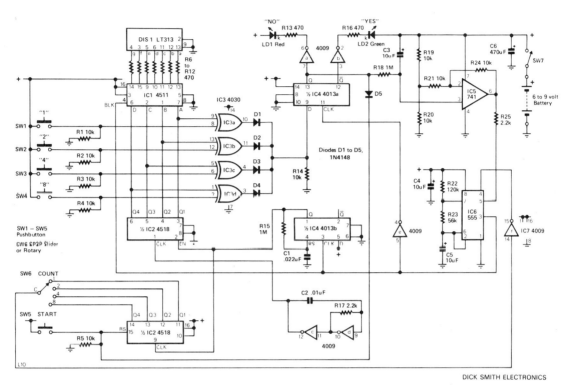

Fig. 17-48. Schematic for Binary Bingo.

PARTS DESCRIPTION		TOTAL QUANTITY
Resistors:		
470 ohm	$R_6, R_7, R_8, R_9, R_{10}, R_{11}, R_{12}$ R_{13}, R_{16}	9
2.2 K	R_{17}, R_{25}	2
10 K	$R_1, R_2, R_3, R_4, R_5, R_{14}, R_{19},$ R_{20}, R_{21}, R_{24}	10
56 K	R_{23}	1
120 K	R_{22}	1
1 Meg	R_{15}, R_{18}	2
Capacitors:		
.01 μF	Ceramic C_2	1
.022 μF	Ceramic C_1	1
10 μF	RB Vertical Electrolytic C_3, C_4, C_5	3
470 μF	RB Vertical Electrolytic C_6	1

PARTS DESCRIPTION		TOTAL QUANTITY
Semiconductors:		
4009	CMOS IC_7	1
4013	CMOS IC_4	1
4030	CMOS IC_3	1
4511	CMOS IC_1	1
4518	CMOS IC_2	1
741	Op-Amp IC_5	1
555	Timer IC IC_6	1
LT303/LT313	7 Seg. Display DIS_1	1
1N4148 Silicon Diodes	D_1, D_2, D_3, D_4, D_5	5
5 mm Red LED	LD_1	1
5 mm Green LED	LD_2	1

PARTS DESCRIPTION		TOTAL QUANTITY
Miscellaneous:		
SP Pushbutton Switch	$SW_1 - SW_5$	5
SP4P Miniature Slide Switch	SW_6	1
DPDT Miniature Slide Switch	SW_7	1
Battery Snap		1
4 AA Battery Carrier		1
Tinned Copper Wire	300 mm	1
Hookup Wire	150 mm	1
PC Board	ZA-1468	1
Optional Components: (Not included in basic kit)		
Zippy Box Case UB1	H-2751	1
AA Cells		4
Basic Kit of Components: Cat K-2668		

Fig. 17-49. Parts list for Binary Bingo.

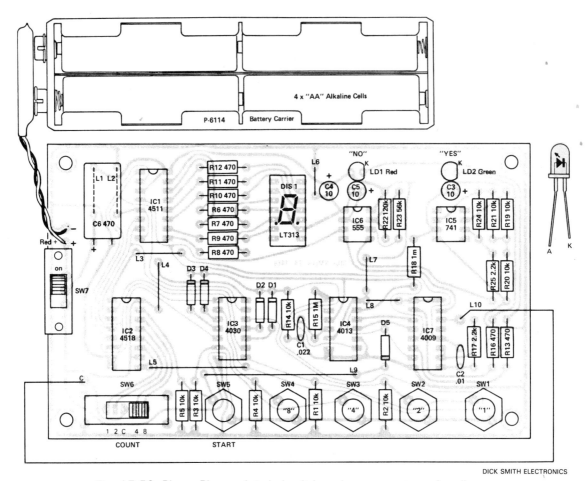

Fig. 17-50 Binary Bingo printed circuit board component overlay diagram.

5. Final housing of the board has to be considered at this point. If you use the Zippy Box method suggested, the display can be inserted. Mount it centered in the 14 hole position. The device itself may only have 10 legs. Pins 1, 7, 8, and 14 are not present and these holes in the board are not used. Do not push the display too hard up onto the board. Only about 1 mm of the legs should go through the foil side of the board. Solder in this position. This gives the display a little extra height so that it is closer to the window of the case. An optional 14 pin IC socket can be used to increase this standoff distance. The display is plugged into this socket. Again, pins 1, 7, 8, and 14 may not be used.

6. UB1 H-2751 Zippy Box can be used to house the board. The dress panel for the lid of the case is shown in Fig. 17-51. The holes and cut-outs for this panel can be marked by using the label as a template. Place it over the lid so that the edges line up with the marked outlines. Using a sharp instrument, carefully mark the center positions of the 5 push buttons, the 2 on-off switches, and the 2 LED holes. Mark each corner of the 3 rectangular cut-outs and remove the label. Scribe a line between each marked point of the cut-outs so that a rectangle is clearly visible.

The holes can now be drilled at the centers marked. Dimensions are 6.5 to 7 mm for the push buttons, 2 x 2 mm for the on-off switches, and 2 x 5 mm for the LEDs. The rectangular sections can be cut out with a piercing or jeweler's saw. A pilot hole will have to be drilled in the center of the cut-out first so the saw blade will fit. Do not cut right up to the scribed line. The final dressing of the hole can be done with a small needle file. If you do not have a saw of this kind, drill a hole in the center of each cut-out. Then use a small needle file to enlarge and shape the hole to final size. All burrs should be removed from each hole with a small file.

The label can now be glued to the lid. Use a rubber based contact adhesive. Position it precisely before pressing down. The holes can be punched through this label by using the rear shank of the drill for the size hole. Push it straight through from the face side so that a clean cut is made. The rectangular cut-outs can be trimmed by using a sharp razor blade or knife. Trim around the edges of the label and lid if needed.

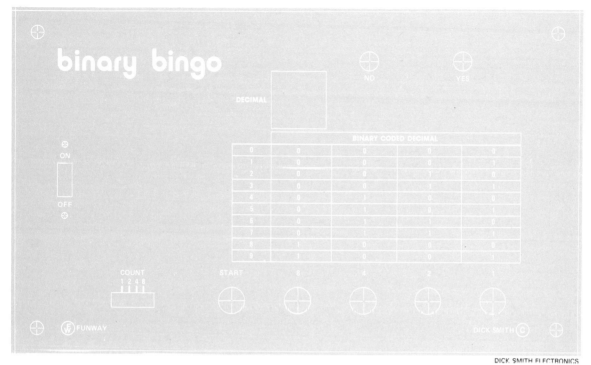

Fig. 17-51. Template for the cover of Binary Bingo.

Final board assembly can now be completed. The S-2060 count switch should drop neatly into the hole system in the board. Solder it with the body flush to the board. If the switch provided does not fit the hole, wire the contacts and mount the body to the lid with screws. Two extra holes will have to be drilled.

The on-off and push-button switches can be mounted on the front panel with the screws and nuts provided. Do not tighten the nuts firmly at this stage. Solder short lengths of tinned copper wire to the contacts of each switch. Now insert the two LEDs in the correct holes. Note that the flat side of the body (k) faces away from the seven segment display. Do not solder at this stage.

Now position the board over these switch wires so that they can be threaded through the holes in the board. If each push button is turned until each contact is over the holes, the board will fit flush. The three wires from the on-off switch will have to be bent at right angles to accommodate the holes in the board. Depending on the push buttons supplied, it may be necessary to bend the end of the contact. This will reduce the overall height of the switch. The lever of SW_6 can then correctly pierce the panel. Securely tighten the push buttons. Solder the wires so that the board and the front panel are parallel. Push the LEDs up to the front panel so that the shoulder of the body is flush with the rear surface. Solder the four leads, and cut off all excess pigtail lengths. This completes all wiring.

7. The battery pack can now be plugged in and the unit tested. If all is well, the total assembly can fit into the Zippy Box. If the unit fails to operate correctly, check all wiring. PC board assembly, and soldering.
8. To make the assembly solid, fix the board to the bottom of the box using four standoffs. The length of these supports will have to be measured carefully so that the board, front panel, and case have the right spacing between them. This extra work is needed only if the unit is to be subjected to rough treatment. Normally, the simple wiring from the switches to the board support the system.
9. The P-6114 battery carrier and 4 AA cells fit between the bottom of the case and the board. Fit a flat piece of foam plastic between the board and the carrier to hold it in place, Fig. 17-52. When the case screws are tightened, this foam will hold the carrier in position.

How it works

The circuit has two basic timing periods. The first and the longest is the decision period. During this cycle, a decimal number is displayed by the seven segment readout. At the same time, the binary equivalent of this number has to be keyed in and held by the four push buttons. During the second period, the display is blanked. At this time the number keyed and the number appearing on the output of the decade counter are compared. If they match, the "yes" LED will light.

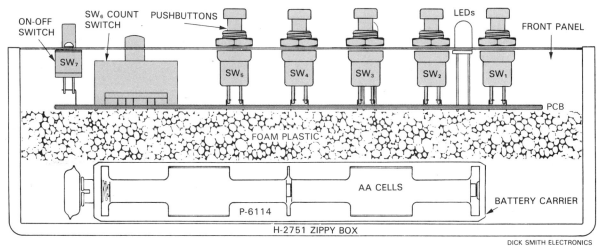

Fig. 17-52. Switches supporting the printed circuit board inside the case.

Likewise, the "no" LED will come on if the two numbers do not match. After comparison, the decade counter is clocked by a free running oscillator. The system returns to the first state to repeat the cycle.

Decade counter clock and display

The decade counter (half of the 4518 CMOS pack), a free running clock made from two inverters (4009 CMOS buffers), the seven segment LED display (LT313), and a variable gating system (4013 and 555) make up a random number generator.

During a short gating period, the decade counter is clocked by the free running clock. It is made from two inverters of the 4009 running at about 20 kHz. At the end of this clock burst period, the last number in the count is stored in the counter. It is present at its outputs Q_1 to Q_4. These outputs are connected to the 4511 decoder/driver for the display.

System clock

A 555 timer is used to create the two timing periods in the operating cycle. It is a low frequency oscillator with a variable mark-space ratio. The longest of these two periods, the decision cycle, is around 3 seconds and pin 3 is high. During this time, a decimal number is displayed and a binary equivalent has to be entered.

For the second and shorter period, around 0.6 seconds, the output is low. This is the active system clocking period.

The decision period can be shortened by using the 555 voltage control pin 5. This is driven by a 741 op-amp that is used as a high input impedance buffer of the voltage present on the 10 μF storage capacitor C_3. Actual voltage can vary somewhere between the supply rails. This will depend on the state of the decision memory flip-flop output pin 13 of IC_4.

Assume that the last answer resulted in a true answer. The "no" LED would be off. Therefore, the Q output, pin 13 would be low. This would result in C_3 being discharged towards the negative rail via R_{18}. This decreasing voltage is applied to the 555 via the op-amp buffer (current limited by R_{25}). This decreases the threshold operating point (normally two-thirds the supply rail). Therefore, the charging period of the timing capacitor C_5 (10 μF) is shortened. This results in reducing the 555 on time and as a result shortens the decision period.

From this action, we can see that if the correct answer is given, the decision time is reduced. The skill required to make these decisions is increased. If a mistake is made, the "yes" LED will be off and pin 13 of the 4013 will be high. This results in an increase in voltage on C_3 and an increase in decision time. The range of this time period may be from about 1 second to 3 seconds. If a shorter period is required, this can be reduced by changing the 555 timing charge component R_{22}. Try moving 100 K down to 68 K.

Correct answer counter

The second half of the dual 4518 is employed as a simple counter. It collects the sequential correct "yes" answers. Outputs Q_1 to Q_4 are used to stop the system clock when a "full card" is reached. This will depend on the position of the 4 position switch. When the output connected via this switch goes high, it is inverted by the 4009 (f) and stops the 555 via the reset input, pin 4. This results in a "winning card" and stops the game. The display is blanked and will remain in this state until the START button is pressed. This restores the counter to zero so that all outputs are low. The 555 is once again enabled. Any "no" answer will reset this counter to zero via D_5.

Comparator

During the decision period, the binary equivalent to the decimal number displayed on the readout has to be entered into the four push buttons SW_1 to SW_4. This number is compared to the binary output of the decade counter using four XOR gates of the 4030. If the numbers match, all outputs of the 4030 will be low and diodes D_1 to D_4 will be reverse biased. This results in the data input, pin 9 of the 4013a being held low by R_{14}. If the switch combination does not match the counter output, at least one of the 4030 outputs will be high. This will pull the D input high.

When the system clock changes to the active period, pin 3 of the 555 goes low. This signal is inverted by the 4009 inverter (e) and transferred to the clock input of the 4013, pin 11. It is on this positive going transition that the 4013 transfers the logic level on the data input, pin 9 to the Q output, pin 13. If the D had a low level from the comparator, a "yes" answer would show.

When the system clock reverts to the decision cycle, this positive going transition at pin 3 of the 555 will clock the 4013b mono via pin 3. As discussed, this short mono period will enable the decade counter clock system and the clock on the answer counter. The display will also be enabled as the blanking input of the 4511 goes high.

To play the game

Switch the unit on and press the START button. Slide the COUNT switch to the 2 position. A number will appear on the display. Now press the switch or switches that would be the binary number equivalent. The table above the four push buttons shows the equivalents. A 1 in the column indicates that the push button is on. 0 is off. So, if the number on the display is 7, the 1, 2 and 4 switches have to be pushed and held. A 9 on the display would require the 8 and 1 buttons to be pushed and held.

After the decision time period, the display will go blank. One of the decison LEDs will light. If you pressed the correct binary equivalent push buttons, the "yes" LED will light.

The next decimal number will come up on the display now. Press the correct push buttons. If your answer is right, the "yes" LED will remain on. The display will go blank but will not come on again if the two answers were correct. This is because the answer counter has increased two places (the number you first entered in the count switch). This means you have a "full card." You have won the game!

Now move the count switch to the 4 or 8 position. Press the start button and play again. Notice that as you get more answers right, the decision time decreases, giving you less time to think. It becomes much easier to make a mistake.

Each time the answer is wrong, the "no" LED comes on. The count resets to zero. You can only win the game with a full card if your answers are given correctly in the order presented.

When a mistake is made, the decision period increases so you have time to think. If you find that after a number of games you become skilled at the operation, the decision time can be shortened.

This game is a good teaching aid for the binary number system. At the same time, it tests and improves your reaction time.

SUMMARY

1. There are two types of integrated circuits: linear and digital.
2. Digital circuits containing semiconductors use on/off or high and low pulse levels to activate them. These two conditions are 0 and 1 in the binary numbering system.
3. Digital logic circuits are on (1) in the valid logic high voltage area and off (0) in the valid logic low voltage areas.
4. Basic gate circuits include:
   a. AND gate.
   b. OR gate.
   c. NOT gate.
   d. NAND gate.
   e. NOR gate.
   f. XOR gate.
   g. XNOR gate.
5. Digital ICs can perform all logic functions.
6. Truth tables show the condition of a gate's output (C), with varying input values of A and B.
7. Some logic families for ICs are:
   a. Resistor-transistor logic (RTL).
   b. Diode-transistor logic (DTL).
   c. Transistor-transistor logic (TTL).
   d. Complementary metal-oxide semiconductor (CMOS).

TEST YOUR KNOWLEDGE, Chapter 17

Please do not write in the text. Place your answers on a separate sheet of paper.

1. The two state system in which circuits are either on or off is the _____ system.
2. Give the binary equivalents for decimal numbers 1 through 10.

3. Complete the following addition problems.
   a. 1101 + 1001.
   b. 010101 + 100110.
   c. 11 + 111.
   d. 111001 + 100001.
4. The _____ _____ _____ is the high value of 1. The _____ _____ _____ is the low value of 0.
5. Electronic switching circuits that decide whether inputs will pass to output or be stopped are called _____ _____.

MATCHING QUESTIONS: Match the following definitions with the correct terms.
   a. AND gate.
   b. OR gate.
   c. NOT gate.
   d. NAND gate.
   e. NOR gate.
   f. XOR gate.
   g. XNOR gate.
6. A negative AND gate.
7. Exclusive NOR gate.
8. Provides an output of 1 if all inputs are 1.
9. Inverts the polarity of the input signal.
10. Exclusive OR gate.
11. Provides an output of 1 when either input is 1.
12. A negative OR gate.
13. Name two logic families that are no longer in wide use.

FOR DISCUSSION

1. What advantages do integrated circuits have over individual transistors?
2. Name the basic types of integrated circuits.
3. What is the difference between a NOT and a NOR logic gate?
4. What are the basic differences between the operational amplifier circuit symbol and the NOT gate?
5. What advances do you think will be made in the computer field in the next 25 years?

Digital Circuits

GOULD INDUSTRIAL AUTOMATION SYSTEMS

Digital circuits are used in programmable controllers. A programmable controller controls procedures done in an industrial process. The unit saves setup time by its use of programming entered in an engineering office.

Electricity and Electronics

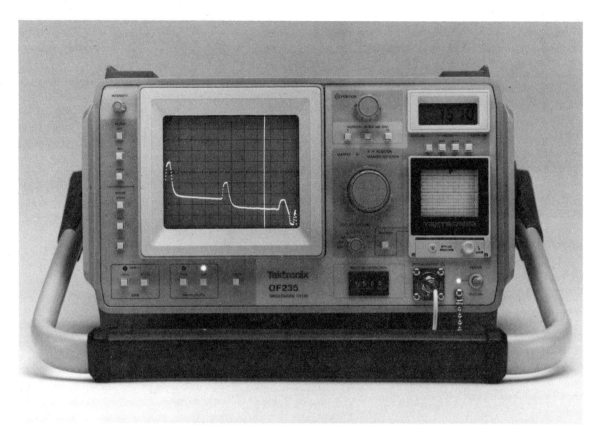

TEKTRONIX

Oscilloscope has a chart recorder for saving the results of a very fast event. An oscillator inside controls the horizontal sweep. A second oscillator known as a "clock" controls the collecting and transmitting of data. This unit can be connected to an optical fiber cable to test cable response.

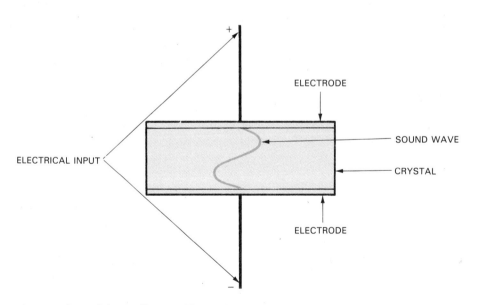

A crystal used in oscillators. Electricity produces stress and sound waves in a piezoelectric crystal, such as one made of quartz. Sound waves and stress in the crystal produce electricity.

Oscillators are the basis of radio and television. They are also the basis of most computer control systems and the systems computers use to communicate over a telephone line. Oscillators are used in test instruments like oscilloscopes. Oscillators can be variable or crystal-controlled. There are many simple circuits that produce ac from dc. The student is encouraged to build and test some of these.

Chapter 18

OSCILLATORS

After studying this chapter, you will be able to:
- *Explain what occurs during an oscillation cycle.*
- *Identify various oscillators.*
- *Discuss and compare the Armstrong oscillator and the Hartley oscillator.*
- *Outline the operation of the crystal oscillator and the power oscillator.*

The pendulum on a grandfather's clock swings back and forth. It marks the time in seconds. Why does the pendulum swing? The pendulum swings to keep time. The main spring, that is used to move the pendulum, is wound with a key. As the spring unwinds, the pendulum swings. How does it keep the correct time? Adjustments are made to the pendulum length so the time required for one complete swing matches one second. The swinging pendulum can be thought of as an OSCILLATOR, Fig. 18-1. An oscillator is an electronic circuit that generates an ac signal at a desired frequency.

18.1 BASIC OSCILLATORS

An oscillating current is one that flows back and forth. It moves first in one direction and then the other.

Follow the voltage amplitude. It starts from its reference line and rises to its peak in one direction and falls to zero. Then it rises to its peak in the opposite direction and returns to zero. One CYCLE has been completed.

As oscillation continues, it repeats these cycles. The time that passes during this cycle is called the PERIOD of cycle. The number of cycles occurring per second is measured and given as the frequency in hertz.

We learned earlier that the electricity used in houses and factories is an alternating or oscillating current. It works at a frequency of 60 hertz. The current is generated by dynamos driven by steam, water, or atomic power.

In our studies of electronics, voltages and currents of much higher frequencies are used. These are generated by a semiconductor (transistor) or an integrated circuit used as an oscillator. The semiconductor or IC does not actually oscillate, but acts as a valve. The valve feeds energy to a tuned circuit to maintain oscillation.

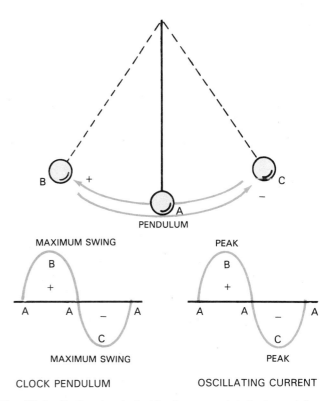

Fig. 18-1. Notice the similarities between the clock pendulum and the oscillating current.

Oscillation was explained in Chapter 11. You may wish to take time now to review this material. The basic block diagram of an oscillator is shown in Fig. 18-2.

Two conditions must exist to sustain oscillation in a tuned circuit:
1. The energy fed back to the tuned circuit must be in phase with the first voltage. The oscillator depends upon REGENERATIVE or POSITIVE feedback.
2. There must be enough feedback voltage amplitude to replace the energy lost by circuit resistance.

THE ARMSTRONG OSCILLATOR

A transistorized Armstrong oscillator is shown in Fig. 18-3. From this circuit, the basic theory of oscillators can be explained. Notice the tuned-tank circuit L_1C_1. This determines the frequency of the oscillator. Follow the sequence of events in this circuit.

Step 1. When the voltage is applied to the circuit, current flows from B−, through the transistor and tickler coil L_2, to B+. L_2 is closely coupled to L_1. The expanding magnetic field of L_2 makes the collector end of L_1 positive. C_1 charges to the polarity shown. The base of Q_1 also collects electrons. It charges C_2 in the polarity shown, Fig. 18-4.

Step 2. When Q_1 reaches its saturation point, there is no longer a change of current in L_2. Magnetic coupling to L_1 drops to zero. The negative charge on the base side of C_2 is no longer opposed by the induced voltage of L_1. The negative charge drives the transistor to cutoff. This rapid decrease in current through the transistor and L_2 causes the base end of L_1 to become negative. This INCREASES the negative bias on Q_1. C_1 discharges through L_1 as the first half-cycle of oscillation. C_2 bleeds off its charge through R_1. See Fig. 18-5.

Step 3. The transistor, Q_1, is held at cutoff until the charge on C_2 is bled off to above cutoff. At that time the transistor starts conduction and the cycle is repeated.

There are a few points to remember in Armstrong oscillator operation.
1. The voltage developed across L_1 first opposes and then adds to the bias developed by the RC_2 combination.
2. The energy added to the tuned-tank circuit L_1C_1 by the tickler coil L_2 is great enough to offset the energy lost in the circuit due to resistance. The coupling between L_1 and L_2 can be adjusted.
3. The combination RC_2 has a somewhat long time constant. It sets the operating bias for the transistor. Q_1 is operated Class C.

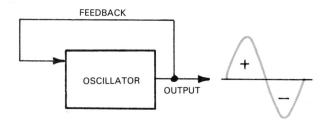

Fig. 18-2. Block diagram of a sine wave oscillator.

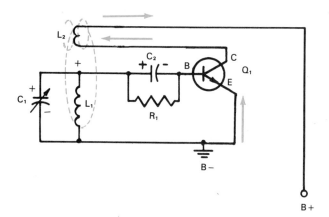

Fig. 18-4. Step 1 in Armstrong oscillator operation. Current flows from B− to B+.

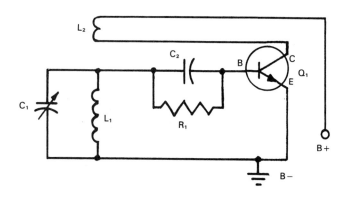

Fig. 18-3. Schematic of an Armstrong oscillator.

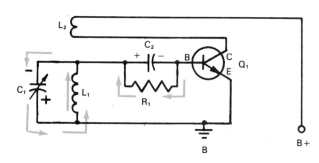

Fig. 18-5. Step 2 in Armstrong oscillator operation. The first half-cycle of oscillation.

Oscillators

Study the voltage waveform on the base of Q_1 in Fig. 18-6. The shaded portion is transistor conduction. At point B the bias is negative. This results from the charge on C_2 plus the induced voltage across L_1. Interval B to C denotes the time that passes while C_2 discharges through R to the cutoff point and conduction begins for the next cycle.

THE HARTLEY OSCILLATOR

An oscillator used commonly in radio receivers and transmitters is the Hartley oscillator. It is more stable than the Armstrong. The operation theory is similar, however. The Hartley oscillator is set apart from other oscillators by the tapped coil, L_1 and L_2 in Fig. 18-7.

The Hartley parts in Fig. 18-7 are labeled in a manner similar to the labeling used for the Armstrong oscillator in Figs. 18-4 and 18-5. The L_1 section of the coil is in series with the emitter-collector circuit. It carries the total collector current.

The current I_E, which includes I_C, is shown by the arrows. When the circuit is turned on, current flow through L_1 induces a voltage at the top of L_2. This makes the base of Q more negative and drives the transistor to saturation. Once saturation is reached, there is no more current change. The coupling between L_1 and L_2 falls to zero. The less negative voltage at the base of Q causes the transistor to decrease conduction. This decrease induces a positive voltage at the top end of L_2. This is reverse bias for the transistor. The transistor is quickly driven to cutoff. The cycle is then repeated. The tank circuit is energized by pulses of current. The transistor switches between saturation and cutoff at the same frequency.

The resting bias condition of the transistor is set by R_B and R_E. The radio frequency choke, or RFC, blocks the rf signal from the power source.

In this circuit, note that coil L_1 is in SERIES with the transistor collector circuit. It is a SERIES FED oscillator.

In Fig. 18-8, a SHUNT FED oscillator is shown. Operation is the same. Note that the dc path for the emitter-collector current is not through coil L_1. The ac signal path, however, is through C and L_1. At point A, the two current components are separated and required to take parallel paths. Both oscillators receive their feedback energy through magnetic coupling.

REVIEW QUESTIONS FOR SECTION 18.1

1. The time that passes during one cycle of an oscillating current is called the _____ of the cycle.
2. What is the purpose of an oscillator?
3. What two conditions must exist to sustain oscillation in a tuned circuit?
4. The _____ oscillator has a tickler coil.
5. The _____ oscillator has a tapped coil.

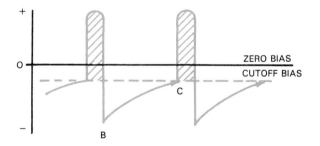

Fig. 18-6. Voltage waveform on base of oscillator. Compare it to steps 1, 2, and 3.

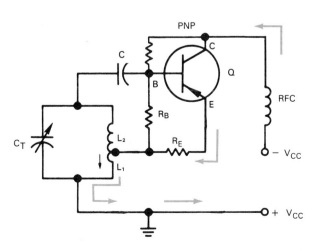

Fig. 18-7. A transistorized Hartley oscillator circuit.

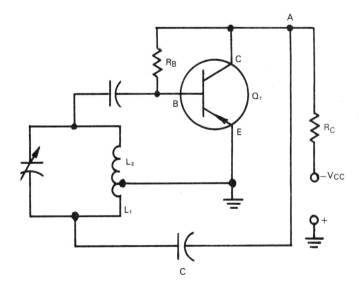

Fig. 18-8. A shunt fed Hartley oscillator.

18.2 OTHER OSCILLATORS

THE COLPITTS OSCILLATOR

Feedback may also be created with an electrostatic field like that found in a capacitor. Replace the tapped coil from the Hartley oscillator with a split stator capacitor. A voltage of proper polarity will be fed back. This will cause the circuit to oscillate. This circuit is called a Colpitts oscillator, Fig. 18-9.

Operation of this oscillator is like the Hartley oscillator. However, the signal is coupled back to C_1 of the tank circuit through coupling capacitor C_3. A changing voltage at the collector appears as a voltage across the tank circuit LC_1C_2. It is in the proper phase to be a regenerative signal. The amount of feedback will depend on the ratio of C_1 to C_2. This ratio is most often fixed. Both capacitors C_1 and C_2 are controlled by a single shaft (ganged capacitor). The frequency of the oscillator is set in the common manner. The tuned-tank consists of L and C_1 and C_2 in series. The circuit is shunt fed. Series fed is not possible due to the blocking of dc by the capacitors.

CRYSTAL OSCILLATORS

A circuit with a stable high frequency is the crystal controlled oscillator. It is used in radio communications, broadcasting stations, and with equipment that needs a fixed frequency with little drift.

You learned earlier that an emf can be made with mechanical pressure and/or distortion of certain crystalline substances. The opposite is also true. A voltage applied to the surface of a crystal will make distortion.

These are called PIEZOELECTRIC EFFECTS. When electrical pressure is applied to a crystal, it will oscillate. Frequency will depend on the size, thickness, and kind of crystal used.

Crystals are used in amateur radio to control the transmitter frequency. In a commercial broadcasting station, crystals used to control transmitter frequency are placed in CRYSTAL OVENS. These are maintained at a constant temperature.

Before using a crystal to stabilize an oscillator, refer to Fig. 18-10. A crystal is placed between two metallic holders. This forms a capacitor C_H. The crystal itself is the dielectric. C_G denotes the series capacitance between the metal holding plates and the air gap between them (as a dielectric). L, C, and R denote the traits of the crystal. Note the likeness of the equivalent crystal circuit to a tuned circuit. It will have a resonant frequency.

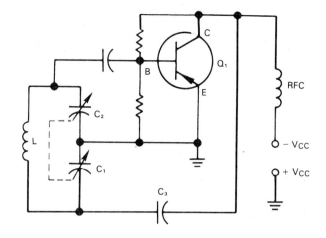

Fig. 18-9. Schematic diagram of a Colpitts oscillator.

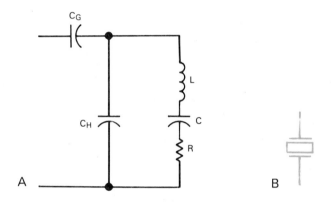

Fig. 18-10. A—An electrical circuit that is equivalent to a crystal. B—The schematic symbol for a crystal.

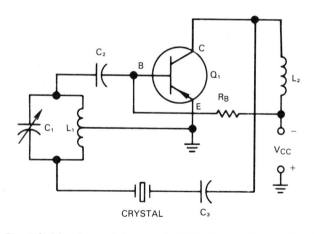

Fig. 18-11. A crystal controlled Hartley oscillator circuit.

A transistorized crystal oscillator circuit is shown in Fig. 18-11. Compare this circuit to Fig. 18-8. It is the same circuit with the crystal added to the FEEDBACK CIRCUIT. The crystal acts as a series resonance circuit. It set the frequency of the feedback currents. The tank circuit must be tuned to this frequency.

Oscillators

In Fig. 18-12, the crystal is used in place of the tuned circuit in a Colpitts oscillator. This is the Pierce oscillator. Compare the circuit to Fig. 18-9.

The amount of feedback needed to energize the crystal again depends on the ratio of C_1 to C_2. These capacitors form a voltage divider across the emitter-base of the transistor. The Pierce circuit is stable under changing circuit conditions.

A SIGNAL GENERATOR is an electronic oscillator that generates various signals for testing, Fig. 18-13.

POWER OSCILLATORS AND CONVERTERS

A power oscillator can convert dc to ac or ac to dc, then transform the signal to a higher voltage and rectify it back to dc.

Fig. 18-14 shows a power oscillator circuit. It is actually a push-pull oscillator. The collector load of each transistor is the primary of the transformer. The ac output is found at the secondary. A proper turns ratio is used if higher or lower voltages are needed.

A slight imbalance in conductivity between Q_1 and Q_2 will start oscillation. This imbalance is always present due to components or temperature. The two transistors are either going toward saturation and cutoff, or vice versa.

Assume Q_1 starts conducting. The voltage at C of Q_1 goes less negative. This makes the base of Q_2 less negative. Q_2 drives toward cutoff. A more negative voltage at C of Q_2 drives Q_1 base more negative. Q_1 reaches saturation. When there is no change in current at saturation, transformer primary reactance drops to zero. Collector voltage toward the value of V_{CC} also decreases. This more negative voltage coupled to the base of Q_2 through R_1 starts Q_2 toward conduction and saturation. The transistors conduct one at a time. The output is combined into a complete cycle at the transformer secondary output.

If the output from the transformer in Fig. 18-14 is connected to a rectifier and filter circuits, the output again becomes dc.

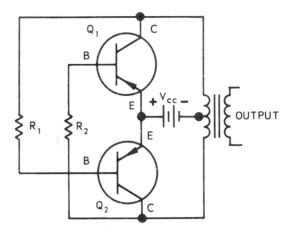

Fig. 18-14. Power oscillator circuit.

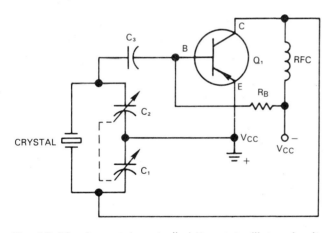

Fig. 18-12. A crystal controlled Pierce oscillator circuit.

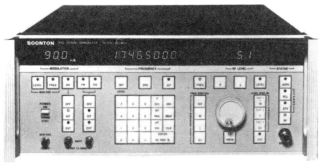

BOONTON ELECTRONICS

Fig. 18-13. This signal generator can produce both AM and FM signals.

REVIEW QUESTIONS FOR SECTION 18.2

1. Why is series feeding impossible in a Colpitts oscillator?
2. What is piezoelectric effect?
3. A _____ oscillator uses a crystal in place of the tuned circuit found in a Colpitts oscillator.
4. Draw the circuit for a power oscillator.

18.3 OSCILLATOR PROJECTS

MORSE CODE OSCILLATOR

The oscillator shown in Fig. 18-15 produces a variable tone. It can be used to send Morse Code. Simple musical tunes can be played on it by varying the tone control knob.

Electricity and Electronics

Fig. 18-15. A completed Morse Code oscillator. It is housed in a shadow cabinet.

Fig. 18-18. A completed Squawker.

The schematic diagram and the parts list are shown in Figs. 18-16 and 18-17. A shadow cabinet houses the oscillator. The cabinet is made of 20 ga. galvanized iron. It is covered with imitation leather.

THE SQUAWKER

The Squawker is a transistorized horn, Fig. 18-18. It uses power transistors in a multivibrator configuration, Fig. 18-19. The amplifier stages also generate audio frequency oscillation by feeding the output of one transistor to the base of the other. Feedback is done by C_1 and C_2. S_1 may be replaced by wires to the horn button. B_1 is the 12 volt battery in the car.

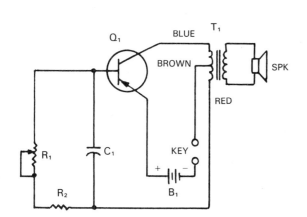

Fig. 18-16. Schematic diagram for the Morse Code oscillator.

C_1 — Capacitor .5 μF
R_1 — Potentiometer 25 kΩ
R_2 — 6800 Ω, 1/2 W
T_1 — Transistor Output Transformer AR 119 or Thordarson TR 27 (pri. = 500 Ω CT, sec. = 3.2 Ω)
Q_1 — RCA transistor SK 3004 or Radio Shack transistor 276-2007
PM — Speaker 1 1/2" Philmore
S_1 — Switch, SPST
Binding Posts or phono jack, knob and small hardware

Fig. 18-17. Parts list for the Morse Code oscillator.

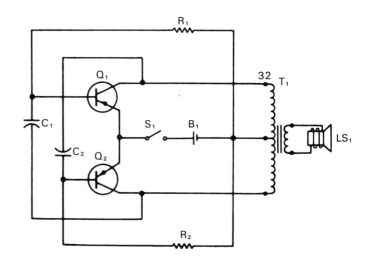

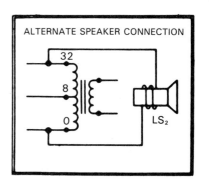

Fig. 18-19. Schematic diagram for the Squawker.

298

B₁ — 12 volt (1 ampere drain) auto or boat battery, or 2 Burgess F4P1 batteries wired in series.
C₁, C₂ — 2.0 μF, electrolytic, 25 volt
LS₁ — 8 ohm tweeter, Lafayette HK-3 or equivalent
Q₁, Q₂ — Sylvania 2N307 power transistor or RCA SK 3009
R₁, R₂ — 200 Ω, 1 W
S₁ — SPST pushbutton switch, normally open
T₁ — Lafayette TR-94 or Triad/Utrad TY-64X
LS₂ — For alternate connection - 45 ohm paging trumpet; University CMIL-45, University MIL-45, or equivalent

Fig. 18-20. Parts list for the Squawker.

Any metal or wooden box may be used to house the horn. It is best to use a weatherproof speaker if the horn is installed on a car. Or it can be protected by a plastic covering. The parts list is given in Fig. 18-20.

SUMMARY

1. An oscillator is an electronic circuit that produces an ac signal at a desired frequency.
2. A cycle is a complete set of positive and negative values for an ac signal.
3. Frequency is the number of cycles per second. It is measured in hertz.
4. An oscillator circuit is usually made up of a wave-producing circuit, amplifier, and feedback circuit. The feedback circuit usually causes variations in oscillator designs.
5. The Armstrong oscillator has a tickler coil for feedback.
6. The Hartley oscillator has a tapped coil in the tank circuit.
7. A Colpitts oscillator has two capacitors in the tank circuit with a "tap" between them.
8. Crystal oscillators use piezoelectric crystals to maintain accurate frequencies.
9. Power oscillators can convert ac to dc or vice versa, then transform the signal to a higher voltage and rectify it back to dc.

TEST YOUR KNOWLEDGE, Chapter 18

Please do not write in the text. Place your answers on a separate sheet of paper.

1. A _____ is a complete set of positive and negative values for an ac signal.
2. What is the period of a cycle?
3. The _____-_____ circuit determines the frequency in the Armstrong oscillator.
4. Outline the steps in the operation of the Armstrong oscillator.
5. What feedback method is used in the Hartley oscillator?
6. What feedback method is used in the Colpitts oscillator?
7. Draw the equivalent electrical circuit of a crystal.
8. A power oscillator can:
   a. Convert dc to ac, or ac to dc.
   b. Transform a signal to a higher voltage.
   c. Rectify a signal back to dc.
   d. All of the above.

FOR DISCUSSION

1. How does an electronic oscillator compare to a clock pendulum?
2. Can a Piezoelectric crystal be used to produce electricity if pressure is applied?
3. Explain the operation of a Hartley oscillator.
4. What is the difference between a series fed oscillator and a shunt fed oscillator?
5. Why are crystal oscillators used in many commercial transmitters?
6. How can a Colpitts oscillator be identified?

Part V Summary

BASIC ELECTRONIC CIRCUITS

IMPORTANT POINTS

1. A power supply is an electronic circuit that provides ac and dc voltages for equipment operation.
2. Power supplies can:
   a. Step up or step down ac line voltage to the required voltage using transformer action.
   b. Change ac voltage to a pulsating dc voltage by either half-wave or full-wave rectification.
   c. Filter pulsating dc voltage to produce a more pure dc for equipment use.
   d. Provide voltage division for equipment use.
   e. With the proper components, regulate output in proportion to the applied load.
3. Most power supplies have a rectifier and filter. Transformers and voltage regulators are also used in some power supplies.
4. Amplifiers control large output voltages using small input voltages.
5. Forward bias provides low resistance to current flow. Reverse bias provides high resistance to current flow.
6. Class A amplifiers are biased so that output current flows for 360 degrees of the input signal cycle. Class B amplifiers are biased so that output current flows during 180 degrees of the input signal cycle. Class C amplifiers are biased so that the output current flows less than 180 degrees of the input signal cycle.
7. Amplifiers can be connected in a number of ways to increase gain.
8. Linear ICs have variable outputs. Digital ICs have only two outputs: on and off.
9. The binary numbering system has only two states: 0 or 1. For example, the binary number for 22 is 10110.
10. An IC is turned on at valid logic high while it is turned off at valid logic low.
11. A bit is a single binary digit (0 or 1). A byte is eight bits.
12. An inverter changes the value of an incoming signal (from 0 to 1, from 1 to 0).
13. An oscillator generates an ac signal at a desired frequency.
14. Oscillators are commonly grouped based on feedback method.
15. The basic parts of an oscillator are the wave producing circuit, amplifier, and feedback circuit.

SUMMARY QUESTIONS

1. What is the percentage of regulation in a power supply with a no load voltage of 20 volts and full load voltage of 13 volts?
2. What is the percentage of ripple for an ac signal with an rms value of 56 volts and peak voltage of 80 volts?
3. An amplifier has a change in output voltage of 15 volts ac and a change in input voltage of 500 millivolts. What is the voltage gain?
4. What is the binary number for 2000?
5. Draw the truth tables for:
   a. NAND gate.
   b. AND gate.
   c. XOR gate.

Part VI

ELECTRONIC COMMUNICATION AND DATA SYSTEMS

19 Radio Wave Transmission
20 Radio Wave Receivers
21 Television
22 Computers

Communication is the process of exchanging information. The information moves from a source through a transmitter to a destination.

There are many ways to communicate. Methods that use electronics include human-to-machine, machine-to-human, and machine-to-machine. In electronic communication, the transmitter is often electromagnetic waves or cable. Transducers are needed by both transmitter and receiver to convert input signals. Transducers are also needed to convert the output signals to other forms of energy. Computers are an excellent example of electronic communication.

Chapter 19 discusses radio wave transmission. Amplitude modulation (AM) and frequency modulation (FM) are explained. Also discussed is the transmission of electromagnetic waves.

Radio wave receivers are presented in Chapter 20. Topics discussed include the history of receivers, AM and FM receivers, the superheterodyne receiver, and transducers (speakers and headphones). There is also a radio construction project.

Chapter 21 explains the operation of televisions (black and white, color, large screen, cable, satellite).

One of the most powerful tools of this century, the computer, is the subject of Chapter 22. Key topics of discussion include the history of the computer, computer operation, and computer components.

Electricity and Electronics

This AM/FM signal generator is just one important device needed to test circuits used for radio wave transmission.

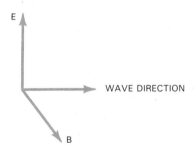

The varying currents in all ac circuits give off electromagnetic waves. A circuit operating at 60 Hz does not give off enough radiation to be a practical transmitter. Frequencies as high as 20 kHz are usually required to transmit information. Waves are transmitted with a combination of a varying electric field (E) and a varying magnetic field (B). See the drawing that shows the E and B fields

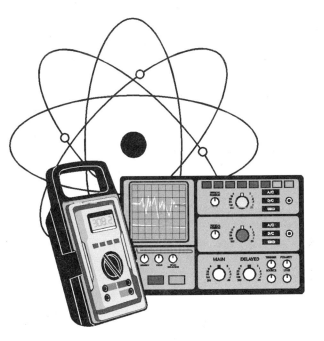

Chapter 19

RADIO WAVE TRANSMISSION

After studying this chapter, you will be able to:
- *Construct the communication model.*
- *Explain the methods of transmitting radio waves.*
- *Discuss amplitude modulation and frequency modulation.*
- *Calculate the percent of modulation.*
- *Calculate wavelength.*
- *Identify types of radio waves.*

Humans have communicated with each other for thousands of years. However, electrical and electronic communication has been present for only the past century. Now one person can communicate with other people and with machines.

19.1 COMMUNICATION MODEL

A basic communication model is shown in Fig. 19-1. All communication has three basic parts: source, transmitter, and destination.

The source is the starting point of the message to be sent. For example, when you are speaking to a friend, you are the source of information. You start the message on its way.

Next, the message is transmitted. During transmission, the message is sent toward its destination.

Messages can be sent in a number of ways. In electronic communication, messages are often sent as electrical signals. These signals or electrical pulses move through a channel. The channel may be electromagnetic waves, wires, air, or some other medium. (A medium means some way of carrying the signal.)

Finally, the message reaches its destination. The DESTINATION is the final stop for the transmitted message. The destination can take many forms. It might be a person or it might be a machine.

At the destination, the message may have to be decoded. Decoding changes the transmitted message into an understandable form. For example, radio broadcast waves reach a radio (destination) as electronic signals. These signals are changed in the stereo receiver to sounds we understand.

Two other important parts of the communication model are feedback and noise, Fig. 19-2. FEEDBACK is a sign that a message has been received. It often is an exchange of signals between the destination and the source.

NOISE is a break in the intended signal. It distorts the message. It commonly enters the communication process during transmission. For example, noise in electronic communication may be caused by lightning breaking an electrical signal.

Fig. 19-1. Study the three parts of the basic communication model.

305

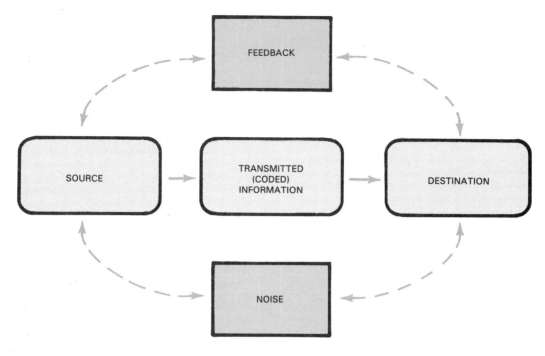

Fig. 19-2. Noise and feedback are also part of the communication model.

One type of communication model is shown in Fig. 19-3. The source of the message is the microphone. The message is transmitted through antennas. The receiver is the destination. It picks up, tunes, amplifies, and reproduces the information.

FREQUENCY SPECTRUM

Earlier study included discussion of waves having certain frequencies. See Fig. 19-4. This is a frequency spectrum. It indicates the frequency bands used by telecommunication areas such as radio, television, and police.

The use of frequencies is strictly controlled. This is because there are many users of these systems. And each user requires a frequency on which to broadcast. There are only so many frequencies available. These must be assigned to meet the needs of the public. At the same time, there must be a minimum of interference between the services. The Federal Communications Commission (FCC) controls frequency assignment in the United States.

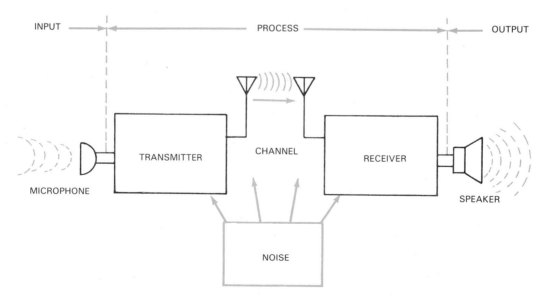

Fig. 19-3. Can you find the source, transmitter, and destination of this communication model?

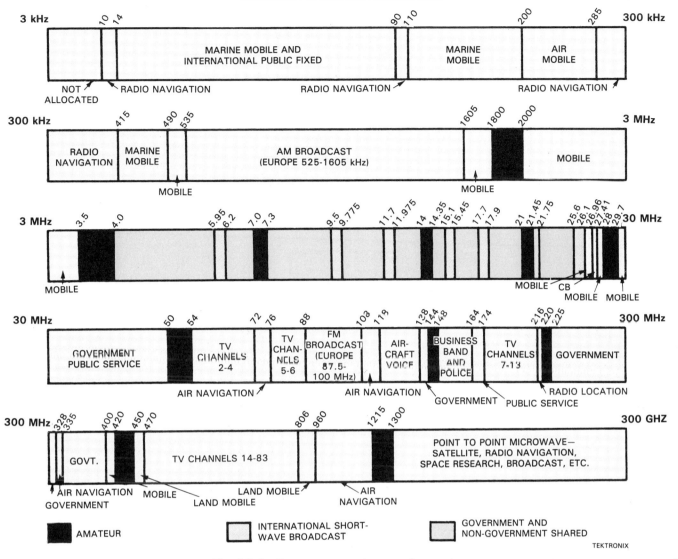

Fig. 19-4. Frequency spectrum assignments.

Waves are also grouped according to frequency, Fig. 19-5. Note the location of some services. AM broadcast stations use the MF range. Television uses the VHF and UHF ranges. A stereo operates in the audio frequency range. Electricity for the home is 60 Hz. It operates at the audio frequency.

FREQUENCY	BAND
20-30,000 Hz	Audio Frequency
Below 30 kHz	VLF Very Low Frequency
30-300 kHz	LF Low Frequency
300-3000 kHz	MF Medium Frequency
3000-30,000 kHz	HF High Frequency
30,000 kHz – 300 MHz	VHF Very High Frequency
300-3000 MHz	UFH Ultra High Frequency

Fig. 19-5. Study this chart of frequency ranges. Can you name one device that operates in each band?

REVIEW QUESTIONS FOR SECTION 19.1

1. Draw a basic communication model. Explain each function.
2. Why are transmitted electromagnetic signals controlled by the FCC?
3. Audio frequencies are in the human hearing range and lie between _____ Hz and _____ Hz.
4. The following popular broadcast bands lie where in the frequency spectrum assigned to telecommunications?
   a. AM.
   b. CB.
   c. TV channels 2 through 4.
   d. TV channels 5 through 6.
   e. TV channels 7 through 13.
   f. TV channels 14 through 83.

19.2 RADIO TRANSMITTER

Any of the oscillators we have discussed produce radio frequency waves. If one was connected to an antenna system, it would send energy into the atmosphere. However, stronger signals are needed. Amplification will increase the amplitude of the oscillator wave so that it will drive a final power amplifier.

A block diagram of a simple continuous wave (CW) transmitter is shown in Fig. 19-6. The first block is the conventional crystal oscillator and then the final power amplifier. A power supply is provided for the oscillator and the final power amplifier.

The oscillator creates an ac sine wave at a desired frequency. This signal is called the CARRIER WAVE. The carrier wave is then amplified by the rf power amplifier to the desired output wattage. A power supply is required to provide the voltages and current needed to work the oscillator and the rf power amplifier. The output is then fed to an antenna. From there the energy is sent into the air as electromagnetic waves.

A CW transmitter sends energy that has no message. At the destination, the signal is picked up. This is the sign that energy has been sent out from the CW transmitter antenna. Therefore, a CW transmitter has only two states, on or off.

How can this type of transmitter be used? By adding a switch, the transmitter can be turned on and off following a code. For example, such a transmitter could be used to send Morse code messages, Fig. 19-7.

The basic switched or keyed CW transmitter can be improved. Place a buffer amplifier between the oscillator and the rf amplifier. The buffer amplifier isolates the oscillator from the rf amplifier. It keeps it from shifting off frequency. It also provides some amplification to the carrier wave.

Many CW transmitters use frequency multipliers to increase the frequency produced by the basic oscillator. These circuits multiply the carrier wave by two (doubler) or three (tripler). These circuits work based on the principle of harmonics in the fundamental carrier frequency created by the oscillator. A fundamental frequency is the basic frequency produced by the oscillator. A harmonic frequency is a multiple of the fundamental frequency.

MICROPHONES

How is a sound wave converted to an electrical wave? Your vocal cords send vibrations in the air. These waves move out to all persons within hearing range. A MICROPHONE will convert these sound waves to electrical audio waves of the same frequency and relative amplitude. Fig. 19-8 shows a microphone built into a video camera.

Microphones are sometimes called transducers. This is because they transform one form of energy (air or mechanical) to electrical energy.

A CARBON MICROPHONE is shown in Fig. 19-9. Granules of carbon are packed in a small container. Electrical connections are made to each side. A transformer and a small battery are joined in series with the carbon. A diaphragm is attached to one side of the container. This diaphragm is called a BUTTON. Sound waves strike the diaphragm and cause the carbon granules to be compressed or pushed together. This varies the resistance of the carbon. Varying resistance causes a varying current to flow through the carbon button and the transformer primary. The output is a current which varies at the same frequency as the sound waves acting on the diaphragm.

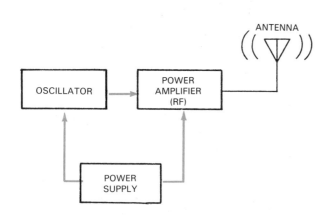

Fig. 19-6. Block diagram representing various stages of basic continuous wave radio transmitter.

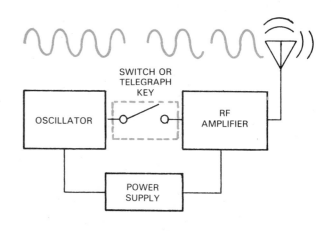

Fig. 19-7. Continuous wave transmitter with telegraph key. Note break in RF waveform indicating an open switch at that point.

Fig. 19-8. This video camera is equipped with a stereo microphone.

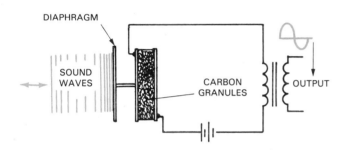

Fig. 19-9. In a carbon microphone, sound waves change the resistance of the circuit.

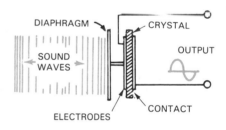

Fig. 19-10. Mechanical pressure produced electrical energy. The crystal microphone takes advantage of piezoelectric effect.

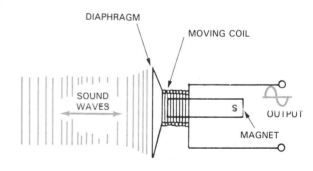

Fig. 19-11. Dynamic microphone. Electrical audio waves are produced by a coil moving in a magnetic field.

The carbon microphone is a very sensitive device. It has a frequency response up to about 4000 Hz. This is useful for voice communication, but not for reproduction of music. It provides a good response for its intended frequencies. It is nondirectional. This means it will pick up sound from all directions.

A second type of microphone uses the piezoelectric effect of certain crystals. It is called a CRYSTAL MICROPHONE. When sound waves strike a diaphragm, mechanical pressure is transferred to the crystal. The flexing or bending of the crystal creates a small voltage between its surfaces. This voltage is the same frequency and relative amplitude as the sound wave.

Crystal microphones have a frequency response up to 10,000 Hz. They are sensitive to shock and vibration. They should be handled with care, Fig. 19-10.

A DYNAMIC MICROPHONE, or moving coil microphone, is sketched in Fig. 19-11. As sound waves strike the diaphragm, they cause the voice coil to move in and out. The voice coil is surrounded by a fixed magnetic field. When the coil moves, a voltage is induced in the coil (Faraday's discovery). This induced voltage causes current to flow at a frequency and amplitude similar to the sound wave causing the motion. It has a frequency response up to 9000 Hz. It is directional. It requires no outside voltage for operation.

A high quality microphone, called a VELOCITY MICROPHONE, is made by suspending a corrugated ribbon of metal in a magnetic field. Sound waves directly striking the ribbon cause the ribbon to vibrate. As the ribbon cuts the magnetic field, a voltage is induced. Proper connections at the ends of the ribbon bring the voltage out to terminals. This voltage varies according to the frequency and amplitude of the incoming sound waves.

The velocity microphone is a somewhat delicate microphone and has a response above 12,000 Hz. When using this microphone, the speaker must speak across its face or stand about 18 inches away. Otherwise, a "booming" effect is created.

MODULATION

When you turn on the radio or TV, you expect to hear music and the voices you understand. The signals of the CW transmitter mean nothing to the average person. To make an understandable message, an audio wave is combined or superimposed on a carrier wave. This is called MODULATION. Sound waves are converted into electrical waves, amplified, and then combined with the radio wave.

In one modulation method used in commercial broadcasting, the radio wave amplitude is made to vary at an audio frequency rate. This is AMPLITUDE MODULATION or AM. In a second method, the radio wave frequency is made to vary at an audio frequency rate. This is called FREQUENCY MODULATION or FM. See Fig. 19-12.

REVIEW QUESTIONS FOR SECTION 19.2

1. An ac sine wave created at a desired frequency is called:
   a. Audio wave.
   b. Carrier wave.
   c. Modulated wave.
   d. None of the above.
2. A CW transmitter signal has a message. True or False?
3. What is a fundamental frequency? What is a harmonic frequency?
4. Sketch a carbon microphone.
5. Define amplitude modulation and frequency modulation.

19.3 AMPLITUDE MODULATION (AM)

Modulation is a process in which an audio wave is combined or superimposed on a carrier wave. Assume a radio transmitter is operating on a frequency of 1000 kHz. A musical tone of 1000 Hz is to be used for modulation. Refer to Fig. 19-13. Using a modulation circuit, the amplitude of the carrier wave is made to vary at the audio signal rate.

Let us look at this process another way. Mixing a 1000 Hz wave with a 1000 kHz wave produces a sum wave and difference wave, which are ALSO IN THE RADIO FREQUENCY RANGE. These two waves will be 1001 kHz and 999 kHz. These are known as SIDEBAND FREQUENCIES. The UPPER SIDEBAND is the higher number. The LOWER SIDEBAND is the lower number, Fig. 19-14.

The sum of the carrier wave and its sidebands is an amplitude modulation wave. The audio tone is present in both sidebands, as either sideband results from modulating a 1000 kHz signal with a 1000 Hz tone.

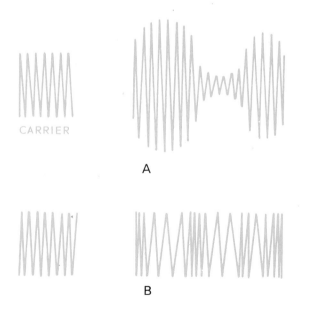

Fig. 19-12. Carrier waves and resulting modulated waves. A—Amplitude modulation, or AM. B—Frequency modulation, or FM.

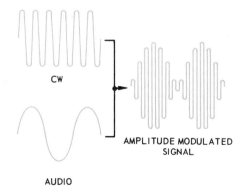

Fig. 19-13. A CW wave, an audio wave, and the resulting amplitude modulated wave.

Radio Wave Transmission

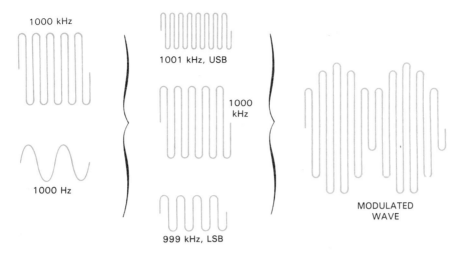

Fig. 19-14. Wave showing formation of sidebands and modulation envelope.

The location of the waves on a frequency base are shown in Fig. 19-15. If a 2000 Hz tone was used for modulation, then sidebands would appear at 998 kHz and 1002 kHz. In order to transmit a 5000 Hz tone of a violin using AM, sidebands of 995 kHz and 1005 kHz would be required. The frequency band width will be 10 kHz to transmit a 5000 Hz musical tone.

There is not enough space in the spectrum for all broadcasters to transmit. And, if all broadcasts contained the same message or operated on the same frequency, the effect would be confusing. Therefore, the BROADCAST BAND for AM radio extends from 535 kHz to 1605 kHz. It is divided into 106 channels, each 10 kHz wide.

A station is licensed to operate at a frequency in one of these channels. The channels are spaced far enough from each other to prevent interference.

In order to improve the fidelity and quality of music within these limitations, a VESTIGIAL SIDEBAND FILTER is used. It removes a large portion of one sideband. Recall that both sidebands contain the same information. By this means, frequencies higher than 5 kHz can be used for modulation. The fidelity is improved.

MODULATION PATTERNS

A radio transmitter is not permitted by law to exceed 100 percent modulation. This means that the modulation signal can not cause the carrier signal to vary over 100 percent of its unmodulated value. Look at the patterns in Fig. 19-16. Notice the amplitude of the modulated waves.

The 100 percent modulation wave variation is from zero to two times the peak value of the carrier wave. Overmodulation is caused when modulation increases the carrier wave to over two times its peak value. At negative peaks the waves cancel each other and leave a straight line of zero value. Overmodulation causes distortion and interference called SPLATTER.

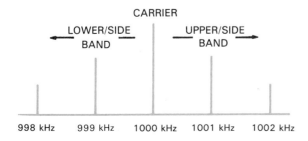

Fig. 19-15. Carrier and sideband locations for modulation tone of 1 kHz and 2 kHz.

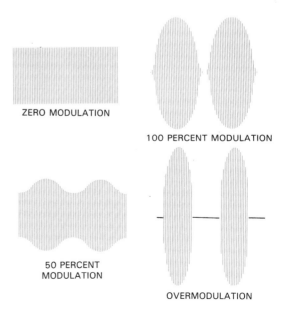

Fig. 19-16. Patterns for 0, 100, and 50 percent of modulation and for overmodulation.

311

Percent of modulation may be computed using the following formula:

$$\% \text{ MODULATION} = \frac{e_{max} - e_{min}}{2e_c} \times 100$$

e_{max} is the maximum amplitude of modulated wave. e_{min} is the minimum amplitude of modulated wave. e_c is the amplitude of unmodulated wave.

SIDEBAND POWER

The dc input power to the final amplifier of a transmitter is the product of voltage and current. To find the power required by a modulator, this formula may be used:

$$P_{audio} = \frac{m^2 P_{dc}}{2}$$

where P_{audio} is the power of the modulator, m is the percentage of modulation (expressed as decimal), and P_{dc} is the input power to the final amplifier.

PROBLEM. What power is required to modulate a transmitter having a dc power input of 500 watts to 100 percent?

$$P_{audio} = \frac{(1)^2\ 500\ \text{watts}}{2} = 250\ \text{watts}$$

This represents a total input power of 750 watts. Notice what happens under 50 percent modulation.

$$P_{audio} = \frac{(.5)^2\ 500\ \text{watts}}{2} = 62.5\ \text{watts}$$

And the total input power is only 562.5 watts.

Where the modulation percentage is reduced to 50 percent, the power is reduced to 25 percent. This is a severe drop in power that decreases the broadcasting range of the transmitter. It is wise to maintain transmitter modulation as close to, but not exceeding, 100 percent.

You may wish to know why the term INPUT POWER has been used. That is because any final amplifier is far from 100 percent efficient.

$$\% \text{ EFF} = \frac{P_{out}}{P_{in}} \times 100$$

If a power amplifier had a 60 percent efficiency and a P_{dc} input of 500 watts, its output power would approach:

$$P_{out} = \%\text{EFF} \times P_{in} = .6 \times 500 = 300\ \text{watts}$$

Ham radio stations are limited by law to 1000 watts input power. Their output power is even less. One key duty of transmitter engineers is to check transmitter input power at the intervals set by law.

A transmitter has 100 percent modulation and power of 750 watts. 500 watts of this power is in the carrier wave and 250 watts is added to produce the sidebands. Therefore, there are 125 watts of power in each sideband or one-sixth of the total power in each sideband. Recall that each sideband contains the same information and each is a radio frequency wave which will radiate as well as the carrier wave. So why waste all this power?

In single sideband transmission, this power is saved. The carrier and one sideband are suppressed. Only one sideband is radiated. At the receiver end the carrier is put back in. The difference signal (the audio signal) is then detected and reproduced.

We will not cover the methods of sideband transmission and reception. But, you may wish to study this communication system on your own.

TRANSISTORIZED TRANSMITTERS

The vacuum tube is no longer key as a power amplifier for transmitters. In recent years, transistors have been developed that can handle large power needs. Transistors can be found in audio circuits, oscillators, and intermediate power amplifiers. See Fig. 19-17.

A hand-held scanner is shown in Fig. 19-18. It is used to pick up radio signals in the citizens band frequencies. Both transistors and ICs are used to receive and transmit. The scanner also contains a rechargeable battery and charger.

A table model citizens band radio is shown in Fig. 19-19. The citizens band frequencies assigned by the FCC are shown in Fig. 19-20.

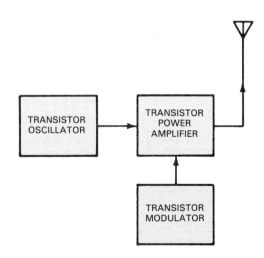

Fig. 19-17. Block diagram of a simple transistorized transmitter.

Fig. 19-18. A 30 channel hand-held scanner.

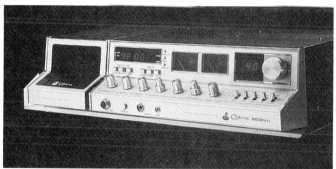

Fig. 19-19. Citizens band radio base station.

REVIEW QUESTIONS FOR SECTION 19.3

1. Define modulation.
2. The sum of the carrier wave and its sidebands results in the _____ _____ wave.
3. A carrier wave has a peak value of 500 volts. A modulating signal causes amplitude variation from 250 volts to 750 volts. What is the percent of modulation?
4. Draw an amplitude modulated wave that has 50 percent modulation.

19.4 FREQUENCY MODULATION (FM)

Recall that in frequency modulation, the radio wave frequency is made to vary at the audio frequency rate. FM radio is a popular method of electronic communication. Frequency modulation allows a high audio sound to be transmitted while still remaining within the space legally assigned to the broadcast station. Also, FM transmits dual channels of sound (stereo) by multiplex systems. The FM band is from 88 MHz to 108 MHz. A block diagram of a FM transmitter is shown in Fig. 19-21.

FM uses a constant amplitude continuous wave signal. The frequency of this wave is varied at an audio rate, Fig. 19-22.

Each station is assigned a CENTER FREQUENCY in the FM band (92.1 to 107.9 MHz). This is the frequency to which a radio is tuned, Fig. 19-23.

40 CHANNEL CB OPERATION (TRANSMIT)

CHANNEL NO.	VCO OUTPUT MHz	TX OSCILLATOR MHz	PLL MIXER OUTPUT MHz	CHANNEL NO.	VCO OUTPUT MHz	TX OSCILLATOR MHz	PLL MIXER OUTPUT MHz
1	26.965	29.515	2.55	21	27.215	29.515	2.30
2	26.975	29.515	2.54	22	27.225	29.515	2.29
3	26.985	29.515	2.53	23	27.255	29.515	2.26
4	27.005	29.515	2.51	24	27.235	29.515	2.28
5	27.015	29.515	2.50	25	27.245	29.515	2.27
6	27.025	29.515	2.49	26	27.265	29.515	2.25
7	27.035	29.515	2.48	27	27.275	29.515	2.24
8	27.055	29.515	2.46	28	27.285	29.515	2.23
9	27.065	29.515	2.45	29	27.295	29.515	2.22
10	27.075	29.515	2.44	30	27.305	29.515	2.21
11	27.085	29.515	2.43	31	27.315	29.515	2.20
12	27.105	29.515	2.41	32	27.325	29.515	2.19
13	27.115	29.515	2.40	33	27.335	29.515	2.18
14	27.125	29.515	2.39	34	27.345	29.515	2.17
15	27.135	29.515	2.38	35	27.355	29.515	2.16
16	27.155	29.515	2.36	36	27.365	29.515	2.15
17	27.165	29.515	2.35	37	27.375	29.515	2.14
18	27.175	29.515	2.34	38	27.385	29.515	2.13
19	27.185	29.515	2.33	39	27.395	29.515	2.12
20	27.205	29.515	2.31	40	27.405	29.515	2.11

Fig. 19-20. Table lists current 40 channels available for CB use.

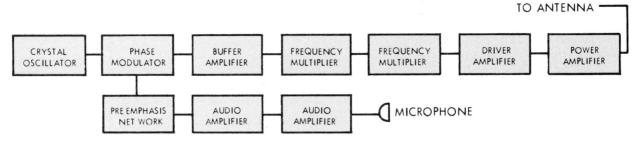

Fig. 19-21. The block diagram of a simplified FM transmitter.

The amount of frequency variation from each side of the center frequency is called the FREQUENCY DEVIATION. It is set by the amplitude or strength of the audio modulating wave.

In Fig. 19-23, a weak audio signal causes the frequency of the carrier wave to vary between 100.01 MHz and 99.99 MHz. The deviation is ± 10 kHz. In the second example a stronger audio signal causes a frequency swing between 100.05 MHz and 99.95 MHz or a deviation of ± 50 kHz. The stronger the modulation signal, the greater the frequency departure and the more the band is filled.

The RATE of frequency deviation depends on the FREQUENCY of the AUDIO MODULATING SIGNAL. See Fig. 19-24.

If the audio signal is 1000 Hz, the carrier wave goes through its greatest deviation 1000 times per second. If the audio signal is 100 Hz, the frequency changes at a rate of 100 times per second. Notice that the modulating frequency does not change the amplitude of the carrier wave.

An FM signal forms sidebands. The number of sidebands produced depends on the frequency and amplitude of the modulating signal. Each sideband is separated from the center frequency by the amount of the frequency of the modulating signal, Fig. 19-25.

The power of the carrier frequency is reduced a great deal by the formation of sidebands, which take power from the carrier. The amount of power taken from the carrier depends on the maximum deviation and the modulating frequency.

Although a station is assigned a center frequency and stays within its maximum deviation, the formation of sidebands determines the bandwidth required for transmission. In FM, the bandwidth is specified by the frequency range between the upper and lower significant sidebands. A SIGNIFICANT SIDEBAND has an amplitude of one percent or more of the unmodulated carrier.

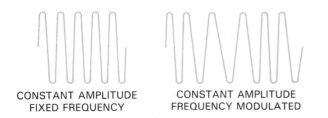

Fig. 19-22. For FM, the frequency of the wave is varied at an audio rate.

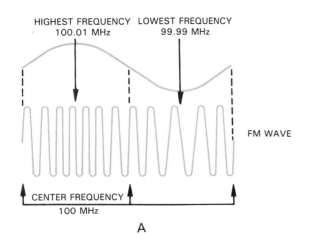

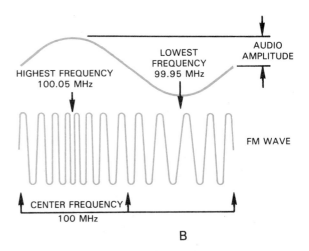

Fig. 19-23. The amplitude of the modulating signal determines the frequency swing from center frequency. A—Weak audio signal. B—Strong audio signal.

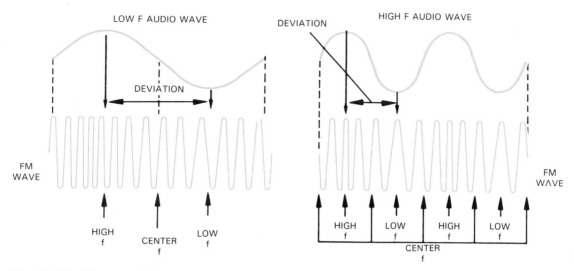

Fig. 19-24. The rate of frequency variation depends on the frequency of the audio modulating signal.

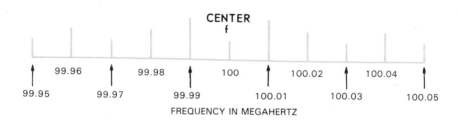

Fig. 19-25. Sidebands generated by a 10 kHz modulating signal on a 100 MHz carrier wave.

NARROW BAND FM

Maximum deviation of a carrier wave can be limited so that the FM wave occupies the same space as an AM wave carrying the same message. This is called NARROW BAND FM. Some distortion occurs in the received signal. This is satisfactory for voice communication but not for high fidelity music.

MODULATION INDEX

Modulation index is the relationship between the MAXIMUM CARRIER DEVIATION and the MAXIMUM MODULATING FREQUENCY:

$$\frac{\text{Maximum carrier deviation}}{\text{Maximum modulating frequency}}$$

Using this index, the number of significant sidebands and the bandwidth of the FM signal can be figured. The complete index may be found in more advanced texts. Examples of the use of the modulation index are given in Fig. 19-26.

If the amplitude of a modulating signal causes a maximum deviation of 10 kHz and the frequency of the modulating signal was 1000 Hz, the index would be:

$$\text{M Index} = \frac{10,000}{1000} = 10$$

The FM signal would have 14 significant sidebands and occupy a bandwidth of 28 kHz.

PERCENT OF MODULATION

The percent of modulation has been randomly set at a maximum deviation of ± 75 kHz for FM radio. The FM sound transmission in television is limited to ± 25 kHz.

MODULATION INDEX	NUMBER OF SIDEBANDS	BANDWIDTH
.5	2	4 × F
1	3	6 × F
5	8	16 × F
10	14	28 × F

Fig. 19-26. Examples of modulation index use. F is the modulating frequency.

REVIEW QUESTIONS FOR SECTION 19.4

1. The _____ frequency is the frequency to which a radio is tuned.
2. Define frequency deviation.
3. On what does the rate of frequency deviation depend?
4. A _____ _____ has an amplitude of at least one percent of the unmodulated carrier.
5. How is the modulation index for FM found?

19.5 THE RADIO WAVE

Think of a wave moving through space as a wave rolling toward the beach. Throw a stone into water. From the point where the stone entered the water, small waves move outward, in circular patterns. The stone creates a disturbance, setting the wave in motion. As the wave moves away from the center (where the stone was dropped) the amplitude of the wave decreases.

Does the water move along as the wave? No. The water moves UP to a crest and DOWN in a hollow or trough as the wave passes. The water has acted as the medium for transmitting the wave.

Scientists know that radio waves travel through space at a speed of 186,000 miles per second (300,000,000 meters per second), or the speed of light. This is the velocity of a radio wave.

Refer to Fig. 19-27. The first radio wave has a frequency of one cycle per second (1 Hz). Starting at point A, the wave will move 186,000 miles by the time it reaches point B. In the second wave, the frequency of the wave is 1000 Hz. The wave will move 186 miles by the time it reaches point B. In the third wave, the frequency is increased to 1,000,000 Hz, or 1 megahertz. The wave will move only .186 miles by the time it reaches point B.

A wave can be described not only by its frequency, then, but also by its length, Fig. 19-28. The distance between the crests of the waves is the WAVELENGTH. As the frequency increases, the wavelength decreases. The Greek letter λ (lambda) stands for wavelength. The mathematical relationship may be stated as:

$$\lambda = \frac{\text{velocity}}{\text{frequency (in Hz)}}$$

$$\lambda \text{ (in miles)} = \frac{186{,}000}{f}$$

and,

$$f = \frac{186{,}000}{\lambda \text{ (in miles)}}$$

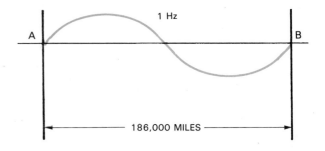
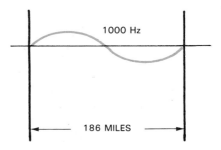
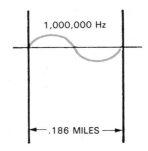

Fig. 19-27. A comparison of the distances traveled in 1 Hz, 1000 Hz, and 1,000,000 Hz waves.

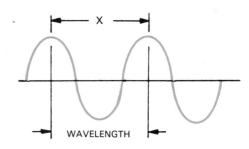

Fig. 19-28. Radio waves are identified by their length and their frequency. X equals one wave.

When using metric measurements, the formula is:

$$\lambda \text{ (in meters)} = \frac{300{,}000{,}000}{f}$$

and

$$f = \frac{300{,}000{,}000}{\lambda \text{ (in meters)}}$$

EXAMPLE 1. An amateur radio transmits on a frequency of 3.9 MHz. What is the wavelength of the radiated waves?

$$\lambda = \frac{186,000}{3.9 \text{ MHz}}$$

3.9 MHz is converted to 3,900,000 Hz.

$$\lambda = \frac{186,000}{3,900,000} = \frac{186}{3900} = .05 \text{ miles}$$

Observe how the math is simplified by using powers of ten.

$$\lambda = \frac{1.86 \times 10^5}{3.9 \times 10^6} = \frac{1.86 \times 10^{-1}}{3.9} = \frac{0.186}{3.9} = .05 \text{ miles}$$

EXAMPLE 2. What is the frequency of a transmitter operating on 40 meters?

$$f = \frac{3 \times 10^8}{4 \times 10} = .75 \times 10^7 = 7,500,000 \text{ Hz} = 7.5 \text{ MHz}$$

A more useful form of this formula is available. Since the foot is the common unit of measure and radio frequencies are usually in megahertz:

$$\lambda \text{ (in feet)} = \frac{984}{f \text{ (in MHz)}}$$

Memorize this formula. It is used a great deal in the design of special frequency antennas.

RADIO WAVE TRAVEL

In the study of electromagnetism, we learned that a conductor carrying an electric current is surrounded by a magnetic field. In an alternating current, the magnetic field expands, collapses, and changes polarity from time to time. Some of these magnetic lines of force also lose the influence of the conductor and travel into space. These radiated electromagnetic waves are perpendicular to the direction in which they travel.

Along with any electromagnetic field is an electrostatic field. It is perpendicular to the electromagnetic field. It is also perpendicular to the direction of radiation or travel. As a result, a radio wave is made up of electromagnetic and electrostatic fields. See Fig. 19-29. The position these waves radiate in respect to the earth is called POLARIZATION. In Fig. 19-30 the waves are radiated from a vertical antenna. Note that the electrostatic or E waves are in the same plane as the antenna yet perpendicular to direction of travel. The vertically polarized waves are perpendicular to the surface of the earth.

In Fig. 19-31, the wave is radiated from a horizontal antenna. It is still perpendicular to the direction of

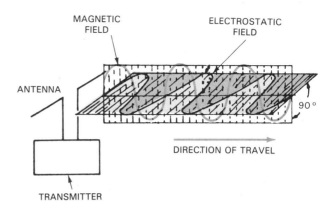

Fig. 19-29. The relationship between electrostatic and electromagnetic waves. They are perpendicular to each other and both are perpendicular to the direction of travel.

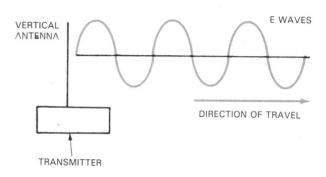

Fig. 19-30. A vertical antenna radiates a vertically polarized wave.

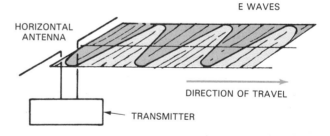

Fig. 19-31. A horizontal antenna radiates a horizontally polarized wave.

travel, but is parallel to the surface of the earth. Generally speaking, the antenna which receives these waves should be positioned in the same way as the transmitting antenna. At high frequencies the polarization changes to some extent as the wave moves.

Does all this mean the transmitting antenna radiates two waves? The answer is found in the fact that without one there cannot be the other. A moving electrostatic field produces a moving electromagnetic field. A moving electromagnetic field produces a moving electrostatic field. These conditions are true whether an actual conductor is present or not.

GROUND WAVES

The radiated waves from an antenna may be divided into two groups. These are ground waves and sky waves.

A GROUND WAVE follows the surface of the earth to the radio receiver. The ground wave has three parts:
1. The surface wave.
2. The direct wave which follows a direct path from the transmitter to the receiver.
3. The ground reflected wave which strikes the ground and is then reflected to the receiver.

The last two waves are combined and called a SPACE wave. The waves that make up the space wave may or may not arrive at the receiver in proper order. They may join together or cancel each other, depending on distances traveled by each wave.

Broadcast stations depend on the surface wave for reliable communications. As the wave travels along the surface of the earth, it will induce currents in the earth's surface. These currents use up the energy contained in the wave. The wave becomes weaker as the distance it travels increases.

An interesting sidelight is that salt water conducts surface waves about 5000 times better than the earth. Overseas communication is very reliable when transmitters are near the coastline. These stations use high power and operate at lower frequencies than the normal broadcast band.

SKY WAVES

The second type of radiated wave is a SKY WAVE. Sky waves use the ionized layer of the earth's atmosphere for transmission. This layer is called the IONOSPHERE. It is located from 40 to 300 miles above the earth's surface. It is believed to consist of large numbers of positive and negative ions. As the sky wave radiates, it strikes the ionosphere. Some of the wave may be absorbed into the ionosphere. But some will bounce off the layer and be sent back to the earth's surface. See Fig. 19-32.

REVIEW QUESTIONS FOR SECTION 19.5

1. Transmitted radio waves travel at:
   a. 186,000 miles per second.
   b. 300,000,000 meters per second.
   c. The speed of light.
   d. All of the above.
2. The wavelength for a 93 megahertz signal is _____ meters.
3. What is polarization?
4. Name the three parts of a ground wave.
5. A space wave is a combination of what waves?
6. Sky waves use the _____ for transmission.

19.6 AM TRANSMITTER PROJECT

In this project, you will make a working AM transmitter. Using it, you will be able to hear yourself over the radio.

The rf carrier range is 550–1500 kHz with an output power of less than 100 milliwatts. Transmitting distance will depend on the environment. However, this distance should be less than 100 feet. Any AM receiver can be used to pick up the transmitted signal. A block diagram of the transmitter is shown in Fig. 19-33.

The circuit used in this AM transmitter was provided by Graymark Enterprises, P.O. Box 54343, Los

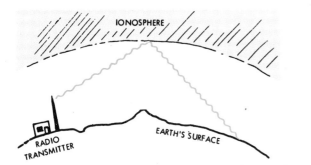

Fig. 19-32. Sky waves bounce off the ionosphere and move back to the earth's surface.

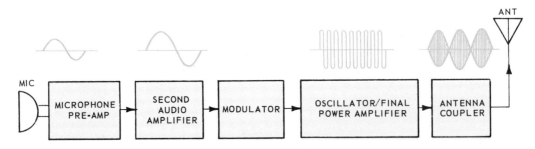

Fig. 19-33. Block diagram of AM transmitter.

Radio Wave Transmission

Angeles, California, 90054. It may be purchased as the Graymark Model 533 AM Transmitter, Fig. 19-34.

See Fig. 19-35 for parts list. The test conditions for the initial operation of the transmitter are:
1. R_1 set at position "1."
2. C_{10} set at position "15" (full CW).
3. Antenna not extended.
4. Mike removed.
5. Power source 9.0 V dc.
6. All readings taken with 11 megohm TVM meter, negative lead connected to common ground ($-$).
7. Tolerance of all readings: $\pm 20\%$.

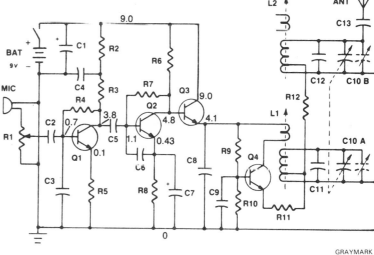

Fig. 19-34. Left. Completed AM transmitter. Right. Schematic for AM transmitter.

QTY.	SYMBOL	PART NO.	DESCRIPTION
1	R1	62389	Potentiometer, with switch, 50K
2	R2, R8	61412	Resistor, 470Ω, ¼-W, 10%
2	R3, R10	61415	Resistor, 4.7KΩ, ¼-W, 10%
1	R4	62390	Resistor, 470KΩ, ¼-W, 10%
1	R5	61420	Resistor, 100Ω, ¼-W, 10%
1	R6	61413	Resistor, 3.9KΩ, ¼-W, 10%
1	R7	62235	Resistor, 560KΩ, ¼-W, 10%
1	R9	61418	Resistor, 3.3KΩ, ¼-W, 10%
1	R11	62391	Resistor, 22Ω, ¼-W, 10%
1	R12	62392	Resistor, 10Ω, ¼-W, 10%
2	C1, C4	62393	Capacitor, electrolytic, 33µF, 16 wv
2	C2, C5	62394	Capacitor, disc, .1µF, 50 wv ("104")
1	C3	62365	Capacitor, disc, .0047µF, 50 wv ("472")
1	C6	62364	Capacitor, disc, .01µF, 50 wv ("103")
1	C7	61528	Capacitor, electrolytic, 4.7µF, 16 wv
2	C8, C9	62395	Capacitor, disc, .047µF, 50 wv ("473")
1	C10	62407	Capacitor, variable (tuning), 266pF
2	C11, C12	62397	Capacitor, disc, 22pF, 50wv ("22")
1	C13	62398	Capacitor, disc, 220pF, 50wv ("221")
2	L1, L2	62399	Coil, oscillator type
4	Q1-Q4	61533	Transistor, NPN silicon, 2SC372-Y
1		62400	Microphone, crystal lapel
1		62401	Antenna, telescopic
1	J1	62402	Jack, microphone, 3.5mm type, NC
1		62403	Battery clip, for 9-volt battery
1		62405	Antenna bracket, steel
2		61548	Knob, black
4		61433	Rubber foot, black
1		62408	Rubber grommet, black
4	MS1	61267	Machine screw, large dia., slot head (3x5mm)
2	MS2	61061	Machine screw, small dia., slot head (2.6x4 mm)
1	MS3	62344	Machine screw, small dia., phillips head (2.6x6mm)
4	MS4	61264	Self-tapping screw
1	WS1	62410	Washer, shoulder, plastic
1	WS2	61258	Washer, flat, fiber
1		62412	Solder lug
35		61357	Soldering pin
1 set		62413	Hookup wire set, stranded, 1-red, 1-yellow
1		61174	Bare wire (buswire), single strand, #22
1		62414	Solder, rosin core, 60/40
1		62388	Printed circuit board
1		62416	Chassis, w/battery holder
1		62417	Cover, aluminum
1		62422	Breadboard panel, with schematic
1		—	Warranty card
1		61943	Instruction manual

Fig. 19-35. Parts list for AM transmitter.

SUMMARY

1. The basic communications process can be shown as a block diagram. Parts include source, transmitter, receiver, and destination. Noise and feedback are also part of this process.
2. The radio spectrum chart shows the frequency assigned to certain communications services. Assignments are made by the FCC.
3. Basic transmitters include continuous wave, modulated continuous wave, amplitude modulation, and frequency modulation.
4. Some basic types of microphones are the carbon, dynamic, crystal, and velocity.
5. Modulation is the process of adding or superimposing audio waves to carrier waves.
6. Modulation percentages refer to the amount a carrier wave has been varied or modulated.
7. FM modulation index is the relationship between maximum carrier deviation and maximum modulating frequency.
8. Radio waves travel at the speed of light (186,000 miles or 300,000,000 meters per second).
9. Two types of transmitted waves are ground waves and sky waves.
10. Ground waves follow the surface of the earth while sky waves are radiated into space.
11. The ionosphere is a layer of charged particles 40 – 300 miles above the earth's surface.

TEST YOUR KNOWLEDGE, Chapter 19

Please do not write in the text. Place your answers on a separate sheet of paper.

1. Name the three basic parts of all communication.
2. _____ is a sign that a message has been received.
3. Noise:
    a. Is a break in the intended signal.
    b. Distorts the message.
    c. Enters the communication process during transmission.
    d. All of the above.
4. The signal created by an oscillator is a _____ wave.
5. What is the purpose of a microphone?
6. _____ is a process in which an audio wave is superimposed on a carrier wave.
7. _____ _____ results from the mixing of two waves.
8. What is the formula for computing percent of modulation?
9. What power is required to modulate a transmitter having a dc input of 250 watts to 50 percent?
10. The amplitude of a modulating signal has a maximum deviation of 15 kHz and a frequency of 1500 Hz. What is the modulating index?
11. What is the wavelength (in meters) of a signal with a 1200 kHz frequency?
12. Make a sketch showing the direction of travel of a wave transmitted from a horizontal antenna.

FOR DISCUSSION

1. Are the waves radiated to your TV antenna horizontally or vertically polarized?
2. Explain the basic block diagram of a transmitter.
3. Discuss the operation of the following microphones:
    a. Carbon microphone.
    b. Dynamic microphone.
    c. Crystal microphone.
4. What is the primary difference between AM and FM?
5. Discuss two basic methods of FM detection.

Radio Wave Transmission

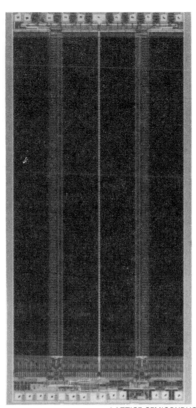

LATTICE SEMICONDUCTOR CORP.

Leads are kept short in digital devices to prevent the radiation of electromagnetic waves to the outside world. The radiated power must be kept below 0.001 mW. The longest lead on this 64K static RAM is 0.1 cm.

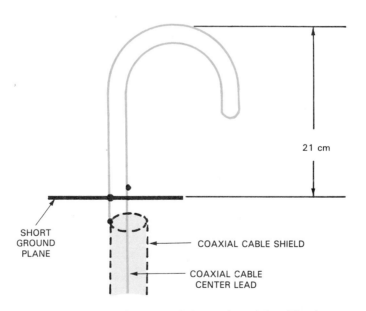

A J-type antenna for transmitting and receiving. The J antenna is used for frequencies above 400 MHz. The magnetic field B is always at right angles to the plane of the J. The electric field E is strongest in the vertical direction and weakest in the horizontal direction.

Electricity and Electronics

MARANTZ

The transducer is the final stage of a radio receiver. A speaker is a type of transducer. It changes electrical energy to sound energy.

Early radio receivers used circuits tuned to the frequency of the signal being sent. This usually required tuning three or four tank circuits at once, which was not easy to do and made radio reception into hard work. When the superheterodyne principle was used, the radio became simple enough for anyone to use. When learning about radio receivers, it is important to pay special attention to the terms "local oscillator" and "intermediate frequency (IF)." These are important terms from the superheterodyne circuit.

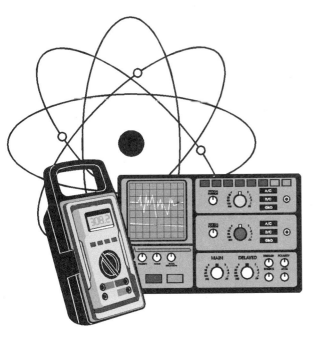

Chapter 20

RADIO WAVE RECEIVERS

After studying this chapter, you will be able to:
- *Discuss the history of radio wave receivers.*
- *List the components and explain the operation of an AM receiver.*
- *List the components and explain the operation of a superheterodyne receiver.*
- *Describe the use of transducers in radio wave receivers.*
- *List the components and explain the operation of an FM receiver.*

One of the greatest inventions of all time is the radio. Like many other inventions, the radio resulted from the work of many scientists. In 1864, James Maxwell theorized that electromagnetic waves existed. In 1887, Heinrich Rudolph Hertz confirmed this theory when he transmitted and received the first radio waves. The first continuous wave (CW) transmitter was developed in 1897 by Guglielmo Marconi.

Two key devices that furthered the development of the radio were the diode and the audion triode amplifier tube. The diode was invented in 1904 by Alexander Fleming. The amplifier tube was developed in 1904 by Lee DeForest.

A key scientist in radio history was E.H. Armstrong. He invented the regenerative radio circuit in 1913 and the superheterodyne radio circuit in 1920. Major Armstrong is also credited with the development of much of the FM radio theory.

In Chapter 19, a basic communication model showing the relationship between the transmitter and receiver was discussed. This model is shown again in Fig. 20-1. In this chapter we will discuss the AM/FM radio receiver.

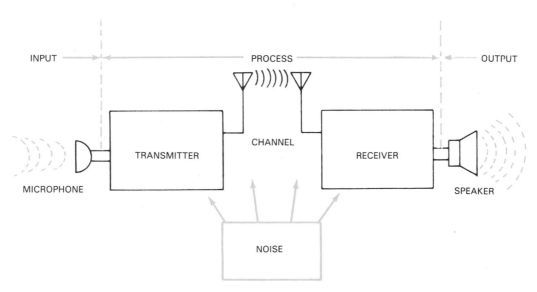

Fig. 20-1. Basic transmitter/receiver communication model.

20.1 AMPLITUDE MODULATED (AM) RECEIVER

A radio wave transmitted through space also carries information, such as voices and music. This information has been combined with the carrier wave at the transmitter by the amplitude modulation process.

These radiated electromagnetic waves cut across and induce a small voltage in the receiving antenna. The small radio frequency voltages are coupled to a tuned circuit in the receiver which selects the signal to be heard. After selection, the voltages may be amplified, then demodulated. DEMODULATION, or DETECTION, is the process of removing the audio portion of a signal from the carrier wave. It is a form of rectification. The audio signal is then amplified until it can drive a loudspeaker.

A block diagram of this receiver is shown in Fig. 20-2. The modulated RF wave is shown as it passes the antenna. The audio wave is shown as the output of the detector. The increase in amplitude between the blocks is the result of stages of amplification. This type of receiver was once quite popular. It was called the TRF or TUNED RADIO FREQUENCY receiver.

For satisfactory operation of the circuit, each stage had to be tuned to the correct incoming frequency. Some early radios had a series of tuning dials on the front panel. Adjusting these to receive a signal required skill and patience. The development of the superheterodyne receiver overcame these obstacles.

THE TUNING CIRCUIT

One function of any radio receiver is selection of the desired radio signal. Details of the "tank circuit" and flywheel action were discussed in Chapter 11.

Fig. 20-3 shows a schematic of the first stage of the TRF receiver. Follow the signal of the antenna to the output of this circuit.

Radio signals of many frequencies from many radio transmitters pass by the antenna. The induced voltage in the antenna causes small currents to alternate from antenna to ground and from ground to antenna through coil L_1. Since L_2 is closely coupled to L_1, the magnetic field created by the antenna current in L_1 transfers energy to L_2. The combination coils L_1 and L_2 are called the ANTENNA COIL. They are wound on a common cylinder of cardboard, Fig. 20-4.

Variable capacitor C_1 is connected across the terminals of L_2 and forms a tank circuit. C_1 may be adjusted to vary the RESONANT frequency of the tank. When

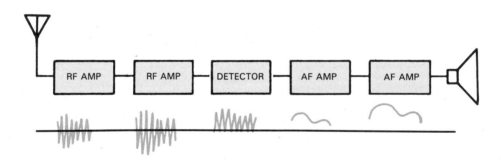

Fig. 20-2. A block diagram of a TRF (tuned radio frequency) receiver.

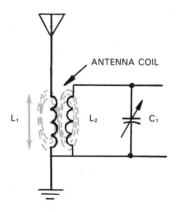

Fig. 20-3. The tuning or station selector section of the radio receiver.

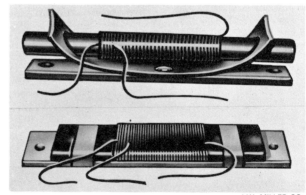

J.W. MILLER CO.

Fig. 20-4. Typical antenna coil used with transistorized receivers.

Radio Wave Receivers

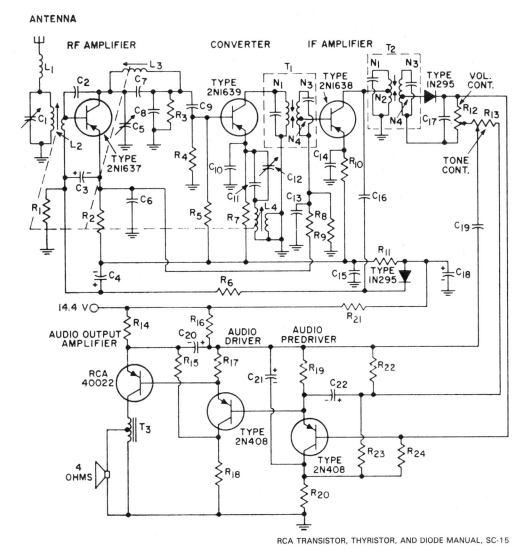

Fig. 20-5. 12 volt automobile radio receiver with RF amplifier.

the resonant frequency of the tank circuit is the same as the incoming signal, high circulating currents develop in the tank. In other words, the tank circuit may be adjusted to give peak response for only a single frequency.

The L_2C_1 combination may cover only a certain range or band of frequencies. Most of our home radios cover only the broadcast band. For other groups of frequencies, such as shortwave, another coil is switched in place of L_2. This coil changes the resonant frequency range of the circuit. This is called a BAND SWITCHING circuit.

The ability of a radio receiver to select a single frequency, and only one frequency, is SELECTIVITY. The ability of a receiver to respond to weak incoming signals is called SENSITIVITY. Both of these traits are useful and wanted. Special circuits and components have been devised to improve selectivity and sensitivity.

RF AMPLIFICATION

The radio frequency (RF) amplifier is most often the first stage to receive the signal from the antenna. The RF amplifier, tuned to incoming signal's frequency, amplifies the signal (provides gain).

Generally, RF amplifiers are narrow band amplifiers that can amplify only the band or frequencies to be picked up by that receiver. An example is the RF amplifier in an AM broadcast band receiver. It can amplify frequencies from 550 to 1500 kHz. An RF amplifier improves the selectivity of a receiver.

A common RF amplifier is shown in Fig. 20-5. The tuning circuit is made up of C_1 and L_2 formed into a tank circuit. The tuned signal is fed into the secondary winding of L_2. The amplifier is the 2N1637 transistor. Some less costly radio receivers do not have RF amplifier stages. However, high quality radio receivers will often have two or more RF amplifier stages.

325

DETECTION

Assume that a modulated radio frequency signal has been amplified by several stages of RF amplification. It now has enough amplitude for detection. Recall that detection is a form of rectification. RF wave is removed, leaving only the AF wave. See Fig. 20-6.

Detection is needed to recover the audio signal from the modulated RF carrier wave.

THE DIODE DETECTOR

One method of detection uses a diode as a unilateral (one direction) conductor.

While the traits of the diode were discussed earlier, they are reviewed here. When the anode of the diode is driven positive, electrons flow through the diode from cathode to anode. The diode conducts. When the anode is driven negative, the diode does not conduct.

Fig. 20-7 shows the basic diode detector. The amplified modulated signal is supplied to the detector by the previous amplifier stage. When the input signal is positive, the diode conducts. A voltage develops across the diode load resistor R_1.

When the incoming signal is negative, there is no current in the diode circuit. The diode has rectified the modulated RF signal into pulses, or waves, of dc voltage. These waves have a frequency and amplitude of the audio wave.

To understand half-wave rectification, look at the input and output wave forms in Fig. 20-8. The dotted curve in the output of the diode denotes the AVERAGE dc value of the rectified voltage. Raising the average dc output voltage will cause it to more closely reproduce the input signal. A filter capacitor of the correct value C_1, is shunted across R_1. This capacitor charges to the peak value of the signal. It also prevents the voltage across R_1 from dropping to zero when the diode is not conducting.

The improved output waveform resulting from filtering is shown in Fig. 20-8. The time constant of R_1C_1 should be long when compared to the RF cycle. This will deter C_1 from discharging to a low value. Likewise, this time constant should be short when compared to the AF cycle. Voltage variations across the R_1C_1 network will follow the audio frequency cycle.

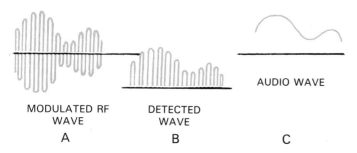

Fig. 20-6. The appearance of waveforms as they are detected. A—The RF carrier within the modulation envelope. B—The rectified RF wave. C—The audio wave when the RF wave has been removed.

REVIEW QUESTIONS FOR SECTION 20.1

1. What is demodulation?
2. Demodulation is a form of:
   a. Rectification.
   b. Detection.
   c. Amplification.
   d. None of the above.
3. What was the drawback of the TRF receiver?
4. The ability of a receiver to choose a single frequency, and only one frequency, is _____.
5. The ability of a receiver to respond to weak incoming signals is _____.
6. _____ _____ are narrow band amplifiers that can amplify only the band or frequencies picked up by that receiver.

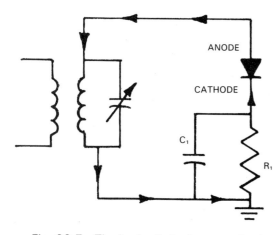

Fig. 20-7. The basic diode detector circuit.

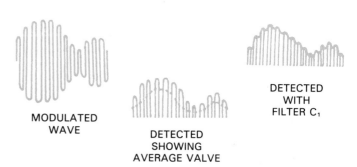

Fig. 20-8. The average value of the dc component is represented by the dotted line. The dc average is raised by adding the filter capacitor.

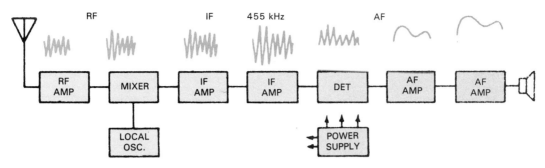

Fig. 20-9. The block diagram of the superheterodyne receiver.

20.2 THE SUPERHETERODYNE RECEIVER

The superheterodyne circuit was developed in answer to the problems of the TRF receiver. HETERODYNING, or mixing of signals, converts all incoming signals to a single intermediate frequency. This signal can be amplified with little loss and distortion.

See Fig. 20-9. The signal picked up by the antenna is fed first to a stage of radio frequency amplification (RF amp). The output of the RF amp is then fed to the MIXER or CONVERTER STAGE. The output of a LOCAL OSCILLATOR is also fed to the converter.

When two signals are mixed together, four signals appear in the output. These will include the two original signals, the sum of the two signals, and the difference between the two signals. For example, a 1000 kHz signal is mixed with a 1455 kHz oscillator signal. Appearing in the output then will be the 1000 kHz original signal, the 1455 kHz original oscillator frequency, a 2455 kHz frequency sum, and a 455 kHz frequency difference.

The BEAT frequency of 455 kHz is key to the study of the superheterodyne receiver. The station is selected using the tuning circuits of the receiver. Turning the tuning knob on the front panel varies the capacitance of the tuning circuit. Attached to the shaft of the tuning capacitor is another tuning capacitor. This adjusts the frequency of the local oscillator, Fig. 20-10.

These capacitors operate in step with each other. They provide a change in oscillator frequency as the tuned frequency is changed. They always maintain a fixed difference or beat frequency of 455 kHz. This is called the INTERMEDIATE FREQUENCY, or IF. The IF output is then amplified by two stages of voltage amplification and fed to the DETECTOR. The detector output is an audio frequency voltage. It is amplified enough to operate the power amplifier and speaker. The waveforms at each stage are shown in Fig. 20-9.

An eight-transistor transistorized AM superheterodyne receiver is shown in Fig. 20-11. This circuit is part of the Graymark 536 AM transistor radio. Each part will be discussed in detail.

J.W. MILLER CO.

Fig. 20-10. A double section tuning capacitor.

MIXER

The MIXER performs three functions, Fig. 20-12. Mixing is just one. This section of the radio converts the incoming signal to a new frequency, called the intermediate frequency. This is done by mixing it with a signal produced by a local oscillator. At the same time, it provides amplification.

The frequency converter, or mixer, is a nonlinear circuit. Signals are combined to produce the sum and the difference frequencies of the original signals. The properties of a mixer can be used to produce a signal of the correct frequency for the IF amplifier. A signal must be produced locally (within the receiver) by an oscillator. This signal and the signal of interest are mixed. The original two and the sum and difference appear at the output of the mixer. An example is shown in Fig. 20-13.

Notice that the signals of many stations could be mixed to a new frequency. However, the IF amplifier will not amplify these because they are not within the IF passband.

The local oscillator is designed to be 455 kHz above the desired signal. The incoming signal must pass through a tuned antenna circuit to get to the mixer.

Electricity and Electronics

SCHEMATIC DIAGRAM

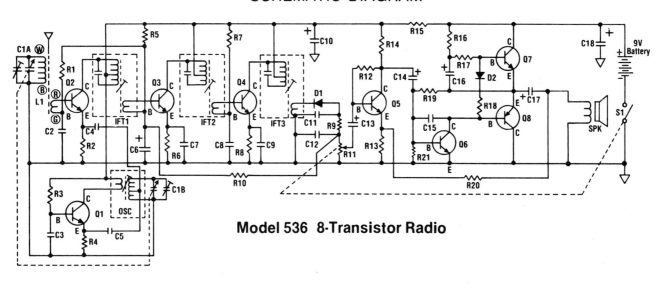

Fig. 20-11. Model 536, eight-transistor radio.

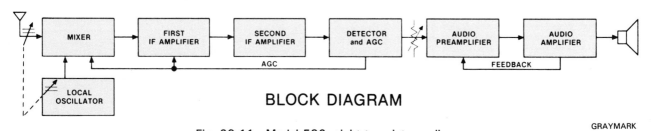

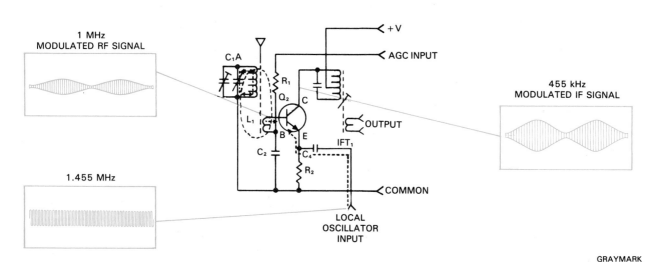

Fig. 20-12. An RF amplifier/mixer/converter.

Radio Wave Receivers

	FREQUENCY (MHz)	ABBREVIATION
INCOMING	1.000	RF
LOCAL OSCILLATOR	1.455	LO
LOCAL OSCILLATOR *PLUS* INCOMING	2.455	LO + RF
LOCAL OSCILLATOR *MINUS* INCOMING	.455	LO − RF = IF

Fig. 20-13. Frequencies of signals being mixed. Their abbreviations are also shown.

This is because there are two possible signals that can be mixed to result in the IF. One is the desired signal, the other is called an IMAGE. See Fig. 20-14.

Images are supressed in front of the mixer by the tuned antenna circuit. The more selectivity in front of the mixer, the less of a problem images will be.

The incoming signal frequency, local oscillator frequency, sum and difference of the two are all presented to the input of the first IF amplifier. The IF amplifier is designed to amplify only the difference frequency (455 kHz). It will reject the other three frequencies. This is because they fall outside of the passband of the IF amplifier.

The partial schematic shown in Fig. 20-12 is the mixer used in the Model 536 receiver. The input circuit is composed of tuning and trimmer capacitors C_{1A}, and core and coil L_1 (also serving as the antenna). This is a high Q resonant circuit. It is designed to select the incoming signal and to reject images. L_1 has a winding joined to the antenna circuit by an inductive couple. This will couple the selected signal to the base of Q_2. Bias current for Q_2 is provided by R_1. C_1 bypasses one end of the coupling winding to ground. The resonant circuit connected to the collector of Q_2 is tuned to the intermediate frequency. The emitter resistor of Q_2, R_2, provides bias voltage. The Q_2 emitter is the point at which the local oscillator output is put into the mixer. It moves through coupling capacitor C_4. Mixer gain is controlled by AGC feedback through resistors R_{10} and R_1.

	FREQUENCY (MHz)
DESIRED RF	1.000
LO	1.455
LO − RF = IF	.455
IMAGE (LO + IF)	1.910
IMAGE − LO = IF	.455

Fig. 20-14. Frequencies of desired and image signals.

LOCAL OSCILLATOR

When the local oscillator is inserted as part of the mixer, oscillator pulling is caused. There is also an increase in distortion products. Although it uses more components, a separate local oscillator creates more stable design. Fig. 20-15 shows a radio that has a separate local oscillator.

Transistor Q_1 is connected as a variable oscillator.

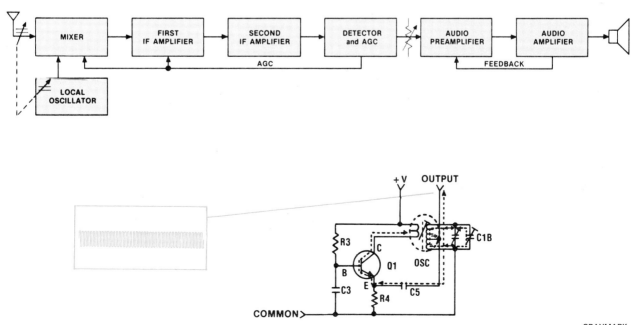

Fig. 20-15. Local oscillator.

Variable capacitor C_{1B} is mechanically coupled to the mixer input circuit. Capacitor C_{1B} is coupled to the tuning capacitor C_{1A}. The frequency of the signal made by the local oscillator will be adjusted to 455 kHz higher than the received signal. The positive feedback needed for oscillation is returned to the Q_1 emitter through C_5.

FIRST IF AMPLIFIER

Intermediate frequency

IF amplifiers are nontunable (except for alignment) radio frequency amplifiers. Because they amplify a fixed frequency, they can be quite useful. Gain and bandwidth can be tailored to meet the needs of the receiver. Some limits to this include:
1. The lower the frequency the easier it is to obtain a narrow bandwidth.
2. The IF frequency will set the image frequency (see mixer discussion). Thus, the IF frequency should be as high as practical. In AM broadcast receivers, 455 kHz has been chosen to be the best frequency. If the intermediate frequency is lower, the image response would suffer. And if higher, it would be harder to reach the proper bandwidth with just two IF stages.

IF amplifiers set selectivity and provide most of the voltage gain in the superheterodyne circuit.

A lower intermediate frequency permits more narrow bandwidth because of ARITHMETICAL SELECTIVITY. To explain this, consider two signals, one at 1.00 MHz and the other at 1.01 MHz. The separation of these two signals is 10 kHz. That is a difference of 1%. But when both signals are converted in the mixer to their intermediate frequencies, the 1.00 MHz signal becomes 455 kHz. The 1.01 MHz signal becomes 465 kHz. The difference of 10 kHz is now 2.2% of the intermediate frequency. That is more than twice the percentage difference. The signals can now be easily separated by the IF amplifier.

The standard intermediate frequencies for receivers that have been in use for several decades are:

AM receivers   455 kHz
FM receivers   10.7 MHz
Television receivers 41-46 MHz

IF amplifiers

There are two IF amplifiers used in the AM receiver shown in Fig. 20-16. Two transistor stages and three resonant circuits are provided to achieve the gain and passband required.

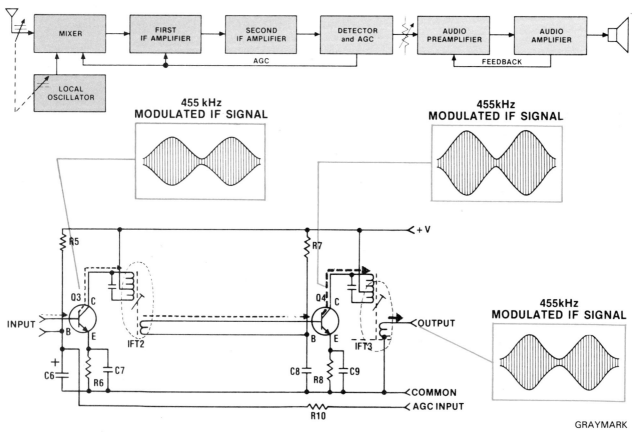

Fig. 20-16. First and second IF amplifiers.

Transistors Q_3 and Q_4 each amplify the IF signal. The transformers IFT_1 (Fig. 20-11), IFT_2, and IFT_3, and their internal capacitors, each have a narrow passband (window). This keeps out unwanted frequencies. The transistor Q_3 is connected to the output of the mixer transistor through a winding with an inductive coupling, transformer IFT_1. The operating bias for transistor Q_3 is developed by resistor R_5 and emitter resistor R_6. The bias current, and thus the collector current, are controlled by the AGC voltage. This voltage is developed at the output of the third IF transformer, IFT_3. Each of three resonant circuits are located in the IF transformers. All are tuned (user adjustable) to 455 kHz, the intermediate frequency. The Q and coupling coefficient are designed to provide the proper bandpass needed.

These factors are crucial. They are an integral part of the design and are not adjustable. Gain is fixed and maximized by proper biasing. The emitter resistor is RF bypassed. If this was not done, the negative feedback developed across the emitter resistor would reduce the gain a great deal.

The second IF stage is like the first IF stage. The gain is fixed and maximized by proper biasing with resistors R_7 and R_8. The emitter resistor is RF bypassed so that the negative feedback developed across the emitter resistor will not reduce the gain.

DETECTOR

The detector circuit shown in Fig. 20-17 is much the same as is used in the simple crystal set. However, because there are two stages of IF amplification before the detector, and more gain is provided by the mixer stage, weak signals can be detected.

The gain in front of the detector allows the signal to overcome the threshold voltage of the detector diode. The detector detects the modulation information from the received signal. This is done by rectifying the amplified signal, then filtering the remaining RF from the signal. The detector diode (D_1) rectifies the RF modulated signal. Note that the envelope of the RF waveform contains the audio information first used to modulate the RF carrier at the transmitter.

The RF is removed (bypassed to ground) by C_{12}. This is because the .02 µF capacitor presents a low impedance path for the RF signal, but very high impedance for audio frequencies.

AUTOMATIC GAIN CONTROL

Automatic gain control, or AGC, keeps the audio output level constant despite the varying strengths of the signals. AGC rectifies and filters the output of the IF stages. This signal is used to control the gain of the

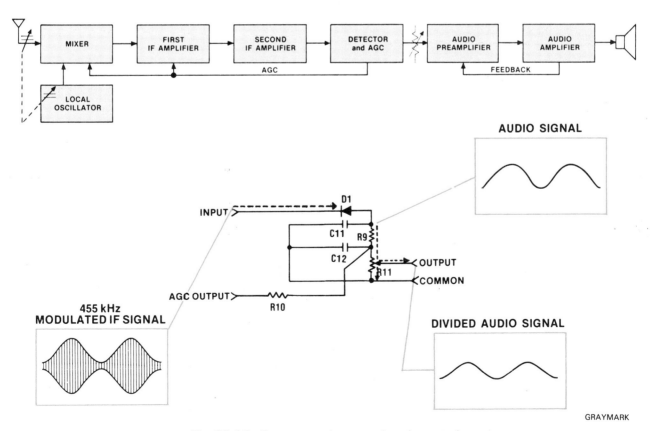

Fig. 20-17. Detector and automatic gain control.

preceding stages. In this manner, input signals with a strength difference of 40 dB can cause as little as 3 dB change in audio output level. Without AGC, the volume would have to be adjusted as each station was tuned in. AGC is also known as AVC (automatic volume control).

Refer again to Fig. 20-17. The AGC voltage is taken from the junction of C_{12} and R_{10}. Resistor R_{10} keeps the audio from the further filtering done by C_6. The AGC voltage appears across C_6. This AGC voltage controls the current into the bases of Q_2 and Q_3. The gain of transistors Q_2 and Q_3 are directly proportional to their collector currents. Detector diode D_1 is connected to supply a negative voltage which increases as signal strength increases. This negative voltage is applied to the bases of Q_2 and Q_3. It controls the gain. Therefore, as the AGC voltage increases, the gain of Q_2 and Q_3 goes down. Note that the detector diode is used both as an audio detector and AGC rectifier.

AUDIO PREAMPLIFIER

The audio preamplifier has a high impedance input. This provides a minimum load on the detector output. Refer to Fig. 20-18. Variable resistor R_{11} is used to select the detector signal at different voltage levels. The preamp gain varies. It is controlled by the negative feedback from the audio amplifier through R_{20}.

AUDIO AMPLIFIER

The audio output from the detector will have an amplitude of several volts. The impedance level will be fairly high. The means that little power will be ready to drive the loudspeaker. In order to match the low impedance loudspeaker to the high impedance detector output, an audio amplifier with a low impedance output is required. In Fig. 20-19, transistors Q_5, Q_6, Q_7, and Q_8 do this.

Transistor Q_5 is biased Class A. It provides enough voltage and power gain to drive the output transistors, Q_7 and Q_8. The output stage is composed of Q_7 (NPN) and Q_8 (PNP). During the positive swing, Q_7 will conduct. During the negative swing, Q_8 will conduct. It is crucial that Q_8 and Q_7 not be on at the same time. This would cause a large current to flow through both transistors. Diode D_2 prevents this event from occurring.

It is best to use all the voltage given by the 9 volt battery. To do this, do not allow the junction of the Q_7 and Q_8 emitter, under NO signal conditions, to idle at 4.5 volts ($\frac{9 \text{ volts}}{2}$). This would make a peak-to-peak audio signal of 9 volts ready for use before the signal is clipped.

If the NO signal voltage is not centered at the halfway point, one side of the signal swing will clip at a

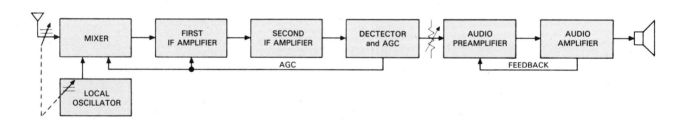

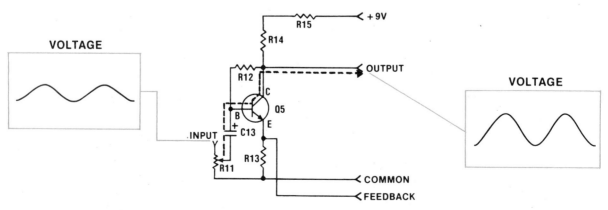

Fig. 20-18. Audio preamplifier.

Radio Wave Receivers

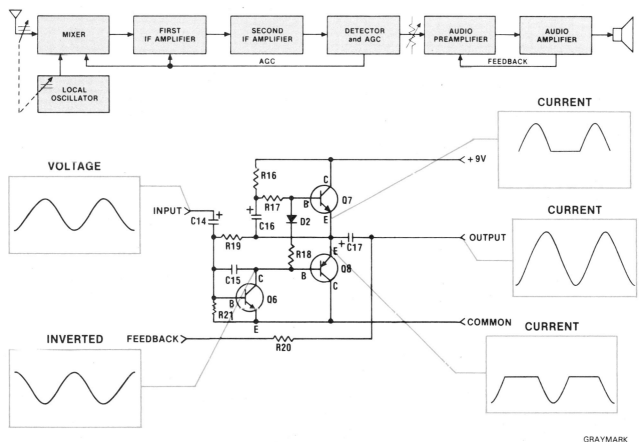

Fig. 20-19. Audio amplifier.

lower voltage. This causes uneven clipping distortion. That limits the power output that will be ready to use. The audio amplifier is Fig. 20-19 achieves halfway idle bias voltage as follows.

Q_7 is an NPN device. A positive voltage applied to its base-emitter junction will cause it to conduct. R_{16} and R_{17} provide bias current that causes Q_7 to turn on. If no other circuit elements are present, the bias current will cause the Q_7 emitter to rise to 9 V minus the collector emitter voltage drop.

To prevent this, the emitter of Q_7 is connected via R_{19} to the base of Q_6. Q_6 is also an NPN transistor. As a higher base current is applied to Q_7, the collector-emitter resistance becomes lower.

The collector of Q_6 is connected via R_{18} and D_2 to the base of Q_7. As Q_6 becomes lower in resistance, it steals base current from Q_7. This increases the collector-emitter resistance of Q_7. The network then becomes a voltage divider. To complete the circuit, think of the effect of the PNP transistor Q_8. If the base-emitter junction of Q_8 becomes forward biased, it will conduct. As stated earlier, it must not be allowed to conduct at the same time as Q_7.

Under working conditions, capacitor C_{14} couples the audio signal to the base of driver/inverter Q_6. When the signal swings positive, Q_6 will conduct more. As a result, Q_8 will also conduct more. While this is happening, more current is stolen from the base of Q_7. Q_7 then conducts less.

On the negative swing of the signal, Q_6 will conduct less. This allows Q_8 to reverse bias. More bias current is then applied to Q_7, causing it to conduct more. As in many circuits, there are two conditions. They are dc, or no signal conditions, and ac, or dynamic conditions. In order to study the circuit, both states should be understood and then combined.

Capacitor C_{17} is used to block dc current from flowing through the speaker.

IC USE IN SUPERHETERODYNE RECEIVERS

Integrated circuits can be used in place of many single parts in a superheterodyne receiver. In the early years of IC production, most devices made were op amps, comparators, and voltage regulators. In the late 1970s and early 1980s, the monolithic (single) IC was developed. It was very useful in communication systems such as radios and televisions. Industry has always worked toward ICs that perform better and are more stable than what is used currently.

ICs such as the LM 1820 from National Semiconductor can replace most transistors, resistors, and capacitors in the mixer oscillator and IF amplifier circuits. ICs cannot replace the tuning components, local oscillator tank circuits, and IF transformers. A linear IC such as the LM 386N can be used as the audio amplifier.

A block diagram of the complete radio is shown in Fig. 20-20. The schematic is shown in Fig. 20-21. The IC LM 1820, houses the major workings for the mixer-oscillator, first IF amplifier stage, second IF amplifier stage, and the automatic gain control (AGC) circuit. The audio amplifier stages are housed in the LM 386. It provides 1/4 watt into an 8 ohm speaker. The operating voltage is 6 volts dc. This can be obtained from four D, C, or AA cells. The V_{cc} is the positive voltage and ground is negative. A complete schematic of LM 1820 is shown in Fig. 20-22. An LM 386 schematic is shown in Fig. 20-23. A layout of a printed circuit for an AM radio is shown in Fig. 20-24.

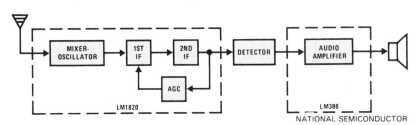

Fig. 20-20. Radio block diagram.

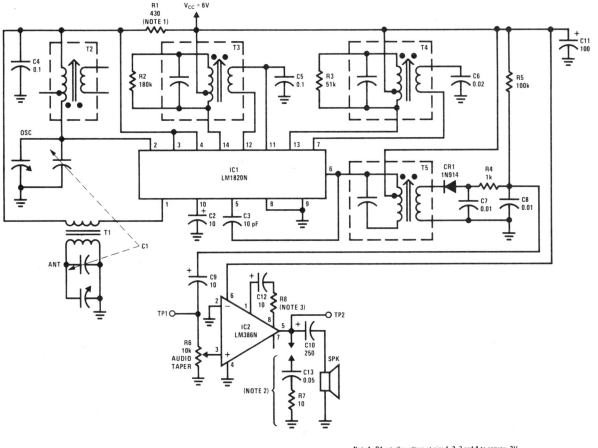

Fig. 20-21. Radio schematic.

Radio Wave Receivers

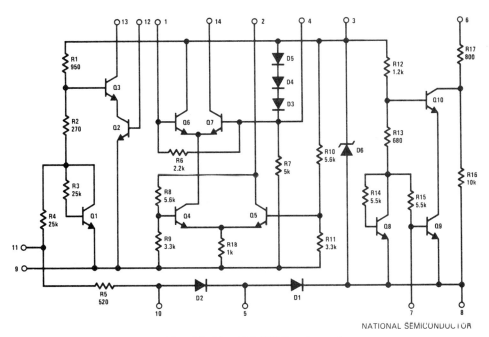

Fig. 20-22. LM 1820 schematic.

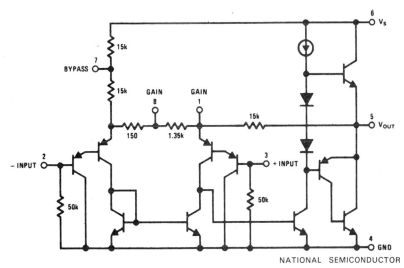

Fig. 20-23. LM 386 schematic.

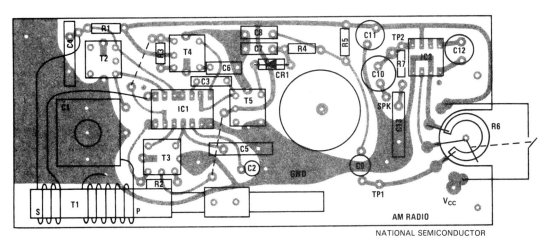

Fig. 20-24. Typical printed circuit board radio layout.

REVIEW QUESTIONS FOR SECTION 20.2

1. Define heterodyning.
2. How does the superheterodyne receiver differ from the TRF receiver?
3. The intermediate frequency in a superheterodyne radio is:
   a. 455 Hz.
   b. 45 kHz.
   c. 455 kHz.
   d. None of the above.
4. What three functions are performed by the mixer?
5. A local oscillator produces an unmodulated signal _____ kHz above the incoming frequency signal.
6. What is the purpose of the AGC circuit in a superheterodyne receiver?

20.3 TRANSDUCERS

The final stage of the radio receiver is a transducer. The transducer converts the electrical energy of audio frequencies into sound energy. Transducer devices are the LOUDSPEAKER and HEADPHONES.

A common type of speaker is shown in Fig. 20-25. The schematic symbol for this speaker is shown in Fig. 20-26. It is made with a permanent magnet. It is called a PM SPEAKER.

In a PM speaker, a strong magnetic field is produced between the poles of a fixed permanent magnet. A small voice coil is hung in the air gap. It is attached to the speaker cone. The audio alternating currents are joined to the voice coil. The action between the fixed field and the moving field causes the voice coil to move back and forth. This motion also causes the speaker cone to move back and forth. The air pressure, in the form of sound energy, changes back and forth, from highest to lowest pressure.

An electrodynamic speaker replaces the permanent magnet with an electromagnet, Fig. 20-27. It works like the PM type. A strong source of direct current must be supplied to the electromagnet. This may come from the power supply. A common practice is to use the field coil of the speaker as the filter choke in the supply.

Due to the popularity of stereos, special sizes and types of speakers have been developed. These provide the best response for certain bands of audio frequencies.

Low frequency speakers are WOOFERS. High frequency speakers are TWEETERS. A MIDRANGE speaker may be used to reproduce intermediate frequencies.

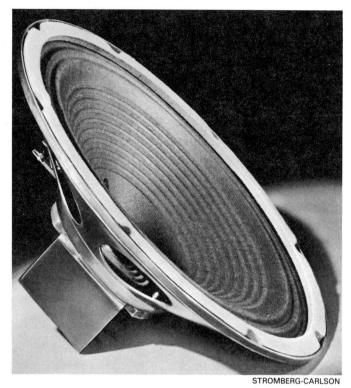

STROMBERG-CARLSON

Fig. 20-25. A typical radio loudspeaker.

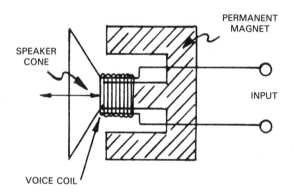

Fig. 20-26. This sketch shows the operating principles of the PM speaker.

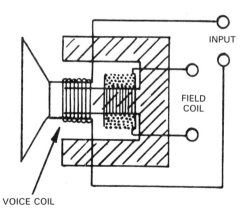

Fig. 20-27. An electrodynamic speaker. The permanent field is replaced with an electromagnetic field.

Special filters and crossover networks allow signals of set frequency ranges to be channeled to the speaker that best reproduces the sound.

A crossover network is shown in Fig. 20-28. Coil L is connected to the woofer. As the frequency of the sound increases, the reactance of L increases at the rate of $X_L = 2\pi fL$. However, the tweeter is connected through C. As the frequency increases, the reactance of C decreases at the rate of $X_C = \frac{1}{2\pi fC}$. Values of L and C may be selected for the desired crossover point. This is usually between 400 and 1200 Hz. Note that at one frequency, X_L equals X_C. The response is equal for both speakers at this crossover frequency.

HEADPHONES

Headphones provide excellent sound reproduction and allow for private listening. Like loudspeakers, there are two basic types of headphones: dynamic and electrostatic. See Fig. 20-29.

Dynamic

Dynamic headphones are divided into two groups. PRESSURE-TYPE dynamic headphones require an air seal around the ears for proper bass response, Fig. 20-30. HEAR-THROUGH or VELOCITY headphones allow the listener to hear outside sounds (telephone, door bell, etc.) with the headphones in place.

Pressure and hear-through headphones use the same basic transducer. Some pressure headphones use a dynamic woofer and a separate tweeter. The tweeter may be electrostatic, ceramic, or dynamic in design. Correct crossover circuits are included in the headphone.

Pressure and hear-through headphones can have two or four channels. A copper voice coil is attached to one side of miniature loudspeaker cone or diaphragm. In quad headphones, there are two driver elements in each earcup.

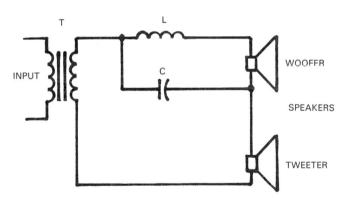

Fig. 20-28. A basic crossover network design.

KOSS CORP.

Fig. 20-30. Pressure-type dynamic headphones.

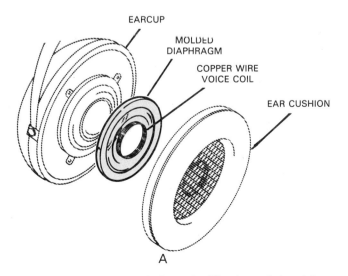

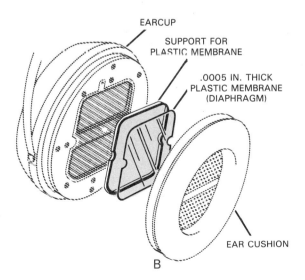

Fig. 20-29. Headphone designs. A—The dynamic headphone uses a voice coil and molded diaphragm. B—With electrostatic headphones, a very thin membrane provides the diaphragm within the driver of the headphone.

The coil is hung in a magnetic field. It acts like a pulsating motor when the electrical form of the music energy flows through its windings. This causes the diaphragm to move the air in a manner similar to the original sound waves made by the musical instruments. See Fig. 20-31.

Sound reproduction quality varies among headphones. Different designs and manufacturing costs affect results. Manufacturers try for a certain "sound," just as the loudspeaker companies do.

Sound waves are made in headphones with either cone-type or element-type drivers. They both perform the same task.

Cone-type drivers are loudspeakers, like those found in transistor radios. They are not designed for headphone use. But, they work well when used with the pressure-type cushions.

Element-type drivers are special loudspeaker structures. They are designed for headphone use. A special element is designed to work with a certain headphone housing. Grouping the driver element and housing into one package results in a driver that performs better and has better quality than the cone-type drivers.

Hear-through headphones are much different from the pressure-type, Fig. 20-32. They are light weight and have porous foam ear cushions.

Recall that the pressure-type headphone requires a closed volume of air. Hear-through headphones, however, vent the back sound waves through the rear of the cup.

The porous cushions provide some acoustic resistance. They help control the sound emitted by the headphone diaphragm. And they provide some acoustic openness. This allows the listener to hear outside sounds when the headphones are in place.

A key advantage of this headphone is its weight. The drawback, however, is a loss in sound quality.

Dynamic headphones use a fairly heavy copper voice coil. It is attached to one side of a miniature loudspeaker cone. In theory, the moving parts act like a piston to reproduce the recorded sound. But the heavy voice coil lags behind the electrical energy. Therefore, the plastic or parchment cone also loses true piston action and distorts the sound wave.

Electrostatic

Electrostatic headphones look like pressure-type headphones. However, their mechanics are very different.

The best electrostatic headphones use light diaphragms of plastic instead of heavy plastic or parchment cones. The diaphragm in electrostatic headphones may be only 1/1000 of an inch thick, and weigh less than the surrounding air.

The diaphragm moves back and forth by controlled charges of static electricity. Lightness and control of the high quality electrostatic allows for excellent sound reproduction, Fig. 20-33.

Fig. 20-32. Hear-through headphones.

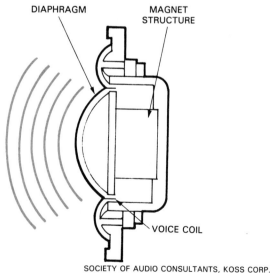

Fig. 20-31. Diaphragm as it moves in a headphone.

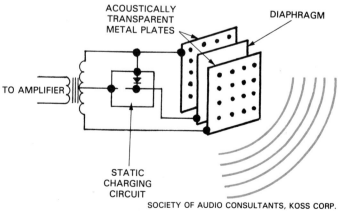

Fig. 20-33. Electrostatic headphones circuit and operation.

Radio Wave Receivers

Frequency response of electrostatics is wider and flatter than dynamics. Using electrostatic headphones makes the drawbacks of disc and tape recordings, power source equipment, and broadcast stations more apparent. The noise and hiss from these sources are more noticeable to the listener.

TONE CONTROLS

Tone controls are used to adjust the amount of high or low frequencies sent out from speakers. Most radios, TVs, and audio amplifiers have these controls. A common circuit is shown in Fig. 20-34.

Capacitor C_1 has a low reactance for high audio frequencies. The value of C_1 is usually .05 μF and R_1 is 50,000 ohms.

A tone switch with three control positions may also be used. The reactance of the capacitor for each position filters or removes the high frequencies. Fig. 20-35 shows the tone switch control system with three capacitors of different values.

ALIGNMENT

The superheterodyne receiver contains several tuned circuits. These include the primary and secondary windings of the IF transformers. These must be tuned to resonance or maximum response for those signals to be passed. Variable trimmer capacitors are connected in parallel with these coils. They provide the adjustment, or ALIGNMENT.

Receivers often need alignment after parts have been replaced in the service shop. The effects of age may be reduced by "peaking." This is aligning the receiver. Tools needed for peaking include the SIGNAL GENERATOR, Fig. 20-36, and an output indicator, such as an ac VOLTMETER.

A signal generator produces either a modulated or unmodulated RF wave. This may be selected by the controls on the panel. When an alignment job is needed, consult the technical manual for that radio or TV. Therefore, this discussion of alignment procedures is general. It refers only to the basic theory.

Refer to Fig. 20-37. The ac voltmeter is connected to the plate of the final power amplifier tube through a .1 μF 600 V capacitor. This is the indicating device. The signal generator is set up to produce a 455 kHz modulated RF wave. It is connected to the input grid of the converter tube through a .001 μF capacitor. The local oscillator must be disabled. This can be done by connecting a "jumper lead" between the rotor and stator plates of the oscillator tuning capacitor. Turn the attenuation control on the generator so that only a faint tone is heard in the speaker. Start with the output of the last IF transformer. Adjust the trimmer capacitor for peak reading on the voltmeter. Use a special alignment tool. Proceed toward the front end of the receiver.

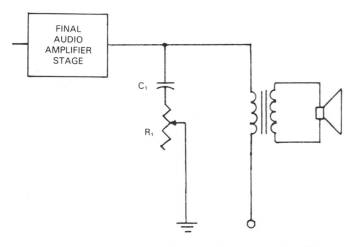

Fig. 20-34. Basic tone control circuit. It removes the higher frequencies from the speaker output.

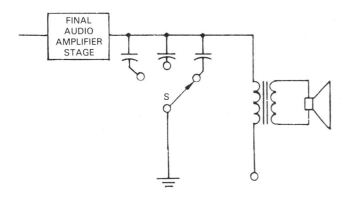

Fig. 20-35. This tone control has three positions. The tone is selected by the listener.

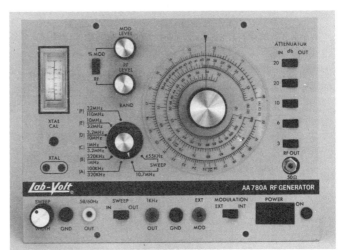

Fig. 20-36. A superheterodyne receiver can be aligned with this signal generator.

Electricity and Electronics

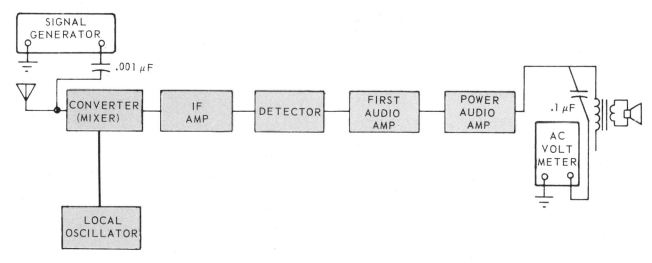

Fig. 20-37. This diagram shows only the signal path through a superheterodyne receiver. Connections for instruments are made for IF alignment.

Adjust, in order, the secondary and primary of all IF transformers for maximum output response. This completes the IF alignment.

The oscillator section of Fig. 20-37 contains a trimmer capacitor and a series padder capacitor, called the 600 padder. To adjust the oscillator, the generator is set at a modulated frequency of 600 kHz. It is attached to the antenna input terminals through a 250 pF capacitor. The receiver is tuned to 600 kHz and the 600 padder is adjusted for maximum response. Now the generator and receiver are tuned to the high frequency end of the band, around 1400 kHz. The trimmer capacitor is carefully adjusted for maximum response.

For complete instructions on alignment of a particular radio, check the manual. You will find antenna trimmer tuning and more precise oscillator adjustments.

REVIEW QUESTIONS FOR SECTION 20.3

1. What is a transducer?
2. Make a sketch showing the operating principles of the PM speaker.
3. Low frequency speakers are _____. High frequency speakers are _____.
4. Name the basic types of headphones.
5. What tools are needed to align a receiver?

20.4 FM RECEIVER

A block diagram of a complete FM receiver is shown in Fig. 20-38. Each block is labeled according to its function in the system.

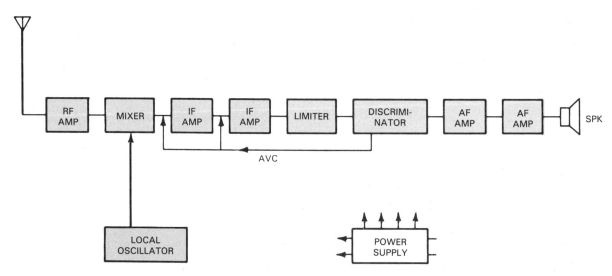

Fig. 20-38. The block diagram of a typical FM radio.

Radio Wave Receivers

The FM receiver is similar to the superheterodyne AM receiver, with three exceptions. The incoming signals to be tuned are from 88 to 108 megahertz. Secondly, the IF frequency used in the FM radio is 10.7 megahertz. The same heterodyne principles apply as with the AM receiver. Thirdly, the detection method an FM receiver uses is different.

FM DETECTION

In the AM radio, the detector is sensitive to amplitude variations. An FM detector must be sensitive to frequency variations and remove this intelligence from the FM wave. The FM detector must produce a varying amplitude and frequency audio signal from an FM wave.

Refer to Fig. 20-39. Assume that a circuit has a peak response at its resonant frequency. All frequencies, other than resonance, will have a lesser response. If the center frequency of an FM wave is on the SLOPE of the resonant response curve, a higher frequency will produce a higher response in voltage. A lower frequency will produce a lower voltage response. The curves in Fig. 20-39 reveal that the amplitude of the output wave is the result of the maximum deviation of the FM signal. The frequency of the audio output depends on the rate of FM signal frequency change.

Discrimination

The discriminator in Fig. 20-40 uses three tuned circuits. In this circuit, L_1C_1 is tuned to the center frequency. L_2C_2 is tuned to above center frequency. L_3C_3 is tuned to below center frequency by an equal amount.

At center frequency, equal voltages are developed across the tuned circuits. D_1 and D_2 conduct equally. The voltages across R_1 and R_2 are equal and opposite in polarity. The circuit output is zero. If the input frequency increases above center, L_2C_2 develops a higher voltage. Then D_1 conducts more than D_2 and unequal voltages develop across R_1 and R_2.

The difference between these voltage drops will be the audio signal. The output, therefore, is a voltage wave varying at the rate of frequency change at the input. Its amplitude depends upon the maximum deviation. The capacitors across the output of the discriminator filter out any remaining radio frequencies.

The discriminator in Fig. 20-41 is a circuit often encountered in FM receivers. L_1 and C_1 are tuned to center frequency. At frequency above resonance, the tuned circuit becomes more inductive. At frequencies below resonance, the circuit becomes more capacitive. The out of phase conditions produce voltages that determine which diode will conduct. The output is an audio wave. In advanced courses, you will study this type of detection in more detail.

Note that each diode in the discriminator must have equal conduction capabilities. This means that the semiconductor diodes used must be in MATCHED PAIRS.

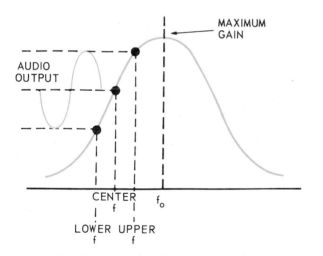

Fig. 20-39. These curves demonstrate slope detection.

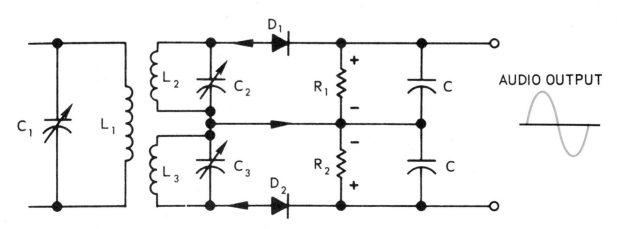

Fig. 20-40. An FM discriminator circuit using semiconductor diodes.

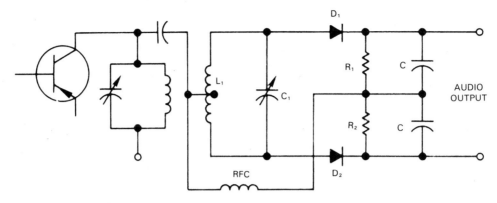

Fig. 20-41. Foster Seeley discriminator using a special transformer designed for this purpose.

Ratio detector

Another type of FM detector is drawn in Fig. 20-42. It is called the RATIO DETECTOR. The diodes are connected in series with the tuned circuit. At center frequencies, both diodes conduct during half-cycles. The voltage across R_1 and R_2 charges C_1 to output voltage. C_1 remains charged because the time constant of $C_1 R_1$ and R_2 is no longer than the period of the incoming waves. C_2 and C_3 also charge to the voltage of C_1. When both D_1 and D_2 are conducting equally, the charge of C_2 equals C_3. They form a voltage divider. At the center point between C_2 and C_3 the voltage is zero.

A frequency shift either below or above center frequency cause one diode to conduct more than the other. As a result, the voltages of C_2 and C_3 become unequal. But they will always total the voltage of C_1. This change of voltage at the junction of C_2 and C_3 is the result of the RATIO of the unequal division of charges between C_2 and C_3. This will vary at an audio rate the same as the rate of change of the FM signal.

Look at the charge on C_1 again. It is the result of the carrier wave amplitude or signal strength. It is charged by half-wave rectification of the FM signal. It is, therefore, a fine point to pick off an automatic volume control voltage to feedback to previous stages to regulate stage gain.

NOISE LIMITING

FM radio receivers are sensitive to and detect frequency variations, not amplitude variations. Most noise and interference in radio reception are amplitude variations. They are called NOISE SPIKES and have little effect on the FM detector. Therefore, FM reception is mostly free of noise and disturbances.

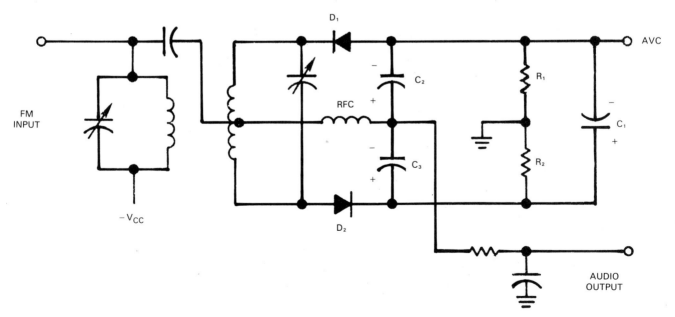

Fig. 20-42. A typical ratio detector circuit using semiconductor diodes.

Radio Wave Receivers

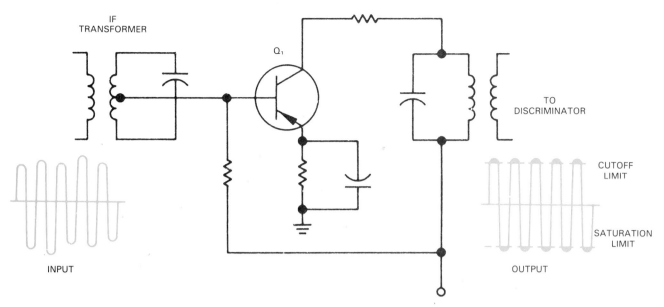

Fig. 20-43. A circuit from a limiter stage before a discriminator.

The FM signal is held at a constant amplitude before detection in a discriminator circuit using a limiter. A schematic of a transistorized LIMITER STAGE is shown in Fig. 20-43.

A limiter is an OVERDRIVEN amplifier stage. If the incoming signal reaches a certain amplitude in voltage, it drives the transistor to cutoff or saturation when the voltage is opposite in polarity. At either of these points, gain cannot increase. The output is confined within these limits. Any noise spikes would be clipped off.

REVIEW QUESTIONS FOR SECTION 20.4

1. What is the difference between the detection methods in an AM receiver and an FM receiver?
2. Draw the block diagram of an FM receiver.
3. Most interferences in radio reception are amplitude variations called _____ _____.
4. The _____ circuit in an FM receiver does not allow signals to go over a certain amplitude after they leave the IF transformer and reach the discriminator circuit.

20.5 RADIO RECEIVER PROJECTS

CRYSTAL RADIO

Crystal radios are often the first electronic project built at home or school. Figs. 20-44 and 20-45 show some of the principles we have studied, such as tuning and detection.

Fig. 20-44. A crystal radio that uses commercially made parts.

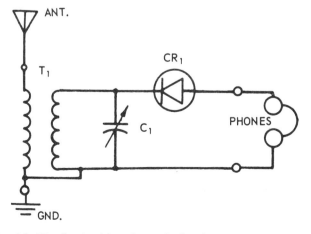

Fig. 20-45. Study this schematic for the crystal radio.

Commercial parts are used in this project, Fig. 20-46. Its construction introduces more complex circuits. Being familiar with these parts will pay off in more advanced projects. The radio may be built on a plastic sheet or a wooden base. It will require a good antenna for proper operation.

FOUR-TRANSISTOR RADIO

A four-transistor radio has good reception on local stations, Fig. 20-47. It is not very costly to build. A schematic for the radio is shown in Fig. 20-48. The parts list is given in Fig. 20-49.

Station selection is done by tuned circuit L_1C_1. The signal is detected by crystal CR_1 and amplified by four transistor stages to drive the loudspeaker and to increase the sensitivity of the detector. The grounded-emitter configuration is used by Q_2, Q_3, and Q_4. Interstage coupling is provided by C_4, C_5, and C_6.

Refer to Fig. 20-47. Notice the use of tiny electrolytic capacitors. The transistor tuning capacitor C_1 and midget speaker LS_1 permit a compact package. All parts have been mounted on a sheet of 1/16 inch plastic, but printed circuitry may be used.

It is best to build the radio first. Use parts mounted on a breadboard. After tests have been done, design a case to hold the radio.

EIGHT-TRANSISTOR SUPERHETERODYNE RADIO

This superheterodyne receiver receives AM broadcast stations transmitting on frequencies between 550 and 1600 kHz. The RF amplifier-converter stage receives the modulated RF signal from the broadcast stations in your area. By tuning the radio, you select the modulated RF signal from one of the broadcast stations. In this stage, it is amplified and combined with the RF signal from the local oscillator in your radio to form an intermediate frequency signal. The IF signal is then amplified in the IF amplifier stage and passed on to the detector stage. The detector stage removes the carrier frequency. This leaves only the AF signal that was originally combined with the carrier frequency at the broadcast station. This is demodulation. This small AF signal is amplified by the first AF amplifier stage. Then by the audio power amplifier stage. It is now strong enough to operate the speaker.

The circuit used in this eight-transistor radio is from Graymark Enterprises. It may be purchased as their Model 536 eight-transistor radio.

The schematic for the radio is shown in Fig. 20-50. The parts list appears as Fig. 20-51.

Order Model 536 from Graymark Enterprises, Inc., P.O. Box 54343, Los Angeles, California 90054. For individual parts, quote part numbers from Fig. 20-51.

T_1 — Antenna Coil: Miller 20A
C_1 — Variable Capacitor 365 pF, Philmore 1946 G
CR_1 — 1N34 Crystal Diode or RCA SK 3087
Headphones 2000 Ω
Binding Posts or Terminals

Fig. 20-46. Parts list for the crystal radio.

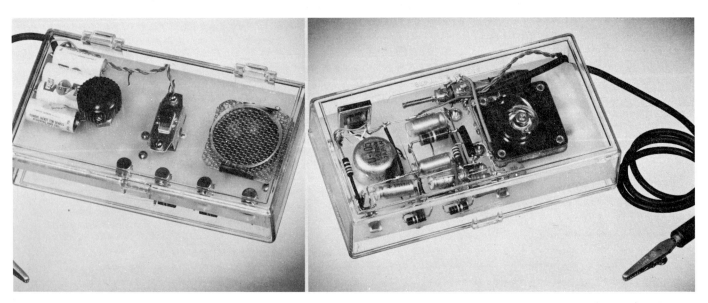

Fig. 20-47. Two views of the four-transistor radio. Left. A pocket size version with a loudspeaker. Right. A rear view of the four-transistor radio showing parts placement.

Radio Wave Receivers

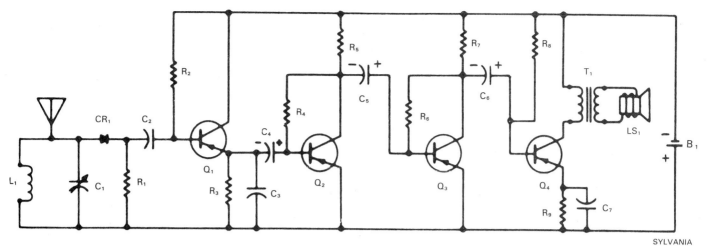

Fig. 20-48. The schematic diagram of the four-transistor radio.

B_1 — 6 volt battery
C_1 — 0.365 pF tuning capacitor
C_2, C_3 — 0.02 μF ceramic disc or paper
C_4, C_5, C_6 — 1.0 μF electrolytic, 15V
C_7 — 25.0 μF electrolytic, 15V
CR_1 — Sylvania 1N64 or 1N34 diode or RCA SK 3087
L_1 — Ferri-loopstick antenna coil

LS_1 — 3-6 ohm loudspeaker
Q_1, Q_2, Q_3, Q_4 — Sylvania 2N 1265, RCA SK 3003
R_1, R_8 — 470 K, 1/2W
R_2, R_4, R_6 — 220 K, 1/2W
R_3, R_5, R_7 — 2.2 K, 1/2W
R_9 — 100 Ω, 1W
T_1 — Argonne AR-133 (pri. = 10 K Ω, sec. = 3.2 Ω, 100 mW)

Fig. 20-49. Parts list for the four-transistor radio.

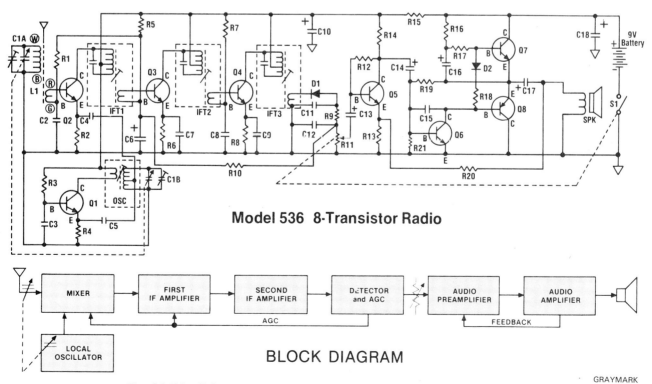

Fig. 20-50. Schematic and block diagrams for the eight-transistor radio.

Electricity and Electronics

QTY	SYMBOL	PART NO.	DESCRIPTION
2	R1, 3	62390	Resistor, 1/4 W, 5%, 470 kΩ
1	R2	62603	Resistor, 1/4 W, 5%, 1.5 k
1	R4	61409	Resistor, 1/4 W, 5%, 5.6 k
1	R5	62608	Resistor, 1/4 W, 5%, 56 k
1	R6	62681	Resistor, 1/4 W, 5%, 1.2 k
2	R7, 12	62235	Resistor, 1/4 W, 5%, 560 k
3	R8, 17, 20	61399	Resistor, 1/4 W, 5%, 1 k
1	R9	62688	Resistor, 1/4 W, 5%, 680 Ω
1	R10	61408	Resistor, 1/4 W, 5%, 27 k
1	R13	62392	Resistor, 1/4 W, 5%, 10 Ω
1	R14	61413	Resistor, 1/4 W, 5%, 3.9 k
1	R15	62689	Resistor, 1/4 W, 5%, 150 Ω
2	R16, 18	62150	Resistor, 1/4 W, 5%, 82 Ω
1	R19	62690	Resistor, 1/4 W, 5%, 12 k
1	R21	62691	Resistor, 1/4 W, 5%, 2.7 k
1	R11	62713	Potentiometer, 5 k, with switch
1	C1	62714	Capacitor, Tuning
3	C2, 3, 5	62207	Capacitor, ceramic, .01 mfd, marked '103'
1	C4	62686	Capacitor, ceramic, .005 mfd, marked '472' or '502'
1	C6	62238	Capacitor, electrolytic, 10 mfd, 10 volt or 16 volt
5	C7, 8, 9, 11, 12	62685	Capacitor, ceramic, .02 mfd, marked '203' or '223'
3	C10, 17, 18	62692	Capacitor, electrolytic, 100 mfd, 10 volt
1	C13	62693	Capacitor, electrolytic, .47 mfd, 50 volt
1	C14	62694	Capacitor, electrolytic, 4.7 mfd, 25 volt
1	C15	61247	Capacitor, ceramic, .001 mfd, marked '102'
1	C16	62695	Capacitor, electrolytic, 47 mfd, 10 volt
6	Q1, 2, 3, 4, 6, 7	63037	Transistor, 2SC1815Y, Silicon, NPN
1	Q5	63038	Transistor, 2SC1815GR, Silicon, NPN
1	Q8	62676	Transistor, 2SA1015Y, Silicon, PNP
1	D1	61517	Diode, 1N60
1	D2	62700	Diode
1	L1	62701	Antenna core with coil
1		62704	Antenna holder (mounted to Printed Circuit Board)
2	OSC	62715	Coil, oscillator, red core
2	IFT1	62716	Transformer, IF, yellow core
2	IFT2	62717	Transformer, IF, white core
2	IFT3	62718	Transformer, IF, black core
1	SPK	62719	Speaker, 8 Ω
1		62403	Battery connector
1		62702	Printed Circuit Board
1		62703	Cabinet, plastic
1		62705	Knob, tuning
1		62721	Knob, volume
2		62711	Screw, machine, flat head, 2.6 x 3 mm
1		63047	Screw, machine, flat head, 2.6 x 6 mm
1		61266	Screw, machine, 1.7 x 3 mm
2		62712	Screw, machine, 2 x 5 mm
4		62683	Screw, self-tapping, 3 x 8 mm
2		62191	Screw, self-tapping, 2.6 x 5 mm
2		61772	Nut, machine, 2 mm
4		62706	Printed Circuit Board, IFT Breadboard
2		62722	Speaker holder
3		62733	Sponge, double-sided adhesive type, 15 x 10 mm
55		61357	Solder pins
1		62707	Solder, rosin core, 600 mm
1		62709	Wire, bare, solid, 22 gauge, 1100 mm
1		62710	Wire stranded, black, 26 gauge, 1100 mm
1		62708	Breadboard
1			Warranty Card
1		62725	Instruction Manual

GRAYMARK

Fig. 20-51. Parts list for the eight-transistor radio.

Radio Wave Receivers

SUMMARY

1. The tuned radio frequency (TRF) receiver picks up a transmitted RF wave, amplifies it, detects or demodulates it, and amplifies the audio wave.
2. The superheterodyne receiver converts the tuned RF signal to an intermediate frequency (IF) signal so that this signal can be further amplified and refined.
3. High frequency rectification of the RF wave is called detection.
4. Selectivity is the ability of a receiver to select a single frequency and reject all others.
5. Sensitivity is the ability of a receiver to respond to weak incoming signals.
6. Some common intermediate frequency signals are 455 kHz and 10.7 MHz.
7. Transducers are devices that convert one form of energy to another. Some common transducers are microphones and loudspeakers.
8. Radios must be properly aligned in order for signals to be properly tuned and passed through the various stages of the receiver.
9. The high frequency circuit used to separate the audio wave is the discriminator or the ratio detector.

TEST YOUR KNOWLEDGE, Chapter 20

Please do not write in the text. Place your answers on a separate sheet of paper.

1. The process of removing the audio wave from the modulated RF wave is:
   a. Demodulation.
   b. Detection.
   c. Both of the above.
   d. None of the above.
2. A _____ _____ circuit changes the resonant frequency range of a circuit.
3. What two characteristics determine the quality of a tuning circuit?
4. The RF amplifier:
   a. Is the first stage that the received signal is fed to from the antenna.
   b. Provides amplification to the incoming signal.
   c. Improves the selectivity of a receiver.
   d. All of the above.
5. What is the purpose of detection?
6. Sketch a block diagram for a superheterodyne receiver.
7. What is the advantage of a separate local oscillator attached to a mixer?
8. What are the standard frequencies for the following receivers?
   a. AM receiver.
   b. FM receiver.
   c. TV receiver.
9. _____ _____ _____ automatically adjusts all volume levels to a constant level.
10. Explain the operation of a PM speaker.
11. What is the purpose of a tone control?
12. What is the purpose of alignment?
13. Name two types of FM detectors.

FOR DISCUSSION

1. How does the tuning circuit select a desired frequency?
2. Explain the process of heterodyning signals.
3. Draw the block diagram of a communication model and explain the function of each stage.
4. How does an AM receiver differ from an FM receiver?
5. Discuss the advantages and disadvantages of AM and FM.

Electricity and Electronics

This camcorder operates in much the same way as a full-sized television camera. The camcorder signal is sent to a videotape rather than through the airwaves.

The principle of television is the breaking of the picture into many dots. Linear amplifier circuits can easily handle the resulting signal. Picture ("video") signals can have frequencies as high as 3.5 MHz. The superheterodyne principle is used, just as in a radio receiver. The video information is sent by AM and the audio information is sent by FM. Some sets have a separate carrier of 38 kHz for stereo sound.

Chapter 21
TELEVISION

After studying this chapter, you will be able to:
- *Explain the steps in the transmission of a television signal.*
- *Discuss the scanning process.*
- *Identify circuits in both black and white and color television receivers and explain their functions.*
- *Discuss a variety of television innovations.*

In just a few years, television has grown from infancy to a giant in the field of communications. Entertainment, education, information, and advertising are available to millions of people. Through satellite links, the effect of television is to bring the entire world closer together.

This text will not address the detailed electronic circuits involved in the production, transmission, and reception of television signals. However, it is important for students interested in the science of electronics to have a basic knowledge of television and related information.

21.1 TELEVISION SIGNALS

TELEVISION CAMERAS

A picture is really an infinite number of dots. In a black and white photo, these dots are varying degrees of black and white. These dots are called PICTURE ELEMENTS. Look at a photo in a newspaper. These elements are clearly visible.

As a television camera looks at a scene in a studio, these picture elements appear to it. The scene is focused on a photosensitive mosaic in the camera. It consists of many photoelectric cells. Each cell responds to the scene by producing a voltage in balance with the strength of the light. These voltages are amplified and used for modulation of the AM carrier wave. This wave is transmitted to the home receiver.

A line drawing of an image orthicon is shown in Fig. 21-1. The scene in front of the camera focuses on the photo cathode through a standard camera lens system.

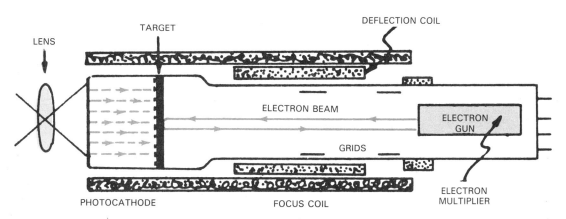

Fig. 21-1. This sketch shows the interior arrangement of the image orthicon tube used in television cameras.

349

The varying degrees of light cause electrons to be emitted on the target side of the cathode. These form an electronic image of the scene. The target plate is operated at a high positive potential. The electrons from the cathode are attracted to the target. The target is made of low resistance glass and has a transparency effect. The electron image appears on both sides of the target plate.

At the right in Fig. 21-1 is the electron gun. It produces a stream of electrons. The speed of the stream is increased by the grids. The beam scans left to right and top to bottom. This is done by the magnetic deflection coils around the tube. The moving electron beam strikes the target plate and the electrons return to the electron multiplier section. The strength of the electron stream returning to the multiplier is balanced with the electron image. It provides the desired signal current for electronic picture reproduction.

SCANNING

Scanning is the point-to-point examination of a picture. In the camera the electron beam scans the electron image. It responds to the point-to-point brilliance of the picture.

The scanning system used in the United States is the interlace system. It consists of 525 scanning lines. The beam starts at the top left hand corner of the picture. It scans, from left to right, the odd numbered lines of the scanning pattern (lines 1, 3, 5, 7, etc.). At the completion of 262 1/2 lines, the electron beam is returned to the starting point by vertical deflection coils. The beam then scans the even numbered lines.

One scan of 262 1/2 lines represents a FIELD. One set of the odd and even numbered fields represents a FRAME. A portion of the interlace scanning pattern is shown in Fig. 21-2.

The frame frequency has been set by the FCC at 30 Hertz. This means your TV receives 30 complete frames per second, or 60 picture fields per second of alternate odd and even lines. This appears as a constant, nonflickering picture to your vision.

In order to produce 525 lines per frame at a frame frequency of 30 Hz, the horizontal deflection oscillator must work at a frequency of 15,750 Hz (525 x 30). The horizontal deflection oscillator causes the beam to move from left to right.

The vertical deflection oscillator must have a frequency of 60 cps. This is a field frequency. The vertical deflection oscillator causes the beam to move from top to bottom.

Closer study of the scanning process reveals that the beam scans as it moves from left to right. After it has read one line, it quickly retraces to the left and starts reading the next line, just the way you are reading this book. The retrace time is very rapid, but still shows a line in the picture. Therefore, the picture must be black during this retrace time. Also, when the beam reaches the bottom of the picture, the beam must be returned to the top to scan again. The picture must also be black during vertical retrace or trace lines would be visible.

The oscillators that make the scanning and retrace voltages for both the horizontal and vertical sweep must produce a waveform as shown in Fig. 21-3. This is a sawtooth waveform. Notice the gradual increase in voltage during the sweep and the rapid decrease during retrace. These voltages are applied to coils that surround the picture tube. These coils are called the DEFLECTION YOKE. An increase in the strength of the magnetic fields in the coils causes the electron beam to move. See Fig. 21-4.

Scanning at the studio and scanning on a television must be in step. In order to do this, a pulse generator triggers the horizontal and vertical oscillators at the studio. This same pulse is also sent over the air and received by the television set. This pulse, known as the SYNCHRONIZATION PULSE or SYNC PULSE, triggers the oscillators in the receiver and keeps them at exactly the same frequency. The HORIZONTAL

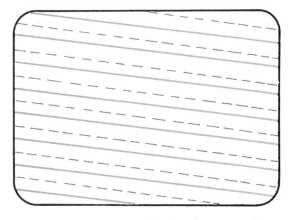

Fig. 21-2. This is a portion of the interlace scanning system used in television. There are actually 525 lines.

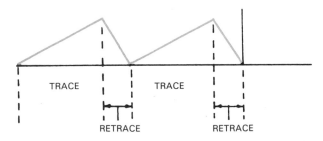

Fig. 21-3. The waveform of the voltages required for scanning and retrace. It is called a sawtooth waveform.

Television

HOLD and VERTICAL HOLD controls on your TV are used to make slight, adjustments so that the oscillators can lock in with the sync pulses.

COMPOSITE VIDEO SIGNALS

The television signal received by TVs contains picture (video) and sound (aural) information. The video information is an AM signal. The audio information is an FM signal. An example of this process is shown in Fig. 21-5.

All video signals are formed in the same way. In that way, a television may be used in any area. These standards are set up by the FCC and are used by all TV broadcasting stations.

The amplitude of the modulation is divided into two parts. The first 75 percent is used to transmit video information. The remaining 25 percent is for the sync pulses. Also, a system of negative transmission has been adopted. This means that the higher amplitudes of video information produce darker areas in the picture. At 75 percent the picture is completely BLACK. In Fig. 21-5 the percentage is shown on the left.

As the beam scans the next black bar, a similar action takes place. At the end of the line, the screen is driven back to the BLACK PEDESTAL or BLANKING LEVEL. During this blanking pulse, beam flyback occurs and a sync pulse is sent in the blacker-than-black or infrablack region (upper 25 percent) for oscillator synchronization. A second line to be scanned would be an exact copy of the first unless the picture is changed. At the bottom of the picture a series of pulses trigger the vertical oscillator and keep it synchronized.

Fig. 21-6 shows a composite (complete) video signal. The video information for one line between the blanking pulses is varying degrees of black and white.

Compare the two video signals in Fig. 21-7. Signal A is made up mostly of bright objects. The average overall brightness of the scene is another form of the information sent to a TV receiver from the transmitter. It may be detected from the composite video signal.

Fig. 21-4. A deflection yoke fits around the neck of the television picture tube.

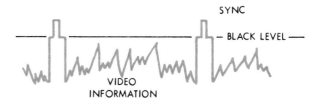

Fig. 21-6. The composite video signal of one line scanned by the television camera.

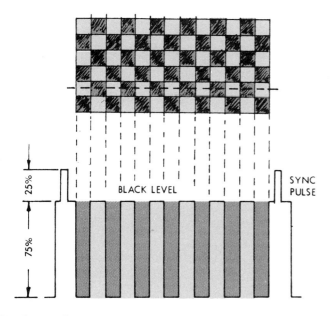

Fig. 21-5. The voltages developed as the camera scans one line across a checkerboard. The sync pulses are transmitted at the end of each line.

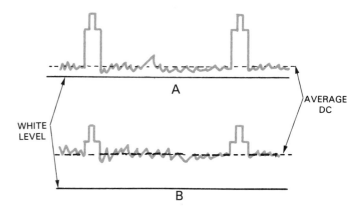

Fig. 21-7. A comparison of a dark and a light picture as they appear in the video signal. A—Light signal. B—Dark signal.

BASIC CATHODE RAY TUBE CONTROLS

The cathode ray tube (CRT) is used to produce images in most television sets. The CRT was discussed in Chapter 13. Refer back to that chapter for review.

Recall the theory of vacuum tube operation. The control grid determines the flow of electrons through the tube. In the CRT this is also true, Fig. 21-8. At zero bias the CRT is at maximum current. The screen, therefore, is bright or white. At cut-off bias, the current is zero and the screen is black. The tube is operated at a selected bias on the control grid. This bias may be controlled by the knob on the TV called the BRIGHTNESS CONTROL.

When no picture is being received on the TV, the scanning electron beam can be seen in the form of lines on the TV screen. This is called the RASTER. Turn a TV to a vacant channel and observe this raster. Now adjust the brightness control from black to bright. The incoming, detected video signal is applied to the grid of the CRT (sometimes to the cathode, depending on polarity of signal). The video signal adds to or subtracts from the bias on the tube. This results in a modulated electron stream that conforms to the picture information in the video signal. The picture is produced on the fluorescent screen.

The sharpness or focus of the electron beam may be adjusted by changing the voltages of the focusing grids. These controls are often found on the rear of the TV. They require adjustments from time to time. Refer to Fig. 21-8 again. Find the centering magnet. Slight adjustments on this magnet will correct a picture that is off center.

NEGATIVE ION EMISSIONS

When the cathode in the cathode ray tube emits electrons, negative ions are also emitted. Since the negative ions have a greater mass than the electrons, they are not deflected as much. This action could cause an ion bombardment in the center of the picture tube. This is called an ION SPOT (brown spot). It is about one inch in diameter.

Many studies have been conducted to find a solution to this problem. One method uses an aluminized picture tube. The tube has an aluminum film on the electron gun side of the screen. This film protects the screen from ion bombardment. It also improves the sharpness of the picture.

REVIEW QUESTIONS FOR SECTION 21.1

1. Briefly explain how a television camera works.
2. _____ is the point-to-point examination of a picture.
3. One scan of 262 1/2 lines represents a _____.
4. What is a deflection yoke and what is its purpose?
5. The _____ _____ triggers the oscillators in the receiver and keeps them at exactly the same frequency.
6. What is a raster?

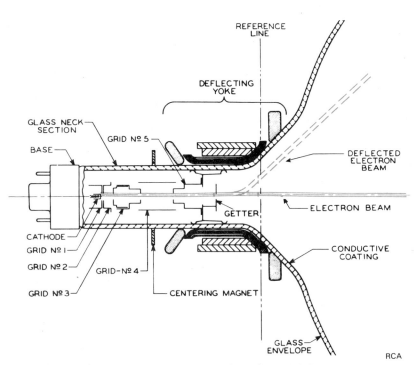

Fig. 21-8. Study this sketch of a CRT. Note the electron gun.

21.2 TELEVISION RECEIVERS

BLACK AND WHITE TV RECEIVER

Fig. 21-9 shows the parts of the television receiver. This block diagram shows the links between the parts of the television. Trace the signal path through the stages. The purpose of each group of components will be apparent. The name of each block reflects is purpose in the circuit.

The RF AMPLIFIER serves a function similar to that in the superheterodyne radio. The incoming television signal is chosen by switching fixed inductors into the tuning circuit. These tuned circuits provide constant gain and selectivity for each television channel. In this stage the video signal, with all its information, is amplified and fed to the mixer.

In the MIXER stage, the incoming video signal is mixed with the signal from a local oscillator to produce an intermediate frequency. The commonly used IF is 45.75 megahertz. When the channel selector switch is turned on to a channel, the tuning circuit is changed. The frequency of the oscillator is also changed, always producing the IF of the correct frequency. A FINE TUNING changes the frequency of the oscillator slightly in order to provide the best response.

The RF amp, mixer, and oscillator are combined in one unit. It is called the TUNER or FRONT END of a television. These units are usually put together in the factory. Adjustments should not be made on these units unless you have the correct instruments and thorough knowledge of procedures.

The PIX-IF AMPLIFIERS amplify the output of the mixer stage: the 45.75 megahertz intermediate frequency, the video and aural information. To provide maximum frequency response for each stage up to 45.75 MHz, each stage must amplify a broad band of frequencies. The voltage gain of each stage is reduced. More stages of IF amplification are required. In this system, the sound is passed through the IF amplifier with the video signal. It is called the intercarrier system.

The output from the last IF stage is fed to the VIDEO DETECTOR or DEMODULATOR. The detection process is the same as in the radio. The video signal used to amplitude modulate the transmitted carrier wave is separated and fed to the next stage.

In the VIDEO AMPLIFIER stages, the demodulated video signal is amplified and fed to the grid (sometimes cathode) of the CRT. This signal modulates or varies the strength of the electron beam and produces the picture on the screen.

The FM sound signal is amplified in the SOUND IF AMPLIFIER. Later in this chapter you will learn that the FM sound of the television program is separated from the video carrier wave by 4.5 megahertz. This produces a 4.5 MHz FM signal at the output of the video detector, which is coupled to the sound IF amplifiers.

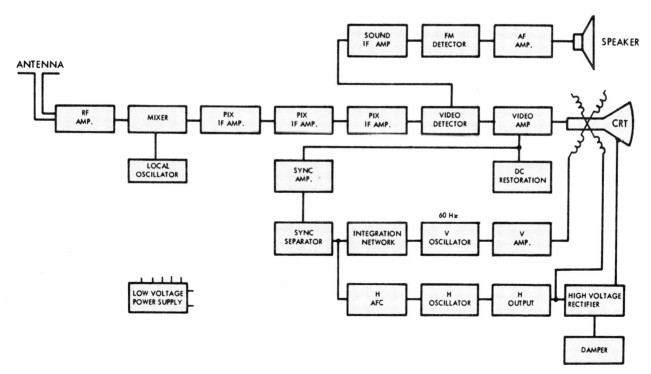

Fig. 21-9. A block diagram of a TV receiver.

The FM AUDIO DETECTOR detects the frequency variations in the modulated signal and converts them to an audio signal.

AF AMPLIFIERS are the same as those used in the conventional radio or audio system. The audio signal is amplified enough to drive the power amplifier and the loudspeaker.

The output of the video amplifier is fed to the SYNC SEPARATOR. This circuit removes the horizontal and vertical sync pulses that were transmitted as part of the composite video signal. These sync pulses trigger the horizontal and vertical oscillators and keep them "in step" with the television camera.

The SYNC AMPLIFIER is a voltage amplifier stage that increases the sync pulses.

In the HORIZONTAL AFC, the horizontal oscillator frequency is compared to the sync pulse frequency. If they are not the same, voltages are developed that change the horizontal oscillator to the same frequency.

The HORIZONTAL OSCILLATOR operates on a frequency of 15,750 Hz. It provides the sawtooth waveform needed for horizontal scanning.

The HORIZONTAL OUTPUT stage correctly shapes the sawtooth waveform for the horizontal deflection coils. It also drives the horizontal deflection coils and provides power for the high voltage rectifier.

The output of the horizontal oscillator shocks the HORIZONTAL OUTPUT TRANSFORMER (HOT). The high ac voltage developed by this autotransformer is rectified by the HIGH VOLTAGE RECTIFIER and is filtered for the anode in the CRT. See Fig. 21-10.

The DAMPER stage dampens out oscillations in the deflection yoke after retrace. It also rectifies part of the ac for a boosted B+ VOLTAGE. Part of this damper voltage is used in the horizontal scanning as reaction scanning.

The output of the sync amplifier is fed through a vertical intergration network to the VERTICAL OSCILLATOR. The integrator is an RC circuit designed with a time constant. The integrator builds up a voltage when a series or groups of closely spaced sync pulses occur.

The transmitted vertical pulse is a wide pulse. It occurs at a frequency of 60 Hz. The vertical oscillator is triggered by this pulse and, therefore, keeps on frequency.

The output of the vertical oscillator provides the sawtooth voltage to the deflection coils. The deflection coils move the beam from the top to the bottom of the screen and the flyback from bottom to top, at the end of each field.

The VERTICAL OUTPUT amplifier is used by the output of the oscillator to provide the proper currents in the deflection yoke for vertical scanning.

Special purpose circuits

The basic television circuit has been improved in many ways. Automatic controls have been designed for many functions which would be tiresome for the TV viewer to manage. A few of these improvements are mentioned in the following paragraphs.

DC RESTORATION. The average darkness or brightness of a TV picture is transmitted as the average value of the dc component of the video signal. If the video amplifiers are RC coupled, the dc value of the signal is lost. In this case, the average value of the video signal is taken from the detector and used to set the bias on the CRT.

AUTOMATIC GAIN CONTROL. The AGC serves the same function as the AVC in the radio receiver. Its purpose is to provide a fairly constant output from the detector by varying the gain of the amplifiers in previous stages. This is done by rectifying the video signal to produce a negative voltage. This voltage is applied to the bias of the previous amplifiers to change their gain.

COLOR TV RECEIVER

Color television was developed in the late 1940s. The system currently used in the United States was pioneered by RCA Laboratories. In March, 1950, a color television demonstration was given in Washington, DC to FCC personnel, reporters and other interested people. As a result of this demonstration, color television development was launched.

TRIAD

Fig. 21-10. A horizontal output transformer (HOT).

Television

An invention that made color television possible was the SHADOW MASK PICTURE TUBE, Fig. 21-11.

The three basic colors used in color television are red, blue, and green, Fig. 21-12. By combining these colors, any color can be produced on the screen.

The first color picture tube produced for retail sale was the DELTA-TYPE tube. It was invented in 1950. It was basically the same tube used in the color television demonstrations given by RCA Laboratories in Washington, DC.

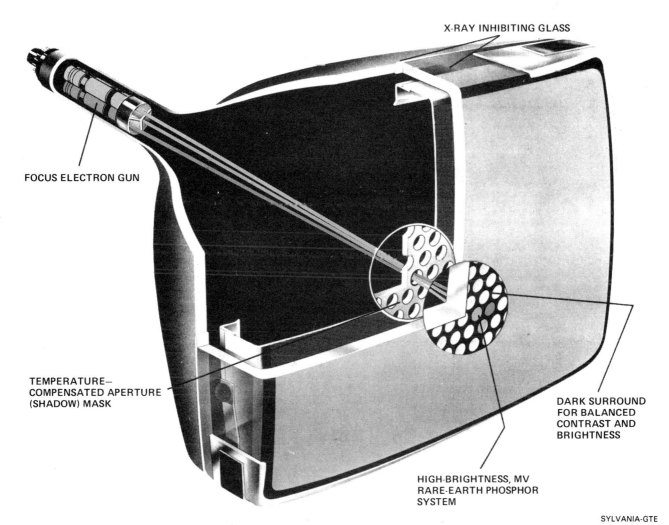

Fig. 21-11. Study this color picture tube.

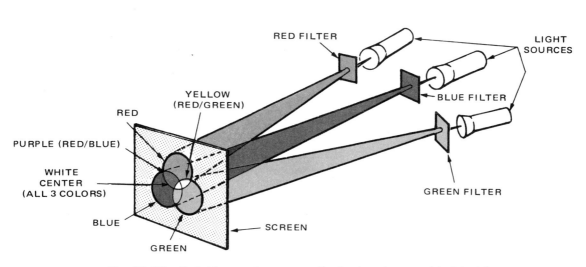

Fig. 21-12. Red, blue, and green are the basic colors used in television.

The delta-type tube uses three electron guns placed in the neck assembly of the picture tube. The electrons are emitted by the three guns toward the screen. The screen is filled with hundreds of thousands of dots containing the colors red, blue, and green. In between the electron gun and the color producing screen, the three electron beams are focused through an aperture or shadow mask, Fig. 21-13. This shadow mask assures that the electron beams strike the dots properly.

A line is made by one complete scan (from left to right) of all three electron beams hitting all the dots across the screen. If all electron beams are adjusted properly, the result will be a white line. A color line can be made by mixing the electron beams.

IN-LINE gun assemblies were invented after the shadow mask tube. Fig. 21-14 shows four in-line gun assemblies with different aperture grill and screen patterns.

A color television receiver is a very complex instrument. It has to be able to produce both color and black and white pictures. The incoming signal must work with black and white sets also. Fig. 21-15 shows a block diagram of a color television receiver.

THE TELEVISION CHANNEL

The FCC has assigned a channel of the radio frequency spectrum for each television channel. There are two types of television channels: very high frequency (VHF) and ultra high frequency (UHF). Each channel is 6 MHz wide.

The VHF channels, 2 to 13, are shown in Fig. 21-16. Examine channel 4, Fig. 21-17.

The basic video carrier frequency is 67.25 MHz. Recall that when an RF carrier wave is amplitude modulated, sideband frequencies appear. These stand for the sum and difference between the carrier frequency and the modulating frequencies.

To send a very clear, sharp picture, frequencies of at least 4 MHz are needed for modulation. These frequencies combine in channel 4 for a band occupancy of 63.25 MHz (67.25 − 4 MHz) to 71.25 MHz (67.25 + 4 MHz). It is a total channel width of 8 MHz. But the FCC allows only 6 MHz. A compromise must be made.

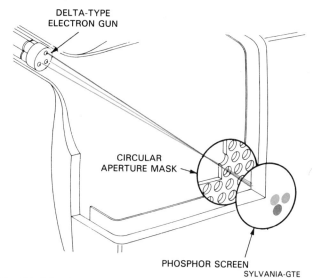

Fig. 21-13. A delta-type gun, shadow mask, and tri-dot screen arrangement.

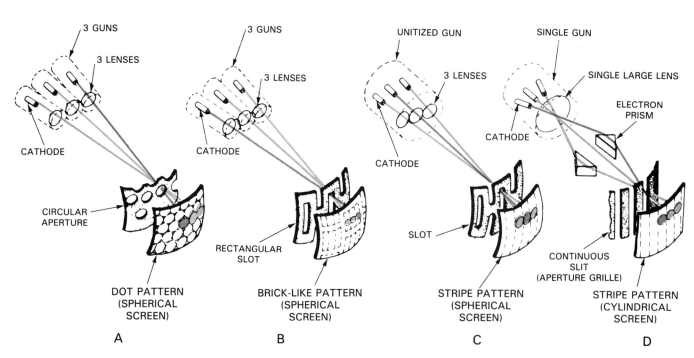

Fig. 21-14. In-line color tubes. A—Three gun with circular aperture. B—Three gun with rectangular aperture. C—Unitized gun with slot aperture. D—Sony Trinitron with one gun and grille aperture.

Television

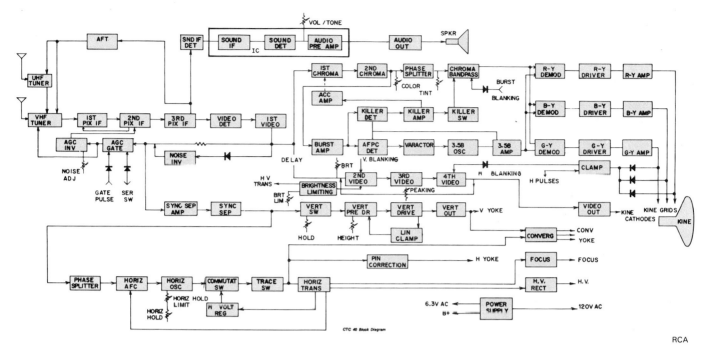

Fig. 21-15. The RCA CTC-40 color receiver.

CHANNEL NUMBER	FREQUENCY BAND (MHz)	VIDEO CARRIER FREQUENCY	AURAL CARRIER FREQUENCY
2	54-60	55.25	59.75
3	60-66	61.25	65.75
4	66-72	67.25	71.75
5	76-82	77.25	81.75
6	82-88	83.25	87.75
7	174-180	175.25	179.75
8	180-186	181.25	185.75
9	186-192	187.25	191.75
10	192-198	193.25	197.75
11	198-204	199.25	203.75
12	204-210	205.25	209.75
13	210-216	211.25	215.75

Fig. 21-16. VHF frequency assignments.

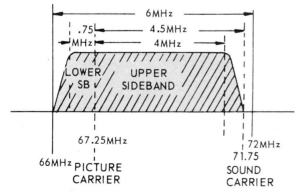

Fig. 21-17. Shown are the locations of the picture carrier and sound for channel 4, VHF. The channel is 6 MHz wide.

In commercial TV broadcasting the upper sideband is transmitted without attenuation. The lower sideband is partly removed by a vestigial-sideband filter at the transmitter. The curve in Fig. 21-17 shows the basic response traits of the TV transmitter. The sound is transmitted as a frequency-modulated signal at a center frequency 4.5 MHz above the video carrier. In channel 4 the sound is at 71.75 MHz.

The ultra high frequency (UHF) television band covers from channel 14 to 83. As in VHF, the bandwidth of each channel is 6 MHz. The same frequencies bandwidths are used for the picture carrier as VHF.

The UHF channels used for commercial TV are set by the FCC. See Fig. 21-18.

CHANNEL NUMBER	FREQUENCY BAND (MHz)
14	470-476
15	476-482
16	482-488
17	488-494
18	494-500
19	500-506
•	
•	
•	
•	
83	884-890

Fig. 21-18. Examples of some ultra high frequency (UHF) channels and where they are located in the frequency band.

Fig. 21-19. A VHS format video recorder with remote control.

GOLDSTAR

REVIEW QUESTIONS FOR SECTION 21.2

1. What is the purpose of an RF amplifer in a television receiver?
2. The output of the mixer stage is amplified by the:
   a. PIX-IF amplifier.
   b. Video amplifier.
   c. AF amplifier.
   d. None of the above.
3. The _____ _____ _____ uses the horizontal oscillator output to shock the horizontal output transformer.
4. What is the purpose of a shadow mask?
5. A Sony Trinitron color tube uses a _____ aperture.
6. Each TV channel bandwidth is:
   a. 6 Hz.
   b. 60 MHz.
   c. 6 MHz.
   d. None of the above.

21.3 TELEVISION INNOVATIONS

VIDEO CASSETTE RECORDERS

Video cassette recorders (VCRs) are very popular and useful. VCRs can be used to play rented videotapes. Or they can be used to show your own movies made with video recorders. There are two types of tape formats: VHS and Beta.

Most VCRs have two HEADS. A video head is a tiny electromagnet that reads information from the recorded tape during playback. It writes information onto the tape during recording. More costly VCRs can have up to seven heads. These extra heads are used for better sound and picture quality. Many VCRs also come with a remote control, Fig. 21-19.

ADVENT

SONY ELECTRONICS CORP.

Fig. 21-20. Top. The three gun projection cabinet used to project a large screen image. Bottom. With new projection TVs, the electron gun assembly is contained in the cabinet.

LARGE SCREEN PROJECTION TV

Most large screen projection TVs use a special electron gun assembly that projects three separate images onto a curved screen, Fig. 21-20. Fig. 21-21 shows how the electron gun is assembled.

Television

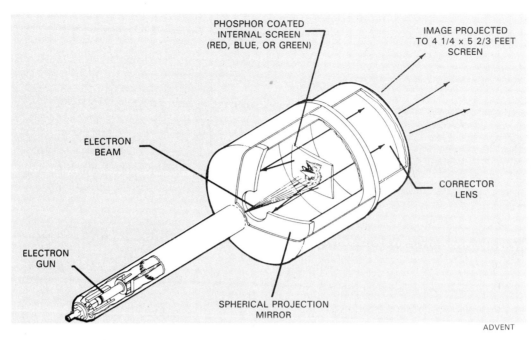

Fig. 21-21. Projection tube for a large screen television.

SATELLITE TV

Arthur C. Clarke first introduced the idea of launching satellites to improve communications. He did this in an article in the Fall, 1945 issue of "Wireless World" magazine. He stated that if satellites could be launched high enough (35,880 kilometers or 22,300 miles) above the equator, they would be in geostationary orbit. GEOSTATIONARY ORBIT means an object rotates with the earth.

The first communications satellite, Telestar I, was launched by the National Aeronautics and Space Adminstration (NASA) in 1962. It was a small, experimental satellite which only operated a few hours each day. This satellite made communication between the United States and Europe possible. In April, 1965, NASA launched the first commercial satellite, Early Bird. This satellite was owned by the International Telecommunications Satellite Organization (INTELSAT), a group created in 1964. Now there are many satellites in orbit allowing for television, telephone, radio, data, and other communications messages.

Rockets place satellites into space. There they are deployed and the circuits are activated, Fig. 21-22. Fig. 21-23 shows an SBS communications satellite now in orbit. This satellite was designed to provide voice, video teleconferencing, data, and electronic mail services to U.S. businesses.

From its synchronous orbit 22,300 miles above the equator, AUSSAT, Australia's first national communication satellite, links that entire country and Papua, New Guinea, through an advanced telecommunications system. See Fig. 21-24. When the satellite

Fig. 21-22. The launching of the Hughes communication Leasat 4 satellite.

Electricity and Electronics

Fig. 21-23. A Satellite Business Systems (SBS) satellite being prepared for launch.
HUGHES COMMUNICATIONS, INC.

is in orbit, the antennas point south, making the spacecraft look upside down if viewed from earth.

Refer to Fig. 21-25. It shows the inside of a satellite. A traveling wave tube amplifier increases the strength of the communication signal for its broadcast back to earth. It is being adjusted by an engineer. The amplifier is onboard a communication satellite. This is the first satellite built to carry both standard traveling wave tubes and new solid state power amplifiers (left). These are expected to be more reliable and have a longer life.

The parts of a satellite are shown in Fig. 21-26. Fig. 21-27 shows satellites in orbit over North America.

Satellite transmission

Once a signal is made by a communication station on earth, it is beamed up to the satellite. The satellite picks up the signal on its receiving antenna. The satellite then amplifies the signal and sends it back down to earth, Fig. 21-28. The signal sent up from the studio to the satellite is a narrow, targeted signal. The signal sent down from the satellite is a wide signal designed to cover a large area of the earth.

The signal transmitted by the satellite is picked up on earth by a receiving dish or parabolic antenna. The dish focuses the received signals into a small area called the FOCAL POINT. The FEEDHORN is located here, Fig. 21-29. Located near the feedhorn is a low noise amplifier (LNA) that amplifies the received signal. The signal is then fed through a piece of electrical coaxial cable to the TV receiver. A coaxial cable has a con-

HUGHES COMMUNICATIONS, INC.
Fig. 21-24. A communications satellite used for communication in Australia.

Television

Fig. 21-25. Looking inside a communications satellite.

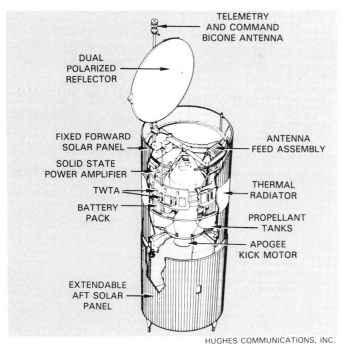

Fig. 21-26. Parts of the Telstar III satellite.

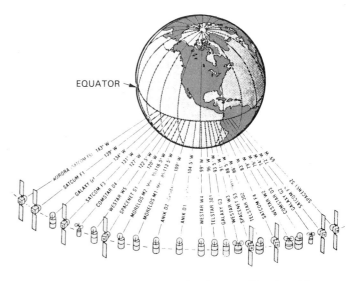

Fig. 21-27. Stationary communication satellites in orbit over North America.

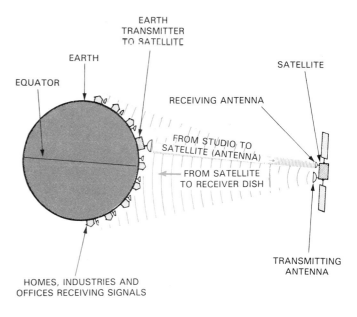

Fig. 21-28. Study the operation of a satellite.

ductor inside another conductor. Each is insulated from the other.

Fig. 21-30 shows common dish designs. Motors are often used to move dishes. This way, more than one communication satellite can be received or the dish can be lined up with a particular satellite.

A satellite receiver is shown in Fig. 21-31. This receiver can be programmed with infrared remote control. Complex circuitry allows the user to store satellite positions, polarity, frequencies, and tuning voltages into memory. The programmed information can then be recalled from the unit's front panel or the remote control unit. Other remote control functions include volume control with mute, direct or scan channel selection, and video fine tuning.

Electricity and Electronics

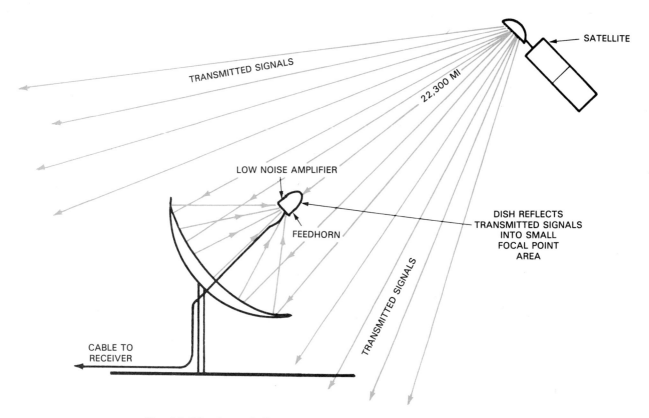

Fig. 21-29. A parabolic or dish antenna. Study the parts.

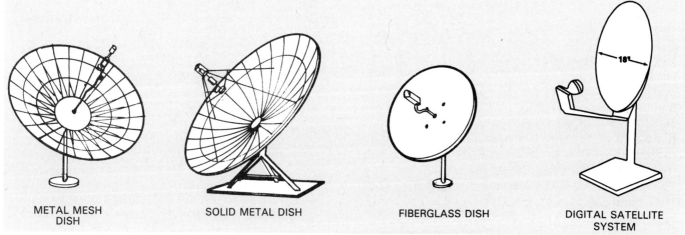

Fig. 21-30. Common dish designs.

REGENCY ELECTRONICS

Fig. 21-31. A satellite receiver.

CABLE TV

Cable television has become very popular in many areas. The viewer pays a monthly fee for the reception of more channels. Cable television is a video distribution system. TV signals are sent through coaxial cable to homes, businesses, schools, and other subscribers.

REVIEW QUESTIONS FOR SECTION 21.3

1. A _____ is an electromagnet that reads information from a recorded tape.
2. How is an image projected onto a large screen TV?
3. Who first proposed the concept of communication satellites?
4. How does a receiving dish work?

SUMMARY

1. A television picture is produced by scanning an image onto a cathode ray tube using a television camera.
2. Scanning is the point-to-point examination of a picture.
3. The scanning system used in the United States is the interlace system. It consists of 525 scanning lines. Scanning starts at the top left hand corner of the picture. It scans from left to right on the odd numbered lines (1,3,5,7, etc.) and then the even numbered lines (2,4,6,8, etc.). All of this takes place 30 times per second.
4. Black and white signals are made by a single color picture tube. Color signals are made by a three color (red, blue, and green) picture tube.
5. The bandwidth of a TV signal is 6 megahertz. The VHF band covers channels 2 through 13. The UHF band covers channels 14 through 83.
6. Video cassette recorders are used for playing movies or tapes.
7. The image on a large screen TV is made by projecting three color images onto a curved screen.
8. Satellite TV is used for communication worldwide. This is done by satellites that orbit the earth in a geostationary position.

TEST YOUR KNOWLEDGE, Chapter 21

Please do not write in the text. Place your answers on a separate sheet of paper.
1. The dots that make up a picture are called _____ _____.
2. Briefly explain the interlace scanning system.
3. What causes an electron beam to move from left to right?
4. The speed of an electron stream from an electron gun is increased by:
   a. Grids.
   b. Target plates.
   c. Scanning.
   d. None of the above.
5. A composite video signal contains the:
   a. Video information.
   b. Sound information.
   c. Both of the above.
   d. None of the above.
6. The _____ _____ _____ is used to produce images in most televisions.
7. What is an ion spot?

MATCHING QUESTIONS: Match the following definitions with the correct terms.
   a. Sync separator.   d. Vertical oscillator.
   b. Sync amplifier.   e. Video amplifier.
   c. Mixer.   f. Horizontal AFC.

8. The output of this device provides the sawtooth voltage to the deflection coils.
9. Removes the horizontal and vertical sync pulses transmitted as part of the composite video signal.
10. Compares the frequencies of the horizontal oscillator and sync pulse.
11. Amplifies the demodulated picture signal and feeds it to the grid of the CRT.
12. Combines the incoming video signal with a local oscillator signal.
13. A voltage amplifier stage in which the sync pulse is increased.
14. What are the three colors used to produce a color TV image?
15. What is the purpose of having more than two heads on a VCR?

FOR DISCUSSION

1. Discuss the function of each of the following controls found in a TV receiver. State the circuit which is regulated by each control.
   a. Horizontal hold.
   b. Brightness.
   c. Vertical hold.
   d. Fine tuning.
   e. Channel selector.
   f. Vertical linearity.
   g. Horizontal linearity.
   h. Height control.
   i. Contrast.
   j. Width control.
2. Explain the process of negative transmission used in television in the U.S.
3. How does the vertical integration network separate the vertical sync pulse?
4. If dc amplifiers were used in the video section, would dc restoration be necessary?

5. Why are both UHF and VHF channels needed for television?
6. Research and discuss the various types of color picture tubes.
7. Discuss what you think television will be like in the year 2000.
8. How did development of the shadow mask picture tube promote the reality of color TV?

SONY ELECTRONICS CORP.

As technology has progressed, appliances such as the television have become small enough to fit comfortably in one hand.

Television

SONY ELECTRONICS CORP.

Televisions are constantly changing. Today's TVs offer flatter and larger picture tubes.

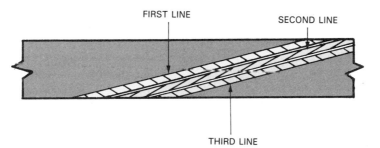

Videotape is scanned on a slanted line. This provides a high speed for the record/play head relative to the tape.

Electricity and Electronics

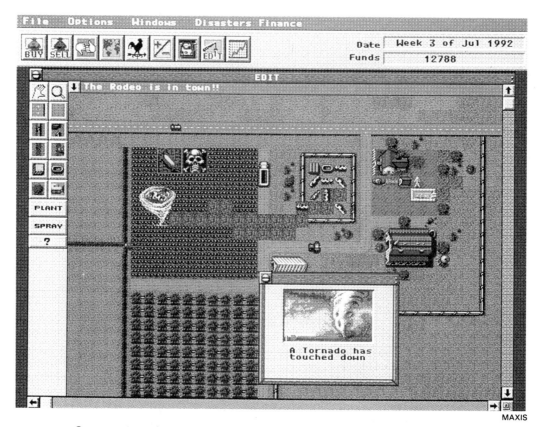

MAXIS

Computers perform a variety of functions in our society. This program simulates owning a farm.

Computers are based on binary arithmetic (1 or 0). Each has memory (RAM and/or ROM), an arithmetic logic unit (ALU), input, output, address buses, and data buses. With suitable interfaces, (circuits between the computer and the outside world), the computer can analyze both analog and digital information. Circuit frequencies as high as 400 MHz are possible. At these frequencies, even the capacitance of a 1/16 in. solder connection can distort the signal. The inductance of a single wire that is bent partially around a transistor or an integrated circuit chip can distort signals. To help prevent distortion, printed circuit board construction is used.

Chapter 22

COMPUTERS

After studying this chapter, you will be able to:
- *Outline the evolution of computers.*
- *List key inventors, inventions, and events in the evolution of computers.*
- *Name the basic pieces of equipment that make up a working computer.*
- *Explain the functions and uses of these various pieces of equipment.*

In this age of information and technology, computers affect our lives in nearly every way. They assist us in education, Fig. 22-1. Computers help predict weather patterns. They assist in the making of music. They are also a form of entertainment, Fig. 22-2. They allow for safer travel in the air and on the highways. They make high-speed communication possible. They also play a key role in our country's defense, Fig. 22-3.

Fig. 22-2. Computer games often serve two functions: they are entertaining and educational.

Fig. 22-1. Drafting students learn to do a great deal of work on personal computers. This work used to be done by hand and was quite time consuming.

Fig. 22-3. This electronic brain is used in a defense system.

367

Electricity and Electronics

Computers help make manufactured products more dependable and less costly. They help us calculate faster, and with much more ease, Fig. 22-4. In addition, they make offices and other work places more efficient, Fig. 22-5. These are only a few of the hundreds of useful tasks computers perform.

Over the past few years, computer science has grown faster than any other technology. Each year computers become more powerful, less costly, and easier to use. See the chart in Fig. 22-6.

Fig. 22-4. This programmable calculator is very useful in computer science and digital electronics applications.

Fig. 22-5. Modern offices are helped by the use of computers.

	1955	1960	1965	1975	1983
COST	$14.54	$2.48	47¢	20¢	7¢
PROCESSING TIME (in seconds)	375	47	29	4	1
TECHNOLOGY	Vacuum tubes Magnetic cores Magnetic tape Magnetic drum	Transistors Channels Faster cores Faster tape	Solid Logic Technology Large, fast disk files New channels Larger, faster core memory Faster tape	Monolithic memory Monolithic logic Virtual Storage Virtual Machine Larger, faster disk files New channels Advanced tape	High-capacity memory chips Very dense logic packaging Extended Architecture Thin film head disk technology Magnetic cartridge mass storage
PROGRAMMING	Stored program	Overlapped input/output Batch processing	Operating system Faster batch processing	Advanced operating systems Virtual Storage Virtual Machine Multiprogramming Batch, plus on-line processing	Full-function advanced operating systems Multiprocessing Advanced networking products Extended Architecture Comprehensive data base facilities Interactive, user-oriented functions

Fig. 22-6. This chart shows the time and cost for a large IBM computer to complete a fixed amount of data processing. Compare the differences in numbers for each year. New technology is the reason for the lowered costs.

Computers

22.1 HISTORY OF COMPUTERS

One of the oldest examples of a computer is Stonehenge. This monument was constructed from 2800 to 1100 BC, Fig. 22-7. Some scientists say it was used as a counting machine. For example, dates and time were set based on the position of shadows created by the stones.

The Chinese abacus is another simple computer, Fig. 22-8. Using movable beads, numbers can be added, substracted, multiplied, and divided. The abacus is still very common in Asia. Some abacus operators compute faster than people using calculators.

In 1617, John Napier invented a pocket-sized device for multiplication. It was called Napier's Bones. Numbers were printed on square rods that could be moved. Answers were found by adding numbers in horizontal, side-by-side sections.

In 1642, a Frenchman names Blaise Pascal invented an arithmetic machine. His father was a tax collector. They used the machine for addition and subtraction.

In 1822, an English inventor named Charles Babbage invented what is now thought of as the first modern computing machine. It was called the Difference Engine. It was designed to add numbers and store the results at each stage in the calculations. However, it never worked due to mechanical problems in machining accuracy.

In 1833, he designed another device called the Analytical Engine. It was the forerunner of the programmable computer. This machine was not built. However, Babbage's ideas and research for these machines were the basis for modern computers.

In 1890, an American inventor, Herman Hollerith, designed an improved process for tabulating U.S. Census returns. Numbers were entered on punch cards. Data could be processed much more quickly. In 1925, Hollerith developed the card which is the model for current computers, Fig. 22-9.

Hollerith went on to form the Tabulating Machine Company. It was one of the firms that later became

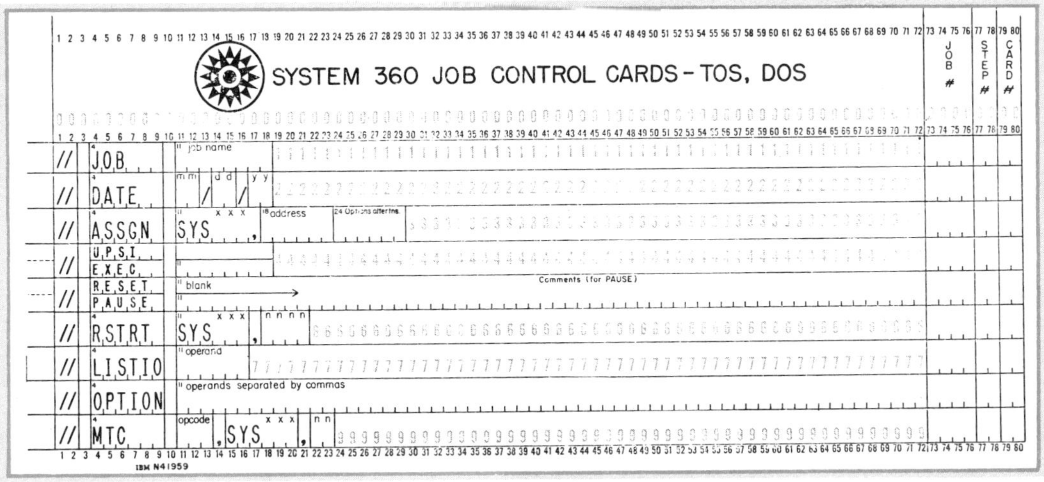

Fig. 22-7. Stonehenge is located in southern Great Britain and is thought to be one of the earliest computing instruments.

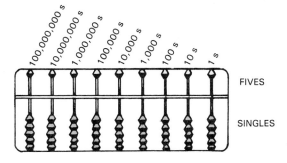

Fig. 22-8. Many arithmetical functions can be done on an abacus.

Fig. 22-9. This IBM computer punch card is a Job Control Card used in the System 360.

part of International Business Machines Corporation, or IBM. The basic Hollerith concept remained the basis of the information processing industry through World War II.

TWENTIETH CENTURY COMPUTER DEVELOPMENTS

Pre-World War II

The 1930s saw progress being made in computer theory. In 1936, Claude Shannon developed a thesis that laid a foundation between symbolic logic and electrical circuits. In 1944, Dr. Howard Aiken of Harvard University and IBM, completed five years of work on the Mark I. It was the largest electromechanical calculator ever built. It had 3300 relays and weighed five tons. Mark I could multiply two, 23-digit numbers in six seconds.

Electromechanical calculators such as the Mark I, however, were too slow for the pace of the postwar world. Users wanted speed. The vacuum tube reponded to that demand. Vacuum tubes could be turned on and off like a controlled switch. They could count thousands of times faster than moving mechanical parts.

Post-World War II

In 1945, John von Neumann developed the architecture of modern computers. He stated that each computer should have four basic pieces of equipment. They are the central processing unit, an input device, an output device, and storage.

Between 1946 and 1952 a number of electronic calculators and computers emerged in rapid order. The world's first large electronic computer was developed in 1946 at the University of Pennsylvania by J. Presper Eckert and John Mauchly. In contained 18,000 vacuum tubes. It was called ENIAC.

The IBM Selective Sequence Electronic Calculator was completed in 1947. It contained both vacuum tubes and relays. It was more than 100 times faster than the Mark I. It was the first IBM computer that stored programs. It could select a calculating sequence by modifying its own stored instructions. In 1946 vacuum tube computers could multiply two, 10-digit numbers in 1/40th of a second. By 1953, the same calculation could be done in 1/2,000th of a second.

Vacuum tube computers are thought of as the first generation of electronic computers. These computers were much faster than the older electromechanical computers. However, vacuum tubes were large and bulky. They required large amounts of voltage and current (power). They gave off large amounts of heat. They were very fragile. And they were not very reliable.

In 1947, the transistor was invented at Bell Laboratories (refer back to Chapter 13). The transistor was 1/200th the size of the vacuum tube. It was faster and could be packaged compactly. Because it was made of a solid substance, it was far more reliable and rugged. During operation it created much less heat than a vacuum tube. Transistors were used in computer circuits starting in the late 1950s. Computers could now multiply two, 10-digit numbers in 1/100,000th of a second. The second generation of computers was made possible by the transistor.

In the early 1950s, there was a demand for faster and cheaper storage and memory devices. As a result, research and development in magnetic storage devices increased. This research created magnetic disks and drums for storage, magnetic cores for memory, and better materials for better magnetic tapes. Also in the 1950s, a number of manufacturers introduced medium and large scale data processing equipment.

In 1958, the integrated circuit (IC) was developed. This began the third generation of computers (refer to Chapter 13). However, it took a few years before the IC was used in a computer. In 1968, the Burroughs Corporation came out with the first computer using ICs. The IC made computers smaller, faster, more reliable, and cheaper to operate than earlier computers. ICs also had built-in memories. This added to their reliability and smaller size.

All of this research and development resulted in the minicomputer and the microcomputer.

MICROCOMPUTERS VS MINICOMPUTERS

A large computer used in a business, industry, or university is called a MAINFRAME computer, Fig. 22-10. A smaller, fixed version of the mainframe computer is the minicomputer. It has less memory than a mainframe. Minicomputers are often used by businesses and industries, Fig. 22-11. The first minicomputer, the PDP-8, was made by Digital Corporation in 1964.

A MICROCOMPUTER is a computer whose CPU is contained on a single chip. Microcomputers are most often portable. They are commonly used for word processing, data analysis, creating graphics, and playing games. See Figs. 22-12 and 22-13.

THE 1970s

The 1970s were eventful years in computer history. During this decade, large scale integration (LSI) of ICs was developed. LSI meant a complete computer circuit could be built on a single silicon chip, Fig. 22-14. Advances were also made in computer memory tech-

Computers

Fig. 22-10. Mainframes perform many complex tasks in industry.

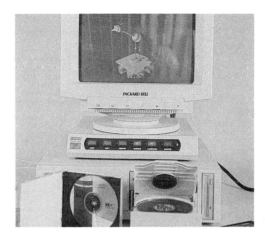

Fig. 22-12. This microcomputer can be used to do word processing and more.

Fig. 22-13. Advances are constantly being made in microcomputer technology. This microcomputer can fit in a briefcase.

Fig. 22-11. Minicomputers are commonly used in small businesses.

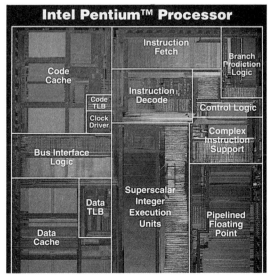

Fig. 22-14. This silicon chip (magnified many times) contains a complete computer circuit.

nology. Products such as floppy disks became available at low costs.

The microprocessor was also developed in the 1970s. A MICROPROCESSOR is a complete processing unit contained on a single IC chip. In 1971, the Intel Manufacturing Company produced the first microprocessor for public sale. It was called the 4004. It had 2250 components. It could add two, 4-bit numbers in 1/11,000,000th of a second.

In November, 1972, Intel came out with an eight bit microprocessor, the 8008. This microprocessor was costly. It had only 16k (16,384) bytes of memory. About one year later, Intel came out with a much better microprocessor IC, the 8080. It was easier to use than the 8008. It had 64k (64,536) bytes of memory. Soon after the 8080 was introduced, Motorola introduced a competing microprocessor chip, the 6800. In addition, Zilog introduced the Z80.

In 1973, the Scelbi Computer Consulting group offered the first make-at-home computer kit. It cost $595 and had an Intel 8008 microprocessor. A company called MITS from Albuquerque, New Mexico then offered a kit in January, 1975. The kit was called the Altair 8800. Two students, Paul Allen and Bill Gates, were hired by MITS to develop a simple software package to be used with the Altair. They invented a program known as BASIC (Beginners All-Purpose Symbolic Instruction Code). It is a very popular language. Later, the students formed their own company, Microsoft. Microsoft is a common name in the computer field.

In September, 1975, IBM announced the first briefcase-sized computer. It was called the 5100 and had a 16k memory, a built-in monitor and tape cartridge storage. In April, 1976, the Apple Computer Company was formed by Stephen Wozniak and Steve Jobs, Fig. 22-15. Close behind, Radio Shack developed its first computer in February, 1977. IBM came out with its first personal computer (PC) in 1981. The microcomputer industry race was on.

Many integrated circuit advances were made in the late 1970s and early 1980s. Very Large Scale Integration (VLSI) manufacturing meant newer and more powerful microprocessors. The Motorola 68000 chip used VLSI. It had about 70,000 components. It was a 16-bit chip.

Memory sizes increased during this period, also. See Fig. 22-16. In 1981, Hewlett Packard developed a 32-bit chip. It had 450,000 components on a silicon chip less than 10 by 10 mm. This chip could multiply two, 32-bit numbers in 1/1,800,000th of a second.

The future of computers is bright and exciting. Superchips are now on the market. They push the upper limits of component density and speed. In the near future, single-chip mainframe central processing unit (CPUs), 4 megabyte, dynamic random access memories, and flat screen color monitors will be available. Gallium arsenide (GaAs) may become a commonly used semiconductor material. GaAs will provide increased microprocessor speed.

In the past twenty-five years, the computer industry has moved from the margins to the center of our lives. The success of the computer is based on its ability to

APPLE COMPUTER INC.

Fig. 22-15. The Apple Computer Company has grown to become a giant in the computer field.

Computers

store and process large amounts of information. It can update and retrieve information and then send it across continents through communication satellites or telephone lines. It can calculate, compare, play, imitate, and monitor. It can do all these things and more with great ease and speed.

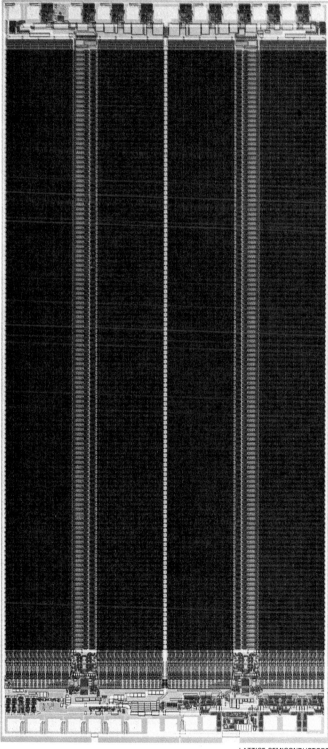

Fig. 22-16. This high speed chip has 64k random access memory (RAM).

REVIEW QUESTIONS FOR SECTION 22.1

1. Who was the inventor of the Difference Engine and the Analytical Engine?
2. What is Herman Hollerith famous for inventing?
3. Who developed the architecture of modern computers?
4. What four pieces of equipment are basic to all computers?
5. _____ _____ computers are thought of as the first generation of electronic computers.
6. The IC began the _____ computer generation.
7. Define microcomputer and microprocessor.

22.2 COMPUTER OPERATION

Modern computers use the same basic equipment pieces devised by John von Neumann in the mid 1940s. He said that computers should be built as a single machine or system. However, they should contain equipment pieces designed and programmed to work as a unit. The basic equipment of any computer includes:

1. An INPUT device for entering data into the system.
2. An OUTPUT device for retrieving data.
3. A CENTRAL PROCESSING UNIT (CPU) that includes an ARITHMETIC/LOGIC UNIT (ALU), a CONTROLLING UNIT (CU), and MEMORY.
4. A STORAGE device.

Fig. 22-17 shows how these pieces work together. Input can be any type of data: numbers, words, symbols, etc. Processing is done in a pre-set sequence.

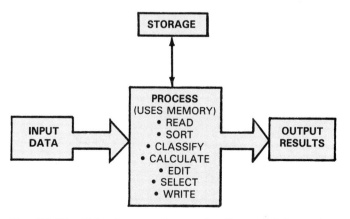

Fig. 22-17. This diagram shows the flow of information through a computer.

Electricity and Electronics

Instructions (the program) are automatically followed by the computer. The programs are stored in the memory. Output can be printed material, data to other equipment, etc.

A computer is a single machine. Engineers, however, often think of it as a number of machines. Each machine is designed and programmed to work as part of the system. Electronics students need to think of the computer system in terms of the constant change of data into electrical signals. These signals are sent back and forth among the machines that make up the total system. Large computers are often joined by cables with telephone lines. Microcomputers and minicomputers tend to be more self-contained.

INPUT/OUTPUT DEVICES

The INPUT or OUTPUT devices either enter data into or retrieve data out of a computer.

An input or output device is started by a program instruction that sends a command to an input or output (I/O) channel. Examples of these commands are ENTER (input) or PRINT (output). Any input or output device that is joined to the outside of the process unit is called PERIPHERAL equipment.

Inputs

Input devices sense and read data. Data can come from keyboards, magnetic or floppy disks, paper, or magnetic ink characters. Data can also come over telephone lines to modems. Modems serve as a link between computers exchanging data. See Fig. 20-18.

Some input devices use manual entry only. A keyboard is an example, Fig. 22-19. Note that the keyboard has letters and other symbols that can be input to the computer.

Some inputs use magnetic recording devices. Examples include floppy disks, magnetic tape, or magnetic ink characters inscribed on paper documents (like checks).

Floppy disks come in a number of sizes. See Fig. 22-20. The construction of a floppy disk is shown in Fig. 22-21. Study this drawing.

Floppy disks used currently have a circle of very thin (.003 in.) polyester film with a recording surface of magnetic oxide. They can store as much as 3,200,000 bytes (26,843,546 bits). That is over 1600 typewritten pages of information.

A disk drive operates like a high-tech stereo turntable, Fig. 22-22. However, a disk drive is much more complex, Fig. 22-23. The disk spins inside the drive at 300 revolutions per second. An electromagnetic head moves across the disk's surface through an opening in the jacket, called the HEAD WINDOW. It moves to position over one of the concentric tracks where the requested data is stored. The head can read information from the disk. It can also write new information for storage. Light passing through the index hole triggers a photoelectric signal. This matches up the recording or reading process with the rotation of the disk.

Fig. 22-19. A keyboard is an example of a manual input device.

INPUT DEVICES MAY INCLUDE:
- KEYBOARDS
- FLOPPY DISKS OR MAGNETIC TAPE
- MAGNETIC INK CHARACTERS
- DATA FROM OTHER COMPUTERS USING MODEMS
- LIGHT PENS

Fig. 22-18. Listed are some examples of inputs into the computer system.

MAXELL CORP.

Fig. 22-20. Floppy disks come in several sizes. Left. 5 1/4 inch disks. Right. 3 1/2 inch disks.

Computers

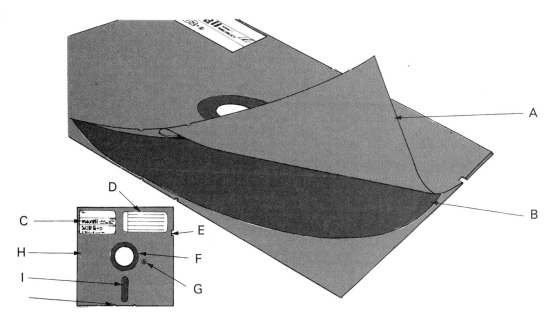

Fig. 22-21. Construction of a floppy disk. A—Liner cushions and cleans the disk. B—Magnetic disk spins inside the sealed jacket. C—Permanent label has information on disk type and product identification number. D—Removable label for user's information. E—Write-enable notch allows user to write new information when notch is exposed and protects information when notch is covered. F—Central window and central hole through which disk is rotated by drive. G—Index window (in jacket) and index hole (in disk) allow light beam in the drive unit to check disk position and rotation. H—Sealed jacket protects the magnetic material inside the disk. I—Head window allows the read/write head of the disk drive to come in contact with the disk. J—Relief notches position the disk and prevent window head distortion.

Fig. 22-22. Typical disk drive.

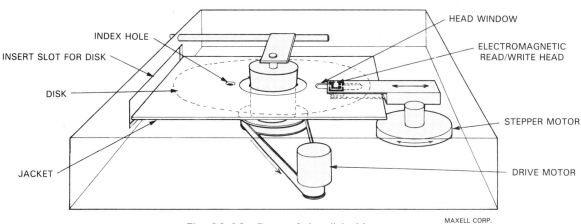

Fig. 22-23. Parts of the disk drive.

375

Electricity and Electronics

Magnetic ink characters are used by banks to read and process daily data transactions. These characters are printed on checks, Fig. 22-24. They are read by a machine with an electromagnetic head. The shape of each character is made up of either an open space or a space covered by magnetic ink. This gives each square a 0 or 1 value.

Another common input process is telecomputing. This is done using modems. Modems allow computers to talk with each other, access information services, and send and receive messages. See Fig. 22-25.

Outputs

The processed results are fed to the output, Fig. 22-26. Output devices record information on paper, magnetic tape, or disks. Output may also take the form of electrical signals. The signals can be sent over transmission lines to other computers. Outputs can also include graphic displays. Output devices vary depending on what a computer system is designed to do.

Computer printers come in a number of types. Many of these printers can print in color or in black and white.

Dot matrix printers are popular, Fig. 22-27. They are fast and inexpensive. They also have a number of print styles.

Dot matrix printers product dot images using pins. These pins are called a PRINTHEAD. The printhead is driven forward, into a ribbon, imprinting an image. The quality of the image will depend on either the number of pins in the printhead or the number of passes made by the printhead. See Fig. 22-28.

Better image quality can be obtained from printers that use THIMBLE or DAISY WHEEL printheads. These printheads are made from high-quality plastic. When forced through the ribbon onto the paper, they produce a letter quality image.

Ink jet printers shoot ink onto the surface of paper using an electric charge, Fig. 22-29. Ink jet printers work best when used with special paper that will absorb the ink. These printers are compact, quiet, and ideal for home offices. They are also ideal for printing graphics, Fig. 22-30.

Plotters are used to make larger prints such as maps, charts, schematic drawings, and architectural plans.

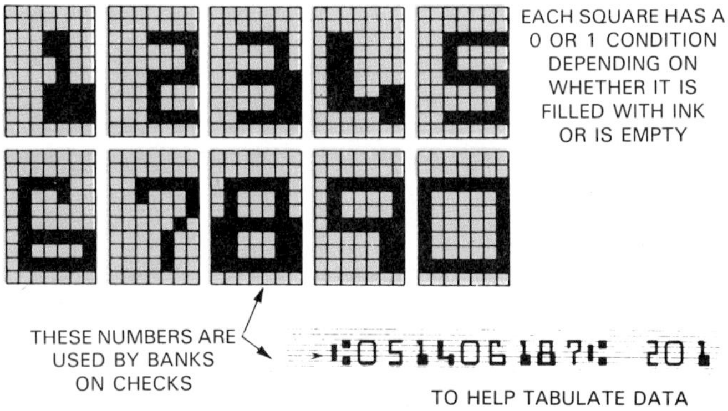

Fig. 22-24. Magnetic ink characters are used by banks and other industries to help tabulate data.

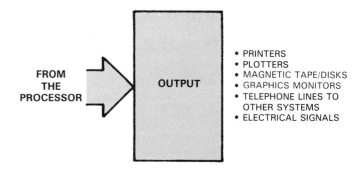

Fig. 22-26. The output devices of the computer provide the results of the processing of data.

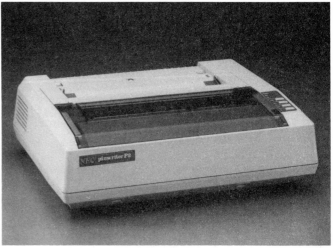

NEC INFORMATION SYSTEMS, INC.

Fig. 22-25. This modem can connect one computer to another using telephone lines.

Fig. 22-27. A dot matrix printer.

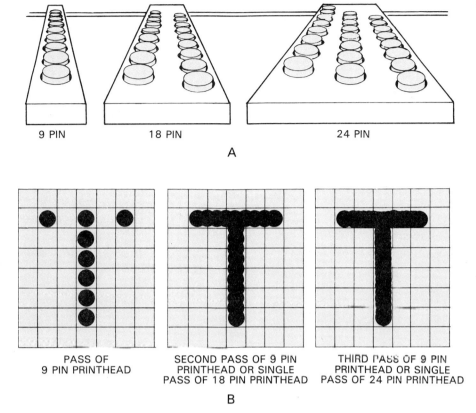

Fig. 22-28. Dot matrix printer operation. A—Printhead designs. B—Images created by dot matrix printer.

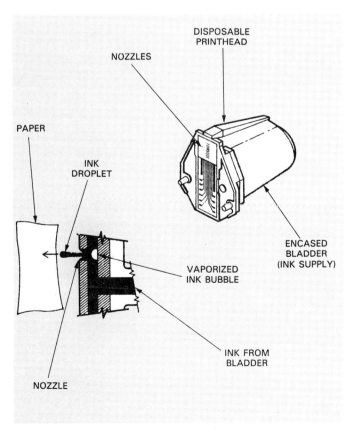

Fig. 22-29. This drawing shows how an ink jet printer operates.

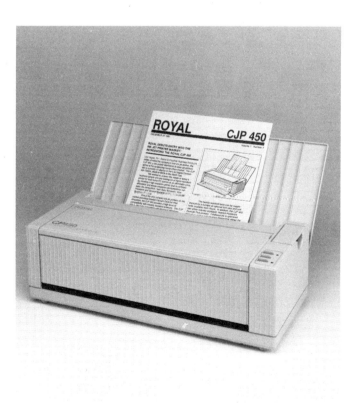

ROYAL CONSUMER BUSINESS PRODUCTS

Fig. 22-30. Graphics turn out nicely on this ink jet printer.

With some plotters, the paper moves back and forth over the printing head or pen. Others print as the printing head or pen moves on a track system. Some plotters create superb color work, Fig. 22-31. Plotters are excellent for use in computers assisted drafting (CAD) applications.

Some printers use lasers to produce text, graphics, and other images. The laser printer shown in Fig. 22-32 is joined to three input devices. Any data is these three devices can be produced as output on the printer. Many laser printers have unlimited fonts (type styles). They provide letter quality printing.

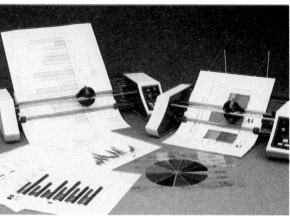
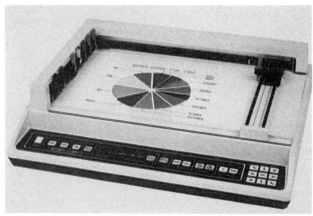

HOUSTON INSTRUMENTS

Fig. 22-31. Plotters can produce colorful, clear output. Left. Vertical plotters are used in computer assisted drafting (CAD). Center. This eight pin plotter produces excellent pie charts. Right. Smaller plotters are ideal for home use.

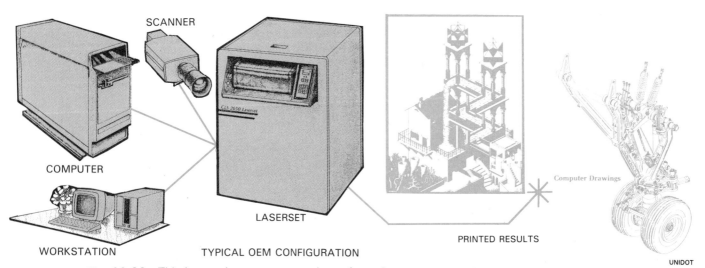

Fig. 22-32. This laser printer can accept input from three sources and produce output from each.

Computers

Magnetic tapes or disks are also a type of output. These are used when information is stored for periods of time. See Fig. 22-33.

Data displayed on a video screen is another form of output. In this case the user may want to see the results, but does not need a copy, Fig. 22-34.

Computer output can be fed into telephone lines or through transmission lines, such as optical fibers. These link the output of one computer to the input of another, or vice versa.

CENTRAL PROCESSING UNIT (CPU)

The central processing unit (CPU) controls the computer system. The computer SYSTEM consists of all the parts that work together to perform computer functions. These functions include directing data, calculating numbers, and performing logic functions. The CPU also has some memory in it. Refer to Fig. 22-35. A CPU built on a single computer chip is called a MICROPROCESSOR. A MICROCOMPUTER is a computer built around a single microprocessor.

Control unit

The control unit (CU) acts like the traffic cop of the CPU. It directs and coordinates data moving throughout the CPU. The program being used (BASIC, Pascal, Fortran) will determine what will be done to the data. The control unit also finds and performs detailed instructions given by the stored program (stored in memory). The control unit then begins sending out data comands. These cause other sections, such as the I/O or arithmetic/logic unit, to perform functions at the proper time.

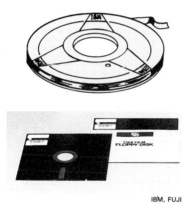

Fig. 22-33. Magnetic tape and floppy disks are two types of outputs.

Fig. 22-34. A monitor is a computer's most used output device.

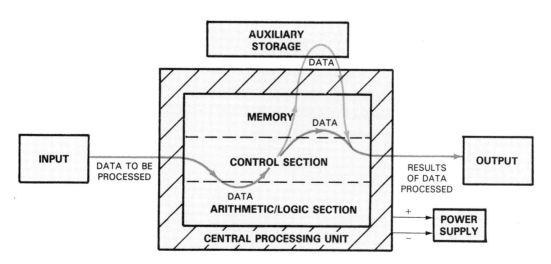

Fig. 22-35. The basic building blocks, or architecture of a central processing unit include the control unit, the arithmetic/logic unit, and memory.

The control unit acts as the "brain" of the CPU. it directs all other sections or devices to do certain things. The control unit does not perform any data processing tasks.

Arithmetic/logic unit

The arithmetic/logic unit (ALU) contains the circuits that perform certain arithmetic functions.

The logic unit carries out the decision making processes that change the sequence of instructions. The logic function tests out the conditions of the arithmetic processes and then takes the correct action.

Memory

Primary computer storage is memory. The memory for instructions and fixed data, which does not change, is called READ ONLY MEMORY (ROM). It contains the information stored by the manufacturer. This information helps the computer operate, recognize, and read what is typed or fed into it.

Fig. 22-36 shows a VLSI integrated circuit with a ROM. The complete IC chip is shown in Fig. 22-37.

Read only memory contents cannot be changed. Even if the power supply is turned off or the computer is unplugged, contents of ROM will be retained. This condition is called NONVOLATILE.

The PROGRAMMABLE READ ONLY MEMORY (PROM) is programmed once after the computer is built. This memory provides ways of updating or changing instructions in ROM.

RANDOM ACCESS MEMORY (RAM) can read, write, and temporarily store data when the computer is operating. New data entered into RAM erases old data. Since the memory is only temporary, data can be lost if a power supply is turned off or interrupted. This makes RAM VOLATILE. Some microcomputers have built-in lithium batteries to provide backup voltage if the power is lost.

STORAGE DEVICES

At times, data may need to be stored on a long term or permanent basis. Storage devices are used for this.

Storage devices include floppy disks, magnetic tape, or magnetic drums. Fig. 22-38 shows a computer system with magnetic tape auxiliary storage units. The key reason for having auxiliary storage is to provide data storage in addition to the internal storage of the computer.

MICROPROCESSORS

A microcomputer is a computer built around a microprocessor. A complete microcomputer consists

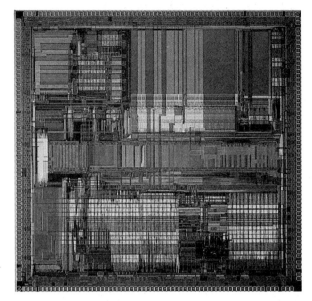
Fig. 22-36. Inside a VSLI integrated circuit.

Fig. 22-37. This is the 32 bit IC shown in Fig. 22-36.

Fig. 22-38. Magnetic tape units are sometimes used for auxiliary storage.

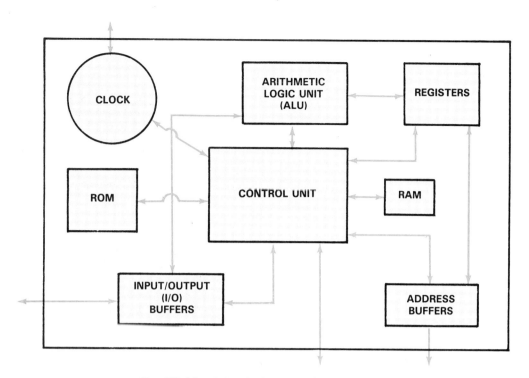

Fig. 22-39. A block diagram of a microprocessor.

of a microprocessor chip, memory interfaces for input and output, and a power supply.

A complete microprocessor is shown in Fig. 22-39. It consists of a control unit, arithmetic/logic unit, ROM, RAM, input/output buffers, registers, address buffers, and a clock.

The control unit fetches and decodes the instructions from ROM. The instructions will determine the action to be taken by the ALU. The control unit and ALU form the "common sense brain" of the microprocessor.

ROM permanently stores the computer operating instructions. RAM stores data the microprocessor needs to perform a task. Once the task is complete, the data is deleted from RAM.

The input/output buffers direct data into or out of the microprocessor. Registers are very fast memory units used to store data for the microprocessor. Address buffers supply the location (address) of an instruction or of the next data item to be read. Address buffers also supply the location in which a data item should be written.

The clock circuit is used to trigger and match up operations.

SOFTWARE

Software is a set of instructions used to run the computer. These instructions provide directions for processing, documenting, and operating. Preprogrammed software can be purchased for word processing,

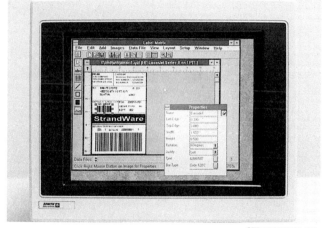

STRANDWARE INC.

Fig. 22-40. There are specialty software programs designed to perform thousands of different chores.

accounting, data base management, graphics, office and project planning, games, etc., Fig. 22-40. Programs can be designed to fit the needs of the computer user, through computer languages such as BASIC.

COMPUTERS IN THE FUTURE

Computers will continue to become more powerful, easier to use, and less expensive.

Integrated circuits will have greater storage. Their computing speeds will increase and their cost will tend to decrease. Gallium arsenide is likely to be a key semiconductor material, along with silicon.

Printers, keyboards, monitors, and disk drives will become more advanced. Printers will provide better quality output. Keyboards will become more "user friendly." Flat-screen monitors will become a reality. Smaller and more powerful floppy disks will be used. Hard drives will continue to grow and become less expensive, Fig. 22-41.

Software will be easier to use and buy. Also, there will be more variety in the marketplace.

Smaller computers will be developed. Briefcase size computers have been common for some time. However, complete hand held computers will be available in the next few years.

Computers have come a long way in the past 25 years. But the future has even more to offer.

Fig. 22-41. The inside of a microcomputers hard drive.

REVIEW QUESTIONS FOR SECTION 22.2

1. Input devices include:
   a. Floppy disks.
   b. Magnetic ink characters.
   c. Keyboards.
   d. All of the above.
2. The processed results of a computer are fed to the _____.
3. The _____ _____ acts like the traffic cop of the CPU.
4. The type of computer memory stored by the manufacturer and that cannot be changed is:
   a. RAM.
   b. ROM.
   c. PROM.
   d. All of the above.
5. What is the difference between nonvolatile and volatile memory?
6. An _____ buffer supplies the location of an instruction or a data item.
7. What is software?

SUMMARY

1. The history of computers dates back to Stonehenge and the abacus.
2. Some early inventors of calculation machines were Pascal, Napier, and Babbage.
3. The Difference Engine, by Charles Babbage, was designed to add numbers and store the sums at each stage of calculation.
4. Herman Hollerith invented the punch card. It was the mainstay of data processing in the first half of the 20th Century.
5. John von Neumann developed the architecture of modern computers.
6. The first electronic computer, ENIAC, used vacuum tubes and was built in 1946.
7. Integrated circuits were developed in 1959. However, 9 years passed before they were used in electronic computers.
8. The first microprocessor was developed in 1971.
9. Microcomputers are computers built around a microprocessor.
10. The four basic pieces of computer equipment are the:
    a. Input device.
    b. Central processing unit (CPU):
       — Control unit (CU).
       — Arithmetic/logic unit (ALU).
       — Memory.
    c. Output device.
    d. Storage.
11. An input device feeds data into a computer.
12. An output device obtains processed data from a computer.
13. The control unit of the CPU supervises, directs, and coordinates the data entering the computer. The arithmetic/logic unit (ALU) performs certain arithmetic functions (calculations). It also makes decisions regarding the sequence of instructions. The memory in the CPU stores operating instructions in ROM, and has working storage RAM.
14. The read only memory (ROM) contains information that helps the computer operate and recognize (read and write) data that is fed into it.
15. The random access memory (RAM) is active memory. It can process, read, and write data. It also provides temporary storate of data, but only when the computer is operating.

Computers

16. Programmable read only memory (PROM) is installed after the computer is built. It has the same basic functions as ROM.
17. A microprocessor is a complete CPU contained on one computer chip.
18. Software is a set of instructions used to operate the computer.

TEST YOUR KNOWLEDGE, Chapter 22

Please do not write in the text. Place your answers on a separate sheet of paper.

1. This man was the inventor of a multiplication device that used movable square rods:
   a. Blaise Pascal.
   b. John Napier.
   c. Charles Babbage.
   d. None of the above.
2. Who developed ENIAC?
3. What was the name of the first computer to store programs?
4. List four disadvantages of vacuum tube computers.
5. The _____ made the second generation of computers possible.
6. What is the difference between a minicomputer and a microcomputer?
7. What is meant by LSI?
8. List three types of input devices and three types of output devices.
9. Name the three major parts of a CPU.
10. The opening in a floppy disk jacket is called a:
    a. Window.
    b. Read window.
    c. Head window.
11. _____ _____ printers must be used with special paper that will absorb ink.
12. The control unit (does, does not) perform any data processing tasks.
13. The _____ unit decides how and when to change the sequence of instructions.
14. Name the three types of memory found in CPU.
15. The main reason for having auxiliary storage is:
    a. To provide storage in addition to the internal storage of the computer.
    b. To provide enough storage to operate the computer.
    c. To provide additional computing power.
    d. All of the above.

FOR DISCUSSION

1. What do you think computers will be like in the year 2050? What will their appearance be? How will they differ from today's computers?
2. Discuss the way in which computers affect your daily life.
3. Is technology affecting our lives in a positive or negative way? Explain your answer.
4. Research a specific event or invention in the evolution of computers.

Modern trains use computers to control their operation and schedule.

Part VI Summary
ELECTRONIC COMMUNICATION AND DATA SYSTEMS

IMPORTANT POINTS

1. A transmitter is an electronic device that sends information toward its destination.
2. Transducers, such as microphones or speakers, convert one form of energy to another (mechanical to electrical, electrical to mechanical).
3. Modulation is the process of adding an audio wave to a carrier wave.
4. In amplitude modulation (AM), the carrier wave amplitude is varied at the audio rate.
5. In frequency modulation (FM), the carrier wave frequency is varied at the audio frequency rate.
6. A receiver is an electronic device which picks up transmitted waves.
7. Ground wave radio waves are transmitted along the surface of the earth to radio receivers.
8. Sky waves are transmitted through the ionosphere to radio receivers.
9. Demodulation, or detection, is the process of removing the audio portion of a signal from the carrier wave.
10. A tuned radio frequency (TRF) receiver picks up a transmitted RF wave, amplifies it, detects or demodulates it, and amplifies the audio wave.
11. A superheterodyne receiver mixes the incoming modulated signal with an unmodulated local oscillator signal. The result is an intermediate frequency signal that carries the message.
12. A television picture is produced by scanning an image onto a cathode ray tube using a television camera.
13. Television uses FM for audio signals and AM for video signals.
14. Black and white television signals are made by a single color picture tube. Color signals are made by a three color (red, blue, green) picture tube.
15. There are two television bands in the U.S.: VHF (very high frequency) band, channels 2 through 13 and UHF (ultra high frequency) band, channels 14 through 83.
16. Satellites handle many communication tasks.
17. Three types of electronic computers are mainframes, microcomputers, and minicomputers.
18. A microcomputer is a computer whose CPU is contained on a single chip.
19. A minicomputer is a small, fixed version of a mainframe computer. It has less memory than a mainframe.
20. A microprocessor is a complete processing unit contained on one chip.
21. Basic computer equipment consists of an input device, an output device, a central processing unit (CPU), and a storage device.
22. Software is a set of instructions used to run the computer.

SUMMARY QUESTIONS

1. Radio waves are grouped according to frequency. AM broadcast stations use the _____ range. TV uses _____ and _____ ranges. A stereo operates in the _____ frequency range.
2. What is the formula for determining percent of modulation?
3. A carrier wave has a peak value of 750 volts. A modulating signal causes amplitude variation from 500 to 1000 volts. What is the percent of modulation?
4. A wave has a frequency of 27 MHz. What is its wave length?
5. Define selectivity and sensitivity.
6. _____ converts all incoming signals to a single intermediate frequency.
7. Name two transducer devices in stereos.
8. An FM _____ must produce a varying amplitude and frequency audio signal from an FM wave.
9. What is scanning?
10. The amplitude of the modulation in a video signal is divided into two parts. The first _____ percent transmits video information. The last _____ percent is for sync pulses.
11. What is the bandwidth of a TV channel?
12. A _____ satellite rotates with the earth.
13. Why is ENIAC famous?
14. During the 1970s:
    a. LSI was developed.
    b. The microprocessor was developed.
    c. BASIC was invented.
    d. All of the above.
15. What three devices are contained in the CPU?

Electricity and Electronics

AT & T

This individual has chosen a career in the electronics industry.

The field of electronics is expanding. In the information age, telecommunications, information handling, and service-related occupations are overtaking production of goods as the major occupational areas in the nation and the world. Anyone with a background in electricity, electronics, and computer theory will be more quickly employed than others. The pay can be two to four times that for jobs in the production of goods.

Chapter 23

CAREER OPPORTUNITIES IN ELECTRONICS

After studying this chapter, you will be able to:
- *Assess your interest in electronics careers.*
- *Distinguish between a goods producing and a service producing business.*
- *Discuss a variety of electronics careers and their educational requirements.*
- *Define entrepreneur and entrepreneurship.*
- *List career information sources.*
- *Outline ideas for a successful job search.*

23.1 FINDING A CAREER

How do you want to earn a living after you complete your education? What vocation will you choose to assure an interesting and rewarding future? What type of education will you need to prepare for your chosen field of study? Do you have the interest, desire, and ability to use the educational opportunities within your reach?

These are questions only you can answer. There are many people, however, willing to help you answer these questions. Parents, school guidance counselors, teachers, and friends in business and industry can supply information gained from years of experience. Do not hesitate to ask them about their careers.

Perhaps you are considering a career in electronics. The electronics industry offers many outstanding career opportunities, Fig. 23-1. Review the following guidelines. They may help you to make decisions regarding a career in this industry.

Fig. 23-1. This engineer has a rewarding career working with communications satellites. Here he is shown conducting tests on an amplifier contained in a satellite.

Electricity and Electronics

- Electronics is a science. If you are interested in scientific subjects in high school, chances are you would enjoy an electronics career.
- Mathematics is the tool of scientists and engineers. Do you like mathematics and do well in it? If your answer is yes, then you may be well-suited to an electronics career.
- Many scientists and engineers began electronics as a hobby. If electronics is one of your hobbies, this may indicate career success in the electronics industry.
- Discuss your abilities and aptitudes with your school guidance counselor. APTITUDES are your talents and natural abilities. Take an inventory test. These tests measure your interest and capacity for learning in specific areas.
- Have you enjoyed this electronics textbook? The fact that you are studying electronics is a good indicator of an interest in this area. Have you taken other classes in electronics? Did you like them? Did you do well?

Give serious thought to your responses to the above questions and guidelines. Your responses may start you on the way to achieving success in your chosen area.

23.2 CAREERS IN THE ELECTRONICS INDUSTRY

Discoveries and developments in science and technology are made every day. Electronics and other fields of service play a major role in daily life.

Think of common devices that have been improved with more complex machines. For example, the communication power of the telephone, television, and radio has been increased through the use of satellites. The safety of boats and airplanes has been improved through the use of sonar and radar devices.

Where else is the use of electronics apparent? Industry is electronically controlled. Robots and machines displace hundreds of production workers every day, Fig. 23-2. In many businesses, computers can complete complex tasks in only a few minutes. Electronics has also entered the field of medicine. Using laser technology, surgeons can now perform surgery without cutting any skin.

With each new device that is developed, many highly-skilled technicians must be trained to maintain and service it. The need for people with electronics training will continue to grow into the next century.

TYPES OF ELECTRONICS INDUSTRIES

In most industries, there are commonly two types of businesses: those that sell goods (products) and those that sell services.

Goods producing businesses sell books, steel, cars, etc. These products can be seen and touched. They are TANGIBLE. Electronic goods include such things as resistors, capacitors, televisions, compact disk players, and computers.

Service producing businesses provide useful labor that results in a need or want being satisfied. Electronic services include installation, maintenance, repair, and update of electronic systems (computers, telephone lines, etc.).

The United States Bureau of Labor Statistics estimates that employment has been shifting from goods producing industries to service producing industries over the past few years. In addition, they predict that by 2000, nearly 4 out of 5 jobs will be in industries that provide services, Fig. 23-3.

CHRYSLER CORP.

Fig. 23-2. With the help of a computer programmer, these robots spot weld body sections of cars in this assembly plant.

Career Opportunities in Electronics

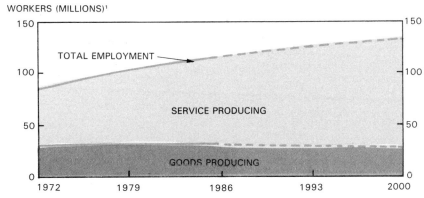

Fig. 23-3. Industries that provide services will account for nearly 4 out of 5 jobs by the year 2000.

CAREERS

In both goods producing and service producing electronics industries, four general classes of workers exist: semiskilled, skilled, technical, and professional. Career opportunities exist in each of these areas. The training needed to perform jobs in each of these areas varies. Likewise, employment outlooks and salaries vary.

Semiskilled workers

Semiskilled workers perform jobs that do not require a high level of training. These workers are most often found working on assembly lines, Fig. 23-4. Most semiskilled jobs are limited to a certain type and number of tasks. These tasks are often simple and repetitive (repeated). In general, semiskilled work is routine. Advance study and training are required to move up from a semiskilled position. There are very few electronics jobs done by semiskilled workers.

Skilled workers

Skilled workers have thorough knowledge of and skill in a particular area. This knowledge and skill are gained through advanced study. Advanced study is obtained through apprenticeship programs or community college programs. Some skilled workers also obtain advance training from the Military Services.

An APPRENTICESHIP is a period of time spent learning a trade from an experienced, skilled worker. This training is done on the job, Fig. 23-5. This is usually combined with special classes or self-study courses. Four years is the usual length of time for an apprenticeship program.

Many community colleges offer courses and programs in the electronics field. Courses can be taken to learn more about a particular subject. Or a program may be followed to gain a particular skill.

Fig. 23-4. This worker examines cable products before they are sent on to the next station.

Fig. 23-5. Many skilled workers learn their trade from on-the-job training.

Electricity and Electronics

The Military Services offer many specialized areas of study in the electronics field. The opportunities for learning a trade are very good in the Military Services.

Many jobs in the electronics field are skilled positions. Some of these include maintenance and construction electricians, assemblers, and quality control inspectors. Assemblers wire and solder various parts for televisions, stereos, printed circuit boards, computers, etc. Quality control inspectors check the finished work of the assemblers.

Technicians

Technicians are specially trained workers capable of doing complex, technical jobs. Many technicians receive their training in two year programs at community colleges.

Technicians work with electronic equipment and assist engineers. They have the training needed to service and repair complex machines and components. Engineers rely on technicians to help them conduct research, test machines and components, and design new devices. Therefore, technicians must keep up to date on developments in the electronics industry.

Careers for technicians include broadcast technicians, Fig. 23-6, robotics technicians, and computer technicians.

Professionals

Nearly all professional workers have four years of college training. Many have more advanced degrees, such as masters and doctorates. Professionals have excellent opportunities for advancement.

One of the best known professional positions in the electronics industry is the engineer, Fig. 23-7. Engineers design and monitor the building of new equipment. Their goal is to design equipment that runs smoothly and does its job. Once this goal is met, technicians are assigned to maintain the equipment.

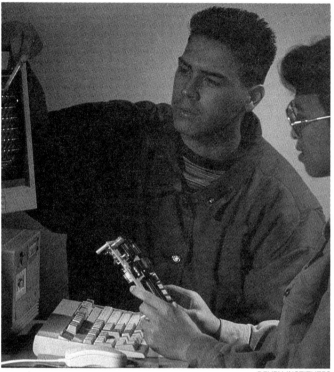

DEVRY INSTITUTES

Fig. 23-7. These research engineers are testing one of the devices they designed.

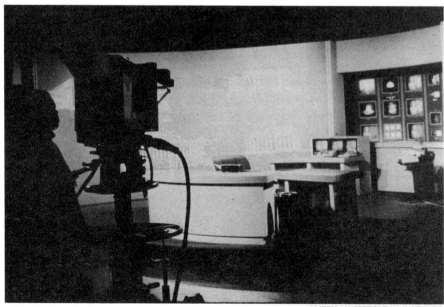

AMERICAN PETROLEUM INSTITUTE

Fig. 23-6. Broadcast technicians maintain and service all the electronics equipment in this television studio.

Engineers must have a solid background in math and science. This background allows them to visualize designs before putting them down on paper.

Teaching is another professional position in the electronics industry. Teachers of electronics have the opportunity to challenge students interested in electronics. They can share their knowledge and interest in electronics with their students, Fig. 23-8. The rewards of teaching are many.

Entrepreneurs

Entrepreneurs own and operate their own businesses. These small businesses make up 97% of businesses in the United States. They provide 58% of the jobs in America.

Entrepreneurships usually start with an idea for filling a hole in the marketplace where a new product or service is needed. Then a business plan is made. This plan outlines goals for the business, along with timetable for meeting those goals. This plan is vital if the business is to succeed.

In addition to a sound business plan, a successful entrepreneur possesses certain skills. The entrepreneur has knowledge of a certain industry, service, or product. This knowledge allows the owner to make smart business decisions about what is being sold. For instance, an appliance service technican needs knowledge of the appliance being serviced. If this is not the case, the business will fail.

The successful entrepreneur also has sound management skills. These skills allow the owner to manage money, time, and employees.

Entrepreneurial skills are also very important for the successful entrepreneur. These skills allow the business owner to control the business and move it in the right direction.

Entrepreneurship opportunities are vast in the electronics industry. With the growth in consumer electronics products, similar growth has occurred in servicing these products. Support of the office products industry also allows for many business opportunities. And servicing of home appliances continues to be a sound business in the electronics industry.

Consulting is yet another growing business in the electronics industry. Consultants work for clients on projects. The specific job they do often depends on what work is needed. The consultant is paid by the client. When the job is completed, the consultant is free to move on to a new job and client.

CAREER INFORMATION SOURCES

An excellent reference on careers in many industries is the *Occupational Outlook Handbook*. It is published

BALL STATE UNIVERSITY

Fig. 23-8. Teaching is a rewarding career for many people.

by the United States Department of Labor and Bureau of Statistics. Most public and secondary school libraries have copies of this book. The book can also be ordered from the Superintendent of Documents, U.S. Government Printing Office, Washington, DC, 20402.

School guidance counselors are another outstanding source of career information. They can help you find information on particular careers, two- and four-year colleges that offer programs in areas that you are interested, training programs through trade schools and the Military Services, etc. These people are well-informed and always ready to help when asked.

EDUCATION

The educational requirements for jobs in the electronics industry vary, Fig. 23-9. However, a minimum of a high school education is a solid foundation on which to build.

Electricity and Electronics

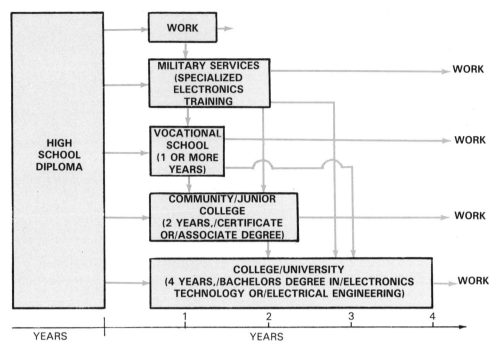

Fig. 23-9. There are many roads to take on your way to an electronics career.

Some high school graduates enter industry directly and receive specialized education in the training programs maintained by large companies. Many of these workers do not stop at this point, however. They continue to study and read to keep abreast of all the changes and new technologies that develop in this industry.

Specialized training may also come from a two-year college offering an associate degree program in technology. Trade schools and the Military Services also offer programs in electronics technology.

Career in science and engineering require a college degree. Advanced degrees are becoming more commonplace as a means of advancing. Many state and private universities offer engineering and electronics degrees. Talk with your school counselor to learn locations and entrance requirements.

The chart in Fig. 23-10 shows major job categories and educational requirements. Study this chart carefully.

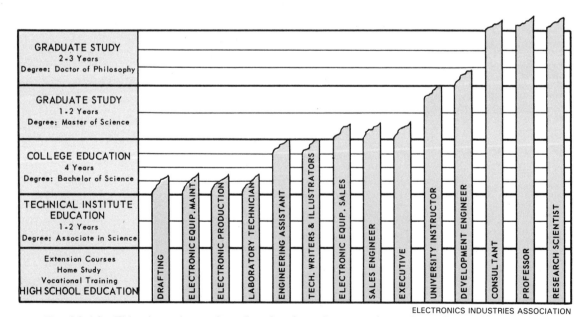

ELECTRONICS INDUSTRIES ASSOCIATION

Fig. 23-10. This chart shows the educational requirements for various electronics careers.

23.3 JOB SEARCH IDEAS

Finding a job is a time-consuming and difficult task. The *Occupational Outlook Handbook* has excellent tips for conducting a job search. Start by talking to your parents, neighbors, teachers, or guidance counselors. These people may know of job openings that have not been advertised. Read the want ads in the newspaper. Look through the Yellow Pages. Companies are grouped according to industry in the Yellow Pages. You may see companies to contact. City, county, or state employment services may also provide useful job leads. Private employment agencies might provide leads also. However, they often charge a fee if a job is found.

Once you have secured an interview, it is important to be prepared for it. Read the following tips for a successful interview.

Preparation:
- Learn about the organization.
- Have a specific job or jobs in mind.
- Review your qualifications for the job.
- Prepare to answer broad questions about yourself.
- Review your resumé.
- Arrive before the scheduled time of your interview.

Personal Appearance:
- Be well groomed.
- Dress appropriately.
- Do not chew gum or smoke.

The Interview:
- Answer each question concisely.
- Be prompt in giving responses.
- Use good manners.
- Use proper English and avoid slang.
- Convey a sense of cooperation and enthusiasm.
- Ask questions about the position and the organization.

Test (if employer gives one):
- Listen carefully to instructions.
- Read each questions carefully.
- Write legibly and clearly.
- Budget your time wisely and don't dwell on one question.

Information to Bring to an Interview:
- Social Security number.
- Driver's license number.
- Resumé. See Fig. 23-11.

```
                James Washington
                123 E. 25th Street
                Clover, Montana 59876
                  (406) 555-1212
```

OBJECTIVE	Briefly state what type of job you are seeking, and what you wish to learn in the job.
EDUCATION	List high school, college, trade school, etc. Give names and addresses, dates of attendance, curriculum studied, and highest grade completed or degree awarded.
EXPERIENCE	List both paid and volunteer work. Include job title, name and address of employer, and dates of employment.
	List any other special skills, knowledge, honors, awards, or memberships that may prove helpful in judging your qualification for a job.

Fig. 23-11. A good resume includes those items listed in the drawing.

- Three references. Get permission from people before using their names. Try to avoid using relatives. For each reference, provide name, address, telephone number, and occupation.

SUMMARY

1. Goods producing businesses produce tangible products.
2. Service producing businesses produce useful labor that results in a need or want being filled.
3. Semiskilled workers perform job tasks that are often simple and repetitive. These workers do not require high levels of training.
4. Skilled workers have thorough knowledge of and skill in a particular area.
5. Apprenticeship is a period of time spend learning a trade from an experienced, skilled worker.
6. Technicians are specially trained workers capable of doing complex, technical jobs.
7. Professionals are highly trained workers. Nearly all have four-year college degrees and many have masters and doctorates.
8. Entrepreneurs are people who own and operate their own businesses.
9. Three skills required by the successful entrepreneur are knowledge of a certain industry, service, or product, management skills, and entrepreneurial skills.

TEST YOUR KNOWLEDGE, Chapter 23

Please do not write in the text. Place your answers on a separate sheet of paper.

1. List three new electronic devices that will require maintenance by specially trained technicians.
2. What are tangible products?
3. In 2000, nearly 4 out of 5 jobs will be in industries that provide _____.
4. An _____ is a period of time spent learning a trade from a skilled, experienced worker.
5. What skills do successful entrepreneurs possess?

FOR DISCUSSION

1. What type of career would you like to have? How much training and education will this career require?
2. What traits do you think are required of successful entrepreneurs?

REFERENCE SECTION

Appendix 1

SCIENTIFIC NOTATION

In the study of electronics, you will work with very small quantities. As a student of mathematics, you understand that multiplication and division of numbers having many zeros and decimal points requires your concentration. Otherwise, errors will occur. The "powers of ten" or scientific notation is a useful method for performing these tasks.

$1 \times 10^0 = 1$
$1 \times 10^1 = 10$
$1 \times 10^2 = 100$
$1 \times 10^3 = 1000$
$1 \times 10^4 = 10,000$
$1 \times 10^5 = 100,000$
$1 \times 10^6 = 1,000,000$
$1 \times 10^{-1} = .1$
$1 \times 10^{-2} = .01$
$1 \times 10^{-3} = .001$
$1 \times 10^{-4} = .0001$
$1 \times 10^{-5} = .00001$
$1 \times 10^{-6} = .000001$

Follow these examples and you will learn to use the scientific notation method.

To express a number:

$47,000 = 47 \times 10^3$
$.000100 = 100 \times 10^{-6}$
$.0025 = 25 \times 10^{-4}$
$3,500,000 = 3.5 \times 10^6$

To multiply, add the exponents of ten:

$47 \times 10^3 \times 25 \times 10^{-4}$
$= 47 \times 25 \times 10^{-1}$

$100 \times 10^{-6} \times 3.5 \times 10^6$
$= 100 \times 3.5$

To divide, subtract exponents:

$$\frac{3.5 \times 10^6}{25 \times 10^{-4}} = \frac{3.5 \times 10^{10}}{25}$$

$$\frac{100 \times 10^{-6}}{25 \times 10^{-4}} = \frac{100 \times 10^{-2}}{25}$$

$6 - (-4) = +10$

$-6 - (-4) = -2$

To square, multiply exponent by 2:

$(25 \times 10^2)^2 = 25^2 \times 10^4$

$(9 \times 10^{-3})^2 = 81 \times 10^{-6}$

To extract square root, divide exponent by 2:

$\sqrt{81 \times 10^4} = 9 \times 10^2$

$\sqrt{225 \times 10^{-8}} = 15 \times 10^{-4}$

See Appendix 3 for using scientific notation in conversions.

Appendix 2
COLOR CODES

A standard color code has been adopted to determine the values of resistors and capacitors. Fig. A2-1 shows those values.

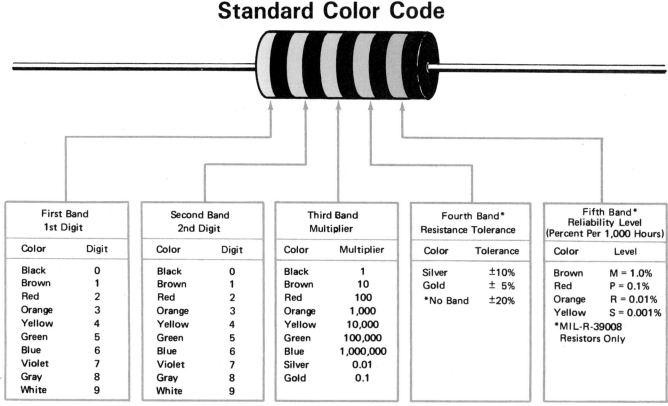

Fig. A2-1. Standard color code for resistors and capacitors is established by the Electronic Industries Association (EIA).

HOW TO READ THE CODE

Secure several resistors from stock. Hold the resistor so that the color bands are to the left, Fig. A2-2. Assume the band colors are (from left to right) brown, black, green, silver.

The brown band is 1. The black band is 0. The green band is the multiplier. Multiply by 10^5. The silver band tells us that it is within the limits of ± 10 percent of the color code value.

Review the following examples:

Yellow, Violet, Brown, Silver .... 470Ω ± 10%
Brown, Black, Red, Gold .... 1000Ω ± 5%
Orange, Orange, Red, None .... 3300Ω ± 20%
Green, Blue, Red, None.......... 5600Ω ± 20%
Red, Red, Green, Silver ...... 2,200,000Ω ± 10%

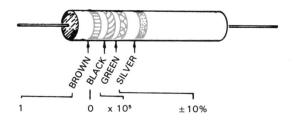

Fig. A2-2. When reading color code, hold the resistor with the bands at the left. This resistor value is 1,000,000 Ω or 1 megohm.

The practice of marking capacitors with numbers is fairly common. Some capacitors can be read directly. Others, however, are coded. There are many special marking systems. A detailed manual describing all codes should be consulted when necessary.

In Fig. A2-3 we will use the code shown in Fig. A2-1. Assume that the upper left-hand corner dot is white. The first significant figure is red. The second significant figure is green. The numbers for these figures then is 25. The multipler is brown. Multiply by 10^1. The value of the capacitor is 250 pF. The tolerance value and class can also be found. However, their meanings will not be covered in this textbook.

Appendix 3
CONVERSIONS

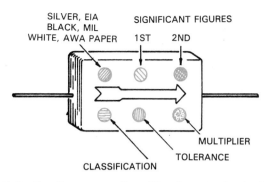

Fig. A2-3. Reading a capacitor. All values are in picofarads.

When using formulas to find unknown values in a circuit, the formula may be given in basic units. But the quantities you wish to use are given in larger or smaller units. A conversion must be made. For example, Ohm's Law states that:

$$I \text{ (in amperes)} = \frac{E \text{ (in volts)}}{R \text{ (in ohms)}}$$

If R were given in megohms such as 2.2 megohms, it would be necessary to change it to 2,200,000 ohms. If E were given in microvolts, such as 500 μV, it would be necessary to change it to .0005 volts.

Scientific notation offers a simple way to make conversions. First, however, look at the common prefixes used in electronics.

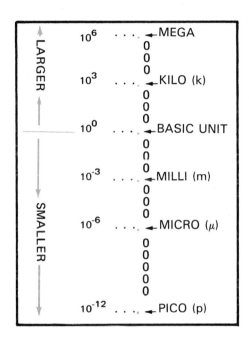

mega = 1,000,000

kilo = 1000

milli = $\frac{1}{1000}$

micro = $\frac{1}{1,000,000}$

pico = $\frac{1}{1,000,000,000,000}$

The prefix pico has now been adopted for micromicro. Compare the meaning of these prefixes to scientific notation in Appendix 1. Study Fig. A3-1.

10,000 ohms = 10,000 x 10^{-3} kilohms

47 kilohms = 47 x 10^3 ohms

950 kilohertz = 950 x 10^3 hertz or

950 x 10^{-3} megahertz

100 milliamperes = 100 x 10^{-3} amperes or

100 x 10^3 microamperes

.01 microfarads = .01 x 10^{-6} farads or

.01 x 10^6 pF

250 pF = 250 x 10^{-6} μF or

250 x 10^{-12} F

8 μF = 8 x 10^{-6} F or

8 x 10^6 pF

75 milliwatts = 75 x 10^{-3} watts

A simple rule will also help you. It states that if the exponent of ten is negative, the decimal point is moved to the left in the answer.

447 x 10^{-3} = .447

250 x 10^{-6} = .00025

If the exponent is positive, the decimal point is moved to the right.

$447 \times 10^3 = 447{,}000$

$250 \times 10^6 = 250{,}000{,}000$

Here is a problem. What is the reactance of a 2.5 mH choke at 1000 KHz?

$X_L = 2 \pi f L$

$= 2 \pi f \text{ (in Hz)} \times L \text{ (in henrys)}$

Two conversions must be made:

$1000 \text{ KHz} = 1000 \times 10^3 \text{ Hz}$

$2.5 \text{ mH} = 2.5 \times 10^{-3} \text{ henrys}$

Therefore:

$X_L = 2 \times 3.14 \times 1000 \times 10^3$

$\times 2.5 \times 10^{-3}$

$= 6.28 \times 2.5 \times 10^3$

(1000 changed to 10^3)

$= 15.6 \times 10^3 = 15{,}600 \text{ ohms}$

Appendix 4
TRIGONOMETRY

Trigonometry is a part of the "tool kit" of the electronic technician. Using it simplifies finding the solutions of alternating current problems. Trigonometry finds many uses in designing and understanding electronic circuits.

Basically, trigonometry is the relationship between the angles and sides of a triangle. These relationships are called functions. They represent the numerical ratio between the two sides of the right triangle.

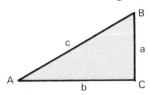

Sine of angle A $= \dfrac{a \text{ opposite side}}{c \text{ hypotenuse}}$

Cosine of angle A $= \dfrac{b \text{ adjacent side}}{c \text{ hypotenuse}}$

Tangent of angle A $= \dfrac{a \text{ opposite side}}{b \text{ adjacent side}}$

There are other functions, but these three are widely used in solving problems in electronics. Using these equations, if two values are known, the third may be found.

For example, in a triangle, side a = 6 and b = 10. Find the tangent of angle A.

$\text{Tan } A = \dfrac{a}{b} = \dfrac{6}{10} = .6$

Look at the table of Natural Trigonometric functions, Fig. A4-1. Find the angle whose tangent is .6. It is 31°.

Try this problem. Angle A is 45° and is 6. What is the value of side c? Use the sine equation.

$\text{Sine } 45° = \dfrac{6}{c}$

Look up the sine of 45°. It is .707. Then,

$c = \dfrac{6}{.707} = 8.5 \text{ (approx.)}$

In electronics, the right triangle used for the previous examples can be labeled other ways. But the problems are worked out in the same manner.

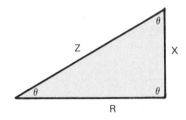

For example, a circuit contains 40Ω resistance (R) and 50Ω reactance (X). What is the circuit impedance (Z) in ohms?

Find angle θ:

$\text{Tan } \theta = \dfrac{X}{R} = \dfrac{50}{40} = 1.25$

Look up in the table:

$\theta = 51° \text{ approx.}$

Find Z:

$\text{Sine } \theta = \dfrac{X}{R}$

Look up in the table:

$.777 = \dfrac{50}{Z}$

$Z = \dfrac{50}{.777} = 64.3Ω$

These applications of trigonometry must be thoroughly understood. The best way to gain this understanding is through practice. Problems have been included in the text for this purpose.

Reference Section

NATURAL TRIGONOMETRIC FUNCTIONS

Angle	Sine	Cosine	Tangent	Angle	Sine	Cosine	Tangent
1°	.0175	.9998	.0175	46°	.7193	.6947	1.0355
2°	.0349	.9994	.0349	47°	.7314	.6820	1.0724
3°	.0523	.9986	.0524	48°	.7431	.6691	1.1106
4°	.0698	.9976	.0699	49°	.7547	.6561	1.1504
5°	.0872	.9962	.0875	50°	.7660	.6428	1.1918
6°	.1045	.9945	.1051	51°	.7771	.6293	1.2349
7°	.1219	.9925	.1228	52°	.7880	.6157	1.2799
8°	.1392	.9903	.1405	53°	.7986	.6018	1.3270
9°	.1564	.9877	.1584	54°	.8090	.5878	1.3764
10°	.1736	.9848	.1763	55°	.8192	.5736	1.4281
11°	.1908	.9816	.1944	56°	.8290	.5592	1.4826
12°	.2079	.9781	.2126	57°	.8387	.5446	1.5399
13°	.2250	.9744	.2309	58°	.8480	.5299	1.6003
14°	.2419	.9703	.2493	59°	.8572	.5150	1.6643
15°	.2588	.9659	.2679	60°	.8660	.5000	1.7321
16°	.2756	.9613	.2867	61°	.8746	.4848	1.8040
17°	.2924	.9563	.3057	62°	.8829	.4695	1.8807
18°	.3090	.9511	.3249	63°	.8910	.4540	1.9626
19°	.3256	.9455	.3443	64°	.8988	.4384	2.0503
20°	.3420	.9397	.3640	65°	.9063	.4226	2.1445
21°	.3584	.9336	.3839	66°	.9135	.4067	2.2460
22°	.3746	.9272	.4040	67°	.9205	.3907	2.3559
23°	.3907	.9205	.4245	68°	.9272	.3746	2.4751
24°	.4067	.9135	.4452	69°	.9336	.3584	2.6051
25°	.4226	.9063	.4663	70°	.9397	.3420	2.7475
26°	.4384	.8988	.4877	71°	.9455	.3256	2.9042
27°	.4540	.8910	.5095	72°	.9511	.3090	3.0777
28°	.4695	.8829	.5317	73°	.9563	.2924	3.2709
29°	.4848	.8746	.5543	74°	.9613	.2756	3.4874
30°	.5000	.8660	.5774	75°	.9659	.2588	3.7321
31°	.5150	.8572	.6009	76°	.9703	.2419	4.0108
32°	.5299	.8480	.6249	77°	.9744	.2250	4.3315
33°	.5446	.8387	.6494	78°	.9781	.2079	4.7046
34°	.5592	.8290	.6745	79°	.9816	.1908	5.1446
35°	.5736	.8192	.7002	80°	.9848	.1736	5.6713
36°	.5878	.8090	.7265	81°	.9877	.1564	6.3138
37°	.6018	.7986	.7536	82°	.9903	.1392	7.1154
38°	.6157	.7880	.7813	83°	.9925	.1219	8.1443
39°	.6293	.7771	.8098	84°	.9945	.1045	9.5144
40°	.6428	.7660	.8391	85°	.9962	.0872	11.4301
41°	.6561	.7547	.8693	86°	.9976	.0698	14.3006
42°	.6691	.7431	.9004	87°	.9986	.0523	19.0811
43°	.6820	.7314	.9325	88°	.9994	.0349	28.6363
44°	.6947	.7193	.9657	89°	.9998	.0175	57.2900
45°	.7071	.7071	1.0000	90°	1.0000	.0000	

Fig. A4-1. The table of natural trigonometric functions.

APPENDIX 5

STANDARD SYMBOLS AND ABBREVIATIONS

The symbols and abbreviations used in this text conform to Military Abbreviations and Contractions and IRE Standards 54 IRE 21S1. They represent accepted practice in trade and industry.

LETTER SYMBOLS

β	beta
I	current
E	voltage
R	resistance
X	reactance
X_L	inductive reactance
X_C	capacitive reactance
Z	impedance
L	inductance
C	capacitor
M	mutual inductance
f	frequency
f_o	resonant frequency
λ	wave length
θ	phase displacement
Δ	a change in
Ω	ohms
Φ	phi-magnetic flux
G	conductance
Q	Q factor
r_p	plate resistance
g_m	transconductance
μ	amplification factor, permeability, micro
A	gain
t	time
ϕ	phase angle
ω	angular velocity
V	vacuum tube or volts in transistor circuits
P	watts
p	pico

STANDARD ABBREVIATIONS

ac	alternating current
AWG	American Wire Gage
amp	ampere
AM	Amplitude Modulation
af	audio frequency
afc	automatic frequency control
avc	automatic volume control
bfo	beat-frequency-oscillator
CRT	Cathode Ray Tube
cw	continuous wave
cps	cycles per second
db	decibel
dc	direct current
DPST	double pole, single throw
DPDT	double pole, double throw
emf	electromotive force
F	farad
FET	Field Effect Transistor
FM	Frequency Modulation
gnd	ground
H	henrys
Hz	hertz
hf	high frequency
hp	horsepower
if	intermediate frequency
JAN	Joint-Army-Navy
kHz	kilohertz
k	kilohm
kV	kilovolt
kWh	kilowatt hour
lf	low frequency
mmf	magnetomotive force
max	maximum
MHz	megahertz
meg	megohm
μ	micro
μA	microampere
μF	microfarad
μH	microhenry
$\mu\mu$F	micromicrofarad
mike	microphone
μV	microvolt
mH	millihenry
mA	millampere
mV	millivolt
mW	milliwatt
min	minimum
osc	oscillator
pF	picofarad
pot	potentiometer
PF	Power Factor
rf	radio frequency
rpm	revolutions per minute
rms	root mean square
SPDT	single pole, double throw
SPST	single pole, single throw
spkr	speaker
sw	switch
TR	Transmit-Receive
uhf	ultra high frequency
VTVM	vacuum tube voltmeter
vhf	very high frequency
vf	video frequency
V	volts
W	watts

Reference Section

COMMON SCHEMATIC SYMBOLS

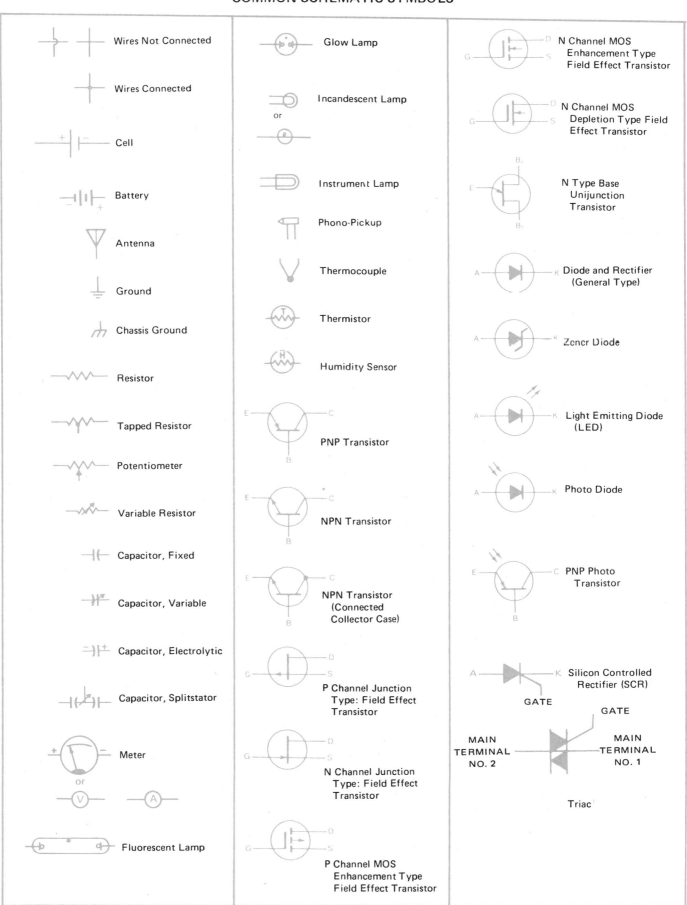

Electricity and Electronics

COMMON SCHEMATIC SYMBOLS

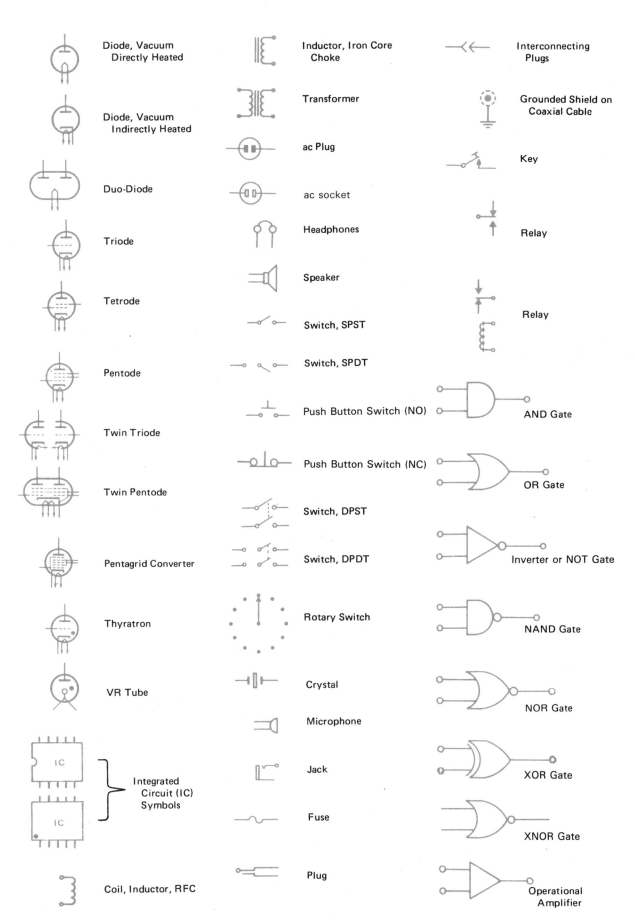

402

Reference Section

× or ·	multiplied by
÷ or :	divided by
+	positive; add
−	negative; subtract
±	plus or minus
= or ::	equals to
≡	identity
≅	is similar to
≠	does not equal
>	is greater than
<	is less than
≥	greater than or equal to
≤	less than or equal to
∴	therefore
∠	angle
Δ	increment or decrement (change in)
⊥	perpendicular to
∥	parallel to
∞	infinity

Electricity and Electronics

Appendix 6

ELECTRODEMONSTRATOR KIT

The electrodemonstrator kit contains the parts needed to build sixteen projects illustrated and described in preceding chapters of this text.

Projects which may be built from the parts include: permanent magnetic fields, electromagnetic field, electromagnet, relay, circuit breaker, door chime, buzzer, series dc motor, shunt dc motor, ac motor, Faraday's experiment, solenoid sucking coil, reactance dimmer, induction coil, series circuits, and parallel circuits.

Each project is assembled on a peg-board base. After the project has been built and tested, it is then disassembled. The parts are reused to make other projects.

Items in the electrodemonstrator kit are listed below. This includes items like sockets, leads, bolts, and nuts that are purchased along with brackets, coils, etc. Construction details for student-made parts are given in the drawings, Fig. A6-1.

THE ELECTRODEMONSTRATOR PARTS LIST

1 — Base, 1/8 x 7 1/2 x 10 Masonite peg-board
4 — Base feet, rubber 3/8
4 — Feet bolts, 3/8 x 8–32 round head with nuts
1 — Base support, wood (Plan, Detail M)
2 — Coils, (Plan, Detail F)
4 — Coil brackets, brass 1/32" (Plan, Detail G)
2 — Coil cores, 1/2 x 2 1/2 mild steel (Plan, Detail E)
1 — Solenoid core, mild steel (Plan, Detail H)
2 — Magnets, 1/2 x 2 1/2 round
1 — Armature bracket brass 1/32" (Plan, Detail J)
1 — dc armature (Plan, Detail A)
1 — ac armature (Plan, Detail B)
1 — Brush support (Plan, Detail L)
1 — Relay armature (Plan, Detail N)
1 — Armature plate (Plan, Detail K)
1 — Circuit breaker stop (Plan, Detail P)
2 — Contact brackets (Plan, Detail R)
1 — Induction coil (Plan, Detail D)
1 — .01 mfd. paper capacitor, Mount on subbases.

1 — Push button switch
1 — Toggle switch
3 — Subbase
6 — Jiffy leads, 8"
3 — Miniature Sockets with bulbs 6 V
1 — Doorbell transformer
1 — 50 Potentiometer with subbase for potentiometer (2W)
1 — Terminal base
3 — Battery holders D cell, Hickok teaching systems,
1 — Chime, hard steel tube or bar
1 — Galvanometer, 500-0-500 μa in meter case,
12 — Bolts, 3/8 x 8-32 round head
12 — Nuts, Hex. 8-32
4 — Stove bolts, 3/16 x 2 1/2
8 — Nuts, 3/16"

Construction Hints:

1. Use the Masonite peg-board base with rubber feet attached at each corner.
2. The two coils, F, are interchangeable in all projects. The coil brackets are bolted to the Masonite base. The saw kerf at each end of coil fits snugly into bracket.
3. The permanent magnets are interchangeable with the iron cores, when used for magnetic field demonstrations or in the dc motor.
4. Induction coil D is made to slide over coil F. Wind with more than 8 layers of No. 30 wire if higher voltages are desired. Capacitor is connected across breaker points.
5. If difficulty is experienced with the ac motor, try starting it by winding a string around the armature shaft and pulling it. Its speed must be synchronized with the 60 hertz line voltage for operation.
6. All motors are constructed by placing armatures between plate R, bolted to base, and armature bracket as the top bearing.
7. The brush support L is constructed so that it may be locked in several positions to demonstrate a change in commutation angle.

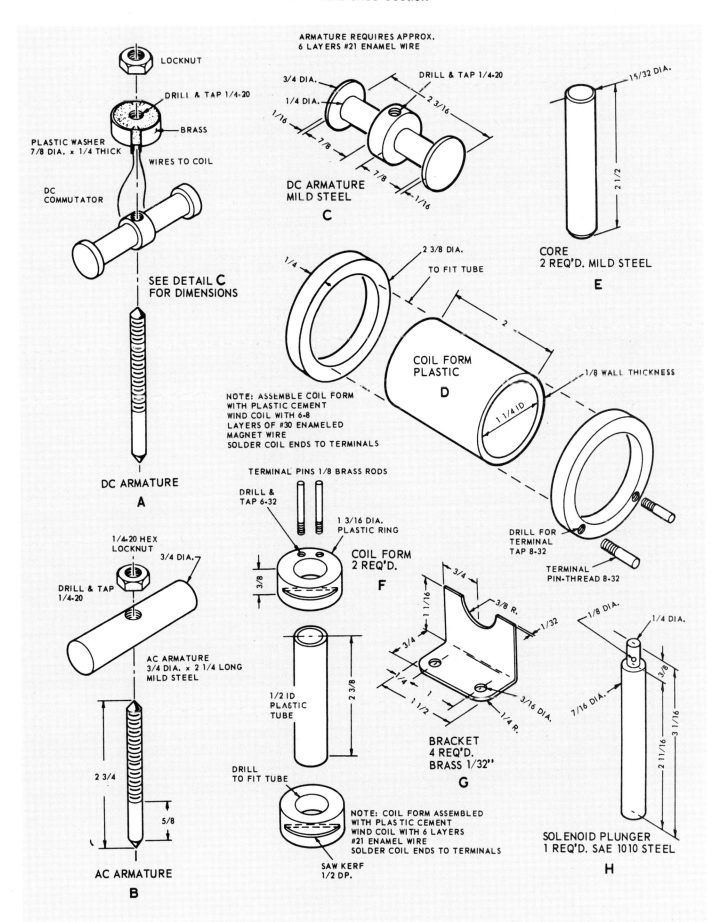

Fig. A6-1. Construction details of electrodemonstrator parts.

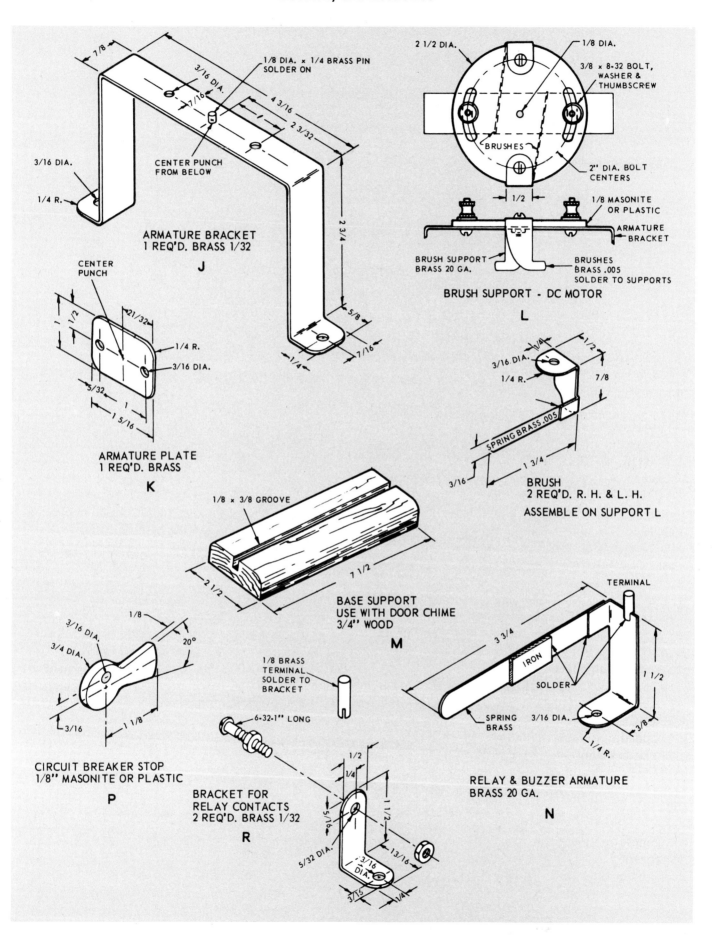

Fig. A6-1. Construction of electrodemonstrator parts (continued).

DICTIONARY OF TERMS

A

ABACUS: An ancient calculating device that contains movable beads used to perform basic mathematic functions.

A-BATTERY: Battery used to supply heater voltage for electron tubes.

AC: Alternating current.

ACCEPTOR CIRCUIT: Series tuned circuit at resonance. Accepts signal at resonant frequency.

ACCEPTER IMPURITY: Impurity added to semiconductor material which creates holes for current carriers.

AC GENERATOR: Generator using slip rings and brushes to connect armature to external circuit. Output is alternating current.

AC PLATE RESISTANCE: Variational characteristic of vacuum tube representing ratio of change of plate voltage to change in plate current, while grid voltage is constant. Its symbol is r_p.

AERIAL: An antenna.

AF AMPLIFIER: Used to amplify audio frequencies.

AGC: Abbreviation for AUTOMATIC GAIN CONTROL. Circuit employed to vary gain or amplifier in proportion to input signal strength so output remains at constant level.

AIR-CORE INDUCTOR: Inductor wound on insulated form without metallic core. Self-supporting coil without core.

ALKALINE BATTERY: A primary cell composed of manganese dioxide for the positive plate, powdered zinc for the negative plate, and caustic alkali for the electrolyte.

ALNICO: Special alloy used to make small permanent magnets.

ALPHA: Greek letter α, representing current gain of a transistor. It is equal to change in collector current caused by change in emitter current for constant collector voltage.

ALPHA CUT-OFF FREQUENCY: Frequency at which current gain drops to .707 of its maximum gain.

ALTERNATOR: An ac generator.

ALTERNATING CURRENT(ac): Current of electrons that moves first in one direction and then in the other.

ALMALGAM: Compound or mixture containing mercury.

ALMALGAMATION: Process of adding small quantity of mercury to zinc during manufacture.

AMMETER: Meter used to measure current.

AMPERE: Electron or current flow representing flow of one coulomb per second past given point in circuit. Its symbol is I.

AMPERE-HOUR: Capacity rating measurement of batteries. A 100 ampere-hour battery will produce, theoretically, 100 amperes for one hour.

AMPERE TURN (IN): Unit of measurement of magnetomotive force. Represents product of amperes times number of turns in coil of electromagnetic. $F = 1.257\ IN$.

AMPLIFICATION: Ability to control a relatively large force by a small force. In a vacuum tube, relatively small variation in grid input signal is accompanied by relatively large variation in output signal.

AMPLIFICATION FACTOR: Expressed as μ(mu). Characteristic of vacuum tube to amplify a voltage. Mu is equal to change in plate voltage as a result of change in grid voltage while plate current is constant.

AMPLIFIER, POWER: Electron tube used to increase power output. Sometimes called a current amplifier.

Class A: An amplifier, so biased that plate current flows during entire cycle of input signal.

Class B: An amplifier, so biased that plate current flows for approximately one-half the cycle of input signal.

Class C: An amplifier, so biased that plate current flows for appreciably less than half of each cycle of applied signal.

Class AB: A compromise between class A and class B.

Class AB$_1$: Same as class AB, only grid is never driven positive and no grid current flows.

Class AB$_2$: Same as AB except that signal drives grid positive and grid current does flow.

AMPLITUDE: Extreme range of varying quantity. Size, height of.

AMPLITUDE DISTORTION: Deviation in amplitude caused by non-linear operation of electron tube when peaks of input signals are reduced or cut off by either excessive input signal or incorrect bias.

AMPLITUDE MODULATION (AM): Varying of the radio wave amplitude at an audio frequency rate. **Grid.** Audio signal is applied in series with the grid of the power amplifier. **Plate.** Audio signal is injected into the plate circuit of the modulated stage.

ANALOG DEVICES: Electronic devices that have variable outputs controlled by variable inputs. These are also called linear integrated circuits.

ANALOG METER: A meter with a continuously variable scale.

AND GATE: A logic gate that is used to determine the presence of yes signals or 1s.

ANGLE OF COMPENSATION: Correcting angle applied to compass reading to compensate for local magnetic influence.

ANGLE OF DECLINATION: See DECLINATION.

ANGULAR PHASE: Position of rotating vector in respect to reference line.

ANGULAR VELOCITY (ω): Speed of rotating vector in radians per second. ω (omega) = $2\pi F$.

ANODE: Positive terminal, such as plate in electron tube.

ANTENNA: Device for radiating or receiving radio waves.

APERTURE MASK: A thin sheet of perforated material placed directly behind the viewing screen in a three-gun color picture tube.

APPARENT POWER: Power apparently used in circuit as product of current times voltage.

ARCBACK: Current flowing opposite direction in a diode, when plate has a high negative voltage.

ARITHMETIC/LOGIC UNIT (ALU): A part of the central processor that performs arithmetic operations, such as subtraction, and logical operations, such as true/false comparisons.

ARMATURE: Revolving part in generator or motor. Vibrating or moving part of relay or buzzer.

ARMATURE REACTION: Effect on main field of generator by armature acting as electromagnet.

ARRL: American Radio Relay League. The association of radio amateurs.

ARTIFICIAL MAGNETS: Manufactured magnets.

ASPECT RATIO: Ratio of width to height of TV picture. This is standardized as 4:3.

A-SUPPLY: Voltages supplied for heater circuits of electron tubes.

AT-CUT CRYSTAL: Crystal cut at approximately a 35 deg. angle with Z axis.

ATOM: Smallest particle that makes up a type of material, called an element.

ATOMIC NUMBER: Number of protons in nucleus of a given atom.

ATOMIC WEIGHT: Mass of nucleus of atom in reference to oxygen, which has a weight of 16.

ATTENUATION: Decrease in amplitude or intensity.

ATTENUATOR: Networks of resistance used to reduce voltage, power, or current to a load.

AUTOTRANSFORMER: Transformer with common primary and secondary winding. Step-up or step-down action is accomplished by taps on common winding.

AVC: Automatic volume control.

AVERAGE VALUE: Value of alternating current or voltage of sine wave form that is found by dividing area under one alternation by distance along X axis between 0 and 180 deg. $E_{avg} = .637\ E_{max}$

AWG: American Wire Gauge—used in sizing wire by numbers.

B

BACK EMF: See COUNTER EMF.

BAND: Group of adjacent frequencies in frequency spectrum.

BAND PASS FILTER: Filter circuit designed to pass currents of frequencies within continuous band and reject or attenuate frequencies above or below the band.

BAND REJECT FILTER: Filter circuit designed to reject currents in continuous band of frequencies, but pass frequencies above or below the band.

BAND SWITCHING: Receiver employing switch to change frequency range of reception.

BANDWIDTH: Band of frequencies allowed for transmitting modulated signal.

BARRIER REGION: Potential difference across a PN junction due to diffusion of electrons and holes across junction.

BASE: Semiconductor between emitter and collector of transistor.

BASS: Low frequency sounds in audio range.

BATTERY: Several voltaic cells connected in series or

Dictionary of Terms

parallel. Usually contained in one case.

B-BATTERY: Group of series cells in one container producing high voltage for plate circuits of electronic devices. Popular batteries include 22 1/2, 45, and 90 volts.

BEAM POWER TUBE: Tube so constructed that electrons flow in concentrated beams from cathode through grids to plate.

BEAT FREQUENCY: The resultant frequency obtained by combining two frequencies.

BEAT FREQUENCY OSCILLATOR: Oscillator whose output is beat with continuous wave to produce beat frequency in audio range. Used in cw reception.

BEL: Unit of measurement of gain equivalent to 10 to 1 ratio of power gain.

BETA: Greek letter β, represents current gain of common-emitter connected transistor. It is equal to ratio of change in collector current to change in base current, while collector voltage is constant.

BIAS,

CATHODE SELF: Bias created by voltage drop across cathode resistor.

FIXED: Voltage supplied by fixed source.

GRID LEAK: Bias created by charging capacitor in grid circuit. Bias level is maintained by leak resistor.

BIAS, FORWARD: Connection of potential to produce current across PN junction. Source potential connected so it opposes potential hill and reduces it.

BIAS, REVERSE: Connection of potential so little or no current will flow across PN junction. Source potential has same polarity as potential hill and adds to it.

BINARY: Number system having base of 2, using only the symbols 0 and 1.

BIT: One binary digit (0 or 1).

BLACK BOX: Box containing unknown and possibly complicated circuit.

BLANKING PULSE: Pulses transmitted by TV transmitter, which are used in receiver to cut off scanning beam during retrace time.

BLEEDER: Resistor connected across power supply to discharge filter capacitors.

BRIDGE CIRCUIT: Circuit with series-parallel groups of components that are connected by common bridge. Bridge frequently is meter in measuring devices.

BRIDGE RECTIFIER: Full-wave rectifier circuit employing four rectifiers in bridge configuration.

BRIGHTNESS: In television. Overall intensity of illumination of picture.

BRUSH: Sliding contact, usually carbon, between commutator and external circuit in dc generator.

B-SUPPLY: Voltages supplied for plate circuits of electron tubes.

BUFFER: Type of amplifier placed between oscillator and power amplifier to isolate from load.

BUS: An electrical pathway on which current flows.

BYPASS CAPACITOR: Fixed capacitor which bypasses unwanted ac to ground.

BYTE: Two binary nibbles, or eight binary bits.

BX CABLE: Cable sheathed in metallic armor.

C

CABLE: May be stranded conductor or group of single conductors insulated from each other.

CABLE TV: A communications system that sends television signals using coaxial cables.

CAPACITANCE: Inherent property of electric circuit that opposes change in voltage. Property of circuit whereby energy may be stored in electrostatic field.

CAPACITIVE COUPLING: Coupling resulting from capacitive effect between components or elements in electron tube.

CAPACITIVE REACTANCE (X_C): Opposition to ac as a result of capacitance.

CAPACITOR: Device which possesses capacitance. Simple capacitor consists of two metal plates separated by insulator.

CAPACITOR INPUT FILTER: Filter employing capacitor as its input.

CAPACITOR MOTOR: Modified version of split-phase motor, employing capacitor in series with its starting winding, to produce phase displacement for starting.

CAPACITY: Ability of battery to produce current over a given length of time. Capacity of a battery is measured in ampere-hours.

CARRIER: Usually radio frequency continuous wave to which modulation is applied. Frequency of transmitting station. In a semiconductor the conducting hole of an electron.

CASCADE: Arrangement of amplifiers where output of one stage becomes input of next, throughout series of stages.

CASCODE: Electron tubes connected so second tube acts as plate load for first. Used to obtain higher input resistance and retain low noise factor.

CATHODE: Emitter in electron tube.

CATHODE FOLLOWER: Single-stage Class A amplifier, output of which is taken from across unbypassed cathode resistor.

CATHODE RAY TUBE (CRT): Vacuum tube in which electrons emitted from cathode are shaped into narrow beam and accelerated to high velocity before striking phosphor coated viewing screen.

CAT WHISKER: Fine wire used to contact a crystal.

C-BATTERY: Battery used to supply grid bias

voltages.
CENTER FREQUENCY: Frequency of transmitted carrier wave in FM when no modulation is applied.
CENTER TAP: Connection made to center of coil.
CENTRAL PROCESSING UNIT: The computer component that directs or controls computer operations. It is composed of an arithmetic logic unit, a control unit, and internal memory.
CHARACTERISTIC CURVE: Graphic representation of characteristics of component, circuit, or device.
CHIP: A complete circuitry package contained on a small piece of silicon. The circuit often consists of diodes, resistors, capacitors, and connector wires.
CHOKE INPUT FILTER: Filter employing choke as its input.
CIRCUIT: A complete pathway on which a current flows.
CIRCUIT BREAKER: Safety device which automatically opens circuit if overloaded.
CIRCULAR MIL: Cross-sectional area of conductor one mil in diameter.
CIRCULAR-MIL-FOOT: Unit conductor one foot long with cross-sectional area of one circular mil.
CIRCULATING CURRENT: Inductive and capacitive currents flowing around parallel circuit.
COAXIAL LINE: Concentric transmission line in which inner conductor is insulated from tubular outer conductor.
COEFFICIENT OF COUPLING (K): Percentage of coupling between coils, expressed as a decimal.
COLLECTOR: The electrode that collects electrons in a tube after they have performed their function.
COLPITTS OSCILLATOR: A basic type of oscillator that is characterized by tapped capacitors in the tank circuit.
COMMON BASE: Transistor circuit, in which base is common to input and output circuits.
COMMON COLLECTOR: Transistor circuit in which collector is common to input and output circuits.
COMMON EMITTER: Transistor circuit in which emitter is common to input and output circuits.
COMMUTATION: Process of reversing current in armature coils and conducting direct circuit to external circuit by means of commutator segments and brushes.
COMMUTATOR: Group of bars providing connections between armature coils and brushes. Mechanical switch to maintain current in one direction in external circuit.
COMPLEMENTARY METAL-OXIDE LOGIC (CMOS): A digital circuit arrangement that uses field effect transistors for its logic circuits.
COMPOUND GENERATOR: Uses both series and shunt windings. A source of energy that converts mechanical energy to electrical energy.
COMPOUND GENERATORS,
  FLAT: When no-load and full-load voltages have same value.
  OVER: Full-load voltage is higher than no-load voltage.
  UNDER: Full-load voltage is less than no-load voltage.
COMPUTER: A machine that accepts data, processes it according to a stored program of instructions, and outputs the results.
CONDENSER: Older name for a capacitor.
CONDUCTANCE: Ability of circuit to conduct current. Symbol is G. It is equal to amperes per volt and is measured in siemens. $G = \frac{1}{R}$.
CONDUCTION BAND: Outermost energy level of atom.
CONDUCTIVITY, N TYPE: Conduction by electrons in N-type crystal.
CONDUCTIVITY, P TYPE: Conduction by holes in a P-type crystal.
CONDUCTOR: A low resistance material along which electric current can easily flow. Conductors have low resistance.
CONTINUOUS WAVE (cw): Uninterrupted sinusoidal rf wave radiated into space, with all wave peaks equal in amplitude and evenly spaced along time axis.
CONTRAST: In television. Relative difference in intensity between blacks and white in reproduced picture.
CONTROL GRID: Grid in vacuum tube closest to cathode. Grid to which input signal is fed to tube.
CONVERTER: Electromechanical system for changing alternating current to direct current.
COORDINATES: Horizontal and vertical distances used to locate point on graph.
COPPER LOSSES: Heat losses in motors, generators and transformers as result of resistance of wire. Sometimes called the I^2R loss.
COULOMB: Quantity of electrons representing approximately 6.24×10^{18} electrons.
COUNTER EMF (cemf): Voltage induced in conductor moving through magnetic field which opposes source voltage.
COUPLING: Percentage of mutual inductance between coils. Also called linkage.
COVALENT BOND: Atoms joined together, sharing each other's electrons to form stable molecule.
CROSSOVER FREQUENCY: Frequency in crossover network at which equal amount of energy is delivered to each of two loudspeakers.
CROSSOVER NETWORK: Network designed to divide audio frequencies into bands for distribution

Dictionary of Terms

to loudspeakers.

CROSS NEUTRALIZATION: Method of neutralization used with push-pull amplifiers, where part of output from each tube is fed back to grid circuit of each opposite tube through capacitor.

CROSS TALK: Leakage from one audio line to another which produces objectional background noise.

CRYSTAL DIODE: Diode formed by small semiconductor crystal and cat whisker.

CRYSTAL LATTICE: Structure of material when outer electrons are joined in covalent bond.

C-SUPPLY: Voltages supplied for grid bias of electron tubes, usually negative voltage.

CURRENT: Transfer of electrical energy in conductor by means of electrons moving constantly and changing positions in vibrating manner. Its symbol is I.

CUTOFF BIAS: Value of negative voltage applied to grid of tube which will cut off current flow through tube.

CW: Abbreviation for continuous wave.

CYCLE: Set of events occurring in sequence. One complete reversal of an alternating current from positive to negative and back to starting point.

D

DAMPER: Tube used in television set as half-wave rectifier to prevent oscillations in horizontal output transformer.

DAMPING: Gradual decrease in amplitude of oscillations in tuned circuit, due to energy dissipated in resistance.

D'ARSONVAL METER: Stationary-magnet moving coil meter.

DATA: The information given to the computer for processing.

DBM: Loss or gain in reference to arbitrary power level of one milliwatt.

DC: Direct current.

DC AMPLIFIER: Directly coupled amplifier that amplifies without loss of dc components.

DC COMPONENT: A dc value of ac wave which has axis other than zero.

DC GENERATOR: Generator with connections to armature through a commutator. Output is direct current.

DECAY: Term used to express gradual decrease in values of current and voltage.

DECIBEL: One-tenth of a Bel.

DECLINATION: Angle between true north and magnetic north.

DECODER: Part of the communications block diagram that changes the coded message into an uncoded message.

DEFLECTION: Deviation from zero of needle in meter. Movement or bending of an electron beam.

DEFLECTION ANGLE: Maximum angle of deflection of electron beam in TV picture tube.

DEGENERATIVE FEEDBACK: Feedback 180 deg. out of phase with input signal so it subtracts from input.

DELTA-TYPE ELECTRON GUN: The electron gun arrangement where the guns are placed 120 deg. apart in a color picture tube.

DEMODULATION: Process of removing modulating signal intelligence from carrier wave in radio receiver.

DEPLETION LAYER: In a semiconductor, region in which mobile carrier charge density is insufficient to neutralize net fixed charge of donors and acceptors. (IRE).

DEPOLARIZER: Chemical agent, rich in oxygen, introduced into cell to minimize polarization.

DETECTION: See DEMODULATION.

DIAPHRAGM: Thin disk, used in an earphone for producing sound.

DIELECTRIC: Insulating material between plates and capacitor.

DIELECTRIC CONSTANT: Numerical figure representing ability of dielectric or insulator to support electric flux. Dry air is assigned number 1.

DIELECTRIC FIELD OF FORCE: See ELECTROSTATIC FIELD.

DIFFUSION: Movement of carriers across semiconductor junction in absence of external force.

DIGITAL CIRCUITS: Electronic circuits that handle digital information (on or off) using switching circuits.

DIGITAL METER: An electronic meter in which the output is displayed as a number rather than on a meter.

DIGITAL INTEGRATED CIRCUIT: A switching type (on or off) integrated circuit.

DIODE: Two-element tube containing cathode and plate.

DIODE DETECTOR: Detector circuit utilizing unilateral conduction characteristics of diode.

DIODE-TRANSISTOR LOGIC (DTL): An arrangement of digital circuits using diodes, resistors, and transistors to produce logic gates.

DIP SWITCH: A microminiature switch made in an in-line assembly (having multiple switches) that is used in computers for attaching peripherals, such as printers.

DIRECT CURRENT (dc): Flow of electrons in one direction.

DISCRIMINATOR: A type of FM detector.

DISTORTION: The deviations in amplitude, phase,

and frequency between input and output signals of amplifier or system.

DOMAIN THEORY: Theory concerning magnetism, assuming that atomic magnets produced by movement of planetary electrons around nucleus have strong tendency to line up together in groups. These groups are called domains.

DONOR IMPURITY: Impurity added to semiconductor material which creates negative electron carriers.

DOPING: Adding impurity to semiconductor material.

DRY CELL: Non-liquid cell, which is composed of zinc case, carbon positive electrode, and paste of ground carbon, manganese dioxide, and ammonium chloride as electrolyte.

DYNAMIC CHARACTERISTICS: Characteristics of tube describing the actual control of grid voltage over plate current when tube is operating as an amplifier.

DYNAMIC PLATE RESISTANCE: See AC PLATE RESISTANCE.

DYNAMIC SPEAKER: Loudspeaker which produces sound as result of reaction between fixed magnetic field and fluctuating field of voice coil.

DYNAMOMETER: Measuring instrument based on opposing torque developed between two sets of current carrying coils.

DYNAMOTOR: Motor-generator combination using two windings on single armature. Used to convert ac to dc.

E

EDISION CELL: Cell using positive electrodes of nickel oxide and negative electrodes of powdered iron. The electrolyte is dilute solution of sodium hydroxide.

EDISION EFFECT: Effect first noticed by Thomas Edison, in which emitted electrons were attracted to positive plate in vacuum tube.

EDDY CURRENT LOSS: Heat loss resulting from eddy current flowing through resistance of core.

EDDY CURRENTS: Induced current flowing in rotating core.

EFFECTIVE RESISTANCE: Ratio between true power absorbed in circuit to square of effective current flowing in circuit.

EFFECTIVE VALUE: That value of alternating current of sine wave form that has equivalent heating effect of a direct current ($.707 \times E_{peak}$).

EFFICIENCY: Ratio between output power and input power.

ELECTRIC FIELD OF FORCE: See ELECTROSTATIC FIELD.

ELECTRICITY: The flow of electrons.

ELECTRODE: Elements in a cell.

ELECTRODYNAMIC SPEAKER: Dynamic speaker that uses electromagnetic fixed field.

ELECTROLYTE: Acid solution in a cell.

ELECTROLYTIC CAPACITOR: Capacitor with positive plate of aluminum and negative plate of dry paste or liquid. Dielectric is thin coat of oxide on aluminum plate.

ELECTROMAGNET: Coil wound on soft iron core. When current runs through coil, core becomes magnetized.

ELECTROMOTIVE FORCE (emf): Force that causes free electrons to move in conductor. Unit of measurement is the volt.

ELECTRON: Negatively charged particle.

ELECTRONICS: The study of the flow of electrons in active devices, such as transistors, semiconductors, diodes, or integrated circuits.

ELECTRON TUBE: Highly evacuated metal or glass shell which encloses several elements.

ELECTROSTATIC FIELD: Space around charged body in which its influence is felt.

ELECTROSTRICTION: Piezoelectric property of some elements in which they change in shape and size when voltage is applied and conversely, they produce voltage when subjected to pressure or stress.

ELEMENT: A distinct substance in which either alone or in combination with other elements, makes up all matter in the universe.

EMISSION: Escape of electrons from a surface.

EMISSION, COLD CATHODE: Phenomena of electrons leaving surface of a material caused by high potential field.

EMISSION, TYPES:
- A0: Continuous wave, no modulation.
- A1: Continuous wave, keyed.
- A2: Telegraphy by keying modulating audio frequency.
- A3: Telephony.
- A4: Facsimile.
- A5: Television.
- F0: Continuous wave, no FM.
- F1: Telegraphy by frequency shift keying.
- F2: Telegraphy by keying modulating audio frequency.
- F3: Telephone—FM.
- F4: Facsimile.
- F5: Television.

EMITTER: In transistor, semiconductor section, either P or N, which emits minority carriers. The cathode in a vacuum tube from which electrons are emitted.

ENCODER: Part of the communications block diagram that changes the information source into coded form.

ENERGY: That which is capable of producing work.

Dictionary of Terms

ENVELOPE: Enclosed waveform made by outlining peaks of modulated RF waves.

EXCITER: Small dc generator used to excite or energize field windings of large alternator.

EXCLUSIVE NOR GATE (XNOR): A logic gate that provides a logic high output (1) only if all inputs are logic high or logic low.

EXCLUSIVE OR GATE (XOR): A logic gate that provides a high output (1) whenever any, but not all, inputs are logic high.

F

FARAD: Unit of measurement of capacitance. A capacitor has a capacitance of one farad when charge of one coulomb raises its potential one volt.
$$C = \frac{Q}{E}$$

FEEDBACK: Transferring voltage from output of circuit back to its input.

FIELD MAGNETS: Electromagnets which make magnetic field motors or generators.

FILAMENT: Heating element in vacuum tube coated with emitting material so it acts also as cathode.

FILTER: Circuit used to attenuate specific band or bands of frequencies.

FLOPPY DISK: A small, flexible disk used with computers to store data. This disk is coated with magnetic oxide.

FLUORESCENT: Property of a phosphor which indicates that radiated light will be extinguished when electron bombardment ceases.

FLUX DENSITY: Number of lines of flux per cross-sectional area of magnetic circuit. Its symbol is B.

FREQUENCY: Number of complete cycles per second measured in hertz (Hz).

FREQUENCY BANDS: Abbreviations and ranges as follows:
  vlf: Very low frequencies 10-30 kHz.
  lf: Low frequencies 30-300 kHz.
  mf: Medium frequencies 300-3000 kHz.
  hf: High frequencies 3-30 MHz.
  vhf: Very high frequencies 30-300 MHz.
  uhf: Ultra high frequenices 300-3000 MHz.
  shf: Super high frequencies 3000-30,000 MHz.
  ehf: Extremely high frequencies 30,000-300,000 MHz.

FREQUENCY DEPARTURE: Instantaneous change from center frequency in FM as result of modulation.

FREQUENCY DISTORTION: Deviation in frequency caused by signals of some frequencies being amplified more than others, or when some frequencies are excluded.

FREQUENCY DOUBLER: Amplifier stage in which plate circuit is tuned to twice the frequency of grid tank circuit.

FREQUENCY METER: Meter used to measure frequency of an ac source.

FREQUENCY MODULATION (FM): Varying frequency of rf carrier wave at an audio rate.

FREQUENCY RESPONSE: Rating of device indicating its ability to operate over specified range of frequencies.

FREQUENCY TRIPLER: Amplifier stage in which plate circuit is tuned to three times the frequency (second harmonic) of grid circuit.

FULL-WAVE RECTIFIER: Rectifier circuit which produces a dc pulse output for each half-cycle of applied alternating current.

FUNDAMENTAL: A sine wave that has the same frequency as complex periodic wave. Component tone of lowest pitch in complex tone. Reciprocal of period of wave.

FUSE: Safety protective device which opens an electric circuit if overloaded. Current above rating of fuse will melt fusible link and open circuit.

G

GAIN: Ratio of output ac voltage to input ac voltage.

GALVANOMETER: Meter which indicates very small amounts of current.

GAS FILLED TUBE: Tubes designed to contain specific gas in place of air, usually nitrogen, neon, argon, or mercury vapor.

GAUSS: Measurement of flux density in lines per square centimeter.

GENERATOR: Rotating electric machine which provides a source of electrical energy. A generator converts mechanical energy to electric energy.

GHOST: In TV, a duplicate image of reproduced picture, caused by multipath reception of reflected signals.

GILBERT: Unit of measurement of magnetomotive force. Represents force required to establish one maxwell in circuit with one Rel of reluctance.

GRID: Grid of fine wire placed between cathode and plate of an electron tube.

GRID BIAS: Voltage between the grid and cathode, usually negative.

GRID CURRENT: Current flowing in grid circuit of electron tube, when grid is driven positive.

GRID DIP METER: A test instrument for measuring resonant frequencies, detecting harmonics, and checking relative field strength of signals.

GRID LEAK DETECTOR: Triode amplifier connected so it functions like a diode detector and an amplifier. Detection takes place in grid circuit.

GRID MODULATION: Modulation circuit where modulating signal is fed to grid of modulated stage.

GRID VOLTAGE: Bias or C voltage applied to grid of a vacuum tube.

GROUND: The common return circuit in electronic equipment whose potential is zero. A connection to earth by means of plates or rods.

H

HALF-WAVE RECTIFIER: Rectifier which permits one-half of an alternating current cycle to pass and rejects reverse current of remaining half-cycle. Its output is pulsating dc.

HARMONIC FREQUENCY: Frequency which is multiple of fundamental frequency.

HARTLEY OSCILLATOR: A basic type of oscillator that has a tapped oscillator coil.

HEATER: Resistance heating element used to heat cathode in vacuum tube.

HEAT SINK: Mass of metal used to carry heat away from component.

HENRY (H): Unit of measurement of inductance. A coil has one henry of inductance if an emf of one volt is induced when current through inductor is changing at rate of one ampere per second.

HERTZ (Hz): Basic unit for frequency. One hertz equals one cycle per second.

HETERODYNE: Process of combining two signals of different frequencies to obtain different frequency.

HOLE: Positive charge. A space left by removed electron.

HOLE INJECTION: Creation of holes in semiconductor material by removal of electrons by strong electric field around point contact.

HORIZONTAL POLARIZATION: An antenna positioned horizontally, so its electric field is parallel to earth's surface.

HORSEPOWER: 33,000 ft. lb. of work per minute or 550 ft. lb. of work per second equals one horsepower. Also 746 watts = 1 HP.

HUM: Form of distortion introduced in an amplifier as result of coupling to stray electromagnetic and electrostatic fields or insufficient filtering.

HYDROMETER: Bulb-type instrument used to measure specific gravity of a liquid.

HYSTERESIS: Property of magnetic substance that causes magnitization to lag behind force that produced it.

HYSTERESIS LOOP: Graph showing density of magnetic field as magnetizing force is varied uniformly through one cycle of alternating current.

HYSTERESIS LOSS: Energy loss in substance as molecule or domains move through cycle of magnetization. Loss due to molecular friction.

I

IF AMPLIFIER: Used to increase power output of audio frequencies.

IMPEDANCE (Z): Total resistance to flow of an alternating current as a result of resistance and reactance.

INDEPENDENTLY EXCITED GENERATOR: A source of electrical energy whose field windings are excited by a separate dc source.

INDIRECTLY HEATED: Electron tube employing separate heater for its cathode.

INDUCED CURRENT: Current that flows as result of induced electromotive force.

INDUCED EMF: Voltage induced in conductor as it moves through magnetic field.

INDUCTANCE: Inherent property of electric circuit that opposes a change in current. Property of circuit whereby energy may be stored in magnetic field.

INDUCTION MOTOR: An ac motor operating on principle of rotating magnetic field produced by out of phase currents. Rotor has no electrical connections, but receives energy by transformer action from field windings. Motor torque is developed by interaction of rotor current and rotating field.

INDUCTIVE CIRCUIT: Circuit in which a noticeable emf is induced while current is changing.

INDUCTIVE REACTANCE (X_L): Opposition to an ac current as a result of inductance.

INPUT: Data placed into a computer. The process of putting data in a computer.

INSTANTANEOUS VALUE: Any value between zero and maximum depending upon instant selected.

INSULATION RESISTANCE: Resistance to current leakage through and over surface of insulating material.

INSULATOR: A material which possesses a high resistance to current flow (electricity).

INTEGRATED CIRCUIT: A packaged electronic circuit containing resistors, transistors, diodes, and capacitors with their interconnected leads. These are usually processed from a chip of silicon.

INTENSITY: Magnetizing force per unit length of magnetic circuit.

INTERELECTRODE CAPACITANCE: Capacitance between metal elements in an electron tube.

INTERLACE SCANNING: Process in television of scanning all odd lines and then all even lines to reproduce picture. Used in United States.

INTERNAL RESISTANCE: Refers to internal resistance of source of voltage or emf. A battery or generator has internal resistance which may be represented as a resistor in series with source.

INTERPOLES: Auxiliary poles located midway between main poles of generator to establish flux or satisfactory commutation.

Dictionary of Terms

INTERRUPTED CONTINUOUS WAVE (icw): Continuous wave radiated by keying transmitter into long and short pulses of energy (dashes and dots), conforming to code such as Morse Code.

INTERSTAGE: Existing between stages, such as an interstage transformer between two stages of amplifiers.

INTRINSIC SEMICONDUCTOR: Semiconductor with electrical characteristics similar to a pure crystal.

IONIZATION: The loss or gain of one or more electrons.

IONIZATION POTENTIAL: Voltage applied to a gas-filled tube at which ionization occurs.

IONOSPHERE: Atmospheric layer from 40 to 350 miles above the earth, containing a high number of positive and negative ions.

IR DROP: See VOLTAGE DROP.

IRON VANE METER: Meter based on principle of repulsion between two concentric vanes placed inside a solenoid.

ISOLATION: Electrical separation between two locations.

J

JOULE: Unit of energy equal to one watt-second.

JUNCTION DIODE: PN junction, having unidirectional current characteristics.

JUNCTION TRANSISTOR: Transistor consisting of thin layer of N or P type crystal between P or N type crystals. Designated as NPN or PNP.

K

KEY: Manually operated switch used to interrupt rf radiation of transmitter.

KEY CLICK FILTER: Filter in keying circuit of a transmitter to prevent surges of current and prevent sparking at key contacts.

KEYING: Process of causing cw transmitter to radiate an rf signal when key contacts are closed.

KEYING,
   CATHODE: Key is inserted in grid and cathode circuits of keyed stage.
   GRID-BLOCK: Keying stage by applying high negative voltage on grid of tube.
   PLATE: Key is inserted in plate circuit of stage to be keyed.

KILO: One thousand times.

KILOGAUSS: One thousand gausses.

KILOWATT-HOUR (kWh): 1000 watts per hour. Common unit of measurement of electrical energy for home and industrial use. Power is priced by the kWh.

KIRCHHOFF'S LAW OF CURRENT: At any junction of conductors in a circuit, algebraic sum of currents is zero.

KIRCHOFF'S LAW OF VOLTAGES: In simple circuit, algebraic sum of voltages around circuit it equal to zero.

L

LAGGING ANGLE: Angle current lags voltage in inductive circuit.

LAMBDA: Greek letter λ. Symbol for wavelength.

LAMINATIONS: Thin sheets of steel used in cores of transformers, motors, and generators.

LARGE SCALE INTEGRATION (LSI): A complete computer circuit built on a single silicon chip with up to 1000 components.

LAWS OF MAGNETISM: Like poles repel; unlike poles attract.

L/C RATIO: Ratio of inductance to capacitance.

LEAD ACID CELL: Secondary cell which uses lead peroxide and sponge lead for plates, and sulfuric acid and water for electrolyte.

LEADING ANGLE: Angle current leads voltage in capacitive circuit.

LECLANCHE CELL: Scientific name for common dry cell.

LEFT-HAND RULE: A method using your left hand, to determine polarity of an electromagnetic field or direction of electron flow.

LENZ'S LAW: Induced emf in any circuit is always in such a direction as to oppose effect that produces it.

LIGHT EMITTING DIODE (LED): Special function diode that, when connected in the forward bias direction, emits light.

LIMITER: A stage or circuit that limits all signals at the same maximum amplitude.

LINEAR: Continuously variable.

LINEAR AMPLIFIER: An amplifier whose output is in exact proportion to its input.

LINEAR DETECTOR: Detector using linear portions of characteristic curve on both sides of knee. Output is proportional to input signal.

LINEAR DEVICE: Electronic device or component whose current-voltage relation is a straight line.

LINEAR INTEGRATED CIRCUIT: An amplifying (variable output) integrated circuit.

LINEARITY: Velocity of the scanning beam. It must be uniform for good linearity.

LINES OF FORCE: Graphic representation of electrostatic and magnetic fields showing direction and intensity.

LITHIUM BATTERY: A primary cell that has a long life.

LIQUID CRYSTAL DISPLAYS: A digital or

alphanumeric display unit that can be used in visual outputs for information.

LOAD: Resistance connected across circuit which determines current flow and energy used.

LOAD LINE: Line drawn on characteristic family of curves of electron tube, when used with specified load resistor, representing plate current at zero and maximum plate voltage.

LOCAL ACTION: Defect in voltaic cells caused by impurities in zinc, such as carbon, iron, and lead. Impurities form many small internal cells which contribute nothing to external circuit. Zinc is wasted away, even when cell is not in use.

LOCAL OSCILLATOR: Oscillator in superheterodyne receiver, output of which is mixed with incoming signal to produce intermediate frequency.

LODESTONE: Natural magnet, so called a "leading stone" or lodestone because early navigators used it to determine directions.

LOUDSPEAKER: Device to convert electrical energy into sound energy.

L-SECTION FILTER: Filter consisting of capacitor and an inductor connected in an inverted L configuration.

M

MAGNET: Substance that has the property of magnetism.

MAGNETIC AMPLIFIER: Transformer type device employing a dc control winding. Control current produces more or less magnetic core saturation, thus varying output voltage of amplifier.

MAGNETIC CIRCUIT: Complete path through which magnetic lines of force may be established under influence of magnetizing force.

MAGNETIC FIELD: Imaginary lines along which magnetic force acts. These lines emanate from north pole and enter south pole, forming closed loops.

MAGNETIC FLUX: Entire quantity of magnetic lines surrounding a magnet. Its symbol is Φ (phi).

MAGNETIC LINE OF FORCE: Magnetic line along which compass needle aligns itself.

MAGNETIC MATERIALS: Materials such as iron, steel, nickel, and cobalt which are attracted to magnet.

MAGNETIC PICKUP: Phono cartridge which produces an electrical output from armature in magnetic field. Armature is mechanically connected to reproducing stylus.

MAGNETIC SATURATION: This condition exists in magnetic material when further increase in magnetizing force produces very little increase influx density. Saturation point has been reached.

MAGNETIZATION: Graph produced by plotting intensity of magnetizing force on X axis and relative magnetism on Y axis.

MAGNETIZING CURRENT: Current used in transformer to produce transformer core flux.

MAGNETOMOTIVE FORCE (F) (mmf): Force that produces flux in magnetic circuit.

MAGNET POLES: Point of maximum attraction on a magnet; designated as north and south poles.

MAJOR CARRIER: Conduction through semiconductor as result of majority of electrons or holes.

MATTER: Anything that occupies space or has mass.

MAXIMUM POWER TRANSFER: This condition exists when resistance of load equals internal resistance of source.

MAXIMUM VALUE: Peak value of sine wave either in positive or negative direction.

MAXWELL: One single line of magnetic flux.

MEGA: Prefix meaning one million times.

MEMORY: The storage facilities of a computer. The term is used to refer to internal storage only, as opposed to auxiliary storage, such as disks or tapes.

MERCURY VAPOR RECTIFIER: Hot cathode diode tube which uses mercury vapor instead of high vacuum.

METALLIC RECTIFIER: Rectifier made of copper oxide, based on principle that electrons flow from copper to copper oxide but not from copper oxide to copper. It is unidirectional conductor.

MHO: Old unit of measurement of conductance.

MICA CAPACITOR: Capacitor made of metal foil plates separated by sheets of mica.

MICRO: Prefix meaning one millionth.

MICROCOMPUTER: A computer whose CPU is contained on a single chip; usually portable.

MICROFARAD (μF): One millionth of a farad.

MICROHENRY (μH): One millionth of a henry.

MICROMHO: One millionth of a mho.

MICROPHONE: Energy converter that changes sound energy into corresponding electrical energy.

MICROPROCESSOR: A single IC that does the processing in a microcomputer.

MICROSECOND: One millionth of a second.

MIL: One thousandth of an inch (.001 inches).

MIL-FOOT: A wire which is one mil in diameter and one foot long.

MILLI: Prefix meaning one thousandth.

MILLIAMMETER: Meter which measures in milliammeter range of currents.

MILLIHENRY (mH): One thousandth of a henry.

MINOR CARRIER: Conduction through semiconductor opposite to major carrier. Example: If electrons are major carrier, then holes are minor carrier.

MINUS: Negative terminal or junction of circuit. Its symbol is $-$.

Dictionary of Terms

MISMATCH: Incorrect matching of load to source.

MIXER: Multi-grid tube used to combine several input signals.

MODEM: A circuit that changes computer data in such a way that it can be transmitted and received over telephone lines.

MODULATED CONTINUOUS WAVE (mcw): Carrier wave amplitude modulated by tone signal of constant frequency.

MODULATION: Process by which amplitude or frequency of sine wave voltage is made to vary according to variations of another voltage or current called modulation signal.

MOLECULE: Smallest division of matter. If further subdivision is made, matter will lose its identity.

MOTOR: Device which converts electrical energy into mechanical energy.

MOTOR REACTION: Opposing force to rotation developed in generator, created by load current.

MOTOR,
  COMPOUND: Uses both series and parallel field coils.
  SERIES: Field coils are connected in series with armature circuit.
  SHUNT: Field coils are connected in parallel with armature circuit.

MU (μ): Greek letter representing the amplification factor of a tube.

MULTIELEMENT TUBE: Electron tube with more elements than cathode, plate, and grid.

MULTIGRID TUBE: Special tube with 4, 5, or 6 grids.

MULTIMETER: Combination volt, ampere, and ohm meter.

MULTIPLIER: Resistance connected in series with meter movement to increase its voltage range.

MULTIUNIT TUBE,
  HEXODE: Six elements with four grids.
  HEPTODE: Seven elements with five grids.
  OCTODE: Eight elements with six grids.
  TWIN DIODE: Two diodes in one envelope.
  TWIN DIODE TRIODE: Diode and triode in one envelope.
  TWIN DIODE TETRODE: Diode and tetrode in one envelope.
  TWIN PENTODE: Two pentodes in one envelope.

MULTIVIBRATORS,
  ASTABLE: A free-running multivibrator.
  BISTABLE: A single trigger pulse switches conduction from one tube to the other.
  CATHODE COUPLED: Both tubes have a common cathode resistor.
  FREE RUNNING: Frequency of oscillation depends on value of circuit components. Continuous oscillation.
  MONOSTABLE: One trigger pulse is required to complete one cycle of operation.
  ONE SHOT: Same as MONOSTABLE.
  PLATE COUPLED: The plates of the tubes and grids are connected by RC networks.

MUTUAL INDUCTANCE (M): Two coils located so that magnetic flux of one coil can link with turns of other coil. The change in flux of one coil will cause an emf in other.

N

NAND GATE: A negative AND logic gate.

NATURAL MAGNET: Magnets found in natural state in form of mineral called magnetite.

NEGATIVE ION: Atom which has gained electrons and is negatively charged.

NETWORK: Two or more components connected in either series or parallel.

NEUTRALIZATION: Process of feeding back voltage from plate of amplifier to grid, 180 deg. out of phase, to prevent self oscillation.

NEUTRON: Particle which is electrically neutral.

NICKEL CADMIUM CELL: Alkaline cell with paste electrolyte hermetically sealed. Used in aircraft.

NOISE: Any desired interference to a signal.

NO LOAD VOLTAGE: Terminal voltage of battery or supply when no current is flowing in external circuit.

NONLINEAR DEVICE: Electronic device or component whose current-voltage relation is not a straight line.

NOR GATE: A negative OR logic gate.

NOT GATE: An inverter that changes the polarity of an incoming signal in the output.

NUCLEUS: Core of the atom.

O

OERSTED: Unit of magnetic intensity equal to one gilbert per centimeter.

OHM: Unit of measurement of resistance. Its symbol is Ω.

OHMMETER: Meter used to measure resistance in ohms.

OHM'S LAW: Mathematical relationship between current, voltage, and resistance discovered by George Simon Ohm.

$$I = \frac{E}{R} \qquad E = IR \qquad R = \frac{E}{I}$$

OHMS PER VOLT: Unit of measurement of sensitivity of a meter.

OPEN CIRCUIT: Circuit broken or load removed. Load resistance equals infinity.

OPERATIONAL AMPLIFIER (OP AMP): A type of linear integrated circuit that is used as basic amplifier circuit.

OR GATE: A logic gate that will provide an output signal if there is a signal on either of its inputs.

OSCILLATOR: An electron tube generator of alternating current voltages.

OSCILLATORS,
  ARMSTRONG: An oscillator using tickler coil for feedback.
  COLPITTS: An oscillator using split tank capacitor as feedback circuit.
  CRYSTAL-CONTROLLED: Oscillator controlled by piezoelectric effect.
  ELECTRON COUPLED OSCILLATOR (ECO): Combination oscillator and power amplifier using electron stream as coupling medium between grid and plate tank circuits.
  HARTLEY: Oscillator using inductive coupling of tapped tank coil for feedback.
  PUSH-PULL: Push-pull circuit using interelectrode capacitance of each tube to feed back energy to grid circuit to sustain oscillations.
  RC OSCILLATORS: Oscillators depending upon charge and discharge of capacitor in series with resistance.
  TRANSITRON: Oscillator using negative transconductance.
  TUNED-GRID TUNED-PLATE: Oscillator using tuned circuits in both grid and plate circuits.
  ULTRAUDION: Oscillator, similar to Colpitts, but employing grid-to-cathode and plate-to-cathode interelectrode capacitance for feedback.

OSCILLOSCOPE: Test instrument, using cathode ray tube, permitting observation of signal.

OUTPUT: Processed results of a computer.

OUTPUT DEVICE: Equipment that records or displays processed results of a computer.

OVERMODULATION: Condition when modulating wave exceeds amplitude of continuous carrier wave, resulting in distortion.

P

PARALLEL CIRCUIT: Circuit which contains two or more paths for electrons supplied by common voltage source.

PARALLEL RESONANCE: Parallel circuit of an inductor and capacitor at frequency when inductive and capacitive reactances are equal. Current in capacitive branch is 180 deg. out of phase with inductive current and their vector sum is zero.

PARASITIC OSCILLATION: Oscillations in circuit resulting from circuit components or conditions, occuring at frequencies other than that desired.

PEAK: Maximum value of sine wave.

PEAK INVERSE VOLTAGE: Value of voltage applied in reverse direction across diode.

PEAK INVERSE VOLTAGE RATING: Inverse voltage diode will withstand without arcback.

PEAK REVERSE VOLTAGE: Same as peak inverse voltage.

PEAK TO PEAK: Measured value of sine wave from peak in positive direction to peak in negative direction.

PEAK VALUE: Maximum value of an alternating current or voltage.

PENTAGRID CONVERTER: Tube with five grids.

PENTAVALENT: Semiconductor impurity having five valence electrons. Donor impurities.

PENTODE: Electron tube with five elements including cathode, plate, control grid, screen grid, and suppressor grid.

PERCENTAGE OF MODULATION: Maximum deviation from normal carrier value as result of modulation expressed as a percentage.

PERCENTAGE OF RIPPLE: Ratio of rms value of ripple voltage to average value of output voltage expressed as a percentage.

PERIOD: Time for one complete cycle.

PERIPHERAL: A device that works as part of a computer but is joined to the outside of the process unit.

PERMEANCE (P): Ability of a material to carry magnetic lines of force. The reciprocal of reluctance.

$$P = \frac{1}{R}$$

PERMANENT MAGNET: Bars of steel and other substances which have been permanently magnetized.

PERMEABILITY: Relative ability of substance to conduct magnetic lines of force as compared with air. Its symbol is μ.

PHASE: Relationship between two vectors in respect to angular displacement.

PHASE DISTORTION: A deviation in phase resulting from the shift of phase of some signal frequencies.

PHASE INVERTER: Device or circuit that changes phase of a signal 180 deg.

PHASE SPLITTER: Amplifier which produces two waves that have exactly opposite polarities from single input wave form.

PHOTOELECTRIC EMISSION: Escape of electrons as a result of light striking the surface of certain materials.

Dictionary of Terms

PHOTONS: Light created when a positive hole and a negative electron combine in the PN junction region of a diode.

PHOTOSENSITIVE: Term used to describe the characteristic of a material which emits electrons from its surface when energized by light.

PHOTOTUBE: Vacuum tube employing photo sensitive material as its emitter or cathode.

PICOFARAD (pF): Same as micromicrofarad.

PICTURE ELEMENT: Small areas or dots of varying intensity from black to white which contain visual image of scene.

PIERCE OSCILLATOR: Crystal oscillator circuit in which crystal is placed between plate and grid circuit of tube.

PIEZOELECTRIC EFFECT: Property in which certain crystalline substances change shape when an emf is impressed upon crystal. Action is also reversible.

PI SECTION FILTER: Filter consisting of two capacitors and an inductor connected in a π configuration.

PITCH: Property of musical tone determined by its frequency.

PLATE: Anode of vacuum tube. Element in tube which attracts electrons.

PLATE DETECTOR: An rf signal is amplified and detected in plate circuit. Tube is biased to approximately cut-off by cathode resistor.

PLATE EFFICIENCY: Ratio between useful output power to dc input power to plate of electron tube.

PLATE MODULATION: Modulation circuit where modulating signal is fed to plate circuit of modulated stage.

PLUS: Positive terminal or junction of circuit. Its symbol is +.

PM SPEAKER: Loudspeaker using a permanent magnet as its field.

PN JUNCTION: Piece of N type and a piece of P type semiconductor material joined together.

POINT CONTACT DIODE: Diode consisting of point and a semiconductor crystal.

POLARITY: Property of device or circuit to have poles such as north and south or positive and negative.

POLARIZATION: Defect in cell caused by hydrogen bubbles surrounding positive electrode and effectively insulating it from chemical reaction. Producing magnetic poles or polarity.

POLES: Number of poles in motor or generator field.

POLYPHASE: Consisting of currents having two or more phases.

POSITIVE ION: Atom which has lost electrons and is positively charged.

POWER: Rate of doing work. In dc circuits, $P = I \times E$.

POWER AMPLIFICATION: Ratio of output power to input grid driving power.

POWER DETECTOR: Detector designed to handle signal voltages having amplitudes greater than one volt.

POWER FACTOR: Relationship between true power and apparent power of circuit.

POWER SUPPLY: Electronic circuit designed to provide various ac and dc voltages for equipment operation. Circuit may include transformers, rectifiers, filters, and regulators.

PREAMPLIFIER: Sensitive, low-level amplifier with sufficient output to drive standard amplifier.

PREEMPHASIS: Process of increasing strength of signals or higher frequencies in FM at transmitter to produce greater frequency swing.

PREFIX: A term added to the beginning of a base word to denote an increase or decrease in value.

PRIMARY CELL: Cell that cannot be recharged.

PRIMARY WINDING: Coil of transformer which receives energy from ac source.

PRINTED CIRCUIT: A circuit made from very thin layers of conductive material (such as copper) adhered to a plastic backing. The circuit is etched to leave active pathways for the circuitry.

PROGRAM: A set of instructions used to run a computer.

PROGRAMMABLE READ ONLY MEMORY (PROM): Memory programmed once after the computer is built.

PROTON: Positively charged particle.

PULSE: Sudden rise and fall of a voltage or current.

PULSE AMPLIFIER: An amplifier used to amplify pulses.

PUSH-PULL AMPLIFIER: Two tubes used to amplify signal in such a manner that each tube amplifies one half cycle of signal. Tubes operate 180 deg. out of phase.

Q

Q: Letter representation for quantity of electricity (coulomb).

Q: Quality, figure of merit; ratio between energy stored in inductor during time magnetic field is being established to losses during same time. $Q = \dfrac{X_L}{R}$

QUANTA: Definite amount of energy required to move an electron to higher energy level.

QUIESCENT: At rest. Inactive.

R

RADIO FREQUENCY CHOKE (RFC): Coil which has high impedance to rf currents.

RADIO SPECTRUM: Division of electromagnetic spectrum used for radio.

RANDOM ACCESS MEMORY (RAM): Active memory that can read, write, and temporarily store data, when the computer is operating.

RASTER: Area of light produced on screen of TV picture tube by electron beam. Contains no picture information.

RATIO DETECTOR: Type of FM detector.

REACTANCE (X): Opposition to alternating current as result of inductance or capacitance.

REACTIVE POWER: Power apparently used by reactive component of circuit.

READ ONLY MEMORY (ROM): Memory for instructions and fixed data that is stored by the manufacturer. ROM cannot be changed.

RECIPROCAL: Reciprocal of number is one divided by the number.

RECTIFIER: Component or device used to convert ac into a pulsating dc.

REGENERATIVE FEEDBACK: Feedback in phase with input signal so it adds to input.

REGULATION: Voltage change that takes place in output of generator or power supply when load is changed.

REGULATION, PERCENTAGE OF: Percentage of change in voltage from no-load in respect to full-load voltage. Expressed as:

$$\frac{E_{no\ load} - E_{full\ load}}{E_{full\ load}} \times 100 = \%$$

REJECT CIRCUIT: Parallel tuned circuit at resonance. Rejects signals at resonant frequency.

REL: Unit of measurement of reluctance.

RELATIVE CONDUCTANCE: Percentage comparison of conductance of material compared to silver which is considered 100 percent.

RELATIVE RESISTANCE: Numerical comparison of resistance of a material compared to silver which is assigned value 1.0.

RELAY: Magnetic switch.

RELAXATION OSCILLATOR: Non-sinusoidal oscillator whose frequency depends upon time required to charge or discharge capacitor through resistor.

RELUCTANCE: Resistance to flow of magnetic lines of force.

REMOTE CUT-OFF TUBE: Tube which gradually approaches its cut-off point at remote bias point, due to special grid construction.

REPULSION-START MOTOR: Motor which develops starting torque by interaction of rotor currents and single-phase stator field.

RESIDUAL MAGNETISM: Magnetism remaining in material after magnetizing force is removed.

RESISTANCE: Quality of electric circuit that opposes flow of current through it.

RESISTOR-TRANSISTOR LOGIC (RTL): An arrangement of digital circuits using resistors and transistors to perform logic function. This is no longer in use.

RESONANT FREQUENCY: Frequency at which tuned circuit oscillates. See TUNED CIRCUIT.

RETENTIVITY: Ability of material to retain magnetism after magnetizing force is removed.

RETMA: Radio Electronics Television Manufacturer's Association.

RETRACE: Process of returning scanning beam to starting point after one line is scanned.

REVERSE CURRENT CUTOUT: Relay which permits current to flow only in one direction.

RF AMPLIFIER: Amplifier used to amplify radio frequencies.

RIPPLE VOLTAGE: An ac component of dc output of power supply due to insufficient filtering.

ROLL-OFF: Gradual attenuation with increase or decrease in frequency of signal.

ROOT-MEAN-SQUARE (RMS) VALUE: The same as effective value. ($.707 \times E_{peak}$)

ROTOR: Rotating part of an ac generator.

ROWLAND'S LAW: Law for magnetic circuits which states that number of lines of magnetic flux is in direct proportion to magnetomotive force and inversely proportional to reluctance of circuit.

$$\Phi = \frac{F}{R}$$

RUBEN (RM) CELL: Mercury cell employing mercuric oxide and zinc. Electrolyte is potassium hydroxide.

RUMBLE: Low-frequency mechanical vibration of a turntable which is transmitted to recorded sound.

S

SATELLITE TV: Television in which the signals are sent via satellites.

SATURATION CURRENT: Current through electron tube when saturation voltage is applied to plate.

SATURATION VOLTAGE: Voltage applied to plate of vacuum tube so all emitted electrons are attracted to plate.

SAWTOOTH GENERATOR: Electron tube oscillator producing sawtooth wave form.

SAWTOOTH WAVE: Wave shaped like teeth of saw.

SCAN: Process of sweeping electron beam across each element of picture in successive order, to reproduce total picture in television.

SCHEMATIC: Diagram of electronic circuit showing electrical connections and identification of various components.

Dictionary of Terms

SCREEN GRID: Second grid in electron tube between control grid and plate that reduces interelectrode capacitance.

SECONDARY CELL: Cell that can be recharged by reversing chemical action with electric current.

SECONDARY EMISSION: Emission of electrons as result of electrons striking plate of electron tube.

SECONDARY WINDING: Coil which receives energy from primary winding by mutual induction and delivers energy to load.

SECOND HARMONIC DISTORTION: Distortion of wave by addition of its second harmonic.

SELECTIVITY: Relative ability of receiver to select desired signal while rejecting all others.

SELF-INDUCTANCE: Emf is self-induced when it is induced in conductor carrying current.

SEMICONDUCTOR: Conductor with resistivity somewhere in range between conductors and insulators.

SEMICONDUCTOR, N TYPE: Semiconductor which uses electrons as major carrier.

SEMICONDUCTOR, P TYPE: Semiconductor which uses holes as major carrier.

SENSITIVITY: Ability of circuit to respond to small signal voltages.

SENSITIVITY OF METER: Indication of loading effect of meter. Resistance of moving coil and multiplier divided by voltage for full scale deflection. Sensitivity equals one divided by current required for full scale deflection. Example: A 100 μA meter movement has sensitivity of $\frac{1}{.0001}$ or 10,000 ohms/volt.

SERIES CIRCUIT: Circuit which contains only one possible path for electrons through circuit.

SERIES GENERATOR: A source of electrical energy whose field windings are connected in series with armature and load.

SERIES PARALLEL: Groups of series cells with output terminals connected in parallel.

SERIES RESONANCE: Series circuit of an inductor, capacitor, and resistor at a frequency when inductive and capacitive reactances are equal and cancelling. Circuit appears as pure resistance and has minimum impedance.

SHADED POLE MOTOR: Motor in which each field pole is split to accommodate a short-circuit copper strap called a shading coil. This coil produces a sweeping movement of field across pole face for starting.

SHADOW MASK: See aperture mask.

SHIELD: Partition or enclosure around components in circuit to minimize effects of stray magnetic and radio frequency fields.

SHORT CIRCUIT: Direct connection across source which provides zero resistance path for current.

SHOT EFFECT: Noise in electron tube results from variation in rate of electron emission from cathode.

SHUNT: To connect across or parallel with circuit or component.

SHUNT: Parallel resistor to conduct excess current around meter moving coil. Shunts are used to increase range of meter.

SHUNT GENERATOR: Source of electrical energy whose field windings are connected across armature in shunt with load.

SIDEBANDS: Frequencies above and below carrier frequency as result of modulation. Lower frequencies are equal to the difference between carrier and modulating frequencies. Upper frequencies are equal to the carrier plus modulating frequencies.

SIDE CARRIER FREQUENCIES: Waves of frequencies equal to sum and difference between carrier wave frequency and modulating wave frequency.

SIEMEN: Unit for conductance.

SILICON: A special quadvalent material found in most transistors. It conducts the flow of electricity from one pathway to another.

SILICON CONTROLLED RECTIFIER (SCR): A three junction device (anode, gate, and cathode) that is usually open until a signal on the gate switches it on.

SILVER OXIDE BATTERY: A compact primary cell used mainly for LCD watches.

SINE WAVE: A graphical representation of a wave whose strength is proportional to the sine of an angle that is a linear function of time or distance.

SINGLE-PHASE MOTOR: Motor which operates on single-phase alternating current.

SINUSOIDAL: Wave varying in proportion to sine of angle.

SKY WAVE: Waves moving toward sky from radio antenna.

SLIP RINGS: Metal rings connected to rotating armature windings in generator. Brushes sliding on these rings provide connections for external circuit.

SOCKET: Device for holding lamp or electron tube.

SOFT TUBE: Gaseous tube.

SOFTWARE: A set of instructions used to run a computer.

SOLENOID: Coil of wire carrying electric current possessing characteristics of magnet.

SOLID-STATE: Electronic devices such as transistors and other solid substances, as opposed to vacuum tubes or electromechanical relays.

SOURCE OF SUPPLY: The device attached to input

SPACE CHARGE: Cloud of electrons around cathode of an electron tube.

SPECIFIC GRAVITY: Weight of liquid in reference to water which is assigned value 1.0.

SPLIT-PHASE MOTOR: Single-phase induction motor, which develops starting torque by phase displacement between field windings.

SQUARE LAW DETECTOR: Detector whose output voltage is proportional to square of effective input voltage.

SQUIRREL CAGE ROTOR: Rotor used in an induction motor made of bars placed in slots of rotor core and all joined together at ends.

STAGE: Section of an electronic circuit, usually containing one electron tube and associated components.

STANDING WAVE: Wave in which ratio of instantaneous value at one point to that at another point does not vary with time. Waves appear on transmission line as a result of reflections from termination of line.

STANDING WAVE RATIO: Ratio of effective voltage at loop of standing wave to effective voltage at node. Also called effective current. Also ratio of characteristic impedance to load impedance.

STATIC CHARACTERISTICS: Characteristics of tube taken with constant plate voltage.

STATIC CHARGE: Charge on body either negatively or positively.

STATIC ELECTRICITY: Electricity at rest as opposed to electric current.

STATOR: Stationary coils of an ac generator.

STEADY STATE: Fixed nonvarying condition.

STORAGE BATTERY: Common name for lead-acid battery used in automotive equipment.

STRATOSPHERE: Atmosphere above troposphere in which temperature is constant and there is no cloud formation.

STYLUS: Phonograph needle or jewel, which follows grooves in a record.

SUBHARMONIC: Frequency below harmonic, usually fractional part of fundamental frequency.

SULFATION: Undesirable condition of lead-acid battery caused by leaving it in discharged condition or by improper care. Sulfates forming on plates make battery partially inactive.

SUPERSONIC: Frequencies above audio frequency range.

SURFACE ALLOY TRANSISTOR: Silicon junction transistor, in which aluminum electrode are deposited in shallow pits etched on both sides of thin silicon crystal, forming P regions.

SUPERHETERODYNE: Radio receiver in which incoming signal is converted to fixed intermediate frequency before detecting audio signal component.

SUPPRESSOR GRID: Third grid in electron tube, between screen grid and plate, to repel or suppress secondary electrons from plate.

SWEEP CIRCUIT: Periodic varying voltage applied to deflection circuits of cathode ray tube to move electron beam at linear rate.

SWITCH: Device for directing or controlling current flow in circuit.

SYNCHRO: Electromechanical device used to transmit angular position of shaft from one position to another without mechanical linkage.

SYNC PULSE: Abbreviation for synchronization pulse, used for triggering an oscillator or circuit.

SYNCHRONOUS: Having same period or frequency.

SYNCHRONOUS MOTOR: Type of ac motor which uses a separate dc source of power for its field. It runs at synchronous speed under varying load conditions.

SYNCHRONOUS VIBRATOR: Vibrator with additional contact points to switch output circuit so current is maintained in one direction through load.

SYSTEM: An assembly of parts linked into an organized whole.

T

TANK CIRCUIT: Parallel resonant circuit.

TAP: Connection made to coil at point other than its terminals.

TELEVISION: Method of transmitting and receiving visual scene by radio broadcasting.

TELEVISION CHANNEL: Allocation in frequency spectrum of 6 MHz assigned to each television station for transmission of picture and sound information.

TETRODE: Electron tube with four elements including cathode, plate, control grid, and screen grid.

THERMAL RUNAWAY: In transistor, regenerative increase in collector current and junction temperature.

THERMIONIC EMISSION: Process in which heat produces energy for release of electrons from surface of emitter.

THERMISTOR: Semiconductor device which changes resistivity with change in temperature.

THERMOCOUPLE METER: Meter based on principle that if two dissimilar metals are welded together and junction is heated, a dc voltage will develop across open ends. Used for measuring radio frequency currents.

THETA (θ): Angle of rotation of vector representing selected instants at which sine wave is plotted. Angular displacement between two vectors.

Dictionary of Terms

THORIATED TUNGSTEN: Tungsten emitter coated with thin layer of thorium.

THREE-PHASE ALTERNATING CURRENT: Combination of three alternating currents have their voltages displaced by 120 deg. or one-third cycle.

THRESHOLD OF SOUND: Minimum frequency at which a sound can be heard.

THYRATHRON: Gas-filled tube in which grid is used to control firing potential.

TICKLER: Coil used to feed back energy from output to input circuit.

TIME CONSTANT (RC): Time period required for the voltage of a capacitor in an RC circuit to increase to 63.2 percent of maximum value or decrease to 36.7 percent of maximum value.

TONE CONTROL: Adjustable filter network to emphasize either high or low frequencies in output of audio amplifier.

TRANSCONDUCTANCE: Grid plate transconductance of vacuum tube expressed as ratio of small change in plate current to small change in grid voltage while plate voltage is held constant. Measured in siemens. Its symbol is g_m.

TRANSDUCER: Device by which one form of energy may be converted to another form, such as electrical, mechanical, or acoustical.

TRANSFER CHARACTERISTIC: Relation between input and output characteristics of device.

TRANSFORMER: Device which transfers energy from one circuit to another by electromagnetic induction.

TRANSFORMERS,
  ISOLATION: Transformer with one-to-one turns ratio.
  STEP-DOWN: Transformer with turns ratio greater than one. The output voltage is less than input voltage.
  STEP-UP: Transformer with turns ratio of less than one. Output voltage is greater than input voltage.

TRANSIENT RESPONSE: Response to momentary signal or force.

TRANSISTOR: Semiconductor device derived from two words, transfer and resistor.

TRANSISTOR-TRANSISTOR LOGIC (TTL): An arrangement of digital circuits using transistors to perform logic functions.

TRANSMISSION LINE: Wire or wires used to conduct or guide electrical energy.

TRANSMITTER: Device for converting intelligence into electrical impulses for transmission through lines or through space from radiating antenna.

TRF: Abbreviation for tuned radio frequency.

TRIAC: A full-wave silicon switch.

TRIODE: Three-element vacuum tube, consisting of cathode, grid, and plate.

TRIVALENT: Semiconductor impurity having three valence electrons. Acceptor impurity.

TROPOSPHERE: Lower part of atmosphere where clouds form and temperature decreases with altitude.

TRUE POWER: Actual power absorbed in circuit.

TUNED AMPLIFIER: Amplifier employing tuned circuits for input and/or output coupling.

TUNED CIRCUIT: Circuit containing capacitance, inductance, and resistance in series or parallel. When energized at specific frequency known as its resonant frequency, an interchange of energy occurs between coil and capacitor.

TURNS RATIO: Comparison of number of turns of primary winding of transformer to number of turns of secondary winding.

U

ULTRA HIGH FREQUENCY (UHF): Television frequenices that cover channels 14-83.

UNIJUNCTION TRANSISTOR (UJT): A three terminal transistor that has an emitter and two bases.

UNITY COUPLING: Two coils positioned so all lines of magnetic flux of one cell cut across all turns of second coil.

UNIVERSAL MOTOR: Series ac motor which operates also on dc. Fractional horsepower ac-dc motor.

UNIVERSAL TIME CONSTANT CHART: Graph with curves representing growth and decay of voltages and currents in RC and RL circuits.

V

VALID LOGIC HIGH: The operating voltage required for a digital circuit to be in the 1, or ON position. Voltage range is usually 2 to 5 volts.

VALID LOGIC LOW: The operating voltage required for a digital circuit to be in the 0 or OFF position. Voltage range is usually 0 to 1.5 volts.

VALVE: British name for vacuum tube.

VARIABLE MU TUBE: Tube with increased range of amplification due to its remote cutoff characteristics.

VECTOR: Straight line drawn to scale, showing direction and magnitude of a force.

VECTOR DIAGRAM: Diagram showing direction and magnitude of several forces, such as voltage and current, resistance, reactance, and impedance.

VELOCITY FACTOR: Speed of propagation of signal along transmission line compared to speed of light.

VERTICAL POLARIZATION: Antenna positioned vertically so its electric field is perpendicular to earth's surface.

Electricity and Electronics

VERY LARGE SCALE INTEGRATION (VLSI): ICs with over 1000 components.

VIBRATOR: Magnetically operated interrupter, similar to buzzer, to change steady state dc to pulsating ac.

VIDEO AMPLIFIER: Used to amplify video frequencies.

VIDEO SIGNAL: Electrical signal from studio camera used to modulate TV transmitter.

VOICE COIL: Small coil attached to speaker cone, to which signal is applied. Reaction between field of voice coil and fixed magnetic field causes mechanical movement of cone.

VOLT: Unit of measurement of electromotive force or potential difference. Its symbol is E in electricity and V in semiconductor circuits.

VOLTAGE: The force or difference in potential which causes electrons to flow.

VOLTAGE DIVIDER: Tapped resistor or series resistors across source voltage to multiply voltages.

VOLTAGE DOUBLER: Rectifier circuit which produces double the input voltage.

VOLTAGE DROP: Voltage measured across resistor. Voltage drop is equal to product of current times resistance in ohms. $E = IR$.

VOLTAGE MULTIPLIER: Rectifier circuits which produce output voltage at multiple greater than input voltage, usually doubling, tripling, or quadrupling.

VOLTAGE REGULATOR TUBE: Cold cathod gas-filled tube which maintains constant voltage drop independent of current, over its operating range.

VOLTAIC CELL: Cell produced by suspending two dissimilar elements in acid solution. Potential difference is developed by chemical action.

VOLTMETER: Meter used to measure voltage.

VOLT-OHM-METER: A portable multimeter.

VR TUBE: Gas-filled, cold cathode tube used for voltage regulation.

VTVM: Vacuum Tube Volt Meter.

VU: Number numerically equal to number of decibels above or below reference volume level. Zero vu represents power level of one milliwatt dissipated in 600 ohm load or voltage of .7746 volts.

W

WATT: Unit of measurement of power.

WATT-HOUR: Unit of energy measurement, equal to one watt per hour.

WATT-HOUR METER: Meter that shows instantaneous rate of power consumption of device or circuit.

WATTLESS POWER: Power not consumed in an ac circuit due to reactance.

WATTMETER: Meter used to measure power in watts.

WAVELENGTH: Distance between point on loop of wave to corresponding point on adjacent wave.

WAVE METER: Meter to measure frequency of wave.

WAVE TRAP: Type of band reject filter.

WEAK-SIGNAL DETECTOR: Unit that detects signal voltages having amplitudes of less than one volt.

WHEATSTONE BRIDGE: Bridge circuit used for precision measurement of resistors.

WORK: When a force moves through a distance, work is done. Work in foot-pounds = Force x Distance.

WORKING VOLTAGE: Maximum voltage that can be steadily applied to capacitor without arc-over.

WOW: Low-frequency flutter resulting from variation in turntable speeds.

WWV and WWVH: National Bureau of Standards Radio Stations in Washington, DC and Hawaii respectively.

X

X AXIS: Optical axis of crystal. Axis through corners of hexagonal crystal. Horizontal axis of graph.

X-CUT CRYSTAL: Cut perpendicular to X axis.

Y

YAGI ANTENNA: Dipole with two or more director elements.

Y AXIS: Vertical axis of graph. Axis drawn perpendicular to faces of hexagonal crystal.

Y-CUT CRYSTAL: Cut perpendicular to Y axis.

YOKE: Coils places around neck of TV picture tube for magnetic deflection of beam.

Z

Z AXIS: Optical axis of crystal.

ZENER DIODE: Silicon diode which makes use of the break down properties of a PN junction. If a reverse voltage across the diode is progressively increased, a point will be reached when the current will greatly increase beyond its normal cutoff value. This voltage point is called the Zener voltage.

ZERO REFERENCE LEVEL: Power level selected as reference for computing gain of amplifier or system.

AUTHORS' ACKNOWLEDGMENTS

We would like to acknowledge and express our appreciation and thanks to those associates who so willingly supplied their assistance and knowledge in the preparation of this text.

1964, 1968, and 1975 editions: Robert Mathison, Gary Van, James Collins, Leon Pearson, and John Lienhart, electronic technicians at Chico State College; Bobbie Fikes and Moore Smalley, graduate students in electronics at San Jose State College; Charles Tyler, San Jose City College, for illustrations; Virtue B. Gerrish (wife of Mr. Gerrish) for typing the original manuscript.

1977 and 1980 editions: Neal S. Hertzog, Marketing Office Manager, Buck Engineering Co.; Gerald J. Kane, Director of Engineering, Buck Engineering Co.; Shirley Miles and Carol Keeton for typing the manuscript; El Mueller, Quasar Electronics Corp.; Carl Giegold for photography; special thanks to Carrie, Ed, Cammie, and Toy Dugger for their special support and patience.

1988 edition: David R. Milson of Dick Smith Electronics for granting permission to use certain projects from *Fun Way into Electronics;* sincere gratitude and love to Carrie Dugger, whose support and help were steadfast during the preparation of this edition, and for typing this manuscript.

Thanks to the many industries whose generous supply of illustrations and product information made this text possible. Special credit is due:

Allen-Bradley, Co.; Advent; Apple Computers; AP Products; AT&T; Ballentine Laboratories, Inc.; Varley and Dexter Laboratories, Beckman Industrial Corp.; Bell Laboratories; Boonton Electronics; British Airways; Bud Radio, Inc.; Burr-Brown; Centralab, Globe Union, Inc., Cincinnati Milicron; Dale Electronics; Daystrom, Weston Instruments Div.; Delco Products, General Motors; Delco-Remy Div., General Motors; DeVilbiss Co.; Dick Smith Electronics; Dynascan Corp.; Electric Storage Battery Corp.; Electronic Industries Association; Electronic Instrument Co. (EICO); ESB Brands, Inc.; First Class Peripherals; J.A. Fleming; John Fluke Mfg. Co.; Fosdick Machine Tool Co.; General Electric Co.; Gould-A/M/I Semiconductors; Graymark Enterprises, Inc.; Gulton Industries, Inc.; Hammarlund Mfg. Co., Inc.; Hayes Microcomputer Products; Heath Company; Heathkit; Hewlett Packard, Inc.; Hickok Teaching Systems, Inc.; Honeywell; Hughes Communications, Inc.; Hughes Optical and Data Systems Group; Hughes Solid State Products; IBM; International Rectifier Corp.; Iwatsu Instruments; Johnson Co.; Kepco, Inc.; Jack Kilby; Koss Corp.; Lab-Volt, Buck Engineering Co.; Lafayette Radio Electronics; Lattice Semiconductors; Lindberg Engineering Co.; Liquid Xtal Displays Inc.; Mallory Battery Co.; Marantz; Maxell Corp.; James Millen Co.; J.W. Miller Co.; Minneapolis-Honeywell Regulator Co.; MJR Co.; Montgomery Ward; Motorola Semiconductor Products, Inc.; Murdock Corp.; National Automatic Tool Co.; National Semiconductor; NEC Information Systems, Inc.; Ohmite Mfg. Co.; Omega Group Co.; Optima Enclosures; Pacific Telephone Co.; Panasonic Battery Sales Division; Perma-Power; Philco Corp.; Plastoid Corp.; Potter & Brumfield, AMF Inc.; Quasar Electronics Corp.; Radio Shack Corporation; Raytheon Company; RCA; Regency Electronics; Science-Electronics, Inc.; Shure Bros., Inc.; Simpson Electric Co.; Siemens; Society of Audio Consultants; Sony Corp. of America; Sprague Products Co.; L.S. Starrett Co.; Stromberg-Carlson; Sublogic Corp.; Superior Electric Co.; Sylvania Electric Products, Inc. (GTE); Tektronix; Texas Instruments Inc.; Triad Transformer Corp.; Tripplett Corp.; Ungar Electric Tools; Union Carbide Consumer Products Co.; Union Switch & Signal Co.; Unisys; United States Bureau of Labor Statistics; United States Office of Education; United Transformer Corp.; Vactec Inc.; Viz Mfg. Co.; Westinghouse Electric Corp.; Welch Scientific Co.; Zenith Data Systems; Zenith Electronics Corp.

INDEX

A

AC-DC supply, 242
Acceptor, circuit, 166, 167
AC meters, 121, 122
AF amplifiers, 354
Alignment, 339, 340
Aklaline cell, 43
Alternating current, 104-108
Alternator, 107
Amalgamation, 43
Ammeters, 64, 112, 113
Ampere (AMP), 28
Amplifier, 247
  circuits, 247-250
  operation, 251, 252
Amplifiers, classes of, 254
Amplifiers and linear integrated circuits, 247-270
Amplitude modulation (AM), 310-313
Amplitude modulated (AM) receiver, 324-326
AM transmitter project, 318, 319
AND gates, 278, 279
Angle of declination or variation, 86
Anode, 202
Antenna coil, 324
Arithmetic/logic unit (ALU), 373, 380
Armature, 92
Armstrong oscillator, 294, 295
Atomic characteristics, 199, 200
Atomic number, 20
Atomic weight, 20
Audio amplifier, 332, 333
Audio preamplifier, 332
Automatic gain control (AGC), 331, 332, 354
Automotive and industrial, 262, 263
Autotransformers, 138

B

Basic
  analog meter movement, 111-113
  cathode ray tube controls, 352
  circuits, 57-69
  electronic devices, 199-214
  magnetic principles, 85-88
  oscillators, 293-295
Battery, 46, 47
Battery capacity, 50
Bias, methods of, 249, 250
Biasing, 248, 249
Binary bingo, 284-289
Binary numbering system, 274-276
Bits, nibbles, and bytes, 277
Black and white TV receiver, 353, 354
Black pedestal, 351
B+ voltage, 354
Breadboard, 59
Bridge rectifier, 234, 235
Broadcast band, 311
Bypassing, 173, 174

C

Cable TV, 363
Capacitance
  and the capacitor, 149, 150
  in ac circuits, 159, 160
  in electrical circuits, 149-163
Capacitor input filter, 236
Capacitors, 222
  can type electrolytic, 151-153
  ceramic, 154
  fixed paper, 150
  mica, 155
  rectangular oil filled, 150
  tantalum, 155, 156
  transient response, 156-159
  trimmer, 155
  tubular electrolytic, 153
  types, 150-156
  variable, 150

Index

Career opportunities, in electronics, 387-394
Careers,
  finding a, 387, 388
  information sources, 391
  in the electronics industry, 388-392
  semiskilled workers, 389
  skilled workers, 389, 390
Cascading, 255
Cathode ray tube controls, basic, 352
Cathode ray tube (CRT), 213
Cathodes, 202, 209
Cell,
  alkaline, 43
  lead acid, 48, 49
  mercury, 44
  nickel-cadmium rechargeable, 45, 46
  secondary, 47, 48
  silver oxide, 44, 45
  zinc-carbon, 43
Center frequency, 313
Central processing unit (CPU), 373, 379, 380
Chemical action, 41-51
Choke input filter, 237
Circuit,
  acceptor, 166, 167
  basic, 62, 63
  breaker, 62, 93
  combination, 63, 76-78
  connecting meters, 63, 64
  fundamentals of, 57-63
  magnetic, 89, 90
  open, 58
  parallel, 62, 74-76
  printed, 58-60
  reject, 169, 170
  series, 62, 71-73
  series-parallel, 63
  tank, 167, 168
CMOS logic, 283
Collector junction, 248
Color coding resistors, 36, 37
Color TV receiver, 354-356
Colpitts oscillator, 296
Combination circuit, 63
Combination circuits (series-parallel), 76-78
Common base (CB), 253
Common collector, 253
Common emitter (CE), 253
Commutation and interpoles, 183
Commutator, 99
Communication model, 305-307
Communication systems, 11, 12
Compound, 19
Compound dc motors, 184, 185
Computer developments, twentieth century, 370
Computer operation, 373-382
Computers, 367-383
  future, 381, 382
  history of, 369, 370
Conduction of electricity, 200, 201
Conductive pathway, 58, 59
Conductors, 24, 29
  conductance, and insulators, 29-31
  insulators, and semiconductors, 27-39
  length of, 33
  surface area of, 32
Connecting meters in a circuit, 63, 64
Constant speed motor, 183
Control grid, 210
Controlling unit (CU), 373
Controls, 60-62
Control unit (CU), 379, 380
Copper losses, 101, 138
Coulomb, 23
Counter EMF, 182
Coupling amplifiers, 255-261
Covalent bond, 200
Crystal oscillator, 296, 297
Crystal radio, 343, 344

D

Damper, 354
DC motors, 183-187
DC restoration, 354
DC versus AC, 99
Deflection yoke, 350
Delta-type tube, 355
Depolarizing agent, 42
Detection, 326
Detector, 331
Dielectric field, 23
Digital
  fundamentals, 274-277
  IC construction projects, 284-289
  meters, 117-119
Digital circuits, 273-290
  other, 284
  voltage logic levels, 276, 277
Diode, 202, 210
  characteristics and ratings, 204
  detector, 326
  transistor logic, 283
Direct coupling, 259
Direct current, pulsating, 99
Direct current, pure, 99
Direction of current, 29
Discriminator, 341
Donor impurities, 201
Door bell and buzzer, 93, 94
Doping, 201, 202
Dry cell, 43

E

Eddy current, 101
Eddy current losses, 138
Education, 391, 392
Eight-transistor superheterodyne radio, 344, 346

Electric
  current, 24, 28, 29
  current and magnetism, 88-91
  motors, 179-192
  voltage and current, 24
Electrical bus, 58
Electrical pressure, 27
Electrical energy
  from heat, 52, 53
  from light, 51, 52
  from mechanical energy, 97-101
  from mechanical pressure, 54
  other sources, 51-54
Electricity and electronics,
  in technology, 10-12
  milestones and events, 12-15
  science of, 19-25
Electricity from magnetism, 54
Electricity, sources of, 41-55
Electrolyte, 42
Electromagnets, 90
Electromotive force (EMF), 27
Electronics industries, types, 388
Electrons, 20
Electrons—protons—neutrons, 20
Electrostatic fields, 23
Elements, 19
Emitter junction, 248
ENIAC, 370
Entrepreneurs, 391
Equivalent resistance, 76

F
Factors affecting resistance, 32, 33
Field, 350
Field effect transistor (FET), 207, 208
Field windings, 179
Filtering action, 172, 173
Filtering circuits, 172-177
Filters, 235-237
Fine tuning, 353
First IF amplifier, 330, 331
Floating ground, 242
FM audio detector, 354
FM detection, 341, 342
FM receiver, 340-343
Forward bias, 248
Four-transistor radio, 344
Frame, 350
Frequency deviation, 314
Frequency modulation (FM), 310, 313-316
Frequency spectrum, 306, 307
Full-wave rectification, 233-235
Fundamentals of a circuit, 57-63
Fundamentals of electricity and electronics technology, 16, 17
Fuse, 61

G
Generator, 54, 97-109
  alternating current, 106
  compound, 102, 103
  construction, 99-101
  independently excited field, 102
  losses, 101
  practical applications, 107
  series, 102
  shunt, 102
  types, 101-104
Geostationary orbit, 359
Gain, 246
  computing, 252
  current, 252
  power, 252
Galvanometer, 53
Gilberts, 89
Ground waves, 318

H
Half-wave rectification, 233
Hartley oscillator, 295
Headphones, 336-339
  dynamic, 337, 338
  electrostatic, 338, 339
Heat sinks, 255
High-pass filter, 174
High retentivity, 90
High voltage rectifier, 354
Horizontal AFC, 354
  oscillator, 354
  output, 354
  output transformer (HOT), 354
Hydrometer, 49
Hysteresis loss, 101
Hysteresis losses, 138

I
IC construction, 218-223
IC use in superheterodyne receivers, 333-335
ICs, common types, 223
IF amplifiers, 330, 331
Ignition system, 139
Impedance, 143
Induced current and voltage, 141, 142
Inductance, 131-134
  and RL circuits, 131-146
  in ac circuits, 141-145
Induction coil, 138, 139
Induction motors, 187-192
Information systems, 10
Input/output devices, 374-379
Inputs, 374-376
Instruments and measurements, 111-125
Insulators, 31
Integrated circuits, 217-224
Integrated circuits, history of, 218
Intercalation, 44

Index

Intermediate frequency (IF), 327, 330
Inverted diode, 233
Ionization, 21
Ionosphere, 318
IR drop, 72
Iron vane meter movement, 119, 120
I^2R loss, 101

J
Job search ideas, 393, 394

K
Kilovolt (kV), 28
Kirchhoff's Current Law, 76
Kirchhoff's Voltage Law, 73

L
Lattice crystalline, 200
Law of charges, 21-23
Laws of magnetism, 86, 87
Local action, 43
Lead acid cells, 48, 49
Left hand rule for a coil, 89
Left hand rule for a conductor, 88
Lenz's Law, 98
Light emitting diode (LED), 118, 205
Linear IC construction projects, 263-269
Linear integrated circuits, 261-263
Liquid crystal display (LCD), 118
Loading a circuit, 122
Loading the tank circuit, 171
Load resistor, 237, 238
Loads, 57, 58
Local oscillator, 329, 330
Lodestones, 85
Logic families, 283, 284
Logic gates, 278-282
Low-pass filters, 174
Low retentivity, 90
L section filter, 237

M
Magnetic
  circuits, 85, 89, 90
  flux, 88
  induction, 97
  poles, 85
  principles, basic, 85-88
  recording tape, 87
  shields, 94
Magnetism, 85-95
Magnetism, third Law of, 88
Material, kind of, 32, 33
Material, temperature of, 33
Matter, 19
Matter, nature of, 19-21
Megaphone amplifier, 266, 267
Megavolt (MV), 28
Memory, 380
Mercury cell, 44

Meters, how to use, 123
Microamp (μA), 29
Microcomputers vs minicomputers, 370-373
Microphones, 308-310
  carbon, 308
  crystal, 309
  dynamic, 309
  velocity, 310
Microprocessors, 380, 381
Microvolt (μV), 28
Milestone and events in electricity, 12-15
Milliamp (mA), 29
Millivolt (mV), 28
Mixer, 327-329, 353
Modulation, 310
  index, 315
  patterns, 311, 312
  percent of, 315
Molecule, 19
Molecule and the atom, 19, 20
Morse code oscillator, 297, 298
Motor operation principles, 179-182
Motor starting circuits, 185, 186
Multimeter, 116-119
Multiplier resistors, 114
Mutual inductance, 134-141

N
NAND gate, 280, 281
Narrow band FM, 315
Negative ion, 21
Negative ion emissions, 352
Nickel-cadmium rechargeable cell, 45, 46
NPN and PNP transistors, 247, 248
Noise limiting, 342
Nomograph, 175-177
NOR gate, 281
NOT gates, 278, 280
Nucleus, 20

O
Ohm, 31
Ohmmeters, 36, 115, 116
Ohm's Law, 31, 32, 64, 65
  and Watt's Law, 67, 68
  for ac circuits, 144, 145
  in series circuits, 72
Open circuit, 58
Operational amplifiers, 262
OR gates, 278, 279
Oscillators, 293-299
  basic, 293-295
  Armstrong, 294, 295
  Colpitts, 296
  crystal, 296, 297
  Hartley, 295
  power, 297
  projects, 297-299
Oscilloscope, 123

Outputs, 376-379
Overloading, 61

P

Paging amplifier, 268, 269
Parallel and series circuits, 158, 159
Parallel circuits, 62, 74-76
Partial short, 58
Peak inverse voltage (PIV), 204
Pentodes, 211, 212
Permeability, 89, 94
Phase displacement, 106
Photoelectric control, 52
Photomasks, 219
Photo voltaic cell, 51
Picture elements, 349
Piezoelectric effect, 54, 296
PIX-IF amplifiers, 353
PM speaker, 336
Point contact diodes, 203
Polarity and meters, 63, 64
Polarization, 42, 317
Portable stereo amplifier, 263, 264
Portable stereo speakers, 265, 266
Positive ion, 21
Potentiometer, 35
Power
  and Watt's Law, 66
  factor, 142, 161
  in a series circuit, 73
  oscillators and converters, 297
  supplies, 231-245
  transformer, 232
Power supply
  construction, 243
  functions, 231-237
  0-15 volt dc low amperage, 243, 244
Prefixes, 28
Primary cell, 42
Primary cells, defected, 42, 43
Printed circuit, 58-60
Printhead, 376
Professionals, 390, 391
Programmable read only memory (PROM), 380
Project,
  AM transmitter, 318, 319
  digital IC construction, 284-289
  linear IC construction, 263-269
  oscillator, 297-299
  owl, 157, 158
  radio receiver, 343-346
Projection TV, large screen, 358, 359
Protons, 20
Push-pull coupling, 260, 261
Pyrometer, 53
Pythagorean theorem, 143

Q

Q of tuned circuit, 170, 171

R

Radio and TV, 262, 263
Radio transmitter, 308-310
Radio wave, 316-318
  receivers, 323-347
  travel, 317
  transmission, 305-320
Random access memory, (RAM), 380
Ratio detector, 342
RC coupling, 256-259
RCL networks, 165-172
RC, parallel circuit, 162
RC time constant, 157
Reactive power, 142, 160, 161
Read only memory (ROM), 380
Reed relay, 93
Regulation, voltage and current, 103, 104
Reject circuit, 169, 170
Relay, 91-94
Relay, reed, 93
Reluctance, 89
Repulsion induction motor, 191
Residual magnetism, 90
Resistance, 31-33
  and capacitance in an ac circuit, 161, 162
  and inductance in an ac circuit, 143, 144
Resistive loads, 57
Resistors, 33-37, 222
Resistors, color coding, 36, 37
Resistor-transistor logic, 283
Respective reaction equations, 44
Resonance, 165, 166
Reverse bias, 248
RF amplification, 325
RF amplifier, 353
RL, parallel circut, 145
RMS, 106

S

Satellite transmission, 360-362
Satellite TV, 359-362
Scanning, 350, 351
Schematics, 58
Science of electricity and electronics, 19-25
Screen grid, 211
Secondary cells, 47, 48
Secondary emission, 211
Self-inductance, 132
Semiconductor diodes, 202, 203
Semiconductors, 37, 38, 199-202
Series and parallel inductance,
Series and parallel rectifier arrangements, 204
Series circuits, 62, 71-73
Series dc motor, 184
Series, parallel and combination circuits, 71-79
Series-parallel circuit, 63
Shaded pole motor, 191, 192

Index

Shadow mask picture tube, 355
Shielding, 94
Short, 58
Shunt, 112
Shunt dc motor, 183, 184
Sideband frequencies, 310
Sideband power, 312
Siemens, 30
Signal generator, 297
Silicon rectifiers, 204
Silver oxide cell, 44, 45
Sine wave, 99
Single battery circuit, 249
Single-phase induction motors, 190, 191
Sky waves, 318
Software, 381
Solenoid, 89
Solenoid sucking coil, 90, 91
Sources of electricity, 41-55
Specific gravity, 49
Speed regulation, 183
Splatter, 311
Squawker, 298, 299
Static electricity, 21-24
Storage devices, 380
Superheterodyne receiver, 327-336
Suppressor grid, 211
Switches, 60
Sync amplifier, 354
Synchronization pulse, 350
Sync separator, 354

T

Tank circuit, 167, 168
Taps, 136, 137
Technicians, 390
Technology of the information age, 9-15
Television, 349-364
  cameras, 349, 350
  channel, 356-358
  innovations, 358-363
  signals, 349-352
Television receivers, 353-358
  black and white, 353, 354
  color, 354-356
Testing diodes, 204
Tetrodes, 211
Three-phase induction motor, 189, 190
Thermal considerations, 254, 255
Thermal runaway, 255
Thermionic emission, 209
Thermionic emitters, 209
Thermistor, 36
Thermocouple, 52
Thermopile, 53
Third Law of Magnetism, 88
Thyristor motor controls, 186, 187

Time constant, 134
Tone controls, 339
Transducers, 336-340
Transformer, 135-138
  coupling, 256
  losses, 138
  phase relationship, 139, 140
  power, 137, 138
  reason for using, 138
Transient responses, 133, 134
Transistor, 206-208
  circuit configurations, 253-255
  precautions, 255
  transistor logic (TTL), 283
Transistorized telephone amplifier, 268
Transistorized transmitters, 312, 313
Transportation systems, 12
Triodes, 210, 211
Truth table, 278
Tuned circuits and RCL networks, 165-177
Tuned circuit filters, 175
Tuned radio frequency (TRF), 324
Tuning circuit, 324, 325
Turns ratio and voltage ratio, 135, 136
Twentieth century computer developments, 370

U

Unidirectional conductor, 51
Universal motor, 187

V

Vacuum tubes, 209-213
Valence electrons, 200
Vector, 105
Vertical oscillator, 354
Vertical output, 354
Vestigial sideband filter, 311
Video amplifier, 353
Video cassette recorders, 358
Video detector (demodulator), 353
Voltage, 24, 27, 28
  and current regulation, 103, 104
  doublers, 241, 242
  drop, 72
  regulation, 237-241
  regulator circuit, 239-241
  regulators and references, 261, 262
  sources, 27, 28, 57
Voltaic cell, 42
Voltmeters, 63, 114-116
Voltmeter sensitivity, 115
Volt-ohm-meter (VOM), 116, 117
Volts, 28

W

Wattless power, 160
Wattmeter, 120, 121
Wheatstone bridge, 119

X

XNOR gate, 282
XOR gate, 281, 282

Z

Zener diode, 238, 239
Zinc-carbon cell, 43